Bayesian Logical Data Analysis for the Physical Sciences

A Comparative Approach with *Mathematica*™ Support

Increasingly, researchers in many branches of science are coming into contact with Bayesian statistics or Bayesian probability theory. By encompassing both inductive and deductive logic, Bayesian analysis can improve model parameter estimates by many orders of magnitude. It provides a simple and unified approach to all data analysis problems, allowing the experimenter to assign probabilities to competing hypotheses of interest, on the basis of the current state of knowledge.

This book provides a clear exposition of the underlying concepts with large numbers of worked examples and problem sets. The book also discusses numerical techniques for implementing the Bayesian calculations, including an introduction to Markov chain Monte Carlo integration and linear and nonlinear least-squares analysis seen from a Bayesian perspective. In addition, background material is provided in appendices and supporting *Mathematica* notebooks are available from www.cambridge.org/052184150X, providing an easy learning route for upper-undergraduate, graduate students, or any serious researcher in physical sciences or engineering.

PHIL GREGORY is Professor Emeritus at the Department of Physics and Astronomy at the University of British Columbia.

T0211226

Bayesian Logical Data Analysis for the Physical Sciences

A Comparative Approach with *Mathematica*™ Support

P. C. Gregory

Department of Physics and Astronomy, University of British Columbia

CAMBRIDGE
UNIVERSITY PRESS

CAMBRIDGE UNIVERSITY PRESS
Cambridge, New York, Melbourne, Madrid, Cape Town,
Singapore, São Paulo, Delhi, Tokyo, Mexico City

Cambridge University Press
The Edinburgh Building, Cambridge CB2 8RU, UK

Published in the United States of America by Cambridge University Press, New York

www.cambridge.org
Information on this title: www.cambridge.org/9780521150125

© P. Gregory 2005

This publication is in copyright. Subject to statutory exception
and to the provisions of relevant collective licensing agreements,
no reproduction of any part may take place without the written
permission of Cambridge University Press.

First published 2005
Reprinted 2006
Paperback edition 2010

A catalogue record for this publication is available from the British Library

Library of Congress Cataloguing in Publication data
Gregory, P. C. (Philip Christopher), 1941–
Bayesian logical data analysis for the physical sciences: a comparative approach with
Mathematica support / P.C. Gregory.
     p.   cm.
Includes bibliographical references and index.
ISBN 978-0-521-84150-4 (hardback)
1. Bayesian statistical decision theory. 2. Physical sciences – Statistical methods.
3. Mathematica (Computer file) I. Title.
QA279.5.G74 2005
519.5′42 – dc22   2004045930

ISBN 978-0-521-84150-4 Hardback
ISBN 978-0-521-15012-5 Paperback

Additional resources for this publication at www.cambridge.org/9780521150125

Cambridge University Press has no responsibility for the persistence or
accuracy of URLs for external or third-party internet websites referred to in
this publication, and does not guarantee that any content on such websites is,
or will remain, accurate or appropriate. Information regarding prices, travel
timetables, and other factual information given in this work is correct at
the time of first printing but Cambridge University Press does not guarantee
the accuracy of such information thereafter.

Contents

Preface

The goal of science is to unlock nature's secrets. This involves the identification and understanding of nature's observable structures or patterns. Our understanding comes through the development of theoretical models which are capable of explaining the existing observations as well as making testable predictions. The focus of this book is on what happens at the interface between the predictions of scientific models and the data from the latest experiments. The data are always limited in accuracy and incomplete (we always want more), so we are unable to employ deductive reasoning to prove or disprove the theory. How do we proceed to extend our theoretical framework of understanding in the face of this? Fortunately, a variety of sophisticated mathematical and computational approaches have been developed to help us through this interface, these go under the general heading of statistical inference. Statistical inference provides a means for assessing the plausibility of one or more competing models, and estimating the model parameters and their uncertainties. These topics are commonly referred to as "data analysis" in the jargon of most physicists.

We are currently in the throes of a major paradigm shift in our understanding of statistical inference based on a powerful theory of extended logic. For historical reasons, it is referred to as Bayesian Inference or Bayesian Probability Theory. To get a taste of how significant this development is, consider the following: probabilities are commonly quantified by a real number between 0 and 1. The end-points, corresponding to absolutely false and absolutely true, are simply the extreme limits of this infinity of real numbers. Deductive logic, which is based on axiomatic knowledge, corresponds to these two extremes of 0 and 1. Ask any mathematician or physicist how important deductive logic is to their discipline! Now try to imagine what you might achieve with a theory of extended logic that encompassed the whole range from 0 to 1. This is exactly what is needed in science and real life where we never know anything is absolutely true or false. Of course, the field of probability has been around for years, but what is new is the appreciation that the rules of probability are not merely rules for manipulating random variables. They are now recognized as uniquely valid principles of logic, for conducting inference about any proposition or hypothesis of interest. Ordinary deductive logic is just a special case in the idealized limit of complete information. The reader should be warned that most books on Bayesian statistics do not make the connection between probability theory and logic. This connection, which is captured in the book by physicist E. T. Jaynes, *Probability Theory – The Logic of Science*,[1] is particularly appealing because of the unifying principles it provides for scientific reasoning.

What are the important consequences of this development? We are only beginning to see the tip of the iceberg. Already we have seen that for data with a high signal-to-noise ratio, a Bayesian analysis can frequently yield many orders of magnitude improvement in model parameter estimation, through the incorporation of relevant prior information about the signal model. For several dramatic demonstrations of this point, have a look at the first four sections of Chapter 13. It also provides a more powerful way of assessing competing theories at the forefront of science by quantifying Occam's razor, and sheds a new light on systematic errors (e.g., Section 3.11). For some problems, a Bayesian analysis may simply lead to a familiar statistic. Even in this situation it often provides a powerful new insight concerning the interpretation of the statistic. But most importantly, Bayesian analysis provides an elegantly simple and rational approach for answering any scientific question for a given state of information.

This textbook is based on a measurement theory course which is aimed at providing first year graduate students in the physical sciences with the tools to help them design, simulate and analyze experimental data. The material is presented at a mathematical level that should make it accessible to physical science undergraduates in their final two years. Each chapter begins with an overview and most end with a summary. The book contains a large number of problems, worked examples and 132 illustrations.

The Bayesian paradigm is becoming very visible at international meetings of physicists and astronomers (e.g., *Statistical Challenges in Modern Astronomy III*, edited by E. D. Feigelson and G. J. Babu, 2002). However, the majority of scientists are still not at home with the topic and much of the current scientific literature still employs the conventional "frequentist" statistical paradigm. This book is an attempt to help new students to make the transition while at the same time exposing them in Chapters 5, 6, and 7 to some of the essential ideas of the frequentist statistical paradigm that will allow them to comprehend much of the current and earlier literature and interface with his or her research supervisor. This also provides an opportunity to compare and contrast the two different approaches to statistical inference. No previous background in statistics is required; in fact, Chapter 6 is entitled "What is a statistic?" For the reader seeking an abridged version of Bayesian inference, Chapter 3 provides a stand-alone introduction on the "How-to of Bayesian inference."

---

[1] Early versions of this much celebrated work by Jaynes have been in circulation since at least 1988. The book was finally submitted for publication in 2002, four years after his death, through the efforts of his former student G. L. Bretthorst. The book is published by Cambridge University Press (Jaynes, 2003, edited by G. L. Bretthorst).

The book begins with a look at the role of statistical inference in the scientific method and the fundamental ideas behind Bayesian Probability Theory (BPT). We next consider how to encode a given state of information into the form of a probability distribution, for use as a prior or likelihood function in Bayes' theorem. We demonstrate why the Gaussian distribution arises in nature so frequently from a study of the Central Limit Theorem and gain powerful new insight into the role of the Gaussian distribution in data analysis from the Maximum Entropy Principle. We also learn how a quantified Occam's razor is automatically incorporated into any Bayesian model comparison and come to understand it at a very fundamental level.

Starting from Bayes' theorem, we learn how to obtain unique and optimal solutions to any well-posed inference problem. With this as a foundation, many common analysis techniques such as linear and nonlinear model fitting are developed and their limitations appreciated. The Bayesian solution to a problem is often very simple in principle, however, the calculations require integrals over the model parameter space which can be very time consuming if there are a large number of parameters. Fortunately, the last decade has seen remarkable developments in practical algorithms for performing Bayesian calculations. Chapter 12 provides an introduction to the very powerful Markov chain Monte Carlo (MCMC) algorithms, and demonstrates an application of a new automated MCMC algorithm to the detection of extrasolar planets.

Although the primary emphasis is on the role of probability theory in inference, there is also focus on an understanding of how to simulate the measurement process. This includes learning how to generate pseudo-random numbers with an arbitrary distribution (in Chapter 5). Any linear measurement process can be modeled as a convolution of nature's signal with the measurement point-spread-function, a process most easily dealt with using the convolution theorem of Fourier analysis. Because of the importance of this material, I have included Appendix B on the Discrete Fourier Transform (DFT), the Fast Fourier Transform (FFT), convolution and Weiner filtering. We consider the limitations of the DFT and learn about the need to zero pad in convolution to avoid aliasing. From the Nyquist Sampling Theorem we learn how to minimally sample the signal without losing information and what prefiltering of the signal is required to prevent aliasing.

In Chapter 13, we apply probability theory to spectral analysis problems and gain a new insight into the role of the DFT, and explore a Bayesian revolution in spectral analysis. We also learn that with non-uniform data sampling, the effective bandwidth (the largest spectral window free of aliases) can be made much wider than for uniform sampling. The final chapter is devoted to Bayesian inference when our prior information leads us to model the probability of the data with a Poisson distribution.

Software support

The material in this book is designed to empower the reader in his or her search to unlock nature's secrets. To do this efficiently, one needs both an understanding of the principles of extended logic, and an efficient computing environment for visualizing and mathematically manipulating the data. All of the course assignments involve the use of a computer. An increasing number of my students are exploiting the power of integrated platforms for programming, symbolic mathematical computations, and visualizing tools. Since the majority of my students opted to use *Mathematica* for their assignments, I adopted *Mathematica* as a default computing environment for the course. There are a number of examples in this book employing *Mathematica* commands, although the book has been designed to be complete without reference to these *Mathematica* examples. In addition, I have developed a *Mathematica* tutorial to support this book, specifically intended to help students and professional scientists with no previous experience with *Mathematica* to efficiently exploit it for data analysis problems. This tutorial also contains many worked examples and is available for download from http://www.cambridge.org/052184150X.

In any scientific endeavor, a great deal of effort is expended in graphically displaying the results for presentation and publication. To simplify this aspect of the problem, the *Mathematica* tutorial provides a large range of easy to use templates for publication-quality plotting.

It used to be the case that interpretative languages were not as useful as compiled languages such as C and Fortran for numerically intensive computations. The last few years have seen dramatic improvements in the speed of *Mathematica*. Wolfram Research now claims[2] that for most of Mathematica's numerical analysis functionality (e.g., data analysis, matrix operations, numerical differential equation solvers, and graphics) Mathematica 5 operates on a par[3] with Fortran or MATLAB code. In the author's experience, the time required to develop and test programs with *Mathematica* is approximately 20 times shorter than the time required to write and debug the same program in Fortran or C, so the efficiency gain is truly remarkable.

---

[2] http://www.wolfram.com/products/mathematica/; newin5/performance/numericallinear.html.
[3] Look up *Mathematica* gigaNumerics on the Web.

Acknowledgements

Most of the Bayesian material presented in this book I have learned from the works of Ed Jaynes, Larry Bretthorst, Tom Loredo, Steve Gull, John Skilling, Myron Tribus, Devinder Sivia, Jim Berger, and many others from the international community devoted to the study of Bayesian inference. On a personal note, I encountered Bayesian inference one day in 1989 when I found a monograph lying on the floor of the men's washroom entitled *Bayesian Spectrum Analysis and Parameter Estimation*, by Larry Bretthorst. I was so enthralled with the book that I didn't even try to find out whose it was for several weeks. Larry's book led me to the work of his Ph.D. supervisor, Edwin T. Jaynes. I became hooked on this simple, elegant and powerful approach to scientific inference. For me, it was a breath of fresh air providing a logical framework for tackling any statistical inference question in an optimal way, in contrast to the recipe or cookbook approach of conventional statistical analysis.

I would also like to acknowledge the proof reading and suggestions made by many students who were exposed to early versions of this manuscript, in particular, Iva Cheung for her very careful proof reading of the final draft. Finally, I am really grateful to my partner, Jackie, and our children, Rene, Neil, Erin, Melanie, Ted, and Laura, for their encouragement over the many years it took to complete this book.

1

Role of probability theory in science

1.1 Scientific inference

This book is primarily concerned with the philosophy and practice of inferring the laws of nature from experimental data and prior information. The role of inference in the larger framework of the scientific method is illustrated in Figure 1.1.

In this simple model, the scientific method is depicted as a loop which is entered through initial observations of nature, followed by the construction of testable hypotheses or theories as to the working of nature, which give rise to the prediction of other properties to be tested by further experimentation or observation. The new data lead to the refinement of our current theories, and/or development of new theories, and the process continues.

The role of deductive inference[1] in this process, especially with regard to deriving the testable predictions of a theory, has long been recognized. Of course, any theory makes certain assumptions about nature which are assumed to be true and these assumptions form the axioms of the deductive inference process. The terms deductive inference and deductive reasoning are considered equivalent in this book. For example, Einstein's Special Theory of Relativity rests on two important assumptions; namely, that the vacuum speed of light is a constant in all inertial reference frames and that the laws of nature have the same form in all inertial frames.

Unfortunately, experimental tests of theoretical predictions do not provide simple yes or no answers. Our state of knowledge is always incomplete, there are always more experiments that could be done and the measurements are limited in their accuracy. Statistical inference is the process of inferring the truth of our theories of nature on the basis of the incomplete information. In science we often make progress by starting with simple models. Usually nature is more complicated and we learn in what direction to modify our theories from the differences between the model predictions and the measurements. It is much like peeling off layers of an onion. At any stage in this iterative process, the still hidden layers give rise to differences from the model predictions which guide the next step.

---

[1] Reasoning from one proposition to another using the strong syllogisms of logic (see Section 2.2.4).

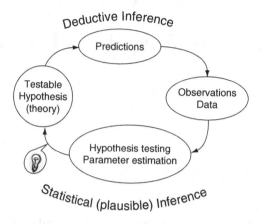

Figure 1.1 The scientific method.

1.2 Inference requires a probability theory

In science, the available information is always incomplete so our knowledge of nature is necessarily probabilistic. Two different approaches based on different definitions of probability will be considered. In conventional statistics, the probability of an event is identified with the long-run relative frequency of occurrence of the event. This is commonly referred to as the "frequentist" view. In this approach, probabilities are restricted to a discussion of random variables, quantities that can meaningfully vary throughout a series of repeated experiments. Two examples are:

1. A measured quantity which contains random errors.
2. Time intervals between successive radioactive decays.

The role of random variables in frequentist statistics is detailed in Section 5.2.

In recent years, a new perception of probability has arisen in recognition that the mathematical rules of probability are not merely rules for calculating frequencies of random variables. They are now recognized as uniquely valid principles of logic for conducting inference about any proposition or hypothesis of interest. This more powerful viewpoint, "Probability Theory as Logic," or Bayesian probability theory, is playing an increasingly important role in physics and astronomy. The Bayesian approach allows us to directly compute the probability of any particular theory or particular value of a model parameter, issues that the conventional statistical approach can attack only indirectly through the use of a random variable statistic. In this book, I adopt the approach which exposes probability theory as an extended theory of logic following the lead of E. T. Jaynes in his book,[2] *Probability Theory –*

---

[2] The book was finally submitted for publication four years after his death, through the efforts of his former student G. Larry Bretthorst.

Table 1.1 *Frequentist and Bayesian approaches to probability.*

Approach	Probability definition
FREQUENTIST STATISTICAL INFERENCE	$p(A)$ = long-run relative frequency with which A occurs in identical repeats of an experiment. "A" restricted to propositions about random variables.
BAYESIAN INFERENCE	$p(A\|B)$ = a real number measure of the plausibility of a proposition/hypothesis A, given (conditional on) the truth of the information represented by proposition B. "A" can be any logical proposition, *not* restricted to propositions about random variables.

The Logic of Science (Jaynes, 2003). The two approaches employ different definitions of probability which must be carefully understood to avoid confusion.

The two different approaches to statistical inference are outlined in Table 1.1 together with their underlying definition of probability. In this book, we will be primarily concerned with the Bayesian approach. However, since much of the current scientific culture is based on "frequentist" statistical inference, some background in this approach is useful.

The frequentist definition contains the term "identical repeats." Of course the repeated experiments can never be identical in all respects. The Bayesian definition of probability involves the rather vague sounding term "plausibility," which must be given a precise meaning (see Chapter 2) for the theory to provide quantitative results. In Bayesian inference, a probability distribution is an encoding of our uncertainty about some model parameter or set of competing theories, based on our current state of information. The approach taken to achieve an operational definition of probability, together with consistent rules for manipulating probabilities, is discussed in the next section and details are given in Chapter 2.

In this book, we will adopt the plausibility definition[3] of probability given in Table 1.1 and follow the approach pioneered by E. T. Jaynes that provides for a unified picture of both deductive and inductive logic. In addition, Jaynes brought

---

[3] Even within the Bayesian statistical literature, other definitions of probability exist. An alternative definition commonly employed is the following: "probability is a measure of the degree of belief that any well-defined proposition (an event) will turn out to be true." The events are still random variables, but the term is generalized so it can refer to the distribution of results from repeated measurements, or, to possible values of a physical parameter, depending on the circumstances. The concept of a coherent bet (e.g., D'Agostini, 1999) is often used to define the value of probability in an operational way. In practice, the final conditional posteriors are the same as those obtained from the extended logic approach adopted in this book.

great clarity to the debate on objectivity and subjectivity with the statement, "the only thing objectivity requires of a scientific approach is that experimenters with the same state of knowledge reach the same conclusion." More on this later.

1.2.1 The two rules for manipulating probabilities

It is now routine to build or program a computer to execute deductive logic. The goal of Bayesian probability theory as employed in this book is to provide an extension of logic to handle situations where we have incomplete information so we may arrive at the relative probabilities of competing hypotheses for a given state of information.

Cox and Jaynes showed that the desired extension can be arrived at uniquely from three "desiderata" which will be introduced in Section 2.5.1. They are called "desiderata" rather than axioms because they do not assert that anything is "true," but only state desirable goals of a theory of plausible inference.

The operations for manipulating probabilities that follow from the desiderata are the sum and product rules. Together with the Bayesian definition of probability, they provide the desired extension to logic to handle the common situation of incomplete information. We will simply state these rules here and leave their derivation together with a precise operational definition of probability to the next chapter.

$$\text{Sum Rule: } p(A|B) + p(\overline{A}|B) = 1 \tag{1.1}$$

$$\text{Product Rule: } p(A, B|C) = p(A|C)p(B|A, C)$$
$$= p(B|C)p(A|B, C), \tag{1.2}$$

where the symbol A stands for a proposition which asserts that something is true. The symbol B is a proposition asserting that something else is true, and similarly, C stands for another proposition. Two symbols separated by a comma represent a compound proposition which asserts that both propositions are true. Thus A, B indicates that both propositions A and B are true and $p(A, B|C)$ is commonly referred to as the joint probability. Any proposition to the right of the vertical bar | is assumed to be true. Thus when we write $p(A|B)$, we mean the probability of the truth of proposition A, given (conditional on) the truth of the information represented by proposition B.

Examples of propositions:

$A \equiv$ "The newly discovered radio astronomy object is a galaxy."
$B \equiv$ "The measured redshift of the object is 0.150 ± 0.005."
$A \equiv$ "Theory X is correct."
$\overline{A} \equiv$ "Theory X is not correct."
$A \equiv$ "The frequency of the signal is between f and $f + df$."

We will have much more to say about propositions in the next chapter.

Bayes' theorem follows directly from the product rule (a rearrangement of the two right sides of the equation):

$$p(A|B, C) = \frac{p(A|C)p(B|A, C)}{p(B|C)}. \qquad (1.3)$$

Another version of the sum rule can be derived (see Equation (2.23)) from the product and sum rules above:

$$\text{Extended Sum Rule: } p(A + B|C) = p(A|C) + p(B|C) - p(A, B|C), \qquad (1.4)$$

where $A + B \equiv$ proposition A is true or B is true or both are true. If propositions A and B are mutually exclusive – only one can be true – then Equation (1.4) becomes

$$p(A + B|C) = p(A|C) + p(B|C). \qquad (1.5)$$

1.3 Usual form of Bayes' theorem

$$p(H_i|D, I) = \frac{p(H_i|I)p(D|H_i, I)}{p(D|I)}, \qquad (1.6)$$

where $H_i \equiv$ proposition asserting the truth of a hypothesis of interest

$I \equiv$ proposition representing our prior information

$D \equiv$ proposition representing data

$p(D|H_i, I) =$ probability of obtaining data D, if H_i and I are true

(also called the likelihood function $\mathcal{L}(H_i)$)

$p(H_i|I) =$ prior probability of hypothesis

$p(H_i|D, I) =$ posterior probability of H_i

$p(D|I) = \sum_i p(H_i|I)p(D|H_i, I)$

(normalization factor which ensures $\sum_i p(H_i|D, I) = 1$).

1.3.1 Discrete hypothesis space

In Bayesian inference, we are interested in assigning probabilities to a set of competing hypotheses perhaps concerning some aspect of nature that we are studying. This set of competing hypotheses is called the *hypothesis space*. For example, a problem of current interest to astronomers is whether the expansion of the universe is accelerating or decelerating. In this case, we would be dealing with a discrete hypothesis space[4] consisting of H_1 ($\equiv$ accelerating) and H_2 ($\equiv$ decelerating). For a discrete hypothesis space, $p(H_i|D,I)$ is called a *probability distribution*. Our posterior probabilities for H_1 and H_2 satisfy the condition that

$$\sum_{i=1}^{2} p(H_i|D,I) = 1. \tag{1.7}$$

1.3.2 Continuous hypothesis space

In another type of problem we might be dealing with a hypothesis space that is continuous. This can be considered as the limiting case of an arbitrarily large number of discrete propositions.[5] For example, we have strong evidence from the measured velocities and distances of galaxies that we live in an expanding universe. Astronomers are continually seeking to refine the value of Hubble's constant, H_0, which relates the recession velocity of a galaxy to its distance. Estimating H_0 is called a parameter estimation problem and in this case, our hypothesis space of interest is continuous. In this case, the proposition H_0 asserts that the true value of Hubble's constant is in the interval h to $h + dh$. The truth of the proposition can be represented by $p(H_0|D,I)dH_0$, where $p(H_0|D,I)$ is a *probability density function* (PDF). The probability density function is defined by

$$p(H_0|D,I) = \lim_{\delta h \to 0} \frac{p(h \leq H_0 < h + \delta h|D,I)}{\delta h}. \tag{1.8}$$

Box 1.1 Note about notation

The term "PDF" is also a common abbreviation for *probability distribution function*, which can pertain to discrete or continuous sets of probabilities. This term is particularly useful when dealing with a mixture of discrete and continuous parameters.

We will use the same symbol, $p(\ldots)$, for probabilities and PDFs; the nature of the argument will identify which use is intended. To arrive at a final numerical answer for the probability or PDF of interest, we eventually need to convert the terms in Bayes' theorem into algebraic expressions, but these expressions can become very complicated in appearance. It is useful to delay this step until the last possible moment.

---

[4] Of course, nothing guarantees that future information will not indicate that the correct hypothesis is outside the current working hypothesis space. With this new information, we might be interested in an expanded hypothesis space.
[5] In Jaynes (2003), there is a clear warning that difficulties can arise if we are not careful in carrying out this limiting procedure explicitly. This is often the underlying cause of so-called paradoxes of probability theory.

Let W be a proposition asserting that the numerical value of H_0 lies in the range a to b. Then

$$p(W|D,I) = \int_a^b p(H_0|D,I)dH_0. \tag{1.9}$$

In the continuum limit, the normalization condition of Equation (1.7) becomes

$$\int_{\Delta H} p(H|D,I)dH = 1, \tag{1.10}$$

where ΔH designates the range of integration corresponding to the hypothesis space of interest.

We can also talk about a *joint probability distribution*, $p(X, Y|D, I)$, in which both X and Y are continuous, or, one is continuous and the other is discrete. If both are continuous, then $p(X, Y|D, I)$ is interpreted to mean

$$p(X, Y|D, I) = \lim_{\delta x, \delta y \to 0} \frac{p(x \leq X < x + \delta x, y \leq Y < y + \delta y|D, I)}{\delta x\, \delta y}. \tag{1.11}$$

In a well-posed problem, the prior information defines our hypothesis space, the means for computing $p(H_i|I)$, and the likelihood function given some data D.

1.3.3 Bayes' theorem – model of the learning process

Bayes' theorem provides a model for inductive inference or the learning process. In the parameter estimation problem of the previous section, H_0 is a continuous hypothesis space. Hubble's constant has some definite value, but because of our limited state of knowledge, we cannot be too precise about what that value is. In all Bayesian inference problems, we proceed in the same way. We start by encoding our prior state of knowledge into a prior probability distribution, $p(H_0|I)$ (in this case a density distribution). We will see a very simple example of how to do this in Section 1.4.1, and many more examples in subsequent chapters. If our prior information is very vague then $p(H_0|I)$ will be very broad, spanning a wide range of possible values of the parameter.

It is important to realize that a Bayesian PDF is a **measure of our state of knowledge** (i.e., ignorance) of the value of the parameter. The actual value of the parameter is not distributed over this range; it has some definite value. This can sometimes be a serious point of confusion, because, in frequentist statistics, the argument of a probability is a random variable, a quantity that can meaningfully take on different values, and these values correspond to possible outcomes of experiments.

We then acquire some new data, D_1. Bayes' theorem provides a means for combining what the data have to say about the parameter, through the likelihood function, with our prior, to arrive at a posterior probability density, $p(H_0|D_1, I)$, for the parameter.

$$p(H_0|D_1, I) \propto p(H_0|I)p(D_1|H_0, I). \tag{1.12}$$

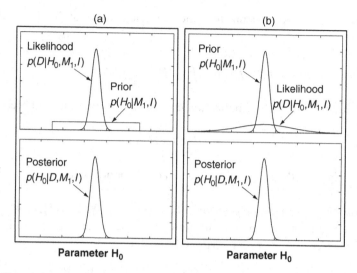

Figure 1.2 Bayes' theorem provides a model of the inductive learning process. The posterior PDF (lower graphs) is proportional to the product of the prior PDF and the likelihood function (upper graphs). This figure illustrates two extreme cases: (a) the prior much broader than likelihood, and (b) likelihood much broader than prior.

Two extreme cases are shown in Figure 1.2. In the first, panel (a), the prior is much broader than the likelihood. In this case, the posterior PDF is determined entirely by the new data. In the second extreme, panel (b), the new data are much less selective than our prior information and hence the posterior is essentially the prior.

Now suppose we acquire more data represented by proposition D_2. We can again apply Bayes' theorem to compute a posterior that reflects our new state of knowledge about the parameter. This time our new prior, I', is the posterior derived from D_1, I, i.e., $I' = D_1, I$. The new posterior is given by

$$p(H_0|D_2, I') \propto p(H_0|I')p(D_2|H_0, I'). \tag{1.13}$$

1.3.4 Example of the use of Bayes' theorem

Here we analyze a simple model comparison problem using Bayes' theorem. We start by stating our prior information, I, and the new data, D.

I stands for:

a) Model M_1 predicts a star's distance, $d_1 = 100$ light years (ly).
b) Model M_2 predicts a star's distance, $d_2 = 200$ ly.
c) The uncertainty, e, in distance measurements is described by a Gaussian distribution of the form

$$p(e|I) = \frac{1}{\sqrt{2\pi}\,\sigma} \exp\left(-\frac{e^2}{2\sigma^2}\right),$$

where $\sigma = 40$ ly.

d) There is no current basis for preferring M_1 over M_2 so we set $p(M_1|I) = p(M_2|I) = 0.5$.

$D \equiv$ "The measured distance $d = 120$ ly."

The prior information tells us that the hypothesis space of interest consists of models (hypotheses) M_1 and M_2. We proceed by writing down Bayes' theorem for each hypothesis, e.g.,

$$p(M_1|D, I) = \frac{p(M_1|I)p(D|M_1, I)}{p(D|I)}; \tag{1.14}$$

$$p(M_2|D, I) = \frac{p(M_2|I)p(D|M_2, I)}{p(D|I)}. \tag{1.15}$$

Since we are interested in comparing the two models, we will compute the *odds ratio*, equal to the ratio of the posterior probabilities of the two models. We will abbreviate the odds ratio of model M_1 to model M_2 by the symbol O_{12}.

$$O_{12} = \frac{p(M_1|D, I)}{p(M_2|D, I)} = \frac{p(M_1|I)}{p(M_2|I)} \frac{p(D|M_1, I)}{p(D|M_2, I)} = \frac{p(D|M_1, I)}{p(D|M_2, I)}. \tag{1.16}$$

The two prior probabilities cancel because they are equal and so does $p(D|I)$ since it is common to both models. To evaluate the likelihood $p(D|M_1, I)$, we note that in this case, we are assuming M_1 is true. That being the case, the only reason the measured d can differ from the prediction d_1 is because of measurement uncertainties, e. We can thus write $d = d_1 + e$ or $e = d - d_1$. Since d_1 is determined by the model, it is certain, and so the probability,[6] $p(D|M_1, I)$, of obtaining the measured distance is equal to the probability of the error. Thus we can write

$$p(D|M_1, I) = \frac{1}{\sqrt{2\pi}\,\sigma} \exp\left(-\frac{(d - d_1)^2}{2\sigma^2}\right)$$

$$= \frac{1}{\sqrt{2\pi}\,40} \exp\left(-\frac{(120 - 100)^2}{2(40)^2}\right) = 0.00880. \tag{1.17}$$

Similarly we can write for model M_2

---

[6] See Section 4.8 for a more detailed treatment of this point.

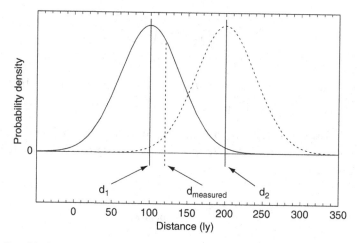

Figure 1.3 Graphical depiction of the evaluation of the likelihood functions, $p(D|M_1, I)$ and $p(D|M_2, I)$.

$$p(D|M_2, I) = \frac{1}{\sqrt{2\pi}\,\sigma} \exp\left(-\frac{(d - d_2)^2}{2\sigma^2}\right)$$

$$= \frac{1}{\sqrt{2\pi}\,40} \exp\left(-\frac{(120 - 200)^2}{2(40)^2}\right) = 0.00135. \tag{1.18}$$

The evaluation of Equations (1.17) and (1.18) is depicted graphically in Figure 1.3. The relative likelihood of the two models is proportional to the heights of the two Gaussian probability distributions at the location of the measured distance. Substituting into Equation (1.16), we obtain an odds ratio of 6.52 in favor of model M_1.

1.4 Probability and frequency

In Bayesian terminology, a *probability* is a representation of our state of knowledge of the real world. A *frequency* is a factual property of the real world that we measure or estimate.[7] One of the great strengths of Bayesian inference is the ability to incorporate relevant prior information in the analysis. As a consequence, some critics have discounted the approach on the grounds that the conclusions are subjective and there has been considerable confusion on that subject. We certainly expect that when scientists from different laboratories come together at an international meeting, their state of knowledge about any particular topic will differ, and as such, they may have arrived at different conclusions. It is important to recognize

---

[7] For example, consider a sample of 400 people attending a conference. Each person sampled has many characteristics or attributes including sex and eye color. Suppose 56 are found to be female. Based on this sample, the frequency of occurrence of the attribute female is $56/400 \equiv 14\%$.

that the only thing objectivity requires of a scientific approach is that *experimenters with the same state of knowledge reach the same conclusion.* Achieving consensus amongst different experimenters is greatly aided by the requirement to specify how relevant prior information has been encoded in the analysis. In Bayesian inference, we can readily incorporate frequency information using Bayes' theorem and by treating it as data. In general, probabilities change when we change our state of knowledge; frequencies do not.

1.4.1 Example: incorporating frequency information

A 1996 newspaper article reported that doctors in Toronto were concerned about a company selling an unapproved mail-order HIV saliva test. According to laboratory tests, the false positive rate for this test was 2.3% and the false negative rate was 1.4% (i.e., 98.6% reliable based on testing of people who actually have the disease).

In this example, suppose a new deadly disease is discovered for which there is no known cause but a saliva test is available with the above specifications. We will refer to this disease by the abbreviation UD, for unknown disease. You have no reason to suspect you have UD but decide to take the test anyway and test positive. What is the probability that you really have the disease? Here is a Bayesian analysis of this situation. For the purpose of this analysis, we will assume that the incidence of the disease in a random sample of the region is $1:10\,000$.

Let $H \equiv$ "You have UD."

$\overline{H} \equiv$ "You do not have UD."

$D_1 \equiv$ "You test positive for UD."

$I_1 \equiv$ "No known cause for the UD,

$p(D_1|H, I_1) = 0.986,$

$p(D_1|\overline{H}, I_1) = 0.023,$

incidence of UD in the population is $1:10^4$."

The starting point for any Bayesian analysis is to write down Bayes' theorem.

$$p(H|D_1, I_1) = \frac{p(H|I_1)p(D_1|H, I_1)}{p(D_1|I_1)}. \tag{1.19}$$

Since $p(D_1|I_1)$ is a normalization factor, which ensures $\sum_i p(H_i|D_1, I_1) = 1$, we can write

$$p(D_1|I_1) = p(H|I_1)p(D_1|H, I_1) + p(\overline{H}|I_1)p(D_1|\overline{H}, I_1). \tag{1.20}$$

In words, this latter equation stands for

$$
\begin{pmatrix} \text{prob. of a} \\ \text{+ test} \end{pmatrix} = \begin{pmatrix} \text{prob. you} \\ \text{have UD} \end{pmatrix} \times \begin{pmatrix} \text{prob. of a +} \\ \text{test when you} \\ \text{have UD} \end{pmatrix}
$$

$$
+ \begin{pmatrix} \text{prob. you} \\ \text{don't have UD} \end{pmatrix} \times \begin{pmatrix} \text{prob. of a +} \\ \text{test when you} \\ \text{don't have UD} \end{pmatrix} \tag{1.21}
$$

$$
= \begin{pmatrix} \text{incidence of} \\ \text{UD in population} \end{pmatrix} \times (\text{reliability of test})
$$

$$
+ \begin{pmatrix} 1 - \dfrac{\text{incidence}}{\text{of UD}} \end{pmatrix} \times (\text{false positive rate})
$$

$$
p(H|D_1, I_1) = \frac{10^{-4} \times 0.986}{10^{-4} \times 0.986 + 0.9999 \times 0.023} = 0.0042.
$$

Thus, the probability you have the disease is 0.4% (not 98.6%).

Question: How would the conclusion change if the false positive rate of the test were reduced to 0.5%?

Suppose you now have a doctor examine you and obtain new independent data D_2, perhaps from a blood test.

$$
I_2 = \text{New state of knowledge} = D_1, I_1 \Rightarrow p(H|D_2, I_2) = \frac{p(H|I_2)p(D_2|H, I_2)}{p(D_2|I_2)},
$$

where $p(H|I_2) = p(H|D_1, I_1)$.

1.5 Marginalization

In this section, we briefly introduce marginalization, but we will learn about important subtleties to this operation in later chapters. Consider the following parameter estimation problem. We have acquired some data, D, which our prior information, I, indicates will contain a periodic signal. Our signal model has two continuous parameters – an angular frequency, ω, and an amplitude, A. We want to focus on the implications of the data for the ω, independent of the signal's amplitude, A.

We can write the joint probability[8] of ω and A given data D and prior information I as $p(\omega, A|D, I)$. In this case ω, A is a compound proposition asserting that the two

---

[8] Since a parameter of a model is not a random variable, the frequentist approach is denied the concept of the probability of a parameter.

propositions are true. How do we obtain an expression for the probability of the proposition ω? We eliminate the uninteresting parameter A by *marginalization*. How do we do this?

For simplicity, we will start by assuming that the parameter A is discrete. In this case, A can only take on the values A_1 or A_2 or A_3, etc. Since we are assuming the model to be true, the proposition represented by $A_1 + A_2 + A_3 + \cdots$, where the $+$ stands for the Boolean 'or', must be true for some value of A_i and hence,

$$p(A_1 + A_2 + A_3 + \cdots | I) = 1. \tag{1.22}$$

Now $\omega, [A_1 + A_2 + A_3 + \cdots]$ is a compound proposition which asserts that both ω and $[A_1 + A_2 + A_3 + \cdots]$ are true. The probability that this compound proposition is true is represented by $p(\omega, [A_1 + A_2 + A_3 + \cdots] | D, I)$. We use the product rule to expand the probability of this compound proposition.

$$p(\omega, [A_1 + A_2 + A_3 + \cdots] | D, I) = p([A_1 + A_2 + A_3 + \cdots] | D, I)$$
$$\times \ p(\omega | [A_1 + A_2 + A_3 + \cdots], D, I) \tag{1.23}$$
$$= 1 \times p(\omega | D, I).$$

The second line of the above equation has the quantity $[A_1 + A_2 + A_3 + \cdots], D, I$ to the right of the vertical bar which should be read as assuming the truth of $[A_1 + A_2 + A_3 + \cdots], D, I$. Now $[A_1 + A_2 + A_3 + \cdots], D, I$ is a compound proposition asserting that all three propositions are true. Since proposition $[A_1 + A_2 + A_3 + \cdots]$ is given as true by our prior information, I, knowledge of its truth is already contained in proposition I. Thus, we can simplify the expression by replacing $p(\omega | [A_1 + A_2 + A_3 + \cdots], D, I)$ by $p(\omega | D, I)$.

Rearranging Equation (1.23), we get

$$p(\omega | D, I) = p(\omega, [A_1 + A_2 + A_3 + \cdots] | D, I). \tag{1.24}$$

The left hand side of the equation is the probability we are seeking, but we are not finished with the right hand side. Now we do a simple expansion of the right hand side of Equation (1.24) by multiplying out the two propositions ω and $[A_1 + A_2 + A_3 + \cdots]$ using a Boolean algebra relation which is discussed in more detail in Chapter 2.

$$p(\omega, [A_1 + A_2 + A_3 + \cdots] | D, I) = p(\{\omega, A_1\} + \{\omega, A_2\} + \{\omega, A_3\} + \cdots | D, I). \tag{1.25}$$

The term $\{\omega, A_1\} + \{\omega, A_2\} + \{\omega, A_3\} + \cdots$ is a proposition which asserts that ω, A_1 is true, or, ω, A_2 is true, or, ω, A_3 is true, etc. We have surrounded each of the ω, A_i terms by curly brackets to help with the interpretation, but normally they are not required because the logical conjunction operation designated by a comma between two propositions takes precedence over the logical "or" operation designated by the $+$ sign.

The extended sum rule, given by Equation (1.5), says that the probability of the sum of two mutually exclusive (only one can be true) propositions is the sum of their individual probabilities. Since the compound propositions ω, A_i for different i are mutually exclusive, we can rewrite Equation (1.25) as

$$p(\omega, [A_1 + A_2 + A_3 + \cdots]|D, I) = p(\omega, A_1|D, I) + p(\omega, A_2|D, I)$$
$$+ p(\omega, A_3|D, I) + \cdots \tag{1.26}$$

Substitution of Equation (1.26) into Equation (1.24) yields:

$$p(\omega|D, I) = \sum_i p(\omega, A_i|D, I). \tag{1.27}$$

Extending this idea to the case where A is a continuously variable parameter instead of a discrete parameter, we can write

$$p(\omega|D, I) = \int dA \, p(\omega, A|D, I). \tag{1.28}$$

The quantity, $p(\omega|D, I)$, is the marginal posterior distribution for ω, which, for a continuous parameter like ω, is a probability density function. It summarizes what D, I (our knowledge state) says about the parameter(s) of interest. The probability that ω will lie in any specific range from ω_1 to ω_2 is given by $\int_{\omega_1}^{\omega_2} p(\omega|D, I)d\omega$.

Another useful form of the marginalization operation can be obtained by expanding Equation (1.28) using Bayes' theorem:

$$p(\omega, A|D, I) = \frac{p(\omega, A|I)p(D|\omega, A, I)}{p(D|I)}. \tag{1.29}$$

Now expand $p(\omega, A|I)$ on the right hand side of Equation (1.29) using the product rule:

$$p(\omega, A|I) = p(\omega|I)p(A|\omega, I). \tag{1.30}$$

Now we will assume the priors for ω and A are independent so we can write $p(A|\omega, I) = p(A|I)$. What this is saying is that any prior information we have about the parameter ω tells us nothing about the parameter A. This assumption is frequently valid and it usually simplifies the calculations. Equation (1.29) can now be rewritten as

$$p(\omega, A|D, I) = \frac{p(\omega|I)p(A|I)p(D|\omega, A, I)}{p(D|I)}. \tag{1.31}$$

Finally, substitution of Equation (1.31) into Equation (1.28) yields:

$$p(\omega|D, I) \propto p(\omega|I) \int dA \, p(A|I)p(D|\omega, A, I). \tag{1.32}$$

This gives the marginal posterior distribution $p(\omega|D, I)$, in terms of the weighted average of the likelihood function, $p(D|\omega, A, I)$, weighted by $p(A|I)$, the prior probability density function for A. This is another form of the operation of marginalizing out the A parameter. The integral in Equation (1.32) can sometimes be evaluated analytically which can greatly reduce the computational aspects of the problem especially when many parameters are involved. A dramatic example of this is given in Gregory and Loredo (1992) which demonstrates how to marginalize analytically over a very large number of parameters in a model describing a waveform of unknown shape.

1.6 The two basic problems in statistical inference

1. **Model selection**: Which of two or more competing models is most probable given our present state of knowledge?

   The competing models may have different numbers of parameters. For example, suppose we have some experimental data consisting of a signal plus some additive noise and we want to distinguish between two different models for the signal present. Model M_1 predicts that the signal is a constant equal to zero, i.e., has no unknown (free) parameters. Model M_2 predicts that the signal consists of a single sine wave of known frequency f. Let us further suppose that the amplitude, A, of the sine wave is a free parameter within some specified prior range. In this problem, M_1 has no free parameters and M_2 has one free parameter, A.

   In model selection, we are interested in the most probable model, independent of the model parameters (i.e., marginalize out all parameters). This is illustrated in the equation below for model M_2.

   $$p(M_2|D, I) = \int_{\Delta A} dA \, p(M_2, A|D, I),\qquad(1.33)$$

   where ΔA designates the appropriate range of integration of A as specified by our prior information, I.

   We can rearrange Equation (1.33) into another useful form by application of Bayes' theorem and the product rule, following the example given in the previous section (Equations (1.28) to (1.32)). The result is

   $$p(M_2|D, I) = \frac{p(M_2|I) \int_{\Delta A} dA \, p(A|M_2, I)p(D|M_2, A, I)}{p(D|I)}.\qquad(1.34)$$

   In model selection, the hypothesis space of interest is discrete (although its parameters may be continous) and M_2 stands for the second member of this discrete space.

2. **Parameter estimation**: Assuming the truth of a model, find the probability density function for each of its parameters.

   Suppose the model M has two free parameters f and A. In this case, we want to solve for $p(f|D, M, I)$ and $p(A|D, M, I)$. The quantity $p(f|D, M, I)$ is called the marginal posterior distribution for f, which, for a continuous parameter like f, is a probability density function as defined by Equation (1.8). In Chapter 3, we will work through a detailed example of both model selection and parameter estimation.

1.7 Advantages of the Bayesian approach

1. Provides an elegantly simple and rational approach for answering, in an optimal way, any scientific question for a given state of information. This contrasts to the recipe or cookbook approach of conventional statistical analysis. The procedure is well-defined:

   (a) Clearly state your question and prior information.
   (b) Apply the sum and product rules. The starting point is always Bayes' theorem.

   For some problems, a Bayesian analysis may simply lead to a familiar statistic. Even in this situation it often provides a powerful new insight concerning the interpretation of the statistic. One example of this is shown in Figure 1.4 and discussed in detail in Chapter 13.
2. Calculates probability of hypothesis directly: $p(H_i|D, I)$.
3. Incorporates relevant prior (e.g., known signal model) information through Bayes' theorem. This is one of the great strengths of Bayesian analysis. For data with a high signal-to-noise ratio, a Bayesian analysis can frequently yield many orders of magnitude improvement in model parameter estimation, through the incorporation of relevant prior information about the signal model. This is illustrated in Figure 1.5 and discussed in more detail in Chapter 13.
4. Provides a way of eliminating nuisance parameters through *marginalization*. For some problems, the marginalization can be performed analytically, permitting certain calculations to become computationally tractable (see Section 13.4).
5. Provides a more powerful way of assessing competing theories at the forefront of science by automatically quantifying *Occam's razor*. Occam's razor is a principle attributed to the medieval philosopher William of Occam (or Ockham). The principle states that one should not make more assumptions than the minimum needed. It underlies all scientific modeling and theory building. It cautions us to choose from a set of otherwise equivalent models of a given phenomenon the simplest one. In any given model, Occam's razor helps us to "shave off" those variables that are not really needed to explain the phenomenon. It was previously thought to be only a qualitative principle. This topic is introduced in Section 3.5.

   The Bayesian quantitative Occam's razor can also save a lot of time that might otherwise be spent chasing noise artifacts that masquerade as possible detections of real phenomena. One example of this is discussed in Section 12.9 on extrasolar planets.
6. Provides a way for incorporating the effects of *systematic errors* arising from both the measurement operation and theoretical model predictions. Figure 1.6 illustrates the effect of a systematic error in the scale of the cosmic ruler (Hubble's constant) used to determine the distance to galaxies. This topic is introduced in Section 3.11.

These advantages will be discussed in detail beginning in Chapter 3. We close with a reminder that in Bayesian inference probabilities are a measure of our state of knowledge about nature, not a measure of nature itself.

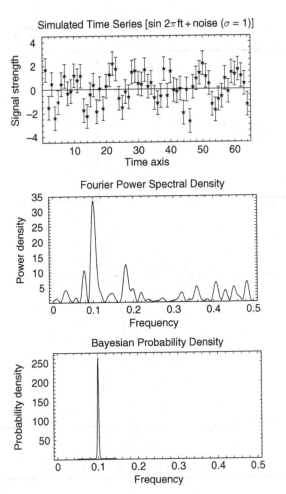

Figure 1.4 The upper panel shows a simulated time series consisting of a single sinusoidal signal with added independent Gaussian noise. A common conventional analysis (middle panel) involves plotting the power spectrum, based on a Discrete Fourier Transform (DFT) statistic of the data. The Bayesian analysis (lower panel) involves a nonlinear processing of the same DFT statistic, which suppresses spurious peaks and the width of the spectral peak reflects the accuracy of the frequency estimate.

1.8 Problems

1. For the example given in Section 1.3.4, compute $p(D|M_1, I)$ and $p(D|M_2, I)$, for a $\sigma = 25$ ly.
2. For the example given in Section 1.4.1, compute the probability that the person has the disease, if the false positive rate for the test $= 0.5\%$, and everything else is the same.

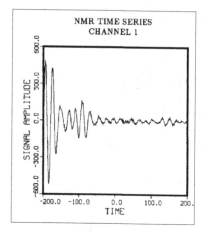
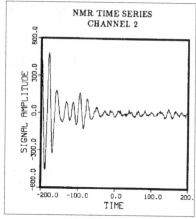
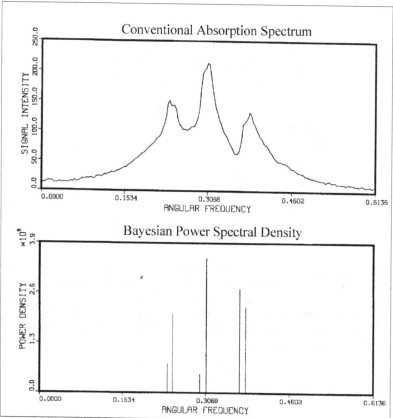

Figure 1.5 Comparison of conventional analysis (middle panel) and Bayesian analysis (lower panel) of the two-channel nuclear magnetic resonance free induction decay time series (upper two panels). By incorporating prior information about the signal model, the Bayesian analysis was able to determine the frequencies and exponential decay rates to an accuracy many orders of magnitude greater than for a conventional analysis. (Figure credit G. L. Bretthorst, reproduced by permission from the American Institute of Physics.)

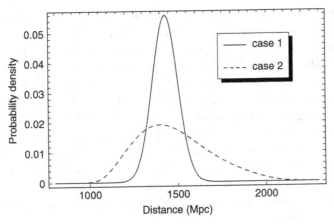

Figure 1.6 The probability density function for the distance to a galaxy assuming: 1) a fixed value for Hubble's constant (H_0), and 2) incorporating a Gaussian prior uncertainty for H_0 of $\pm 14\%$.

3. In Section 1.4.1, based on the saliva test result and the prior information, the probability that the person had the unknown disease (UD) was found to be 0.42%. Subsequently, the same person received an independent blood test for UD and again tested positive. If the false negative rate for this test is 1.4% and the false positive rate is 0.5%, what is the new probability that the person has UD on the basis of both tests?

4. **Joint and marginal probability distributions**

   (Refer to the example on this topic in the *Mathematica* tutorial.)

   (a) Suppose we are interested in estimating the parameters X and Y of a certain model M, where both parameters are continuous as opposed to discrete. Make a contour plot of the following posterior joint probability density function given by:

   $$p(X, Y|D, M, I) = A_1 \, \exp\left(-\frac{(x - x_1)^2 + (y - y_1)^2}{2\sigma_1^2}\right)$$
   $$+ A_2 \, \exp\left(-\frac{(x - x_2)^2 + (y - y_2)^2}{2\sigma_2^2}\right),$$

   where $A_1 = 4.82033$, $A_2 = 4.43181$, $x_1 = 0.5$, $y_1 = 0.5$, $x_2 = 0.65$, $y_2 = 0.75$, $\sigma_1 = 0.2$, $\sigma_2 = 0.04$, where $0 \le x \le 1$ and $0 \le y \le 1$. Your contour plot should cover the interval $x = 0 \rightarrow 1, y = 0 \rightarrow 1$. In *Mathematica*, this can be accomplished with **ContourPlot**.

   (b) Now make a 3-dimensional plot of $p(X, Y|D, M, I)$. In *Mathematica*, this can be accomplished with **Plot3D**.

   (c) Now compute the marginal probability distributions $p(X|D, M, I)$ and $p(Y|D, M, I)$. The prior information is
   $I \equiv$ "X and Y are only non-zero in the interval $0 \rightarrow 1$, and uniform within that interval."
   Check that the integral of $p(X|D, M, I)$ in the interval $0 \rightarrow 1$ is equal to 1.

(d) In your 3-dimensional plot of part (b), probability is represented by a height along the z-axis. Now imagine a light source located a great distance away along the y-axis illuminating the 3-dimensional probability density function. The shadow cast by $p(X, Y|D, M, I)$ on the plane defined by $y = 0$, we will call the *projected probability density function* of X. Compute and compare the projected probability density function of X with the marginal distribution on the same plot. To accomplish this effectively, both density functions should be normalized to have an integral $= 1$ in the interval $x = 0 \rightarrow 1$.

Note: the location of the peak of the marginal does not correspond to the location of the projection peak although they would if the joint probability density function were a single multi-dimensional Gaussian.

(e) Plot the normalized marginal and projected probability density functions for Y on one graph.

2
Probability theory as extended logic

2.1 Overview

The goal of this chapter is to provide an extension of logic to handle situations where we have incomplete information so we may arrive at the relative probabilities of competing propositions (theories, hypotheses, or models) for a given state of information. We start by reviewing the algebra of logical propositions and explore the structure (syllogisms) of deductive and plausible inference. We then set off on a course to come up with a quantitative theory of plausible inference (probability theory as extended logic) based on the three desirable goals called *desiderata*. This amounts to finding an adequate set of mathematical operations for plausible inference that satisfies the desiderata. The two operations required turn out to be the product rule and sum rule of probability theory. The process of arriving at these operations uncovers a precise operational definition of plausibility, which is determined by the data. The material presented in this chapter is an abridged version of the treatment given by E. T. Jaynes in his book, *Probability Theory – The Logic of Science* (Jaynes, 2003), with permission from Cambridge University Press.

2.2 Fundamentals of logic

2.2.1 Logical propositions

In general, we will represent propositions by capital letters $\{A, B, C, \text{etc.}\}$. A proposition asserts that something is true.

$$\text{e.g., } A \equiv \text{“The age of the specimen is } \geq 10^6 \text{ years.”}$$

The denial of a proposition is indicated by a bar:

$$\overline{A} \equiv \text{“}A\text{ is false.”}$$

We will only be concerned with two-valued logic; thus, any proposition has a truth value of either

$$\left.\begin{array}{ccc} \text{True} & \text{or} & \text{False} \\ 1 & \text{or} & 0 \end{array}\right\} \leftarrow \text{Truth value.}$$

21

2.2.2 Compound propositions

$A, B \equiv$ asserts both A and B are true

(*logical product* or *conjunction*)

$A, \overline{A} \equiv$ impossible statement, truth value $=$ F or zero

$A + B \equiv$ asserts A is true or B is true or both are true

(*logical sum* or *disjunction*)

$A, \overline{B} + B, \overline{A} \equiv$ asserts either A is true or B is true but both are not true

(*exclusive form of logical sum*)

2.2.3 Truth tables and Boolean algebra

Consider the two compound propositions $A = \overline{B, C}$ and $D = \overline{B} + \overline{C}$. Are the propositions A and D equal? Two propositions are equal if they have the same truth value. We can verify that $A = D$ by constructing a truth table which lays out the truth values for A and D for all the possible combinations of the truth values of the propositions B and C on which they are based (Table 2.1).

Since A and D have the same truth value for all possible truth values of propositions B and C, then we can write

$A = D$ (which means they are logically equivalent).

We have thus established the relationship

$$\overline{B, C} = \overline{B} + \overline{C} \text{ and } \neq \overline{B}, \overline{C}. \tag{2.1}$$

In addition, the last two columns of the table establish the relationship

$$\overline{B, C} = \overline{B + C}. \tag{2.2}$$

Boole (1854) pointed out that the propositional statements in symbolic logic obey the rules of algebra provided one interprets them as having values of 1 or 0 (Boolean algebra). There are no operations equivalent to subtraction or division. The only operations required are multiplications ('and') and additions ('or').

Table 2.1

B	C	B, C	$A = \overline{B, C}$	$D = \overline{B} + \overline{C}$	$B + C$	$\overline{B + C}$	$\overline{B}, \overline{C}$
T	T	T	F	F	T	F	F
T	F	F	T	T	T	F	F
F	T	F	T	T	T	F	F
F	F	F	T	T	F	T	T

Box 2.1   Worked exercise:

construct a truth table to show $A, (B + C) = A, B + A, C$.

A	B	C	$B + C$	$A, (B + C)$	A, B	A, C	$A, B + A, C$
T	T	T	T	T	T	T	T
T	F	F	F	F	F	F	F
T	T	F	T	T	T	F	T
T	F	T	T	T	F	T	T
F	T	T	T	F	F	F	F
F	F	F	F	F	F	F	F
F	T	F	T	F	F	F	F
F	F	T	T	F	F	F	F

Since $A, (B + C)$ and $A, B + A, C$ have the same truth value for all possible truth values of propositions A, B and C, then we can write
$A, (B + C) = A, B + A, C$. (This is a *distributivity identity*.)

One surprising result of Boolean algebra manipulations is that a given statement may take several different forms which don't resemble one another.

For example, show that $D = A + B, C = (A + B), (A + C)$.
In the proof below, we make use of the relationships $\overline{X, Y} = \overline{X} + \overline{Y}$ (on line 1), and $\overline{X, Y} = \overline{X} + \overline{Y}$ (on line 3), from Equations (2.1) and (2.2).

$$D = A + B, C = \overline{\overline{A}} + \overline{\overline{B, C}} = \overline{\overline{A}, \overline{B, C}}$$

$$D = \overline{\overline{A}, (\overline{B} + \overline{C})}$$

$$\overline{D} = \overline{A}, \overline{B} + \overline{A}, \overline{C} = \overline{(A + B)} + \overline{(A + C)}$$

$$\overline{D} = \overline{(A + B), (A + C)}$$

$$D = (A + B), (A + C)$$

or $A + B, C = (A + B), (A + C)$.

This can also be verified by constructing a truth table.

Basic Boolean Identities

Idempotence:	A, A	$=$	A		
	$A + A$	$=$	A		
Commutativity:	A, B	$=$	B, A		
	$A + B$	$=$	$B + A$		
Associativity:	$A, (B, C)$	$=$	$(A, B), C$	$=$	A, B, C
	$A + (B + C)$	$=$	$(A + B) + C$	$=$	$A + B + C$
Distributivity:	$A, (B + C)$	$=$	$A, B + A, C$		
	$A + (B, C)$	$=$	$(A + B), (A + C)$		
Duality:	If $C = A, B$,	then	$\overline{C} = \overline{A} + \overline{B}$		
	If $D = A + B$,	then	$\overline{D} = \overline{A}, \overline{B}$		

By the application of these identities, one can prove any number of further relations, some highly non-trivial. For example, we shall presently have use for the rather elementary "theorem":

$$\text{If } \overline{B} = A, D$$
$$A, \overline{B} = A, A, D = A, D = \overline{B} \tag{2.3}$$
$$\text{then } A, \overline{B} = \overline{B}.$$

Also, we can show that:

$$B, \overline{A} = \overline{A}. \tag{2.4}$$

Proof of the latter follows from

$$B = \overline{\overline{A}, D} = \overline{A} + \overline{D}$$
$$B, \overline{A} = \overline{A}, \overline{A} + \overline{A}, \overline{D} = \overline{A} + \overline{A}, \overline{D} = \overline{A}. \tag{2.5}$$

Clearly, Equation (2.5) is true if $\overline{A}$ is true and false if $\overline{A}$ is false, regardless of the truth of $\overline{D}$.

2.2.4 Deductive inference

Deductive inference is the process of reasoning from one proposition to another. It was recognized by Aristotle (fourth century BC) that deductive inference can be analyzed into repeated applications of the *strong syllogisms*:

1. If A is true, then B is true (major premise)

   $\underline{\hspace{4em} A \text{ is true} \hspace{4em} \text{(minor premise)}}$
   Therefore B is true   (conclusion)

2. If A is true, then B is true

   $\underline{\hspace{4em} B \text{ is false} \hspace{4em}}$
   Therefore A is false

In Boolean algebra, these strong syllogisms can be written as:

$$A = A, B. \qquad (2.6)$$

This equation says that the truth value of proposition A, B is equal to the truth value of proposition A. It does not assert that either A or B is true. Clearly, if B is false, then the right hand side of the equation equals 0, and so A must be false. On the other hand, if B is known to be true, then according to Equation (2.6), proposition A can be true or false. It is also written as the implication operation $A \Rightarrow B$.

2.2.5 Inductive or plausible inference

In almost all situations confronting us, we do not have the information required to do deductive inference. We have to fall back on weaker syllogisms:

If A is true, then B is true

B is true
_____
Therefore A becomes more plausible

Example

$A \equiv$ "It will start to rain by 10 AM at the latest."

$B \equiv$ "The sky becomes cloudy before 10 AM."

Observing clouds at 9:45 AM does not give us logical certainty that rain will follow; nevertheless, our common sense, obeying the weak syllogism, may induce us to change our plans and behave as if we believed that it will rain, if the clouds are sufficiently dark.

This example also shows the major premise: "If A then B" expresses B only as a logical consequence of A and not necessarily as a causal consequence (i.e., the rain is not the cause of the clouds).

Another weak syllogism:

If A is true, then B is true

A is false
_____
Therefore B becomes less plausible

2.3 Brief history

The early work on probability theory by James Bernoulli (1713), Rev. Thomas Bayes (1763), and Pierre Simon Laplace (1774), viewed probability as an extension of logic to the case where, because of incomplete information, Aristotelian deductive reasoning is unavailable. Unfortunately, Laplace failed to give convincing arguments to show why the Bayesian definition of probability uniquely required the sum and product rules for manipulating probabilities. The frequentist definition of probability was introduced to satisfy this point, but in the process, eliminated the interpretation of probability as extended logic. This caused a split in the subject into the Bayesian and frequentist camps. The frequentist approach dominated statistical inference throughout most of the twentieth century, but the Bayesian viewpoint was kept alive notably by Sir Harold Jeffreys (1891–1989).

In the 1940s and 1950s, G. Polya, R. T. Cox and E. T. Jaynes provided the missing rationale for Bayesian probability theory. In his book *Mathematics and Plausible Reasoning*, George Polya dissected our "common sense" into a set of elementary desiderata and showed that mathematicians had been using them all along to guide the early stages of discovery, which necessarily precede the finding of a rigorous proof. When one added (see Section 2.5.1) the consistency desiderata of Cox (1946) and Jaynes, the result was a proof that, if degrees of plausibility are represented by real numbers, then there is a unique set of rules for conducting inference according to Polya's desiderata which provides for an operationally defined scale of plausibility. The final result was just the standard product and sum rules of probability theory, given axiomatically by Bernoulli and Laplace! The important new feature is that these rules are now seen as uniquely valid principles of logic in general, making no reference to "random variables", so their range of application is vastly greater than that supposed in the conventional probability theory that was developed in the early twentieth century. With this came a revival of the notion of probability theory as extended logic.

The work of Cox and Jaynes was little appreciated at first. Widespread application of Bayesian methodology did not occur until the 1980s. By this time computers had become sufficiently powerful to demonstrate that the methodology could outperform standard techniques in many areas of science. We are now in the midst of a "Bayesian Revolution" in statistical inference. In spite of this, many scientists are still unaware of the significance of the revolution and the frequentist approach currently dominates statistical inference. New graduate students often find themselves caught between the two cultures. This book represents an attempt to provide a bridge.

2.4 An adequate set of operations

So far, we have discussed the following logical operations:

A, B logical product (conjunction)

$A + B$ logical sum (disjunction)

$A \Rightarrow B$ implication

$\overline{A}$ negation

By combining these operations repeatedly in every possible way, we can generate any number of new propositions, such as:

$$C \equiv (A + \overline{B}), (\overline{A} + A, \overline{B}) + \overline{A}, B, (A + B). \qquad (2.7)$$

We now consider the following questions:

1. How large is the class of new propositions?
2. Is it infinite or finite?
3. Can every proposition defined from A and B be represented in terms of the above operations, or are new operations required?
4. Are the four operations already over-complete?

Note: two propositions are not different from the standpoint of logic if they have the same truth value. C, in the above equation, is logically the same statement as the implication $C = (B \Rightarrow \overline{A})$. Recall that the implication $B \Rightarrow \overline{A}$ can also be written as $B = \overline{A}, B$. This does not assert that either A or B is true; it only means that A, B is false, or equivalently that $(\overline{A} + \overline{B})$ is true.

Box 2.2   Worked exercise:

expand the right hand side (RHS) of proposition C given by Equation (2.7), and show that it can be reduced to $(\overline{A} + \overline{B})$.

$$\text{RHS} = A, \overline{A} + \overline{A}, \overline{B} + A, A, \overline{B} + A, \overline{B}, B + \overline{A}, A, B + \overline{A}, B, B$$

Drop all terms that are clearly impossible (false), e.g., $A, \overline{A}$. Adding any number of impossible propositions to a proposition in a logical sum does not alter the truth value of the proposition. It is like adding a zero to a function; it doesn't alter the value of the function.

$$= \overline{A}, \overline{B} + A, \overline{B} + \overline{A}, B$$
$$= \overline{A}, (B + \overline{B}) + A, \overline{B} = \overline{A} + A, \overline{B}$$
$$= \overline{A} + \overline{\overline{\overline{A}, \overline{B}}} = \overline{A, \overline{A}, \overline{B}} = \overline{A, (\overline{A} + B)} = \overline{A, B}$$
$$= \overline{A} + \overline{B}.$$

2.4.1 *Examination of a logic function*

Any logic function $C = f(A, B)$ has only two possible values, and likewise for the independent variables A and B. A logic function with n variables is defined on a discrete space consisting of only $m = 2^n$ points. For example, in the case of $C = f(A, B)$, $m = 4$ points; namely those at which A and B take on the values {TT,TF,FT,FF}. The number of independent logic functions $= 2^m = 16$. Table 2.2 lists these 16 logical functions.

Table 2.2 *Logic functions of the two propositions A and B.*

A, B	TT	TF	FT	FF	
$f_1(A, B)$	T	F	F	F	$= A, B$
$f_2(A, B)$	F	T	F	F	$= A, \overline{B}$
$f_3(A, B)$	F	F	T	F	$= \overline{A}, B$
$f_4(A, B)$	F	F	F	T	$= \overline{A}, \overline{B}$
$f_5(A, B)$	T	T	T	T	
$f_6(A, B)$	T	T	T	F	
$f_7(A, B)$	T	T	F	T	
$f_8(A, B)$	T	F	T	T	
$f_9(A, B)$	F	T	T	T	
$f_{10}(A, B)$	T	T	F	F	
$f_{11}(A, B)$	T	F	T	F	
$f_{12}(A, B)$	F	T	T	F	
$f_{13}(A, B)$	F	T	F	T	
$f_{14}(A, B)$	F	F	T	T	
$f_{15}(A, B)$	T	F	F	T	
$f_{16}(A, B)$	F	F	F	F	$= A, \overline{A}$

We can show that $f_5 \rightarrow f_{16}$ are logical sums of $f_1 \rightarrow f_4$.

Example 1:

$$
\begin{aligned}
f_1 + f_3 + f_4 &= A, B + \overline{A}, B + \overline{A}, \overline{B} \\
&= B + \overline{A}, \overline{B} = (B + \overline{A}), (B + \overline{B}) \\
&\quad \text{last step is a distributivity identity} \\
&= B + \overline{A} \\
&= f_8.
\end{aligned}
\tag{2.8}
$$

Example 2:

$$
\begin{aligned}
f_2 + f_4 &= A, \overline{B} + \overline{A}, \overline{B} \\
&= (A + \overline{A}), \overline{B} = \overline{B} \\
&= f_{13}.
\end{aligned}
\tag{2.9}
$$

This method (called "reduction to disjunctive normal form" in logic textbooks) will work for any *n*. Thus, one can verify that the three operations:

$$
\left\{
\begin{array}{ccc}
\text{conjunction,} & \text{disjunction,} & \text{negation} \\
\text{logical product,} & \text{logical sum,} & \text{negation} \\
AND & OR & NOT
\end{array}
\right\}
$$

suffice to generate all logic functions, i.e., form an adequate set. But the logical sum $A + B$ is the same as denying that they are both false: $A + B = \overline{\overline{A}, \overline{B}}$. Therefore AND and NOT are already an adequate set.

Is there a still smaller set? Answer: Yes.

NAND, defined as $\overline{AND}$ which is represented by $A \uparrow B$.

$$A \uparrow B \equiv \overline{A, B} = \overline{A} + \overline{B}$$
$$\overline{A} = A \uparrow A$$
$$A, B = (A \uparrow B) \uparrow (A \uparrow B)$$
$$A + B = (A \uparrow A) \uparrow (B \uparrow B).$$

Every logic function can be constructed from NAND alone.

The NOR operator is defined by:

$$A \downarrow B \equiv \overline{A + B} = \overline{A}, \overline{B}$$

and is also powerful enough to generate all logic functions.

$$\overline{A} = A \downarrow A$$
$$A + B = (A \downarrow B) \downarrow (A \downarrow B)$$
$$A, B = (A \downarrow A) \downarrow (B \downarrow B).$$

2.5 Operations for plausible inference

We now turn to the extension of logic for a common situation where we lack the axiomatic information necessary for deductive logic. The goal according to Jaynes, is to arrive at a useful mathematical theory of plausible inference which will enable us to build a robot (write a computer program) to quantify the plausibility of any hypothesis in our hypothesis space of interest based on incomplete information. For example, given 10^7 observations, determine (in the light of these data and whatever prior information is at hand) the relative plausibilities of many different hypotheses about the causes at work.

We expect that any mathematical model we succeed in constructing will be replaced by more complete ones in the future as part of the much grander goal of developing a theory of common sense reasoning. Experience in physics has shown that as knowledge advances, we are able to invent better models, which reproduce more features of the real world, with more accuracy. We are also accustomed to finding that these advances lead to consequences of great practical value, like a computer program to carry out useful plausible inference following clearly defined principles (rules or operations) expressing an idealized common sense.

The rules of plausible inference are deduced from a set of three *desiderata* (see Section 2.5.1) rather than *axioms*, because they do not assert anything is true, but only state what appear to be desirable goals. We would definitely want to revise the operation of our robot or computer program if they violated one of these elementary desiderata. Whether these goals are attainable without contradiction and whether they determine any unique extension of logic are a matter of mathematical analysis. We also need to compare the inference of a robot built in this way to our own reasoning, to decide whether we are prepared to trust the robot to help us with our inference problems.

2.5.1 The desiderata of Bayesian probability theory

I. Degrees of plausibility are represented by real numbers.

II. The measure of plausibility must exhibit qualitative agreement with rationality. This means that as new information supporting the truth of a proposition is supplied, the number which represents the plausibility will increase continuously and monotonically. Also, to maintain rationality, the deductive limit must be obtained where appropriate.

III. Consistency

(a) *Structural consistency*: If a conclusion can be reasoned out in more than one way, every possible way must lead to the same result.

(b) *Propriety*: The theory must take account of all information, provided it is relevant to the question.

(c) *Jaynes consistency*: Equivalent states of knowledge must be represented by equivalent plausibility assignments. For example, if $A, B|C = B|C$, then the plausibility of $A, B|C$ must equal the plausibility of $B|C$.

2.5.2 Development of the product rule

In Section 2.4 we established that the logical product and negation (AND, NOT) are an adequate set of operations to generate any proposition derivable from $\{A_1, \ldots, A_N\}$. For Bayesian inference, our goal is to find operations (rules) to determine the plausibility of logical conjunction and negation that satisfy the above desiderata. Start with the plausibility of A, B:

Let $(A, B|C) \equiv$ plausibility of A, B supposing the truth of C.

Remember, we are going to represent plausibility by real numbers (desideratum I). Now $(A, B|C)$ must be a function of some combination of $(A|C)$, $(B|C)$, $(B|A, C)$, $(A|B, C)$.

There are 11 possibilities:

$$(A, B|C) = F_1[(A|C), (A|B, C)]$$

$$(A, B|C) = F_2[(A|C), (B|C)]$$

$$(A, B|C) = F_3[(A|C), (B|A, C)]$$

$$(A, B|C) = F_4[(A|B, C), (B|C)]$$

$$(A, B|C) = F_5[(A|B, C), (B|A, C)]$$

$$(A, B|C) = F_6[(B|C), (B|A, C)]$$
$$(A, B|C) = F_7[(A|C), (A|B, C), (B|C)]$$
$$(A, B|C) = F_8[(A|C), (A|B, C), (B|A, C)]$$
$$(A, B|C) = F_9[(A|C), (B|C), (B|A, C)]$$
$$(A, B|C) = F_{10}[(A|B, C), (B|C), (B|A, C)]$$
$$(A, B|C) = F_{11}[(A|C), (A|B, C), (B|C), (B|A, C)]$$

Box 2.3   Note on the use of the " = " sign

1. In Boolean algebra, the equals sign is used to denote equal truth value. By definition, $A = B$ asserts that A is true if and only if B is true.
2. When talking about plausibility, which is represented by a real number, $(A, B|C) = ()()\ldots$ means equal numerically.
3. $\equiv$ means equal by definition.

Now let us examine these 11 different functions more closely. Since the order in which the symbols A and B appear has no meaning (i.e., $A, B = B, A$) it follows that

$$F_1[(A|C), (A|B, C)] = F_6[(B|C), (B|A, C)]$$
$$F_3[(A|C), (B|A, C)] = F_4[(A|B, C), (B|C)]$$
$$F_7[(A|C), (A|B, C), (B|C)] = F_9[(A|C), (B|C), (B|A, C)]$$
$$F_8[(A|C), (A|B, C), (B|A, C)] = F_{10}[(A|B, C), (B|C), (B|A, C)]$$

This reduces the number of equations dramatically from 11 to 7. The seven functions remaining are $F_1, F_2, F_3, F_5, F_7, F_8, F_{11}$.

If any function leads to an absurdity in even one example, it must be ruled out, even if for other examples it would be satisfactory. Consider

$$(A, B|C) = F_2[(A|C), (B|C)].$$

Suppose $A \equiv$ next person will have blue left eye.

$\quad\quad B \equiv$ next person will have brown right eye.

$\quad\quad C \equiv$ prior information concerning our expectation that the left and right eye colors of any individual will be very similar.

Now $(A|C)$ could be very plausible as could $(B|C)$, but $(A, B|C)$ is extremely implausible. We rule out functions of this form because they have no way of taking such influence into account. Our robot could not reason the way humans do, even qualitatively, with that functional form.

Similarly, we can rule out F_1 for the extreme case where the conditional (given) information represented by proposition C is that "A and B are independent." In this extreme case,

$$(A|B, C) = (A|C).$$

Therefore,

$$(A, B|C) = F_1[(A|C), (A|B, C)] = F_1[(A|C), (A|C)], \qquad (2.10)$$

which is clearly absurd because F_1 claims that the plausibility of $A, B|C$ depends only on the plausibility of $A|C$.

Other extreme conditions are $A = B, A = C, C = \overline{A}$, etc. Carrying out this type of analysis, Tribus (1969) shows that all but one of the remaining possibilities can exhibit qualitative violations with common sense in some extreme case. There is only one survivor which can be written in two equivalent ways:

$$\begin{aligned}(A, B|C) &= F[(B|C), (A|B, C)] \\ &= F[(A|C), (B|A, C)].\end{aligned} \qquad (2.11)$$

In addition, desideratum II, qualitative agreement with common sense, requires that $F[(A|C), (B|A, C)]$ must be a continuous monotonic function of $(A|C)$ and $(B|A, C)$. The continuity assumption requires that if $(A|C)$ changes only infinitesimally, it can induce only an infinitesimal change in $(A, B|C)$ or $(\overline{A}|C)$.

Now use desideratum III: "Consistency"
Suppose we want $(A, B, C|D)$

1. Consider B, C to be a single proposition at first; then we can apply Equation (2.11):

$$\begin{aligned}(A, B, C|D) &= F[(B, C|D), (A|B, C, D)] \\ &= F\{F[(C|D), (B|C, D)], (A|B, C, D)\}.\end{aligned} \qquad (2.12)$$

2. Consider A, B to be a single proposition at first:

$$\begin{aligned}(A, B, C|D) &= F[(C|D), (A, B|C, D)] \\ &= F\{(C|D), F[(B|C, D), (A|B, C, D)]\}.\end{aligned} \qquad (2.13)$$

For consistency, 1 and 2 must be equal.
Let $x \equiv (A|B, C, D), y \equiv (B|C, D), z \equiv (C|D)$, then:

$$F\{x, F[y, z]\} = F\{F[x, y], z\}. \qquad (2.14)$$

This equation has a long history in mathematics and is called the "the Associativity Equation." Aczél (1966) derives the general solution (Equation (2.15) below) without assuming differentiability; unfortunately, the proof fills 11 pages of his book. R. T. Cox (1961) provided a shorter proof, but assumed differentiability.

The solution is

$$w\{F[x, y]\} = w\{x\}w\{y\}, \tag{2.15}$$

where $w\{x\}$ is any positive continuous monotonic function.

In the case of just two propositions, A, B given the truth of C, the solution to the associativity equation becomes

$$w\{(A, B|C)\} = w\{(A|B, C)\}w\{(B|C)\}$$
$$= w\{(B|A, C)\}w\{(A|C)\}. \tag{2.16}$$

For simplicity, drop the $\{\}$ brackets, but it should be remembered that the argument of w is a plausibility.

$$w(A, B|C) = w(A|B, C)w(B|C)$$
$$= w(B|A, C)w(A|C). \tag{2.17}$$

Henceforth this will be called the product rule. Recall that at this moment, $w()$ is any positive, continuous, monotonic function.

Desideratum II: Qualitative correspondence with common sense imposes further restrictions on $w\{x\}$

Suppose A is certain given C. Then $A, B|C = B|C$ (i.e., same truth value). By our primitive axiom that propositions with the same truth value must have the same plausibility,

$$(A, B|C) = (B|C)$$
$$(A|B, C) = (A|C). \tag{2.18}$$

Therefore, Equation (2.17), the solution to the associativity equation, becomes

$$w(B|C) = w(A|C)w(B|C). \tag{2.19}$$

This is only true when $A|C$ is certain.

Thus we have arrived at a new constraint on $w()$; it must equal 1 when the argument is certain.

For the next constraint, suppose that A is impossible given C. This implies

$$A, B|C = A|C$$
$$A|B, C = A|C.$$

Then

$$w(A, B|C) = w(A|B, C)w(B|C) \tag{2.20}$$

becomes

$$w(A|C) = w(A|C)w(B|C). \tag{2.21}$$

This must be true for any $(B|C)$. There are only two choices: either $w(A|C)=0$ or $+\infty$.

1. $w(x)$ is a positive, increasing function $(0 \rightarrow 1)$.
2. $w(x)$ is a positive, decreasing function $(\infty \rightarrow 1)$.

They do not differ in content.

Suppose $w_1(x)$ represents impossibility by $+\infty$. We can define $w_2(x) = 1/w_1(x)$ which represents impossibility by 0. Therefore, there is no loss of generality if we adopt:

$$0 \leq w(x) \leq 1.$$

Summary:

Using our desiderata, we have arrived at our present form of the product rule:

$$w(A, B|C) = w(A|C)w(B|A, C) = w(B|C)w(A|B, C).$$

At this point we are <u>still not</u> referring to $w(x)$ as the probability of x. $w(x)$ is any continuous, monotonic function satisfying:

$$0 \leq w(x) \leq 1,$$

where $w(x) = 0$ when the argument x is impossible and 1 when x is certain.

2.5.3 Development of sum rule

We have succeeded in deriving an operation for determining the plausibility of the logical product (conjunction). We now turn to the problem of finding an operation to determine the plausibility of negation. Since the logical sum $A + \overline{A}$ is always true, it follows that the plausibility that A is false must depend on the plausibility that A is true. Thus, there must exist some functional relation

$$w(\overline{A}|B) = S(w(A|B)). \tag{2.22}$$

Again, using our desiderata and functional analysis, one can show (Jaynes, 2003) that the monotonic function $w(A|B)$ obeys

$$w^m(A|B) + w^m(\overline{A}|B) = 1$$

for positive m. This is known as the sum rule.

The product rule can equally well be written as

$$w^m(A, B|C) = w^m(A|C)w^m(B|A, C) = w^m(B|C)w^m(A|B, C).$$

But then we see that the value of m is actually irrelevant; for whatever value is chosen, we can define a new function

$$p(x) \equiv w^m(x)$$

and our rules take the form

$$p(A, B|C) = p(A|C)p(B|A, C) = p(B|C)p(A|B, C)$$
$$p(A|B) + p(\overline{A}|B) = 1$$

This entails no loss of generality, for the only requirement we imposed on the function $w(x)$ is that $w(x)$ is a continuous, monotonic, increasing function ranging from $w = 0$ for impossibility to $w = 1$ for certainty. But if $w(x)$ satisfies this, so does $w^m(x)$, $0 < m < \infty$.

Reminder: We are still not referring to $p(x)$ as a probability.

We showed earlier that conjunction, A, B, and negation, $\overline{A}$, are an adequate set of operations, from which all logic functions can be constructed. Therefore, it ought to be possible, by repeated applications of the product and sum rules, to arrive at the plausibility of any proposition. To show this, we derive a formula for the logical sum $A + B$.

$$\begin{aligned}
p(A + B|C) &= 1 - p(\overline{A + B}|C) = 1 - p(\overline{A}, \overline{B}|C) \\
&= 1 - p(\overline{A}|C)p(\overline{B}|\overline{A}, C) \\
&= 1 - p(\overline{A}|C)[1 - p(B|\overline{A}, C)] \\
&= 1 - p(\overline{A}|C) + p(\overline{A}|C)p(B|\overline{A}, C) \\
&= p(A|C) + p(\overline{A}, B|C) \\
&= p(A|C) + p(B|C)p(\overline{A}|B, C) \\
&= p(A|C) + p(B|C)[1 - p(A|B, C)] \\
&= p(A|C) + p(B|C) - p(B|C)p(A|B, C) \\
\Rightarrow p(A + B|C) &= p(A|C) + p(B|C) - p(A, B|C).
\end{aligned} \tag{2.23}$$

This is a very useful relationship and is called the *extended sum rule*.

Starting with our three desiderata, we arrived at a set of rules for plausible inference:

product rule:

$$p(A, B|C) = p(A|C)p(B|A, C) = p(B|C)p(A|B, C),$$

sum rule:

$$p(A|B) + p(\overline{A}|B) = 1.$$

We have in the two rules formulae for the plausibility of the conjunction, A, B, and negation, $\overline{A}$, which are an adequate set of operations to generate any proposition derivable from the set $\{A_1, \dots, A_N\}$.

Using the product and sum rules, we also derived the **extended sum rule**

$$p(A + B|C) = p(A|C) + p(B|C) - p(A, B|C). \tag{2.24}$$

For mutually exclusive propositions $p(A, B|C) = 0$, so Equation (2.24) becomes

$$p(A + B|C) = p(A|C) + p(B|C). \tag{2.25}$$

We will refer to Equation (2.25) as the *generalized sum rule*.

2.5.4 Qualitative properties of product and sum rules

Check to see if the product and sum rules predict the strong (deductive logic) and weak (inductive logic) syllogisms.

Strong syllogisms:

(a)	(b)	
$A \Rightarrow B$	$A \Rightarrow B$	major premise (prior information)
<u>A true</u>	<u>B false</u>	minor premise (data)
B true	A false	conclusion

Example:

- Let $A \equiv$ "Corn was harvested in Eastern Canada in AD 1000."
- Let $B \equiv$ "Corn seed was available in Eastern Canada in AD 1000."
- Let $I \equiv$ "Corn seed is required to grow corn, so if corn was harvested, the seed must have been available." This is our prior information or major premise.

In both cases, we start by writing down the product rule:

Syllogism (a)

$$p(A, B|I) = p(A|I)p(B|A, I)$$
$$\rightarrow p(B|A, I) = \frac{p(A, B|I)}{p(A|I)}$$

Prior info. $I \equiv$ "$A, B = A$"

$$\rightarrow p(A, B|I) = p(A|I)$$
$$\rightarrow p(B|A, I) = 1$$
i.e., B is true if A is true

Data: $A = 1$ (true)

$$\rightarrow p(B|A, I) = 1$$
Certainty

Syllogism (b)

$$p(A, \overline{B}|I) = p(\overline{B}|I)p(A|\overline{B}, I)$$
$$p(A|\overline{B}, I) = \frac{p(A, \overline{B}|I)}{p(\overline{B}|I)}$$

Prior info. $I \equiv$ "$A, B = A$"

$$\rightarrow p(A, \overline{B}|I) = 0$$
since B could not be false if A is
true according to the information I.

Data: $\overline{B} = 1$ (B false)

$$\rightarrow p(A|\overline{B}, I) = 0$$
Impossibility

Weak syllogisms:

(a)	(b)	
$A \Rightarrow B$	$A \Rightarrow B$	prior information
<u>B true</u>	<u>A false</u>	data
A more plausible	B less plausible	conclusion

Start by writing down the product rule in the form of Bayes' theorem:

<table>
<tr><td><u>Weak Syllogism (a)</u></td><td><u>Weak Syllogism (b)</u></td></tr>
</table>

$$p(A|B,I) = \frac{p(A|I)p(B|A,I)}{p(B|I)} \qquad\qquad p(B|\overline{A},I) = \frac{p(B|I)p(\overline{A}|B,I)}{p(\overline{A}|I)}$$

Prior info. $I \equiv$ "$A, B = A$"	Syllogism (a) gives $p(A|B,I) \geq p(A|I)$ based on the same prior information.

$$\rightarrow p(B|A,I) = 1$$

$$\rightarrow 1 - p(\overline{A}|B,I) \geq 1 - p(\overline{A}|I)$$

and $p(B|I) \leq 1$
since I says nothing about the truth of B.

$$\rightarrow p(\overline{A}|B,I) \leq p(\overline{A}|I) \text{ or } \frac{p(\overline{A}|B,I)}{p(\overline{A}|I)} \leq 1$$

Substituting into Bayes' theorem   Substituting into Bayes' theorem
$$\rightarrow p(A|B,I) \geq p(A|I) \qquad\qquad\qquad \rightarrow p(B|\overline{A},I) \leq p(B|I)$$

A more plausible B less plausible

2.6 Uniqueness of the product and sum rules

Corresponding to every different choice of continuous monotonic function $p(x)$ there seems to be a different set of rules. Nothing given so far tells us what numerical value of plausibility should be assigned at the beginning of the problem. To answer both issues, consider the following: suppose we have N mutually exclusive and exhaustive propositions $\{A_1, \ldots, A_N\}$.

- *Mutually exclusive* $\equiv$ only one can be true, i.e., $p(A_i, A_j|B) = 0$ for $i \neq j$

$$p(A_1 + \cdots + A_N|B) = \sum_{i=1}^{N} p(A_i|B)$$

- *Exhaustive* $\equiv$ the true proposition is contained in the set

$$\sum_{i=1}^{N} p(A_i|B) = 1 \text{(certain)}$$

This information is not enough to determine the individual $p(A_i|B)$ since there is no end to the variety of complicated information that might be contained in B. Development of new methods for translating prior information to numerical values of $p(A_i|B)$ is an ongoing research problem. We will discuss several valuable approaches to this problem in later chapters.

Suppose our information B is indifferent regarding the $p(A_i|B)$'s. Then the only possibility that reflects this state of knowledge is:

$$p(A_i|B) = \frac{1}{N}, \qquad\qquad\qquad (2.26)$$

where N is the number of mutually exclusive propositions. This is called the *Principle of Indifference*.

In this one particular case, which can be generalized, we see that information B leads to a definite numerical value for $p(A_i|B)$, but not the numerical value of the plausibility $(A_i|B)$.

Instead of saying $p(A_i|B)$ is an arbitrary, monotonic function of the plausibility $(A_i|B)$, it is much more useful to turn this around and say: "the plausibility $(A_i|B)$ is an arbitrary, monotonic function of $p()$, defined in $0 \leq p() \leq 1$."

It is $p()$ that is rigidly fixed by the data, not $(A_i|B)$.

The p's define a particular scale on which degrees of plausibility can be measured. Out of the possible monotonic functions we pick this particular one, not because it is "correct," but because it is more convenient. p is the quantity that obeys the product and sum rules and the numerical value of p is determined by the available information. From now on we will refer to them as probabilities.

Jaynes (2003) writes, "This situation is analogous to that in thermodynamics, where out of all possible empirical temperature scales T, which are monotonic functions of each other, we finally decide to use the Kelvin scale T; not because it is more 'correct' than others, but because it is more convenient; i.e., the laws of thermodynamics take their simplest form $[dU = Tds - PdV, dG = -SdT + VdP]$ in terms of this particular scale. Because of this, numerical values of Kelvin temperatures are directly measurable in experiments."

With this operational definition of probability, we can readily derive another intuitively pleasing result. In this problem, our prior information is:

$I \equiv$ "An urn is filled with 10 balls of identical size, weight and texture, labeled $1, 2, \ldots, 10$. Three of the balls (numbers 3, 4, 7) are red and the others are green. We are to shake the urn and draw one ball blindfolded."

Define the proposition:

$$E_i \equiv \text{"the } i\text{th ball is drawn"}, 1 \leq i \leq 10.$$

Since the prior information is indifferent to these ten possibilities, Equation (2.26) applies.

$$p(E_i|I) = \frac{1}{10}, \quad 1 \leq i \leq 10. \tag{2.27}$$

The proposition $R \equiv$ "that we draw a red ball" is equivalent to "we draw ball 3, 4, or 7." This can be written as the logical sum statement:

$$R = E_3 + E_4 + E_7. \tag{2.28}$$

It follows from the extended sum rule that

$$p(R|I) = p(E_3 + E_4 + E_7|I) = \frac{3}{10}, \tag{2.29}$$

in accordance with our intuition.

More generally, if there are N such balls, and the proposition R is defined to be true for any specified subset of M of them, and false on the rest, then we have

$$p(R|I) = \frac{M}{N}. \tag{2.30}$$

2.7 Summary

Rather remarkably, the three desiderata of Section 2.5.1 have enabled us to arrive at a theory of extended logic together with a particular scale for measuring the plausibility of any hypothesis conditional on given information. We have shown that the rules for manipulating plausibility are the product and sum rules. The particular scale of plausibility we have adopted, now called probability, is determined by the data in a way that agrees with our intuition. We also showed that in the limit of complete information (certainty), the theory gives the same conclusions as the strong syllogisms of deductive inference.

The main constructive requirement which determined the product and sum rules was the desideratum of structural consistency, "If a conclusion can be reasoned out in more than one way, every possible way must lead to the same result." This does not mean that our rules have been proved consistent,[1] only that any other rules which represent degrees of plausibility by real numbers, but which differ in content from the product and sum rules, will lead to a violation of one of our desiderata.

Apart from the justification for probability as extended logic, the value of this approach to solving inference problems is being demonstrated on a regular basis in a wide variety of areas leading both to new scientific discoveries and a new level of understanding. Modern computing power permits a simple comparison of the power of different approaches in the analysis of well-understood simulated data sets. Some examples of the power of Bayesian inference will be brought out in later chapters.

2.8 Problems

1. Construct a truth table to show

$$\overline{A}, \overline{B} = \overline{A + B}.$$

2. Construct a truth table to show

$$A + (B, C) = (A + B), (A + C).$$

---

[1] According to Gödel's theorem, no mathematical system can provide a proof of its own consistency.

3. With reference to Table 2.2, construct a truth table to show that

$$f_8(A, B) = \overline{A} + B.$$

4. Based on the available evidence, the probability that Jones is guilty is equal to 0.7, the probability that Susan is guilty is equal to 0.6, and the probability that both are guilty is equal to 0.5. Compute the probability that Jones is guilty and/or Susan is guilty.

5. The probability that Mr. Smith will make a donation is equal to 0.5, if his brother Harry has made a donation. The probability that Harry will make a donation is equal to 0.02. What is the probability that both men will make a donation?

3

The how-to of Bayesian inference

3.1 Overview

The first part of this chapter is devoted to a brief description of the methods and terminology employed in Bayesian inference and can be read as a stand-alone introduction on how to do Bayesian analysis.[1] Following a review of the basics in Section 3.2, we consider the two main inference problems: parameter estimation and model selection. This includes how to specify credible regions for parameters and how to eliminate nuisance parameters through marginalization. We also learn that Bayesian model comparison has a built-in "Occam's razor," which automatically penalizes complicated models, assigning them large probabilities only if the complexity of the data justifies the additional complication of the model. We also learn how this penalty arises through marginalization and depends both on the number of parameters and the prior ranges of these parameters.

We illustrate these features with a detailed analysis of a toy spectral line problem and in the process introduce the Jeffreys prior and learn how different choices of priors affect our conclusions. We also have a look at a general argument for selecting priors for location and scale parameters in the early phases of an investigation when our state of ignorance is very high. The final section illustrates how Bayesian analysis provides valuable new insights on systematic errors and how to deal with them.

I recommend that Sections 3.2 to 3.5 of this chapter be read twice; once quickly, and again after seeing these ideas applied in the detailed example treated in Sections 3.6 to 3.11.

3.2 Basics

In Bayesian inference, the viability of each member of a set of rival hypotheses, $\{H_i\}$, is assessed in the light of some observed data, D, by calculating the probability of each hypothesis, given the data and any prior information, I, we may have regarding the

---

[1] The treatment of this topic is a revised version of Section 2 of a paper by Gregory and Loredo (1992), which is reproduced here with the permission of the Astrophysical Journal.

hypotheses and data. Following a notation introduced by Jeffreys (1961), we write such a probability as $p(H_i|D,I)$, explicitly denoting the prior information by the proposition, I, to the right of the bar. At the very least, the prior information must specify the class of alternative hypotheses being considered (hypothesis space of interest), and the relationship between the hypotheses and the data (the statistical model).

The basic rules for manipulating Bayesian probabilities are the sum rule,

$$p(H_i|I) + p(\overline{H}_i|I) = 1, \tag{3.1}$$

and the product rule,

$$p(H_i, D|I) = p(H_i|I)p(D|H_i, I)$$
$$= p(D|I)p(H_i|D, I). \tag{3.2}$$

The various symbols appearing as arguments should be understood as propositions; for example, D might be the proposition, "N photons were counted in a time T." The symbol $\overline{H}_i$ signifies the negation of H_i (a proposition that is true if one of the alternatives to H_i is true), and (H_i, D) signifies the logical conjunction of H_i and D (a proposition that is true only if H_i *and* D are both true). The rules hold for any propositions, not just those indicated above.

Throughout this work, we will be concerned with exclusive hypotheses, so that if one particular hypothesis is true, all others are false. For such hypotheses, we saw in Section 2.5.3 that the sum and product rules imply the generalized sum rule,

$$p(H_i + H_j|I) = p(H_i|I) + p(H_j|I). \tag{3.3}$$

To say that the hypothesis space of interest consists of n mutually exclusive hypotheses means that for the purpose of the present analysis, we are assuming that one of them is true and the objective is to assign a probability to each hypothesis in this space, based on D, I. We will use normalized prior probability distributions, unless otherwise stated, such that

$$\sum_i p(H_i|I) = 1. \tag{3.4}$$

Here a "+" within a probability symbol stands for logical disjunction, so that $H_i + H_j$ is a proposition that is true if either H_i or H_j is true.

One of the most important calculating rules in Bayesian inference is Bayes' theorem, found by equating the two right hand sides of Equation (3.2) and solving for $p(H_i|D, I)$:

$$p(H_i|D, I) = \frac{p(H_i|I)p(D|H_i, I)}{p(D|I)}. \tag{3.5}$$

Bayes' theorem describes a type of learning: how the probability for each member of a class of hypotheses should be modified on obtaining new information, D. The probabilities for the hypotheses in the absence of D are called their *prior probabilities*, $p(H_i|I)$, and those including the information D are called their *posterior probabilities*, $p(H_i|D, I)$. The quantity $p(D|H_i, I)$ is called the *sampling probability* for D, or the *likelihood* of H_i, and the quantity $p(D|I)$ is called the *prior predictive probability* for D, or the *global likelihood* for the entire class of hypotheses.

All of the rules we have written down so far show how to manipulate known probabilities to find the values of other probabilities. But to be useful in applications, we additionally need rules that assign numerical values or functions to the initial *direct probabilities* that will be manipulated. For example, to use Bayes' theorem, we need to know the values of the three probabilities on the right side of Equation (3.5). These three probabilities are not independent. The quantity $p(D|I)$ must satisfy the requirement that the sum of the posterior probabilities over the hypothesis space of interest is equal to 1.

$$\sum_i p(H_i|D, I) = \frac{\sum_i p(H_i|I)p(D|H_i, I)}{p(D|I)} = 1. \tag{3.6}$$

Therefore,

$$p(D|I) = \sum_i p(H_i|I)p(D|H_i, I). \tag{3.7}$$

That is, the denominator of Bayes' theorem, which does not depend on H_i, must be equal to the sum of the numerator over H_i. It thus plays the role of a normalization constant.

3.3 Parameter estimation

We frequently deal with problems in which a particular model is assumed to be true and the hypothesis space of interest concerns the values of the model parameters. For example, in a straight line model, the two parameters are the intercept and slope. We can look at this problem as a hypothesis space that is labeled, not by discrete numbers, but by the possible values of two continuous parameters. In such cases, the quantity of interest (see also Section 1.3.2) is a *probability density function* or PDF. More generally, 'PDF' is an abbreviation for a probability distribution function which can apply to both discrete and continuous parameters. For example, given some prior information, M, specifying a parameterized model with one parameter, θ, $p(\theta|M)$ is the prior density for θ, which means that $p(\theta|M)d\theta$ is the prior probability that the true value of the parameter is in the interval $[\theta, \theta + d\theta]$. We use the same symbol, $p(\ldots)$, for probabilities and PDFs; the nature of the argument will identify which use is intended.

Bayes' theorem, and all the other rules just discussed, hold for PDFs, with all sums replaced by integrals. For example, the global likelihood for model M can be calculated with the continuous counterpart of Equation (3.7),

$$p(D|M) = \int d\theta \; p(\theta|M)p(D|\theta, M) = \mathcal{L}(M). \qquad (3.8)$$

In words, the global likelihood of a model is equal to the weighted average likelihood for its parameters. We will utilize the global likelihood of a model in Section 3.5 where we deal with model comparison and Occam's razor.

If there is more than one parameter, multiple integrals are used. If the prior density and the likelihood are assigned directly, the global likelihood is an uninteresting normalization constant. The posterior PDF for the parameters is simply proportional to the product of the prior and the likelihood.

The use of Bayes' theorem to determine what one can learn about the values of parameters from data is called *parameter estimation*, though strictly speaking, Bayesian inference does not provide estimates for parameters. Rather, the Bayesian solution to the parameter estimation problem is the full posterior PDF, $p(\theta|D, M)$, and not just a single point in parameter space. Of course, it is useful to summarize this distribution for textual, graphical, or tabular display in terms of a "best-fit" value and "error bars." Possible summaries of the best-fit values are the *posterior mode* (most probable value of θ) or the *posterior mean*,

$$\langle \theta \rangle = \int d\theta \; \theta \; p(\theta|D, M). \qquad (3.9)$$

If the mode and mean are very different, the posterior PDF is too asymmetric to be adequately summarized by a single estimate. An allowed range for a parameter with probability content C (e.g., $C = 0.95$ or 95%) is provided by a *credible region*, or highest posterior density region, R, defined by

$$\int_R d\theta \; p(\theta|D, M) = C, \qquad (3.10)$$

with the posterior density inside R everywhere greater than that outside it. We sometimes speak picturesquely of the region of parameter space that is assigned a large density as the "posterior bubble." In practice, the probability (density function) $p(\theta|D, M)$ is represented by a finite list of values, p_i, representing the probability in discrete intervals of θ.

A simple way to compute the credible region is to sort these probability values in descending order. Then starting with the largest value, add successively smaller p_i values until adding the next value would exceed the desired value of C. At each step keep track of the corresponding θ_i value. The credible region is the range of θ that just includes all the θ_i values corresponding to the p_i values added. The boundaries of the credible region are obtained by sorting these θ_i values and taking the smallest and largest values.

3.4 Nuisance parameters

Frequently, a parameterized model will have more than one parameter, but we will want to focus attention on a subset of the parameters. For example, we may want to focus on the implications of the data for the frequency of a periodic signal, independent of the signal's amplitude, shape, or phase. Or we may want to focus on the implications of spectral data for the parameters of some line feature, independent of the shape of the background continuum. In such problems, the uninteresting parameters are known as *nuisance parameters*. As always, the full Bayesian inference is the full joint posterior PDF for all of the parameters; but its implications for the parameters of interest can be simply summarized by integrating out the nuisance parameters. Explicitly, if model M has two parameters, θ and ϕ, and we are interested only in θ, then it is a simple consequence of the sum and product rules (see Section 1.5) that,

$$p(\theta|D, M) = \int d\phi \, p(\theta, \phi|D, M). \tag{3.11}$$

For historical reasons, the procedure of integrating out nuisance parameters is called *marginalization*, and $p(\theta|D, M)$ is called the *marginal posterior PDF* for θ. Equation (3.8) for the global likelihood is a special case of marginalization in which *all* of the model parameters are marginalized out of the joint prior distribution, $p(D, \theta|M)$.

The use of marginalization to eliminate nuisance parameters is one of the most important technical advantages of Bayesian inference over standard frequentist statistics. Indeed, the name "nuisance parameters" originated in frequentist statistics because there is no general frequentist method for dealing with such parameters; they are indeed a "nuisance" in frequentist statistics. Marginalization plays a very important role in this work. We will see a detailed example of marginalization in action in Section 3.6.

3.5 Model comparison and Occam's razor

Often, more than one parameterized model will be available to explain a phenomenon, and we will wish to compare them. The models may differ in form or in number of parameters. Use of Bayes' theorem to compare competing models by calculating the probability of each model as a whole is called *model comparison*. Bayesian model comparison has a built-in "Occam's razor:" Bayes' theorem automatically penalizes complicated models, assigning them large probabilities only if the complexity of the data justifies the additional complication of the model. See Jeffreys and Berger (1992) for a historical account of the connection between Occam's (Ockham's) razor and Bayesian analysis.

Model comparison calculations require the explicit specification of two or more specific alternative models, M_i. We take as our prior information the proposition that one of the models under consideration is true. Symbolically, we might write this as $I = M_1 + M_2 + \cdots + M_N$, where the "+" symbol here stands for disjunction ("or"). Given this information, we can calculate the probability for each model with Bayes' theorem:

$$p(M_i|D, I) = \frac{p(M_i|I)p(D|M_i, I)}{p(D|I)}. \tag{3.12}$$

We recognize $p(D|M_i, I)$ as the global likelihood for model M_i, which we can calculate according to Equation (3.8). The term in the denominator is again a normalization constant, obtained by summing the products of the priors and the global likelihoods of all models being considered. Model comparison is thus completely analogous to parameter estimation: just as the posterior PDF for a parameter is proportional to its prior times its likelihood, so the posterior probability for a model as a whole is proportional to its prior probability times its global likelihood.

It is often useful to consider the ratios of the probabilities of two models, rather than the probabilities directly. The ratio,

$$O_{ij} = p(M_i|D, I)/p(M_j|D, I), \tag{3.13}$$

is called the *odds ratio* in favor of model M_i over model M_j. From Equation (3.12),

$$\begin{aligned}
O_{ij} &= \frac{p(M_i|I)}{p(M_j|I)} \frac{p(D|M_i, I)}{p(D|M_j, I)} \\
&\equiv \frac{p(M_i|I)}{p(M_j|I)} B_{ij},
\end{aligned} \tag{3.14}$$

where the first factor is the prior odds ratio, and the second factor is called the *Bayes factor*. Note: the normalization constant in Equation (3.12) drops out of the odds ratio; this can make the odds ratio somewhat easier to work with. The odds ratio is also conceptually useful when one particular model is of special interest. For example, suppose we want to compare a constant rate model with a class of periodic alternatives, and will thus calculate the odds in favor of each alternative over the constant model.

If we have calculated the odds ratios, O_{i1}, in favor of each model over model M_1, we can find the probabilities for each model in terms of these odds ratios as follows:

$$\sum_{i=1}^{N_{mod}} p(M_i|D, I) = 1, \tag{3.15}$$

where N_{mod} is the total number of models considered. Dividing through by $p(M_1|D,I)$, we have

$$\frac{1}{p(M_1|D,I)} = \sum_{i=1}^{N_{\mathrm{mod}}} O_{i1}. \qquad (3.16)$$

Comparing Equation (3.16) to the expression for O_{i1}, given by

$$O_{i1} = p(M_i|D,I)/p(M_1|D,I), \qquad (3.17)$$

we have the result that

$$p(M_i|D,I) = \frac{O_{i1}}{\sum_{j=1}^{N_{\mathrm{mod}}} O_{j1}}, \qquad (3.18)$$

where of course $O_{11} = 1$. If there are only two models, the probability of M_2 is given by

$$p(M_2|D,I) = \frac{O_{21}}{1+O_{21}} = \frac{1}{1+\frac{1}{O_{21}}}. \qquad (3.19)$$

In this work, we will assume that we have no information leading to a prior preference for one model over another, so the prior odds ratio will be unity, and the odds ratio will equal the Bayes factor, the ratio of global likelihoods. A crucial consequence of the marginalization procedure used to calculate global likelihoods is that the Bayes factor automatically favors simpler models unless the data justify the complexity of more complicated alternatives. This is illustrated by the following simple example.

Imagine comparing two models: M_1 with a single parameter, θ, and M_0 with θ fixed at some default value θ_0 (so M_0 has no free parameters). To calculate the Bayes factor B_{10} in favor of model M_1, we will need to perform the integral in Equation (3.8) to compute $p(D|M_1,I)$, the global likelihood of M_1. To develop our intuition about the Occam penalty, we will carry out a back-of-the-envelope calculation for the Bayes factor. Often the data provide us with more information about parameters than we had without the data, so that the likelihood function, $\mathcal{L}(\theta) = p(D|\theta,M_1,I)$, will be much more "peaked" than the prior, $p(\theta|M_1,I)$. In Figure 3.1 we show a Gaussian-looking likelihood centered at $\hat{\theta}$, the *maximum likelihood* value of θ, together with a flat prior for θ. Let $\Delta\theta$ be the characteristic width of the prior. For a flat prior, we have that

$$\int_{\Delta\theta} d\theta \, p(\theta|M_1,I) = p(\theta|M_1,I)\Delta\theta = 1. \qquad (3.20)$$

Therefore, $p(\theta|M_1,I) = 1/\Delta\theta$.

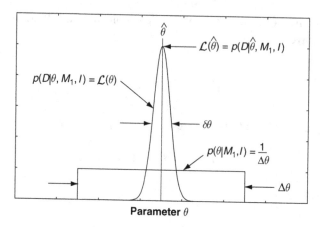

Figure 3.1 The characteristic width $\delta\theta$ of the likelihood peak and $\Delta\theta$ of the prior.

The likelihood has a characteristic width[2] which we represent by $\delta\theta$. The characteristic width is defined by

$$\int_{\Delta\theta} d\theta \, p(D|\theta, M_1, I) = p(D|\hat{\theta}, M_1, I) \times \delta\theta. \tag{3.21}$$

Then we can approximate the global likelihood (Equation (3.8)) for M_1 in the following way:

$$\begin{aligned} p(D|M_1, I) &= \int d\theta \, p(\theta|M_1, I) p(D|\theta, M_1, I) = \mathcal{L}(M_1) \\ &= \frac{1}{\Delta\theta} \int d\theta \, p(D|\theta, M_1, I) \\ &\approx p(D|\hat{\theta}, M_1, I) \frac{\delta\theta}{\Delta\theta} \end{aligned} \tag{3.22}$$

or alternatively, $\mathcal{L}(M_1) \approx \mathcal{L}(\hat{\theta}) \dfrac{\delta\theta}{\Delta\theta}.$

Since model M_0 has no free parameters, no integral need be calculated to find its global likelihood, which is simply equal to the likelihood for model M_1 for $\theta = \theta_0$,

$$p(D|M_0, I) = p(D|\theta_0, M_1, I) = \mathcal{L}(\theta_0). \tag{3.23}$$

Thus the Bayes factor in favor of the more complicated model is

$$B_{10} \approx \frac{p(D|\hat{\theta}, M_1, I)}{p(D|\theta_0, M_1, I)} \frac{\delta\theta}{\Delta\theta} = \frac{\mathcal{L}(\hat{\theta})}{\mathcal{L}(\theta_0)} \frac{\delta\theta}{\Delta\theta}. \tag{3.24}$$

---

[2] If the likelihood function is really a Gaussian and the prior is flat, it is simple to show that $\delta\theta = \sigma_\theta \sqrt{2\pi}$, where σ_θ is the standard deviation of the posterior PDF for θ.

The likelihood ratio in the first factor can never favor the simpler model because M_1 contains it as a special case. However, since the posterior width, $\delta\theta$, is narrower than the prior width, $\Delta\theta$, the second factor penalizes the complicated model for any "wasted" parameter space that gets ruled out by the data. The Bayes factor will thus favor the more complicated model only if the likelihood ratio is large enough to overcome this penalty.

Equation (3.22) has the form of the best-fit likelihood times the factor that penalizes M_1. In the above illustrative calculation we assumed a simple Gaussian likelihood function for convenience. In general, the actual likelihood function can be very complicated with several peaks. However, one can always write the global likelihood of a model with parameter θ, as the maximum value of its likelihood times some factor, Ω_θ:

$$p(D|M, I) \equiv \mathcal{L}_{max}\Omega_\theta. \tag{3.25}$$

The second factor, Ω_θ, is called the *Occam factor* associated with the parameters, θ. It is so named because it corrects the likelihood ratio usually considered in statistical tests in a manner that quantifies the qualitative notion behind "Occam's razor:" simpler explanations are to be preferred unless there is sufficient evidence in favor of more complicated explanations. Bayes' theorem both quantifies such evidence and determines how much additional evidence is "sufficient" through the calculation of global likelihoods.

Suppose M_1 has two parameters θ and ϕ, then following Equation (3.22), we can write

$$p(D|M_1, I) = \iint d\theta d\phi \, p(\theta|M_1, I)p(\phi|M_1, I)p(D|\theta, \phi, M_1, I)$$
$$\approx p(D|\hat{\theta}, \hat{\phi}, M_1, I)\frac{\delta\theta}{\Delta\theta}\frac{\delta\phi}{\Delta\phi} = \mathcal{L}_{max}\Omega_\theta\Omega_\phi. \tag{3.26}$$

The above equation assumes independent flat priors for the two parameters. It is clear from Equation (3.26) that the total Occam penalty, $\Omega_{total} = \Omega_\theta\Omega_\phi$, can become very large. For example, if $\delta\theta/\Delta\theta = \delta\phi/\Delta\phi = 0.01$ then $\Omega_{total} = 10^{-4}$. Thus for the Bayes factor in Equation (3.24) to favor M_1, the ratio of the maximum likelihoods,

$$\frac{p(D|\hat{\theta}, \hat{\phi}, M_1, I)}{p(D|M_0, I)} = \frac{\mathcal{L}_{max}(M_1)}{\mathcal{L}_{max}(M_0)}$$

must be $\geq 10^4$. Unless the data argue very strongly for the greater complexity of M_1 through the likelihood ratio, the Occam factor will ensure we favor the simpler model. We will explore the Occam factor further in a worked example in Section 3.6.

In the above calculations, we have specifically made a point of identifying the Occam factors and how they arise. In many instances we are not interested in the value of the Occam factor, but only in the final posterior probabilities of the competing models.

Because the Occam factor arises automatically in the marginalization process, its effect will be present in any model selection calculation.

3.6 Sample spectral line problem

In this section, we will illustrate many of the above points in a detailed Bayesian analysis of a toy spectral line problem. In a real problem, as opposed to the hypothetical one discussed below, there could be all sorts of complicated prior information. Although Bayesian analysis can readily handle these complexities, our aim here is to bring out the main features of the Bayesian approach as simply as possible. Be warned; even though it is a relatively simple problem, our detailed solution, together with commentary and a summary of the lessons learned, will occupy quite a few pages.

3.6.1 Background information

In this problem, we suppose that two competing grand unification theories have been proposed. Each one is championed by a Nobel prize winner in physics. We want to compute the relative probability of the truth of each theory based on our prior (background) information and some new data. Both theories make definite predictions in energy ranges beyond the reach of the present generation of particle accelerators. In addition, theory 1 uniquely predicts the existence of a new short-lived baryon which is expected to form a short-lived atom and give rise to a spectral line at an accurately calculable radio wavelength. Unfortunately, it is not feasible to detect the line in the laboratory. The only possibility of obtaining a sufficient column density of the short-lived atom is in interstellar space. Prior estimates of the line strength expected from the Orion nebula according to theory 1 range from 0.1 to 100 mK.

Theory 1 also predicts the line will have a Gaussian line shape of the form

$$T \exp \left\{ \frac{-(\nu_i - \nu_o)^2}{2\sigma_L^2} \right\} \quad \text{(abbreviated by } Tf_i), \quad (3.27)$$

where the signal strength is measured in temperature units of mK and T is the amplitude of the line. The frequency, ν_i, is in units of channel number and $\nu_o = 37$. The width of the line profile is characterized by σ_L, and $\sigma_L = 2$ channel numbers. The predicted line shape is shown in Figure 3.2.

Data:

To test this prediction, a new spectrometer was mounted on the James Clerk Maxwell telescope on Mauna Kea and the spectrum shown in Figure 3.3 was obtained. The spectrometer has 64 frequency channels with neighboring channels separated by

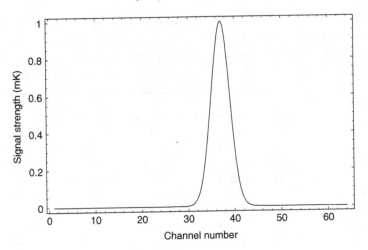

Figure 3.2 Predicted spectral shape according to theory 1.

0.5 σ_L. All channels have Gaussian noise characterized by $\sigma = 1\,\text{mK}$. The noise in separate channels is independent. The data are given in Table 3.1.

Let D be a proposition representing the data from the spectrometer.

$$D \equiv D_1, D_2, \ldots, D_N; \quad N = 64 \qquad (3.28)$$

where D_1 is a proposition that asserts that "the data value recorded in the first channel was d_1."

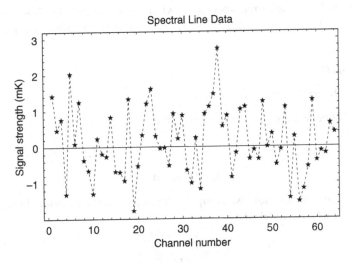

Figure 3.3 Measured spectrum.

Table 3.1 *Spectral line data consisting of 64 frequency channels (#) obtained with a radio astronomy spectrometer. The output voltage from each channel has been calibrated in units of effective black body temperature expressed in mK. The existence of negative values arises from receiver channel noise which gives rise to both positive and negative fluctuations.*

#	mK	#	mK	#	mK	#	mK
1	1.420	17	−0.937	33	0.248	49	0.001
2	0.468	18	1.331	34	−1.169	50	0.360
3	0.762	19	−1.772	35	0.915	51	−0.497
4	−1.312	20	−0.530	36	1.113	52	−0.072
5	2.029	21	0.330	37	1.463	53	1.094
6	0.086	22	1.205	38	2.732	54	−1.425
7	1.249	23	1.613	39	0.571	55	0.283
8	−0.368	24	0.300	40	0.865	56	−1.526
9	−0.657	25	−0.046	41	−0.849	57	−1.174
10	−1.294	26	−0.026	42	−0.171	58	−0.558
11	0.235	27	−0.519	43	1.031	59	1.282
12	−0.192	28	0.924	44	1.105	60	−0.384
13	−0.269	29	0.230	45	−0.344	61	−0.120
14	0.827	30	0.877	46	−0.087	62	−0.187
15	−0.685	31	−0.650	47	−0.351	63	0.646
16	−0.702	32	−1.004	48	1.248	64	0.399

Question: Which theory is more probable?

Based on our current state of information, which includes just the above prior information and the measured spectrum, what do we conclude about the relative probabilities of the two competing theories and what is the posterior PDF for the line strength?

Hypothesis space:

$M_1 \equiv$ "Theory 1 correct, line exists"

$M_2 \equiv$ "Theory 2 correct, no line predicted"

3.7 Odds ratio

To answer the above question, we compute the odds ratio (abbreviated simply by the *odds*) of model M_1 to model M_2.

$$O_{12} = \frac{p(M_1|D, I)}{p(M_2|D, I)}. \tag{3.29}$$

From Equation (3.14) we can write

$$O_{12} = \frac{p(M_1|I)\, p(D|M_1,I)}{p(M_2|I)\, p(D|M_2,I)}$$

$$\equiv \frac{p(M_1|I)}{p(M_2|I)} B_{12} \tag{3.30}$$

where $p(M_1|I)/p(M_2|I)$ is the prior odds, and $p(D|M_1,I)/p(D|M_2,I)$ is the global likelihood ratio, which is also called the Bayes factor.

Based on the prior information given in the statement of the problem, we assign the prior odds = 1, so our final odds is given by,

$$O_{12} = \frac{p(D|M_1,I)}{p(D|M_2,I)} \quad \text{(the Bayes factor).} \tag{3.31}$$

To obtain $p(D|M_1,I)$, the global likelihood of M_1, we need to marginalize over its unknown parameter, T. From Equation (3.8), we can write

$$p(D|M_1,I) = \int dT\ p(T|M_1,I)p(D|M_1,T,I). \tag{3.32}$$

In the following section we will consider what form of prior to use for $p(T|M_1,I)$. In Section 3.7.2 we will show how to evaluate the likelihood, $p(D|M_1,T,I)$.

3.7.1 Choice of prior $p(T|M_1,I)$

We need to evaluate the global likelihood of model M_1 for use in the Bayes factor. One of the items we need in this calculation is $p(T|M_1,I)$, the prior for T. Choosing a prior is an important part of any Bayesian calculation and we will have a lot to say about this topic in Section 3.10 and other chapters, e.g., Chapter 8, and Sections 9.2.3, 13.3 and 13.4. For this example, we will investigate two common choices: the uniform prior and the Jeffreys prior.[3]

Uniform prior
Suppose we chose a uniform prior for $p(T|M_1,I)$ in the range $T_{\min} \leq T \leq T_{\max}$

$$p(T|M_1,I) = \frac{1}{\Delta T}, \tag{3.33}$$

where $\Delta T = T_{\max} - T_{\min}$.

There is a problem with this prior if the range of T is large. In the current example $T_{\max} = 100$ and $T_{\min} = 0.1$. To illustrate the problem, we compare the probability that

---

[3] If the lower limit on T extended all the way to zero, we would not be able to use a Jeffreys prior because of the infinity at $T = 0$. A modified version of the form, $p(T|M_1,I) = 1/\{(T+a)\ln[(a+T_{\max})/a]\}$, where a is a constant, eliminates this singularity. This *modified Jeffreys* behaves like a uniform prior for $T < a$ and a Jeffreys for $T > a$.

T lies in the upper decade of the prior range (10 to 100 mK) to the lowest decade (0.1 to 1 mK). This is given by

$$\frac{\int_{10}^{100} p(T|M_1, I)dT}{\int_{0.1}^{1} p(T|M_1, I)dT} = 100. \tag{3.34}$$

We see that in this case, a uniform prior implies that the line strength is 100 times more probable to be in the top decade of the predicted range than the bottom, i.e., it is much more probable that T is strong than weak. Usually, expressing great uncertainty in some quantity corresponds more closely to a statement of scale invariance or equal probability per decade. In this situation, we recommend using a Jeffreys prior which is scale invariant.

Jeffreys prior
The form of the prior which represents equal probability per decade (scale invariance) is given by $p(T|M_1, I) = k/T$, where $k = \text{constant}$.

$$\int_{0.1}^{1} p(T|M_1, I)dT = k \int_{0.1}^{1} \frac{dT}{T} = k \ln 10 = \int_{10}^{100} p(T|M_1, I)dT. \tag{3.35}$$

We can evaluate k from the requirement that

$$\int_{T_{\min}}^{T_{\max}} p(T|M_1, I)dT = 1 = k \ln\left(\frac{T_{\max}}{T_{\min}}\right) \tag{3.36}$$

$$\frac{1}{k} = \ln\left(\frac{T_{\max}}{T_{\min}}\right). \tag{3.37}$$

Thus, the form of the Jeffreys prior is given by

$$p(T|M_1, I) = \frac{1}{T \ln(T_{\max}/T_{\min})}. \tag{3.38}$$

A convenient way of summarizing the above comparison between the uniform and Jeffreys prior is to plot the probability of each distribution per logarithmic interval or $p(\ln T|M_1, I)$. This can be obtained from the condition that the probability in the interval T to $T + dT$ must equal the probability in the transformed interval $\ln T$ to $\ln T + d \ln T$.

$$p(T|M_1, I)dT = p(\ln T|M_1, I)d \ln T$$
$$p(T|M_1, I) = p(\ln T|M_1, I)\frac{d \ln T}{dT} = \frac{1}{T}p(\ln T|M_1, I) \tag{3.39}$$
$$p(\ln T|M_1, I) = T \times p(T|M_1, I).$$

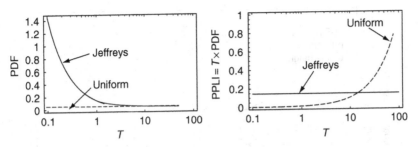

Figure 3.4 The left panel shows the probability density function (PDF), $p(T|M_1, I)$, for the uniform and Jeffreys priors. The right panel shows the probability per logarithmic interval (PPLI), $T \times p(T|M_1, I)$.

Figure 3.4 compares plots of the probability density function (PDF), $p(T|M_1, I)$ (left panel), and the probability per logarithmic interval (PPLI), $T \times p(T|M_1, I)$ (right panel), for the uniform and Jeffreys priors.

3.7.2 Calculation of $p(D|M_1, T, I)$

Let d_i represent the measured data value for the ith channel of the spectrometer. According to model M_1,

$$d_i = Tf_i + e_i, \qquad (3.40)$$

where e_i is an error term. Our prior information indicates that this error is caused by receiver noise which has a Gaussian distribution with a standard deviation of σ. Also, from Equation (3.27), we have

$$f_i = \exp\left\{\frac{-(\nu_i - \nu_o)^2}{2\sigma_L^2}\right\}. \qquad (3.41)$$

Assuming M_1 is true, then if it were not for the error e_i, d_i would equal Tf_i. Let $E_i \equiv$ "a proposition asserting that the ith error value is in the range e_i to $e_i + de_i$." In this case, we can show (see Section 4.8) that $p(D_i|M_1, T, I) = p(E_i|M_1, T, I)$. If all the E_i are independent[4] then

$$\begin{aligned}
p(D|M_1, T, I) &= p(D_1, D_2, \ldots, D_N|M_1, T, I) \\
&= p(E_1, E_2, \ldots, E_N|M_1, T, I) \\
&= p(E_1|M_1, T, I)p(E_2|M_1, T, I)\ldots p(E_N|M_1, T, I) \qquad (3.42) \\
&= \prod_{i=1}^{N} p(E_i|M_1, T, I)
\end{aligned}$$

---

[4] We deal with the effect of correlated errors in Section 10.2.2.

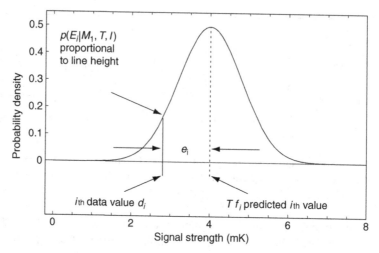

Figure 3.5 Probability of getting a data value d_i a distance e_i away from the predicted value is proportional to the height of the Gaussian error curve at that location.

where $\prod_{i=1}^{N}$ stands for the product of N of these terms. From the prior information, we can write

$$p(E_i|M_1, T, I) = \frac{1}{\sigma\sqrt{2\pi}}\exp\left\{-\frac{e_i^2}{2\sigma^2}\right\}$$
$$= \frac{1}{\sigma\sqrt{2\pi}}\exp\left\{-\frac{(d_i - Tf_i)^2}{2\sigma^2}\right\}. \tag{3.43}$$

It is apparent that $p(E_i|M_1, T, I)$ is a probability density function since e_i, the value of the error for channel i, is a continuous variable. The factor $(\sigma\sqrt{2\pi})^{-1}$ in the above equation ensures that the integral over e_i from $-\infty$ to $+\infty$ is equal to 1. In Figure 3.5, $p(E_i|M_1, T, I)$ is shown proportional to the height of the Gaussian error curve at the position of the actual data value d_i.

Combining Equations (3.42) and (3.43), we obtain the probability of the entire data set

$$p(D|M_1, T, I) = \prod_{i=1}^{N}\frac{1}{\sigma\sqrt{2\pi}}\exp\left\{-\frac{(d_i - Tf_i)^2}{2\sigma^2}\right\}$$
$$= (2\pi)^{-N/2}\sigma^{-N}\exp\left\{-\frac{\sum_i(d_i - Tf_i)^2}{2\sigma^2}\right\}. \tag{3.44}$$

In Section 3.7.4, we will need the maximum value of the likelihood given by Equation (3.44). Since we now know all the quantities in Equation (3.44) except T, we can readily compute the likelihood as a function of T in the prior range $0.1 \leq T \leq 100$. The likelihood has a maximum $= 8.520 \times 10^{-37}$ (called the *maximum likelihood*) at $T = 1.561$ mK.

What we want is $p(D|M_1, I)$, the global likelihood of M_1, for use in Equation (3.31). We now evaluate $p(D|M_1, I)$, given by Equation (3.32), for the two different priors discussed in Section 3.7.1, where we argued that the Jeffreys prior matches much more closely the prior information given in this particular problem. Nevertheless, it is interesting to explore what effect the choice of a uniform prior would have on our conclusions. For this reason, we will do the calculations for both priors.

Uniform prior case:

$$p(D|M_1, I) = \frac{(2\pi)^{-N/2}\sigma^{-N}}{\Delta T}\exp\left\{\frac{-\sum d_i^2}{2\sigma^2}\right\}\int_{T_{\min}}^{T_{\max}} dT \exp\left\{\frac{T\sum d_i f_i}{\sigma^2}\right\}\exp\left\{-\frac{T^2\sum f_i^2}{2\sigma^2}\right\}$$

$$= 1.131 \times 10^{-38}.$$

$$(3.45)$$

According to Equation (3.25), we can always write the global likelihood of a model as the maximum value of its likelihood times an Occam factor, Ω_T, which arises in this case from marginalizing T.

$$p(D|M_1, I) = \mathcal{L}_{\max}(M_1) \times \Omega_T$$
$$= \text{maximum value of } [p(D|M_1, T, I)] \times \text{Occam factor} \qquad (3.46)$$
$$= 8.520 \times 10^{-37}\,\Omega_T.$$

Comparison of the results of Equations (3.45) and (3.46) leads directly to a value for the Occam factor, associated with our prior uncertainty in the T parameter, of $\Omega_T = 0.0133$.

Jeffreys prior case:

$$p(D|M_1, I) = \frac{(2\pi)^{-N/2}\sigma^{-N}}{\ln(T_{\max}/T_{\min})}\exp\left\{\frac{-\sum d_i^2}{2\sigma^2}\right\}$$
$$\times \int_{T_{\min}}^{T_{\max}} dT\,\frac{\exp\left\{\frac{T\sum d_i f_i}{\sigma^2}\right\}\exp\left\{-\frac{T^2\sum f_i^2}{2\sigma^2}\right\}}{T} \qquad (3.47)$$
$$= 1.239 \times 10^{-37}.$$

In this case the Occam factor associated with our prior uncertainty in the T parameter, based on a Jeffreys prior, is 0.145. Note: the Occam factor based on the Jeffreys prior is a factor of ≈ 10 less of a penalty than for the uniform prior for the same parameter.

3.7.3 Calculation of $p(D|M_2, I)$

Model M_2 assumes the spectrum is consistent with noise and has no free parameters so in analogy to Equation (3.40), we can write

$$d_i = 0 + e_i \tag{3.48}$$

where e_i = Gaussian noise with a standard deviation of σ. Assuming M_2 is true, then if it were not for the noise e_i, d_i would equal 0.

$$p(D|M_2, I) = (2\pi)^{-N/2} \, \sigma^{-N} \, \exp\left\{ -\frac{\sum d_i^2}{2\sigma^2} \right\}$$
$$= 1.133 \times 10^{-38}. \tag{3.49}$$

Since this model has no free parameters, there is no Occam factor, so the global likelihood is also the maximum likelihood, $\mathcal{L}_{max}(M_2)$, for M_2.

3.7.4 Odds, uniform prior

Substitution of Equations (3.45) and (3.49) into Equation (3.31) leads to an odds ratio for the uniform prior case given by

$$\text{odds} = \frac{1}{\Delta T} \int_{T_{min}}^{T_{max}} dT \, \exp\left\{ \frac{T \sum d_i f_i}{\sigma^2} \right\} \exp\left\{ -\frac{T^2 \sum f_i^2}{2\sigma^2} \right\}. \tag{3.50}$$

For $T_{min} = 0.1 \, \text{mK}$ and $T_{max} = 100 \, \text{mK}$, the odds $= 0.9986$ and

$$p(M_1|D, I) = \frac{1}{1 + \frac{1}{\text{odds}}} = 0.4996. \tag{3.51}$$

Although the ratio of the maximum likelihoods for the two models favors model M_1, by a factor of $\mathcal{L}_{max}(M_1)/\mathcal{L}_{max}(M_2) = 8.520 \times 10^{-37}/1.131 \times 10^{-38} \approx 75$, the ratio of the global likelihoods marginally favors M_2 because of the Occam factor which penalizes M_1 for its extra complexity.

3.7.5 Odds, Jeffreys prior

Substitution of Equations (3.47) and (3.49) into Equation (3.31) leads to an odds ratio for the Jeffreys prior case, given by

$$\text{odds} = \frac{1}{\ln(T_{max}/T_{min})} \int_{T_{min}}^{T_{max}} dT \, \frac{\exp\left\{ \frac{T \sum d_i f_i}{\sigma^2} \right\} \exp\left\{ -\frac{T^2 \sum f_i^2}{2\sigma^2} \right\}}{T}. \tag{3.52}$$

For $T_{min} = 0.1 \, \text{mK}$ and $T_{max} = 100 \, \text{mK}$, the odds $= 10.94$, and $p(M_1|D, I) = 0.916$.

As noted earlier in this chapter, we consider the Jeffreys prior to be much more consistent with the large uncertainty in signal strength which was part of the prior information of the problem. On this basis, we conclude that for our current state of information, $p(M_1|D,I) = 0.916$ and $p(M_2|D,I) = 0.084$.

3.8 Parameter estimation problem

Now that we have solved the model selection problem leading to a significant preference for M_1, which argues for the existence of the short-lived baryon, we would like to compute $p(T|D,M_1,I)$, the posterior PDF for the signal strength. Again we will compute the result for both choices of prior for comparison, but consider the Jeffreys result to be more reasonable for the current problem.

Again, start with Bayes' theorem:

$$p(T|D,M_1,I) = \frac{p(T|M_1,I)p(D|M_1,T,I)}{p(D|M_1,I)}$$
$$\propto p(T|M_1,I)p(D|M_1,T,I).$$
(3.53)

We have already evaluated $p(D|M_1,T,I)$ in Equation (3.44). All that remains is to plug in our two different choices for the prior $p(T|M_1,I)$.

Uniform prior case:

$$p(T|D,M_1,I) \propto \exp\left\{\frac{T\sum d_i f_i}{\sigma^2}\right\}\exp\left\{-\frac{T^2\sum f_i^2}{2\sigma^2}\right\}$$
(3.54)

Jeffreys prior case:

$$p(T|D,M_1,I) \propto \frac{1}{T}\exp\left\{\frac{T\sum d_i f_i}{\sigma^2}\right\}\exp\left\{-\frac{T^2\sum f_i^2}{2\sigma^2}\right\}.$$
(3.55)

Figure 3.6 shows the posterior PDF for the signal strength for both the uniform and Jeffreys priors. As we saw earlier, the uniform prior favors stronger signals.

In our original spectrum, the line strength was comparable to the noise level. How do the results change as we increase the line strength? Figure 3.7 shows a simulated spectrum for a line strength equal to five times the noise σ together with the estimated posterior PDF for the line strength. The increase in line strength has a dramatic effect on the odds which rise to a whopping 1.6×10^{12} for the uniform prior and 5.3×10^{12} for the Jeffreys prior.

3.8.1 Sensitivity of odds to T_{max}

Figure 3.8 is a plot of the dependence of the odds on the assumed value of T_{max} for both uniform and Jeffreys priors. We see that under the uniform prior, the odds are

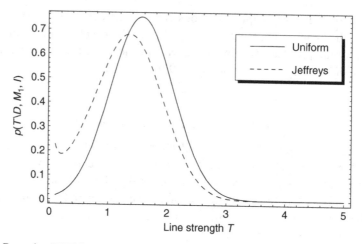

Figure 3.6 Posterior PDF for the line strength, T, for uniform and Jeffreys priors.

much more strongly dependent on the prior range of T than for the Jeffreys case. In both cases, the Occam's razor penalizing M_1 compared to M_2 for its greater complexity increases as the prior range for T increases. Model complexity depends not only on the number of free parameters but also on their prior ranges.

In this problem, we assumed that both the center frequency and line width were accurately predicted by M_1; the only uncertain quantity was the line strength. Suppose the center frequency and/or line width were uncertain as well. In this case, to compute the odds ratio, we would have to marginalize over the prior ranges for these parameters as well, giving rise to additional Occam's factors and a subsequent lowering of the odds. This agrees with our intuition: the more uncertain our prior information about the expected properties of the line, the less significance we attach to any bump in the spectrum.

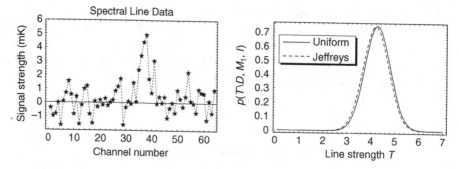

Figure 3.7 The left panel shows a spectrum with a stronger spectral line. The right panel shows the computed posterior PDF for the line strength.

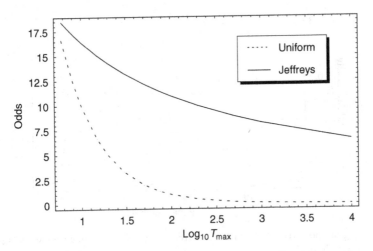

Figure 3.8 The odds ratio versus upper limit on the predicted line strength (T_{max}) for the uniform and Jeffreys priors.

3.9 Lessons

1. In the model selection problem, we are interested in the global probabilities of the two models independent of the most probable model parameters. This was achieved using Bayes' theorem and marginalizing over model M_1's parameter T, the signal strength (model M_2 had no parameters pertaining to the spectral line data). An Occam's razor automatically arises each time a model parameter is marginalized, penalizing the model for prior parameter space that gets ruled out by the data. The larger the prior range that is excluded by the likelihood function, $p(D|M_1, T, I)$, the greater the Occam penalty as can be seen from Figure 3.8. Recall that the global likelihood for a model is the weighted average likelihood for its parameter(s). The weighting function is the prior for the parameter. Thus, the Occam penalty can be very different for two different choices of prior (uniform and Jeffreys). The results are always conditional on the truth of the prior which must be specified in the analysis, and there is a need to seriously examine the consequences of the choice of prior.

2. When the prior range for a parameter spans many orders of magnitude, a uniform prior implies that it is much more probable that the true value of the parameter is in the upper decade. Often, a large prior parameter range can be taken to mean we are ignorant of the scale, i.e., small values of the parameter are equally likely to large values. For these situations, a useful choice is a Jeffreys prior, which corresponds to equal probability per decade (scale invariance). Note: when the range of a prior is a small fraction of the central value, then the conclusion will be the same whether a uniform or Jeffreys prior is used. In the spectrum problem just analyzed, we started out with very crude prior information on the line strength predicted by M_1. Now that we have incorporated the new experimental information D, we have arrived at a posterior probability for the line strength, $p(T|D, M_1, I)$. Were we to obtain more data, D_2, we would set our new prior $p(T|M_1, I_2)$ equal to our current posterior $p(T|D, M_1, I)$, i.e., $I_2 = D, I$. The question of whether to use a Jeffreys or uniform prior would no longer be relevant.

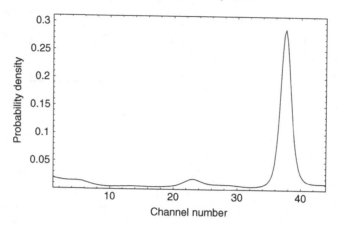

Figure 3.9 Marginal posterior PDF for the line frequency, where the line frequency is expressed as a spectrometer channel number.

3. If the location and line width were also uncertain, we would have to marginalize over these parameters as well, giving rise to other Occam factors which would decrease the odds still further. For example, if the prior range for the expected channel number of the spectral line were increased from less than 1 to 44 channels, the odds would decrease from ≈ 11 to 1, assuming a uniform prior for the line location. We can also compute the marginal posterior PDF for the line frequency for this case which is shown in Figure 3.9. This permits us to update our knowledge of the line frequency given the data and assuming the theory is correct. For further insights on this matter, see the discussion on systematic errors in Section 3.11.

4. Once we established that model M_1 was more probable, we were able to apply Bayes' theorem again, to compute the posterior PDF for the line strength. Note: no Occam factors arise in parameter estimation. Parameter estimation can be viewed as model selection where the competing models all have exactly the same complexity so the Occam penalties are identical and cancel out in the analysis. It can happen that the $p(T|D, M_1, I)$ can be very small for values of T close to zero. One might be tempted to rule out M_2 because it predicts $T = 0$, thus bypassing the model selection problem. This is not wise, however, because the model selection analysis includes Occam factors that could rule out M_1 compared to the simpler M_2. As we noted, these Occam factors do not appear in the parameter estimation analysis.

5. In this toy problem, the spectral line data assume that any background continuum radiation or instrumental DC level has been subtracted off, which can only be done to a certain accuracy. It would be better to parameterize this DC level and marginalize over this parameter so that the effect of our uncertainty in this quantity (see Section 3.11) will be included in our final odds ratio and spectral line parameter estimates. A still more complicated version of this problem is if M_1 simply predicts a certain prior range for the optical depth of the line but

leaves unanswered whether the line will be seen in emission or absorption against the background continuum. In this problem, a Bayesian solution is still possible but will involve a more complicated model of the spectral line data.

3.10 Ignorance priors

In the analysis of the spectral line problem of Section 3.7.1, we considered two different forms of prior (uniform and Jeffreys) for the unknown line temperature parameter. We learned that there was a strong reason for picking the Jeffreys prior in this problem. What motivated a consideration of these particular priors in the first place? In this section we will attempt to answer this question.

As we study any particular phenomenon, our state of knowledge changes. When we are well into the study, our prior for the analysis of new data will be well defined by our previous posterior. But in the earliest phase, our state of "ignorance" will be high. It is therefore useful to have arguments to aid us in selecting an appropriate form of prior to use in such situations. Of course, if we are completely ignorant we cannot even state the problem of interest, and in that case we have no use for a prior. Let us suppose our state of knowledge is sufficient to pose the problem but not much more. For example, we might be interested in the location of the highest point on the equator of Pluto. Are there any general arguments to help us select a suitable prior? In Section 2.6 we saw how to use the *Principle of Indifference* to arrive at a probability distribution for a discrete set of hypotheses.

In the discussion that follows, we will consider a general argument that suggests the form of priors to use for two types of continuous parameters. We will make a distinction between *location parameters*, and *scale parameters*. For example, consider the location of an event in space. To describe this, we must locate the event with respect to some origin and specify the size (scale) of our units of space (e.g., ft, m, light years). The location of an event can be either a positive or negative quantity depending on our choice of origin but the scale (size of our space units) is always a positive quantity. We will first consider a prior for a location parameter.

Suppose we are interested in evaluating $p(X|I)$, where $X \equiv$ "a proposition asserting that the location of the tallest tree along the shore of Lake Superior is between x and $x + dx$." In this statement of the problem, x is measured with respect to a particular survey stake. We will represent the probability density by the function $f(x)$.

What if we consider a different statement of the problem in which the only change is that the origin of our distance measurement has been shifted by an amount c and we are interested in $p(X'|I)$ where $x' = x + c$? If a shift of location (origin) can make the problem appear in any way different, then it must be that we had some kind of prior knowledge about location. In the limit of complete ignorance, the choice of prior would be invariant to a shift in location. Although we are not completely ignorant it still might be useful, in the earliest phase of an investigation, to adopt a prior which is invariant to a shift in location. What form of prior does this imply? If we define our state of ignorance to mean that the above two statements of the problem are equivalent, then the desideratum of consistency demands that

$$p(X|I)dX = p(X'|I)dX' = p(X'|I)d(X + c) = p(X'|I)dX. \tag{3.56}$$

From this it follows that

$$f(x) = f(x') = f(x + c). \tag{3.57}$$

The solution of this equation is $f(x) = $ constant, so

$$p(X|I) = \text{constant}. \tag{3.58}$$

In the Lake Superior problem, it is apparent that we have knowledge of the upper (x_{max}) and lower (x_{min}) bounds of x, so the constant $= 1/(x_{max} - x_{min})$. If we are ignorant of these limits then we refer to $p(X|I)$ as an improper prior, meaning that it is not normalized. An improper prior is useable in parameter estimation problems but is not suitable for model selection problems, because the Occam factors depend on knowing the prior range for each model parameter.

Now consider a problem where we are interested in the mean lifetime of a newly discovered aquatic creature found in the ocean below the ice crust on the moon Europa. We call the lifetime a scale parameter because it can only have positive values, unlike a location parameter which can assume both positive and negative values. Let $T \equiv$ "the mean lifetime is between τ and $\tau + d\tau$." What form of prior probability density, $p(T|I)$, should we use in this case? We will represent the probability density by the function $g(\tau)$.

What if we consider a different statement of the problem in which the only change is that the time is measured in units differing by a factor β? Now we are interested in $p(T'|I)$ where $\tau' = \beta\tau$. If we define our state of ignorance to mean that the two statements of the problems are equivalent, then the desideratum of consistency demands that

$$p(T|I)dT = p(T'|I)dT' = p(T'|I)d(\beta T) = \beta p(T'|I)dT. \tag{3.59}$$

From this it follows that

$$g(\tau) = \beta g(\tau') = \beta g(\beta\tau). \tag{3.60}$$

The solution of this equation is $g(\tau) = \text{constant}/\tau$, so

$$p(T|I) = \frac{\text{constant}}{\tau}. \tag{3.61}$$

This form of prior is called the *Jeffreys prior* after Sir Harold Jeffreys who first suggested it. If we have knowledge of the upper (τ_{max}) and lower (τ_{min}) bounds of τ then we can evaluate the normalization constant. The result is

$$p(T|I) = \frac{1}{\tau \ln(\tau_{max}/\tau_{min})}. \tag{3.62}$$

Returning to the spectral line problem, we now see another reason for preferring the choice of the Jeffreys prior for the temperature parameter, because it is a scale parameter. In Section 9.2.3, we will discover yet another powerful argument for selecting the Jeffreys prior for a scale parameter.

3.11 Systematic errors

In scientific inference, we encounter at least two general types of uncertainties which are broadly classified as random and systematic. Random uncertainties can be reduced by acquiring and averaging more data. This is the basis behind signal averaging which is discussed in Section 5.11.1. Of course, what appears random for one state of information might later be discovered to have a predictable pattern as our state of information changes.

Some typical examples of systematic errors include errors of calibration of meters and rulers,[5] and stickiness and wear in the moving parts of meters. For example, over time an old wooden meter stick may shrink by as much as a few mm. Some potential systematic errors can be detected by careful analysis of the experiment before performing it and can then be eliminated either by applying suitable corrections or through careful experimental design. The remaining systematic errors can be very subtle, and are detected with certainty only when the same quantity is measured by two or more completely different experimental methods. The systematic errors are then revealed by discrepancies between the measurements made by the different methods.

Bayesian inference provides a powerful way of looking and dealing with some of these subtle systematic errors. We almost always have some prior information about the accuracy of our "ruler." Clearly, if we had no information about its accuracy (in contrast to its repeatability), we would have no logical grounds to use it at all except as a means for ordering events. In this case, we would be expecting no more from our ruler and we would have no concern about a systematic error. What this implies is that we require at least some limited prior information about our ruler's scale to be concerned about a systematic error.

As we have seen, a unique feature of the Bayesian approach is the ability to incorporate prior information and see how it affects our conclusions. In the case of the ruler accuracy, the approach taken is to introduce the scale of the ruler into the calculation as a parameter, i.e., we parameterize the systematic error. We can then treat this as a nuisance parameter and marginalize (integrate over) this parameter to obtain our final inference about the quantity of interest. If the uncertainty in the accuracy of our scale is very large, this will be reflected quantitatively in a larger uncertainty in our final inference.

In a complex measurement, many different types of systematic errors can occur, which in principle, can be parameterized and marginalized. For example, consider the

---

[5] One important ruler in astronomy is the Hubble relation relating redshift or velocity to distance.

following modification to the spectral line problem of Section 3.6. Even if we know the predicted frequency of the spectral line accurately, the observed frequency depends on the velocity of the source with respect to the observer through the Doppler effect. The observed frequency of the line, f_o, is related to the emitted frequency, f_e by

$$f_o = f_e\left(1 + \frac{v}{c}\right) \quad \text{for } \frac{v}{c} \ll 1, \tag{3.63}$$

where v is the line of sight component of the velocity of the line emitting region and c equals the velocity of light. In our search for a spectral line, we may be examining a small portion of the Orion nebula and only know the distribution of velocities for the integrated emission from the whole nebula, which may be dominated by turbulent and rotational motion of its parts. The unknown factor v introduces a systematic error in our frequency scale. In this case, we might choose to parameterize the systematic error in v by a Gaussian with a mean and σ equal to that of the Orion nebula as a whole.

From the Bayesian viewpoint, we can even consider uncertain scales that arise in a theoretical model as introducing a systematic error on the same footing, for the purposes of inference, as those associated with a measurement. In the above example, we may know the velocity of the source accurately but the theory may be imprecise with regard to its frequency scale.

Of course, the exact form by which we parameterize a systematic error is constrained by our available information, and just as our theories of nature are updated as our state of knowledge changes, so in general will our understanding of these systematic errors.

It is often the case that we can obtain useful information about a systematic error from the interaction between measurements and theory in Bayesian inference. In particular, we can compute the marginal posterior for the parameter characterizing our systematic error as was done in Figure 3.9. This and other points raised in this section are brought out by the problems at the end of this chapter.

The effect of marginalizing over any parameter, whether or not it is associated with a systematic error, is to introduce an Occam factor which penalizes the model for any prior parameter space that gets ruled out by the data through the likelihood function. The larger the prior range that is excluded by the likelihood function, the greater the Occam penalty. It is thus possible to rule out a valid model by employing an artificially large prior for some systematic error or model parameter. Fortunately, Bayesian inference requires one to specify one's choice of prior so its effect on the conclusions can readily be assessed.

3.11.1 Systematic error example

In 1929, Edwin Hubble found a simple linear relationship between the distance of a galaxy, x, and its recessional velocity, v, of the form $v = H_0 x$, where H_0 is known as *Hubble's constant*. Hubble's constant provides the scale of our ruler for astronomical distance determination. An error in H_0 leads to a systematic error in distance determination. A modern value of $H_0 = 70 \pm 10 \, \text{km s}^{-1} \, \text{Mpc}^{-1}$. Note: astronomical distances are commonly measured in Mpc (a million parsecs). Suppose a particular galaxy has a measured recessional velocity $v_m = (100 \pm 5) \times 10^3 \, \text{km s}^{-1}$. Determine the posterior PDF for the distance to the galaxy assuming:

1) A fixed value of $H_0 = 70 \, \text{km s}^{-1} \, \text{Mpc}^{-1}$.
2) We allow for uncertainty in the value of Hubble's constant. We assume a Gaussian probability density function for H_0, of the form

$$p(H_0|I) = k \exp\left\{ -\frac{(H_0 - 70)^2}{2 \times 10^2} \right\}, \tag{3.64}$$

where k is a normalization constant.
3) We assume a uniform probability density function for H_0, given by

$$p(H_0|I) = \begin{cases} 1/(90 - 50), & \text{for } 50 \leq H_0 \leq 90 \\ 0, & \text{elsewhere.} \end{cases} \tag{3.65}$$

4) We assume a Jeffreys probability density function for H_0, given by

$$p(H_0|I) = \begin{cases} [H_0 \ln(90/50)]^{-1}, & \text{for } 50 \leq H_0 \leq 90 \\ 0, & \text{elsewhere.} \end{cases} \tag{3.66}$$

As usual, we can write

$$v_m = v_{\text{true}} + e \tag{3.67}$$

where v_{true} is the true recessional velocity and e represents the noise component of the measured velocity, v_m. Assume that the probability density function for e can be described by a Gaussian with mean 0 and $\sigma = 5 \times 10^3 \, \text{km s}^{-1}$. To keep the problem simple, we also assume the error in v is uncorrelated with the uncertainty in H_0.

Through the application of Bayes' theorem, as outlined in earlier sections of this chapter, we can readily evaluate the posterior PDF, $p(x|D, I)$, for the distance to the galaxy. The results for the four cases are given below and plotted in Figure 3.10.

Case 1:

$$p(x|D, I) \propto p(x|I) \, p(D|x, I) = p(x|I) \frac{1}{\sqrt{2\pi}\sigma} \exp\left\{ -\frac{e^2}{2\sigma^2} \right\}$$

$$= p(x|I) \frac{1}{\sqrt{2\pi}\sigma} \exp\left\{ -\frac{(v_m - v_{\text{true}})^2}{2\sigma^2} \right\} \tag{3.68}$$

$$= p(x|I) \frac{1}{\sqrt{2\pi}\sigma} \exp\left\{ -\frac{(v_m - H_0 x)^2}{2\sigma^2} \right\}.$$

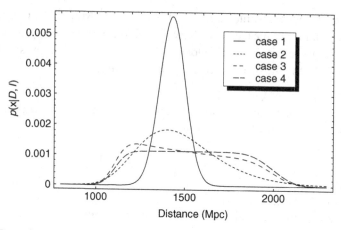

Figure 3.10 Posterior PDF for the galaxy distance, x: 1) assuming a fixed value of Hubble's constant (H_0), 2) incorporating a Gaussian prior uncertainty for H_0, 3) incorporating a uniform prior uncertainty for H_0, and 4) incorporating a Jeffreys prior uncertainty for H_0.

Case 2:

In this case, I incorporates a Gaussian prior uncertainty in the value of H_0.

$$p(x|D, I) = \int_{-\infty}^{\infty} dH_0 \, p(x, H_0|D, I)$$

$$\propto p(x|I) \int_{-\infty}^{\infty} dH_0 \, p(H_0|x, I) \, p(D|x, H_0, I)$$

$$= p(x|I) \int_{-\infty}^{\infty} dH_0 \, p(H_0|I) \, p(D|x, H_0, I) \tag{3.69}$$

$$= p(x|I) \int_{-\infty}^{\infty} dH_0 \, k \exp\left\{ -\frac{(H_0 - 70)^2}{2 \times 10^2} \right\}$$

$$\times \frac{1}{\sqrt{2\pi}\sigma} \exp\left\{ -\frac{(v_m - H_0 x)^2}{2\sigma^2} \right\}.$$

Case 3:

In this case, I incorporates a uniform prior uncertainty in the value of H_0.

$$p(x|D, I) \propto p(x|I) \int_{50}^{90} dH_0 \, p(H_0|I) \, p(D|x, H_0, I)$$

$$= p(x|I) \int_{50}^{90} dH_0 \, \frac{1}{(90 - 50)} \frac{1}{\sqrt{2\pi}\sigma} \exp\left\{ -\frac{(v_m - H_0 x)^2}{2\sigma^2} \right\}. \tag{3.70}$$

Case 4:

In this case, I incorporates a Jeffreys prior uncertainty in the value of H_0.

$$p(x|D, I) \propto p(x|I) \int_{50}^{90} dH_0 \, p(H_0|I) \, p(D|x, H_0, I)$$

$$= p(x|I) \int_{50}^{90} dH_0 \frac{1}{H_0 \ln(90/50)} \frac{1}{\sqrt{2\pi}\sigma} \exp\left\{ -\frac{(v_m - H_0 x)^2}{2\sigma^2} \right\}. \tag{3.71}$$

Equations (3.68), (3.69), (3.70), and (3.71) have been evaluated assuming a uniform prior for $p(x|I)$, and are plotted in Figure 3.10. Incorporating the uncertainty in the scale of our astronomical ruler can lead to two effects. Firstly, the posterior PDF for the galaxy distance is broader. Secondly the mean of the PDF is clearly shifted to a larger value. The means of the PDFs for the four cases are 1429, 1486, 1512, and 1556 km s^{-1}, respectively.

It may surprise you that $p(x|D, I)$ becomes asymmetric when we allow for the uncertainty in H_0. One way to appreciate this is to approximate the integral by a weighted summation over a discrete set of choices for H_0. For each choice of H_0, $p(x|D, I)$ is a symmetric Gaussian offset by a distance Δx given by

$$\Delta x = \frac{v_m}{H_0 + \Delta H_0} - \frac{v_m}{H_0} = \left(-\frac{\Delta H_0}{H_0 + \Delta H_0} \right) \frac{v_m}{H_0}. \tag{3.72}$$

For $\Delta H_0 = +20$ km s^{-1} Mpc^{-1}, the bracketed term in Equation (3.72) is equal to -0.22. For $\Delta H_0 = -20$ km s^{-1} Mpc^{-1}, this term is equal to $+0.4$. Thus, the set of discrete Gaussians is more spread out on one side than the other, which accounts for the asymmetry.

3.12 Problems

1. Redo the calculation of the odds for the spectral line problem of Section 3.6 for the case where there is a systematic uncertainty in the line center of ± 5 channels.

2. The prior information is the same as that given for the spectral line problem in Section 3.6 of the text. The measured spectrum is given in Table 3.2. The spectrum consists of 64 frequency channels. Theory predicts the spectral line has a Gaussian shape with a line width $\sigma_L = 2$ frequency channels. The noise in each channel is known to be Gaussian with a $\sigma = 1.0$ mK and the spectrometer output is in units of mK.

   (a) Plot a graph of the raw data.
   (b) Compute the posterior probability of $M_1 \equiv$ "theory 1 is correct, the spectral line exists," for the two cases: (1) Jeffreys prior for the signal strength, and (2) uniform prior. For this part of the problem, assume that the theory predicts that the spectral line is in channel 24. The prior range for the signal strength is 0.1 to 100 mK. In *Mathematica* you can use the command **NIntegrate** to do the numerical integration required in marginalizing over the line strength.

The how-to of Bayesian inference

Table 3.2 *Spectral line data consisting of 64 frequency channels obtained with a radio astronomy spectrometer. The output voltage from each channel has been calibrated in units of effective black body temperature expressed in mK. The existence of negative values arises from receiver channel noise which gives rise to both positive and negative fluctuations.*

ch. #	mK	ch. #	mK	ch. #	mK	ch. #	mK
1	0.25	17	−0.42	33	0.44	49	−1.56
2	−0.19	18	1.43	34	0.05	50	−0.64
3	0.25	19	−1.33	35	0.59	51	0.48
4	−0.56	20	0.06	36	0.94	52	1.79
5	−0.41	21	0.82	37	−0.10	53	0.07
6	−0.94	22	0.42	38	0.57	54	1.30
7	0.84	23	3.76	39	0.40	55	0.29
8	−0.30	24	1.10	40	−0.97	56	−0.23
9	−2.06	25	1.31	41	2.20	57	−0.50
10	−1.39	26	1.86	42	0.15	58	0.93
11	0.07	27	0.32	43	−0.37	59	−1.28
12	1.80	28	−1.14	44	−0.67	60	−1.98
13	−1.02	29	1.24	45	−0.05	61	1.85
14	−0.46	30	−0.29	46	−0.20	62	0.89
15	0.29	31	0.02	47	0.65	63	0.65
16	−0.36	32	−1.52	48	−1.24	64	0.28

(c) Explain your reasons for preferring one or the other of the two priors.

(d) On the assumption that the model predicting the spectral line is correct, compute and plot the posterior probability (density function) for the line strength for both priors.

(e) Summarize the posterior probability for the line strength by quoting the most probable value and the (+) and (−) error bars that span the 95% credible region (see the last part of Section 3.3 for a definition of credible region). The credible region can be evaluated by computing the probability for a discrete grid of closely spaced line temperature values. Sort these (probability, temperature) pairs in descending order of probability and then sum the probabilities starting from the highest until they equal 95%. As each term is added, keep track of the upper and lower temperature bounds of the terms included in the sum. *Mathematica* command **Sort[yourdata, OrderedQ[{#2, #1] &];**, will sort the file "yourdata" in descending order according to the first item in each row of the data list.

(f) Repeat the calculations in (b) and (d), only this time, assume that the prior prediction on the location of the spectral line frequency is uncertain; it is predicted to occur somewhere between channels 1 and 50. Assume a uniform prior for the unknown line center.[6] This will involve computing a two-dimensional likelihood distribution in the variables line frequency and line strength for a discrete set of values of these parameters, and then using a summation operation to approximate integration[7] (you will probably find **NIntegrate** too slow in two dimensions), for marginalizing over both parameters to obtain the global likelihood for computing the odds. For this purpose, you can use a line frequency interval of 1 channel and a signal strength interval of 0.1 mK for 100 intervals. Although this only spans the prior range 0.1 to 10 mK the PDF will be so low beyond 10 mK that it will not contribute significantly to the integral.

(g) Calculate and plot the marginal posterior probabilities for the line frequency.

(h) What additional Occam factor is associated with marginalizing over the prior line frequency range?

3. Plot $p(x|D, I)$ for case 4 (Jeffreys prior) in Section 3.11.1, assuming

$$p(H_0|I) = \begin{cases} \frac{1}{H_0 \ln(80/60)}, & \text{for } 60 \le H_0 \le 80 \\ 0, & \text{elsewhere.} \end{cases} \tag{3.73}$$

Box 3.1

Equation (3.69) can be evaluated using *Mathematica*.
The evaluation will be faster if you compute a **Table** of values for $p(x, H_0|D, I)$ at equally spaced intervals in x, and use **NIntegrate** to integrate over the given range for H_0.

$p(x|D, I) \propto$ **Table** [

$\left\{ x, \text{NIntegrate}\left[\frac{1}{H_0 \sqrt{2\pi}\, \sigma} \exp\left(-\frac{(v_m - xH_0)^2}{2\sigma^2} \right), \{H_0, 60, 80\} \right] \right\},$

$\{x, 800, 2200, 50\}]$

---

[6] Note: when the frequency range of the prior is a small fraction of center frequency, the conclusion will be the same whether a uniform or Jeffreys prior is assumed for the unknown frequency.
[7] A convenient way to sum elements in a list is to use the *Mathematica* command **Plus@@list**.

4

Assigning probabilities

4.1 Introduction

When we adopt the approach of probability theory as extended logic, the solution to any inference problem begins with Bayes' theorem:

$$p(H_i|D, I) = \frac{p(H_i|I)p(D|H_i, I)}{p(D|I)}. \tag{4.1}$$

In a well-posed problem, the prior information, I, defines the hypothesis space and provides the information necessary to compute the terms in Bayes' theorem.

In this chapter we will be concerned with how to encode our prior information, I, into a probability distribution to use for $p(D|H_i, I)$. Different states of knowledge correspond to different probability distributions. These probability distributions are frequently called *sampling distributions*, a carry-over from conventional statistics literature. Recall that in inference problems, $p(D|H_i, I)$ gives the probability of obtaining the data, D, that we actually got, under the assumption that H_i is true. Thus, $p(D|H_i, I)$ yields how likely it is that H_i is true,[1] and hence it is referred to as the likelihood and frequently written as $\mathcal{L}(H_i)$.

For example, we might have two competing hypotheses H_1 and H_2 that each predicts different values of some temperature, say 1 K and 4.5 K, respectively. If the measured value is 1.2 ± 0.4 K then it is clear that H_1 is more likely to be true. In precisely this type of situation we can use $p(D|H_i, I)$ to compute quantitatively the relative likelihood of H_1 and H_2. We saw how to do that in one case (Section 3.6) where the likelihood was the product of N independent Gaussian distributions.

4.2 Binomial distribution

In this section, we will see how a particular state of knowledge (prior information I) leads us to the choice of likelihood, $p(D|H_i, I)$, which is the well-known binomial distribution (derivation due to M. Tribus, 1969). In this case, our prior information is as follows:

---

[1] Conversely, if we know that H_i is true, then we can directly calculate the probability of observing any particular data value. We will use $p(D|H_i, I)$ in this way to generate simulated data sets in Section 5.13.

$I \equiv$ "Proposition E represents an event that is repeated many times and has two possible outcomes represented by propositions, Q and $\overline{Q}$, e.g., tossing a coin. The probability of outcome Q is constant from event to event, i.e., the probability of getting an outcome Q in any individual event is independent of the outcome for any other event."

In the Boolean algebra of propositions we can write E as

$$E = Q + \overline{Q}, \tag{4.2}$$

where $Q + \overline{Q}$ is the logical sum. Then the possible outcomes of n events can be written as

$$E_1, E_2, \ldots, E_n = (Q_1 + \overline{Q_1}), (Q_2 + \overline{Q_2}), \ldots, (Q_n + \overline{Q_n}), \tag{4.3}$$

where $Q_i \equiv$ "outcome Q occurred for the ith event." If the multiplication on the right is carried out, the result will be a logical sum of 2^n terms, each a product of n logical statements, thereby enumerating all possible outcomes of the n events. For $n = 3$ we find:

$$
\begin{aligned}
E_1, E_2, E_3 = {} & Q_1, Q_2, Q_3 + Q_1, Q_2, \overline{Q_3} + Q_1, \overline{Q_2}, Q_3 + \overline{Q_1}, Q_2, Q_3 \\
& + Q_1, \overline{Q_2}, \overline{Q_3} + \overline{Q_1}, Q_2, \overline{Q_3} + \overline{Q_1}, \overline{Q_2}, Q_3 + \overline{Q_1}, \overline{Q_2}, \overline{Q_3}.
\end{aligned} \tag{4.4}
$$

The probability of the particular sequence $Q_1, \overline{Q_2}, \overline{Q_3}$ can be obtained from repeated applications of the product rule.

$$
\begin{aligned}
p(Q_1, \overline{Q_2}, \overline{Q_3}|I) &= p(Q_1|I)\, p(\overline{Q_2}, \overline{Q_3}|Q_1, I) \\
&= p(Q_1|I)\, p(\overline{Q_2}|Q_1, I)\, p(\overline{Q_3}|Q_1, \overline{Q_2}, I).
\end{aligned} \tag{4.5}
$$

Information I leads us to assign the same probability for outcome Q for each event independent of what happened earlier or later, so Equation (4.5) becomes

$$
\begin{aligned}
p(Q_1, \overline{Q_2}, \overline{Q_3}|I) &= p(Q_1|I)\, p(\overline{Q_2}|I)\, p(\overline{Q_3}|I) \\
&= p(Q|I)\, p(\overline{Q}|I)\, p(\overline{Q}|I) \\
&= p(Q|I)\, p(\overline{Q}|I)^2.
\end{aligned} \tag{4.6}
$$

Thus, the probability of a particular outcome depends only on the number of Q's and $\overline{Q}$'s in it and not on the order in which they occur. Returning to Equation (4.4), we note that:

one outcome, the first, contains three Q's,
three outcomes contain two Q's,
three outcomes contain only one Q,
and one outcome contains no Q's.

More generally, we are going to be interested in the number of ways of getting an outcome with r Q's in n events or trials. In each event, it is possible to obtain a Q, so the question becomes in how many ways can we select r Q's from n events where their order is irrelevant, which is given by ${}^n C_r$.

$$^nC_r = \frac{n!}{r!(n-r)!} = \binom{n}{r}. \tag{4.7}$$

For example, $\binom{3}{2} = \frac{3!}{2!1!} = 3,$ $Q, Q, \overline{Q}$ $Q, \overline{Q}, Q$ $\overline{Q}, Q, Q.$

Thus, the probability of getting r Q's in n events is the probability of any one sequence with r Q's and $(n-r)$ $\overline{Q}$'s, multiplied by ^nC_r, the multiplicity of ways of obtaining r Q's in n events or trials. Therefore, we conclude that in n trials, the probability of seeing the outcome of r Q's and $(n-r)$ $\overline{Q}$'s is

$$p(r|n, I) = \frac{n!}{r!(n-r)!} p(Q|I)^r p(\overline{Q}|I)^{n-r}. \tag{4.8}$$

This distribution is called the *binomial distribution*.

Note the similarity to the *binomial expansion*

$$(x+y)^n = \sum_{r=0}^{n} \frac{n!}{r!(n-r)!} x^r y^{n-r}. \tag{4.9}$$

Referring back to Equation (4.4), in the algebra of propositions, we can interpret E^n to mean E carried out n times and write it in a form analogous to Equation (4.9):

$$E^n = (Q + \overline{Q})^n.$$

Example:

$I \equiv$ "You pick up one of two coins which appear identical. One, coin A, is known to be a fair coin, while coin B is a weighted coin with $p(\text{head}) = 0.2$." From this information and from experimental information you will acquire from tossing the coin, compute the probability that you picked up coin A.

$D \equiv$ "3 heads turn up in 5 tosses."

What is the probability you picked coin A?

$$\text{Let odds} = \frac{p(A|D, I)}{p(B|D, I)}$$

$$= \frac{p(A|I)p(D|A, I)}{p(B|I)p(D|B, I)} = \frac{\frac{1}{2} p(D|A, I)}{\frac{1}{2} p(D|B, I)}. \tag{4.10}$$

To evaluate the likelihoods $p(D|A, I)$ and $p(D|B, I)$, we use the binomial distribution, given by

$$p(r|n, I) = \frac{n!}{r!(n-r)!} p(\text{head}|A, I)^r p(\text{tail}|A, I)^{n-r},$$

where $p(r|n, I)$ is the probability of obtaining r heads in n tosses and $p(\text{head}|A, I)$ is the probability of obtaining a head in any single toss assuming A is true. Now

$$p(D|A, I) = \binom{n}{r} p(\text{head}|A, I)^r p(\text{tail}|A, I)^{n-r} = \binom{5}{3}(0.5)^3(0.5)^2$$

and $p(D|B, I) = \binom{5}{3}(0.2)^3(0.8)^2 \rightarrow \text{odds} = 6.1 = \frac{p(A|D,I)}{1-p(A|D,I)}$ and so $p(A|D, I) = 0.86$.

Thus, the probability you picked up coin $A = 0.86$, based on our current state of knowledge.

4.2.1 Bernoulli's law of large numbers

The binomial distribution allows us to compute $p(r|n, I)$, where r is, for example, the number of heads occurring in n tosses of a coin. According to Bernoulli's law of large numbers, the *long-run frequency of occurrence* tends to the probability of the event occurring in any single trial, i.e.,

$$\lim_{n \to \infty} \frac{r}{n} = p(\text{head}|I). \tag{4.11}$$

We can easily demonstrate this using the binomial distribution. If the probability of a head in any single toss is $p(\text{head}|I) = 0.4$, Figure 4.1 shows a plot of $p(r/n|n, I)$ versus the fraction r/n for a variety of different choices of n ranging from 20 to 1000.

Box 4.1 *Mathematica* evaluation of binomial distribution:

Needs["Statistics `DiscreteDistributions`"]

The line above loads a package containing a wide range of discrete distributions of importance to statistics, and the following line computes the probability of r heads in n trials where the probability of a head in any one trial is p.

PDF[BinomialDistribution[n, p], r]

$\rightarrow$ answer $= 0.205$ $(n = 10,\ p = 0.5,\ r = 4)$

Notice as n increases, the PDF for the frequency becomes progressively more sharply peaked, converging on a value of 0.4, the probability of a head in any single toss.

Although Bernoulli was able to derive this result, his unfulfilled quest lay in the inverse process: what could one say about the probability of obtaining a head, in a single toss, given a finite number of observed outcomes? This turns out to be a straightforward problem for Bayesian inference as we see in the next section.

4.2.2 The gambler's coin problem

Let $I \equiv$ "You have acquired a coin from a gambling table. You want to determine whether it is a biased coin from the results of tossing the coin many times. You specify the bias of the coin by a proposition H, representing the probability of a head

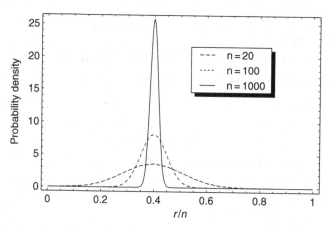

Figure 4.1 A numerical illustration of Bernoulli's law of large numbers. The PDF for the frequency of heads, r/n, in n tosses of a coin is shown for three different choices of n. As n increases, the distribution narrows about the probability of a head in any single toss $= 0.4$.

occurring in any single toss. *A priori*, you assume that H can have any value in the range $0 \rightarrow 1$ with equal probability. You want to see how $p(H|D, I)$ evolves as a function of the number of tosses."

Let $D \equiv$ "You toss the coin 50 times and record the following results: (a) 2 heads in the first 3 tosses, (b) 7 heads in the first 10 tosses, and (c) 33 heads in 50 tosses."

From the prior information, we determine that our hypothesis space H is continuous in the range $0 \rightarrow 1$. As usual, our starting point is Bayes' theorem:

$$p(H|D, I) = \frac{p(H|I)p(D|H, I)}{p(D|I)}. \tag{4.12}$$

Since we are assuming a uniform prior for $p(H|I)$, the action will all be in the likelihood term $p(D|H, I)$, which, in this case, is given by the binomial distribution:

$$p(r|n, I) = \frac{n!}{r!(n-r)!} H^r (1-H)^{n-r}. \tag{4.13}$$

Note: the symbol H is being employed in two different ways. In Equation (4.13), it is acting as an ordinary algebraic variable standing for possible numerical values in the range 0 to 1. When it appears as an argument of a probability or PDF, e.g., $p(H|D, I)$, it acts as a proposition (obeying the rules of Boolean algebra) and asserts that the true value lies in the numerical range H to $H + dH$.

Figure 4.2 shows the results from Equation (4.13) as a function of H in the range $0 \rightarrow 1$. From the figure, we can clearly see how the evolution of our state of knowledge of the coin translates into a progressively more sharply peaked posterior PDF. From this simple example, we can see how Bayes' theorem solves the inverse problem: find $p(H|D, I)$ given a finite number of observed outcomes represented by D.

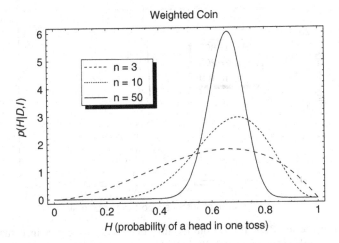

Figure 4.2 The posterior PDF for the bias of a coin determined from: (a) 3 tosses, (b) 10 tosses, and (c) 50 tosses.

4.2.3 Bayesian analysis of an opinion poll

Let $I \equiv$ "A number of political parties are seeking election in British Columbia. The questions to be addressed are: (a) what is the fraction of decided voters that support the Liberals, and (b) what is the probability that the Liberals will achieve a majority of at least 51% in the upcoming election, assuming the poll will be representative of the population at the time of the election?"

Let $D \equiv$ "In a poll of 800 decided voters, 18% supported the New Democratic Party versus 55% for the Liberals, 19% for Reform BC and 8% for other parties."

Let the proposition $H \equiv$ "The fraction of the voters that will support the Liberals is between H and $H + dH$." In this problem our hypothesis space of interest is continuous in the range 0 to 1, so $p(H|D, I)$ is a probability density function.

Based only on the prior information as stated, we adopt a flat prior $p(H|I) = 1$.

Let $r =$ the number of respondents in the poll that support the Liberals. As far as this problem is concerned, there are only two outcomes of interest; a voter either will or will not vote for the Liberals. We can therefore use the binomial distribution to evaluate the likelihood function $p(D|H, I)$. Given a particular value of H, the binomial distribution gives the probability of obtaining $D = r$ successes in n samples, where in this case, a success means support for the Liberals.

$$p(D|H, I) = \frac{n!}{r!(n-r)!} H^r (1-H)^{n-r}. \tag{4.14}$$

In this problem $n = 800$, and $r = 440$. From Bayes' theorem we can write

$$p(H|D, I) = \frac{p(H|I)p(D|H, I)}{p(D|I)} = \frac{p(D|H, I)}{p(D|I)} = \frac{p(D|H, I)}{\int_0^1 dH \, p(D|H, I)}. \tag{4.15}$$

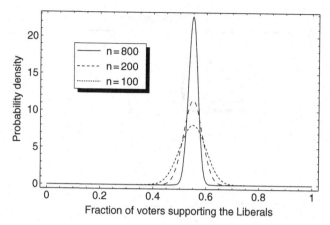

Figure 4.3 The posterior PDF for H, the fraction of voters in the province supporting the Liberals based on polls of size $n = 100, 200, 800$ decided voters.

Figure 4.3 shows a graph of the posterior probability of H for a variety of poll sizes including $n = 800$. The 95% *credible region*[2] for H is $55^{+3.4}_{-3.5}\%$. A frequentist interpretation of the same poll would express the uncertainty in the fraction of decided voters supporting the Liberals in the following way: "The poll of 800 people claims an accuracy of $\pm 3.5\%$, 19 times out of 20." We will see why when we deal with frequentist confidence intervals in Section 6.6.

The second question, concerning the probability that the Liberals will achieve a majority of at least 51% of the vote, is addressed as a model selection problem. The two models are:

1. Model $M_1 \equiv$ "the Liberals will achieve a majority." The parameter of the model is H, which is assumed to have a uniform prior in the range $0.51 \leq H \leq 1.0$.
2. Model $M_2 \equiv$ "the Liberals will not achieve a majority." The parameter of the model is H, which is assumed to have a uniform prior in the range $0 \leq H < 0.51$.

From Equation (3.14) we can write

$$\text{odds} = O_{12} = \frac{p(M_1|I)}{p(M_2|I)} B_{12}, \tag{4.16}$$

---

[2] Note: a Bayesian credible region is not the same as a frequentist confidence interval. For a uniform prior for H the 95% confidence interval has essentially the same value as the 95% credible region, but the interpretation is very different. The recipe for computing a credible region was given at the end of Section 3.3.

where

$$B_{12} = \frac{p(D|M_1, I)}{p(D|M_2, I)}$$

$$= \frac{\int_{H=0.51}^{1} dH\, p(H|M_1, I)\, p(D|M_1, H, I)}{\int_{H=0}^{0.51} dH\, p(H|M_2, I)\, p(D|M_2, H, I)} \qquad (4.17)$$

$$= \frac{\int_{H=0.51}^{1} dH(1/0.49)\, p(D|M_1, H, I)}{\int_{H=0}^{0.51} dH(1/0.51)\, p(D|M_2, H, I)}$$

$$= 87.68.$$

Based on I, we have no prior reason to prefer M_1 over M_2, so $O_{12} = B_{12}$. The probability that the Liberal party will win a majority is then given by (see Equation (3.18))

$$p(M_1|D, I) = \frac{1}{(1 + 1/O_{12})} = 0.989. \qquad (4.18)$$

Again, we emphasize that our conclusions are conditional on the assumed prior information, which includes the assumption that the poll will be representative of the population at the time of the election. Now that we have set up the equations to answer the questions posed above, it is a simple exercise to recompute the answers assuming different prior information, e.g., suppose the prior lower bound on H were 0.4 instead of 0.

4.3 Multinomial distribution

When we throw a six-sided die there are six possible outcomes. This motivates the following question: Is there a generalization of the binomial distribution for the case where we have more than two possible outcomes? Again we can use probability theory as extended logic to derive the appropriate distribution starting from a statement of our prior information.

$I \equiv$ "Proposition E represents an event that is repeated many times and has m possible outcomes represented by propositions, $O_1, O_2, \ldots, O_m$. The outcomes of individual events are logically independent, i.e., the probability of getting an outcome O_i in event j is independent of what outcome occurred in any other event." $E = O_1 + O_2 + O_3 + \cdots + O_m$, then for the event E repeated n times:

$$E^n = (O_1 + O_2 + \cdots + O_m)^n.$$

The probability of any particular E^n having

O_1	occurring	n_1	times
O_2	occurring	n_2	times
$\vdots$	$\vdots$	$\vdots$	$\vdots$
O_m	occurring	n_m	times

is $p(E^n|I) = p(O_1|I)^{n_1}\, p(O_2|I)^{n_2} \ldots p(O_m|I)^{n_m}$.

Next we need to find the number of sequences having the same number of $O_1, O_2, \ldots, O_m$ (multiplicity) independent of the order. We can readily guess at the form of multiplicity by rewriting Equation (4.7) setting the denominator $r!(n-r)! = n_1!n_2!$.

$$\text{multiplicity for the two-outcome case} = \frac{n!}{r!(n-r)!} = \frac{n!}{n_1!n_2!}, \qquad (4.19)$$

where n_1 stands for the number of A's and n_2 for the number of $\overline{A}$'s. Now in the current problem, we have m possible outcomes for each event, so,

$$\text{multiplicity for the } m\text{-outcome case} = \frac{n!}{n_1!n_2!\ldots n_m!}, \qquad (4.20)$$

where $n = \sum_{i=1}^{m} n_i$.

Therefore, the probability of seeing the outcome defined by $n_1 n_2 \ldots n_m$ where $n_i \equiv$ "Outcome O_i occurred n_i times" is

$$p(n_1, n_2, \ldots, n_m | E^n, I) = \frac{n!}{n_1!n_2!\ldots n_m!} \prod_{i=1}^{m} p(O_i|I)^{n_i}. \qquad (4.21)$$

This is called the *multinomial distribution*.

Compare this with the *multinomial expansion*:

$$(x_1 + x_2 + \cdots + x_m)^n = \sum \frac{n!}{n_1!n_2!\ldots n_m!} x_1^{n_1} x_2^{n_2} \ldots x_m^{n_m}, \qquad (4.22)$$

where the sum is taken over all possible values of n_i, subject to the constraint that $\sum_{i=1}^{m} n_i = n$.

4.4 Can you really answer that question?

Let $I \equiv$ "A tin contains N buttons, identical in all respects except that M are black and the remainder are white."

What is the probability that you will a pick a black button on the first draw assuming you are blindfolded? The answer is clearly M/N. What is the probability that you will a pick a black button on the second draw if you know that a black button was picked on the first and not put back in the tin (sampling without replacement)?
Let $B_i \equiv$ "A black button was picked on the ith draw."
Let $W_i \equiv$ "A white button was picked on the ith draw."
Then

$$p(B_2|B_1, I) = \frac{M-1}{N-1},$$

because for the second draw there is one less black button and one less button in total.

Now, what is the probability of picking a black button on the second draw $p(B_2|I)$ when we are not told what color was picked on the first draw? In this case the answer might *appear* to be indeterminate, but as we shall show, questions of this kind can be answered using probability theory as extended logic.

We know that either B_1 or W_1 is true, which can be expressed as the Boolean equation $B_1 + W_1 = 1$. Thus we can write:

$$B_2 = (B_1 + W_1), B_2 = B_1, B_2 + W_1, B_2.$$

But according to Jaynes consistency (see Section 2.5.1), equivalent states of knowledge must be represented by equivalent plausibility assignments. Therefore

$$p(B_2|I) = p(B_1, B_2|I) + p(W_1, B_2|I)$$
$$= p(B_1|I)p(B_2|B_1, I) + p(W_1|I)p(B_2|W_1, I)$$
$$= \left(\frac{M}{N}\right)\left(\frac{M-1}{N-1}\right) + \left(\frac{N-M}{N}\right)\left(\frac{M}{N-1}\right) \tag{4.23}$$
$$= \frac{M}{N}.$$

In like fashion, we can show

$$p(B_3|I) = \frac{M}{N}.$$

The probability of black at any draw, if we do not know the result of any other draw, is always the same.

The method used to obtain this result is very useful.

1. Resolve the quantity whose probability is wanted into mutually exclusive sub-propositions:[3]

$$B_3 = (B_1 + W_1), (B_2 + W_2), B_3$$
$$= B_1, B_2, B_3 + B_1, W_2, B_3 + W_1, B_2, B_3 + W_1, W_2, B_3.$$

2. Apply the sum rule.
3. Apply the product rule.

If the sub-propositions are well chosen (i.e., they have a simple meaning in the context of the problem), their probabilities are often calculable.

While we are on the topic of sampling without replacement, let's introduce the hypergeometric distribution (see Jaynes, 2003). This gives the probability of drawing r

---

[3] In his book, *Rational Descriptions, Decisions and Designs*, M. Tribus refers to this technique as *extending the conversation*. In many problems, there are many pieces of information which do not seem to fit together in any simple mathematical formulation. The technique of extending the conversation provides a formal method for introducing this information into the calculation of the desired probability.

black buttons (blindfolded) in n tries from a tin containing N buttons, identical in all respects except that M are black and the remainder are white.

$$p(r|N,M,n) = \frac{\begin{pmatrix} M \\ r \end{pmatrix} \begin{pmatrix} N-M \\ n-r \end{pmatrix}}{\begin{pmatrix} N \\ n \end{pmatrix}}, \tag{4.24}$$

where

$$\begin{pmatrix} M \\ r \end{pmatrix} = \frac{M!}{r!(M-r)!} \text{ etc.} \tag{4.25}$$

Box 4.2 *Mathematica* **evaluation of hypergeometric distribution:**

Needs["Statistics 'DiscreteDistributions' "]

PDF[HypergeometricDistribution [n, n_{succ}, n_{tot}], r]

gives the probability of r successes in n trials corresponding to sampling without replacement from a population of size n_{tot} with n_{succ} potential successes.

4.5 Logical versus causal connections

We now need to clear up an important distinction between a logical connection between two propositions and a causal connection. In the previous problem with M black buttons and $N-M$ white buttons, it is clear that $p(B_j|B_{j-1},I) < p(B_j|I)$ since we know there is one less black button in the tin when we take our next pick. Clearly, what was drawn on earlier draws can affect what will happen in later draws. We can say there is some kind of partial causal influence of B_{j-1} on B_j.

Now suppose we ask the question what is the probability $p(B_{j-1}|B_j,I)$? Clearly in this case what we get on a later draw can have no effect on what occurs on an earlier draw, so it may be surprising to learn that $p(B_{j-1}|B_j,I) = p(B_j|B_{j-1},I)$. Consider the following simple proof (Jaynes, 2003). From the product rule we write

$$p(B_{j-1},B_j|I) = p(B_{j-1}|B_j,I)p(B_j|I) = p(B_j|B_{j-1},I)p(B_{j-1}|I).$$

But we have just seen that $p(B_j|I) = p(B_{j-1}|I) = M/N$ for all j, so

$$p(B_{j-1}|B_j,I) = p(B_j|B_{j-1},I), \tag{4.26}$$

or more generally,

$$p(B_k|B_j,I) = p(B_j|B_k,I), \quad \text{for all } j,k. \tag{4.27}$$

How can information about a later draw affect the probability of an earlier draw? Recall that in Bayesian analysis, probabilities are an encoding of our state of knowledge about some question. Performing the later draw does not physically affect the number M_j of black buttons in the tin at the jth draw. However, information about the result of a later draw has the same effect on our state of knowledge about what could have been taken on the jth draw, as does information about an earlier draw. Bayesian probability theory is concerned with all *logical* connections between propositions independent of whether there are causal connections.

Example 1:

$I \equiv$ "A shooting has occurred and the police arrest a suspect on the same day."
$A \equiv$ "Suspect is guilty of shooting."
$B \equiv$ "A gun is found seven days after the shooting with suspect's fingerprints on it."
Clearly, B is not a partial cause of A but still we conclude that

$$p(A|B, I) > p(A|I).$$

Example 2:

$I \equiv$ "A virulent virus invades Montreal. Anyone infected loses their hair a month before dying."
$A \equiv$ "The mayor of Montreal lost his hair in September."
$B \equiv$ "The mayor of Montreal died in October."
Again, in this case, $p(A|B, I) > p(A|I)$.

Although a logical connection does not imply a causal connection, a causal connection does imply a logical connection, so we can certainly use probability theory to address possible causal connections.

4.6 Exchangeable distributions

In the previous section, we learned that information about the result of a later draw has the same effect on our state of knowledge about what could have been taken on the jth draw, as does information about an earlier one. Every draw has the same relevance to every other draw regardless of their time order. For example, $p(B_j|B_{j-1}, B_{j-2}, I) = p(B_j|B_{j+1}, B_{j+2}, I)$, where again B_j is the proposition asserting a black button on the jth draw. The only thing that is significant about the knowledge of outcomes of other draws is the number of black or white buttons in these draws, not their time order. Probability distributions of this kind are called *exchangeable distributions*. It is clear that the hypergeometric distribution is exchangeable since for $p(r|N, M, n)$ we are not required to specify the exact sequence of the r black button outcomes. The hypergeometric distribution takes into account the changing contents of the tin. The result of any draw changes the probability of a black on any other draw. If the number, N, of buttons in the tin is much larger than the number of draws n, then this probability changes very little. In the limit as $N \to \infty$, the hypergeometric distribution simplifies to the binomial distribution, another exchangeable distribution.

The multinomial distribution, discussed in Section 4.3, can be viewed as a generalization of the binomial distribution to the case where we have m possible outcomes, not just two. From its form given in Equation (4.21), which we repeat here,

$$p(n_1, n_2, \ldots, n_m | E^n, I) = \frac{n!}{n_1! n_2! \ldots n_m!} \prod_{i=1}^{m} p(O_i | I)^{n_i},$$

we can see that this is another exchangeable distribution because the probability depends only on the numbers of different outcomes $(n_1, n_2, \ldots, n_m)$ observed and not on their order.

Worked example:

A spacecraft carrying two female and three male astronauts makes a trip to Mars. The plan calls for three of the astronauts to board a detachable capsule to land on the planet, while the other two remain behind in orbit. Which three will board the capsule is decided by a lottery, consisting of picking names from a box. The first person selected is to be the captain of the capsule. The second and third names selected become capsule support crew. What is the probability that the captain is female if we know that at least one of the support crew members is female? Let F_i stand for the proposition that the ith name selected is female, and M_i if the person is male.

Let $F_{\text{later}} \equiv$ "We learn that at least one of the crew members is female."

$$F_{\text{later}} = F_2 + F_3.$$

This information reduces the number of females available for the first draw by at least one. To solve the problem we will make use of Bayes' theorem and abbreviate F_{later} by F_{L}.

$$p(F_1 | F_{\text{L}}, I) = \frac{p(F_1 | I) p(F_{\text{L}} | F_1, I)}{p(F_{\text{L}} | I)}. \tag{4.28}$$

To evaluate two of the terms on the right, it will be convenient to work with denials of F_{L}. From the sum rule, $p(F_{\text{L}} | F_1, I) = 1 - p(\overline{F_{\text{L}}} | F_1, I)$. Since $F_{\text{L}} = F_2 + F_3$, we have that $\overline{F_{\text{L}}} = \overline{F_2}, \overline{F_3} = M_2, M_3$, according to the duality identity of Boolean algebra (Section

2.2.3). In words, the denial of at least one female in draws 2 and 3 is a male on both draws. Therefore,

$$p(F_L|F_1,I) = 1 - p(M_2,M_3|F_1,I) = 1 - p(M_2|F_1,I)p(M_3|M_2,F_1,I)$$

$$= 1 - \left(\frac{3}{4}\right)\left(\frac{2}{3}\right) = \frac{1}{2}. \tag{4.29}$$

Similarly, we can write $p(F_L|I) = 1 - p(M_2,M_3|I)$. By exchangeability, $p(M_2,M_3|I)$ is the same as the probability of a male on the first two draws given only the conditional information I, i.e., not F_1, I. Therefore,

$$p(F_L|I) = 1 - p(M_2,M_3|I) = 1 - p(M_1,M_2|I) = 1 - p(M_1|I)p(M_2|M_1,I)$$

$$= 1 - \left(\frac{3}{5}\right)\left(\frac{2}{4}\right) = \frac{7}{10}. \tag{4.30}$$

Substituting Equations (4.29) and (4.30) into Equation (4.28), we obtain

$$p(F_1|F_L,I) = \frac{\left(\frac{2}{5}\right)\left(\frac{1}{2}\right)}{\left(\frac{7}{10}\right)} = \frac{2}{7}. \tag{4.31}$$

The property of exchangeability has allowed us to evaluate the desired probability in a circumstance where we were given less precise information, namely, a female will be picked at least once on the second and third draws. Note: the result for $p(F_1|F_L,I)$ is different from these two cases:

$$p(F_1|I) = \frac{2}{5}$$

and

$$p(F_1|F_2,I) = \left(\frac{2-1}{5-1}\right) = \frac{1}{4}.$$

4.7 Poisson distribution

In this section[4] we will see how a particular state of prior information, I, leads us to choose the well-known *Poisson distribution* for the likelihood. Later, in Section 5.7.2, we will derive the Poisson distribution as a limiting "low count rate" approximation to the binomial distribution.

Prior information: $I \equiv$ "There is a positive real number r such that, given r, the probability that an event, or count, will occur in the time interval $(t, t+dt)$ is $= r\,dt$. Furthermore, knowledge of r makes any information about the occurrence or

---

[4] Section 4.7 is based on a paper by E. T. Jaynes (1990).

non-occurrence of the event in any other time interval (that does not include $(t, t + dt)$) irrelevant to this probability."

Let $q(t) =$ probability of no count in time interval $(0, t)$.

Let $E \equiv$ "no count in $(0, t + dt)$".

E is the conjunction of two propositions A and B given by

$$E = [\text{"no count in } (0, t)\text{"}], [\text{"no count in } (t, t + dt)\text{"}] = A, B.$$

From the product rule, $p(E|I) = p(A, B|I) = p(A|I)p(B|A, I)$.

It follows that

$$p(E|I) = q(t + dt) = q(t)(1 - r\,dt) \quad \text{or} \quad \frac{dq}{dt} = -r\,q(t).$$

The solution for the evident initial condition $q(0) = 1$ is $q(t) = \exp(-r\,t)$.

Now consider the probability of the proposition:

$C \equiv$ "In the interval $(0, t)$, there are exactly n counts which happen at times $(t_1, t_2, \ldots, t_n)$ with infinitesimal tolerances $(dt_1, \ldots, dt_n)$, where $(0 < t_1 < t_2 \ldots < t_n < t)$"

This is the conjunction of $2n + 1$ propositions

$$C = [\text{"no count in } (0, t_1)\text{"}], (\text{"count in } dt_1\text{"}),$$

$$[\text{"no count in } (t_1, t_2)\text{"}], (\text{"count in } dt_2\text{"}), \ldots,$$

$$[\text{"no count in } (t_{n-1}, t_n)\text{"}], (\text{"count in } dt_n\text{"}), [\text{"no count in } (t_n, t)\text{"}].$$

By the product rule and the independence of different time intervals,

$$\begin{aligned}
p(C|r, I) &= \exp(-r\,t_1) \cdot (r\,dt_1) \cdot \exp(-r(t_2 - t_1)) \cdot (r\,dt_2) \ldots \\
&\quad \times \exp(-r(t_n - t_{n-1})) \cdot (r\,dt_n) \cdot (\exp - r(t - t_n)) \qquad (4.32) \\
&= \exp(-r\,t)r^n\,dt_1 \ldots dt_n.
\end{aligned}$$

The probability (given r) that in the interval $(0, t)$ there are exactly n counts, whatever the times, is given by

$$\begin{aligned}
p(n|r, t, I) &= \exp(-rt)r^n \int_0^t dt_n \ldots \int_0^{t_4} dt_3 \int_0^{t_3} dt_2 \int_0^{t_2} dt_1 \\
&= \exp(-rt)r^n \int_0^t dt_n \ldots \int_0^{t_4} dt_3 \int_0^{t_3} \frac{t_2}{1!} dt_2 \\
&= \exp(-rt)r^n \int_0^t dt_n \ldots \int_0^{t_4} \frac{t_3^2}{2!} dt_3 \\
&= \exp(-rt)\frac{(rt)^n}{n!} \qquad \text{Poisson distribution.}
\end{aligned} \qquad (4.33)$$

We will return to the Poisson distribution again in Chapter 5. Some sample Poisson distributions are shown in Figure 5.6, and its relationship to the binomial and Gaussian distributions is discussed in Section 5.7.2, together with some typical examples. Chapter 14 is devoted to Bayesian inference with Poisson sampling.

4.7.1 Bayesian and frequentist comparison

Let's use the Poisson distribution to clarify a fundamental difference between the Bayesian and frequentist approaches to inference. Consider how the probability of n_1 counts in a time interval $(0, t_1)$ changes if we learn that n_2 counts occurred in the interval $(0, t_2)$ where $t_2 > t_1$. According to I, the occurrence or non-occurrence of counts in any other time intervals that do not include the interval $(0, t_1)$ is irrelevant to the probability of interest. Since $(0, t_2)$ contains $(0, t_1)$ it is contributing information which we can incorporate through Bayes' theorem which we write now.

$$
\begin{aligned}
p(n_1 | n_2, r, t_1, t_2, I) &= \frac{p(n_1 | r, t_1, t_2, I) \; p(n_2 | n_1, r, t_1, t_2, I)}{p(n_2 | r, t_1, t_2, I)} \\
&= \frac{p(n_1 | r, t_1, I) \; p(n_1, (n_2 - n_1) | n_1, r, t_1, t_2, I)}{p(n_2 | r, t_2, I)}.
\end{aligned}
\tag{4.34}
$$

Using the product rule, we can expand the second term in the numerator of Equation (4.34):

$$
\begin{aligned}
p(n_1, (n_2 - n_1) | n_1, r, t_1, t_2, I) &= p(n_1 | n_1, r, t_1, t_2, I) \times p(n_2 - n_1 | n_1, r, t_1, t_2, I) \\
&= 1 \times p(n_2 - n_1 | n_1, r, t_1, t_2, I) \\
&= \exp[-r(t_2 - t_1)] \frac{[r(t_2 - t_1)]^{n_2 - n_1}}{(n_2 - n_1)!}.
\end{aligned}
\tag{4.35}
$$

The other terms in Equation (4.34) can readily be evaluated by reference to Equation (4.33). Substituting into Equation (4.34) and simplifying, we obtain

$$
\begin{aligned}
p(n_1 | n_2, r, t_1, t_2, I) &= \frac{n_2!}{n_1! \, (n_2 - n_1)!} \frac{\exp[-rt_1] \; \exp[-r(t_2 - t_1)]}{\exp[-rt_2]} \\
&\quad \times \frac{[rt_1]^{n_1} \; [r(t_2 - t_1)]^{n_2 - n_1}}{[rt_2]^{n_2}} \\
&= \binom{n_2}{n_1} \left(\frac{t_1}{t_2}\right)^{n_1} \left(1 - \frac{t_1}{t_2}\right)^{n_2 - n_1} \left\{ \begin{matrix} t_1 < t_2 \\ n_1 < n_2 \end{matrix} \right\}.
\end{aligned}
\tag{4.36}
$$

The result is rather surprising because the new posterior does not even depend on r. The point is that r does not determine n_1; it only gives probabilities for different values of n_1. If we know the actual value over the interval that includes $(0, t_1)$, then this takes precedence over anything we could infer from r.

In frequentist random variable probability theory, one might think that r is the sole relevant quantity, and thus arrive at a different conclusion, namely,

$$p(n_1|n_2, r, t_1, t_2, I) = p(n_1|r, t_1, I) = \left[\exp(-rt_1)\frac{(rt_1)^{n_1}}{n_1!}\right]. \qquad (4.37)$$

What if we used the measured n_2 counts in the time interval t_2 to compute a new estimate of $r' = n_2/t_2$ and then used Equation (4.37) to compute $p(n_1|r', t_1, I)$. Would we get the same result as predicted by Equation (4.36)? The two distributions are compared in Figure 4.4 for $n_2 = 10$ counts, $t_2 = 10$ s and $t_1 = 8$ s. The two curves are clearly very different. In addition, the probability distribution given by Equation (4.37) predicts a tail extending well beyond 10 counts which makes no physical sense given that we know only 10 counts will occur in the longer interval t_2 which contains t_1. From the frequentist point of view, replacing r by r' would make little sense regarding long-run performance if the original r were estimated on the basis of the counts in a much longer time span than t_2. However, for any non-zero value of r, Equation (4.37) predicts there is a finite probability that n_1 can exceed the actual measured value n_2 in the larger interval, which is clearly impossible.

In frequentist theory, a probability represents the percentage of time that something will happen in a very large number of identical repeats of an experiment, i.e., the long-run relative frequency. As we will learn in Section 6.6 and Chapter 7, frequentist theory says nothing directly about the probability of any estimate derived from a single data set. The significance of any frequentist result can only be interpreted with reference to a population of hypothetical data sets. From this point of view, the

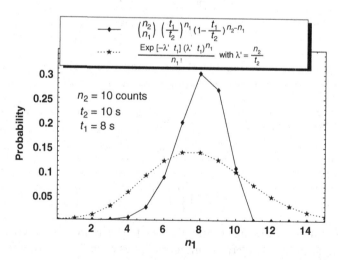

Figure 4.4 A comparison of the predictions for $p(n_1|n_2, r, t_1, t_2, I)$ based on Equations (4.36) and (4.37) where we set $r' = n_2/t_2$. The assumed values are $t_1 = 8$ s, $t_2 = 10$ s and $n_2 = 10$ counts.

frequentist procedure represented by Equation (4.37) is not intended to be optimum in the individual case that we are considering here. In contrast, Bayesian probability theory does apply to the individual case, where the goal is to reason as best we can on the basis of our current state of information. In a Bayesian analysis, only the data that were actually measured, combined with relevant prior information, are considered, hypothetical data sets play no role.

4.8 Constructing likelihood functions

In this section, we amplify on the process of arriving at the likelihood function, $p(D|M, \theta, I)$, for use in a Bayesian analysis, where

$$p(D|M, \theta, I) = \text{probability of obtaining data } D, \text{if model } M$$
$$\text{and background (prior) information } I \text{ are true}$$
$$\text{(also called the likelihood function } \mathcal{L}(M)).$$

The parameters of model M are collectively designated by the symbol θ. We can write $D = Y_1, Y_2, \ldots, Y_N = \{Y_i\}$, where

- $Y_i \equiv$ "A proposition asserting that the ith data value is in the infinitesimal range y_i to $y_i + dy_i$."
- $Z_i \equiv$ "A proposition asserting that the M model prediction for the ith data value is in the range z_i to $z_i + dz_i$."
- $E_i \equiv$ "A proposition asserting that the ith error value is in the range e_i to $e_i + de_i$."

As usual, we can write

$$y_i = z_i + e_i. \tag{4.38}$$

In the simplest case (see Section 4.8.1) the predicted value, z_i, is given by a deterministic model, $m(x_i|\theta)$, which is a function of some independent variable(s) x_i, like position or time. More generally, the value of z_i itself may be uncertain because of statistical uncertainties in $m(x_i|\theta)$, and/or uncertainties in the value of the independent variable(s) x_i. We will represent the probability distribution for proposition Z_i by the function

$$p(Z_i|M, \theta, I) = f_Z(z_i). \tag{4.39}$$

We can also represent the probability distribution for proposition E_i by another function given by

$$p(E_i|M, \theta, I) = f_E(e_i). \tag{4.40}$$

Our next step is to compute $p(Y_i|M, \theta, I)$. Now Y_i depends on propositions Z_i and E_i. To evaluate $p(Y_i|M, \theta, I)$, we first extend the conversation (Tribus, 1969) to include these propositions by writing down the joint probability distribution

$p(Y_i, Z_i, E_i|M, \theta, I)$. We can then solve for $p(Y_i|M, \theta, I)$ by using the marginalizing operation as follows:

$$p(Y_i|M, \theta, I) = \iint dZ_i \, dE_i \, p(Y_i, Z_i, E_i|M, \theta, I)$$

$$= \iint dZ_i \, dE_i \, p(Z_i|M, \theta, I) \, p(E_i|M, \theta, I) \, p(Y_i|Z_i, E_i, M, \theta, I),$$

$$\text{(4.41)}$$

where we assume Z_i and E_i are independent.

Since $y_i = z_i + e_i$,

$$p(Y_i|Z_i, E_i, M, \theta, I) = \delta(y_i - z_i - e_i) \tag{4.42}$$

$$\rightarrow p(Y_i|M, \theta, I) = \int dz_i \, f_Z(z_i) \int de_i \, f_E(e_i) \delta(y_i - z_i - e_i). \tag{4.43}$$

The presence of the delta function in the second integral serves to pick out the value of the integrand at $e_i = y_i - z_i$, so we have:

$$p(Y_i|M, \theta, I) = \int dz_i \, f_Z(z_i) \, f_E(y_i - z_i). \tag{4.44}$$

The right hand side of the equation is the convolution integral.[5] We now evaluate our equation for $p(Y_i|M, \theta, I)$ for two useful general cases.

4.8.1 Deterministic model

In this case, we assume that for any specific choice of the model parameters there is no uncertainty in the predicted value, z_i. We will refer to models of this kind as deterministic models. Given the model and the values of any of its parameters, then $f_Z(z_i) = \delta(z_i - m(x_i|\theta))$. In this case, Equation (4.44) becomes

$$p(Y_i|M, \theta, I) = f_E(y_i - m(x_i|\theta)) = p(E_i|M, \theta, I). \tag{4.45}$$

Thus, the probability of the ith data value is simply equal to the probability of the ith error term. If the errors are all independent,[6] then

$$p(D|M, \theta, I) = p(Y_1, Y_2, \ldots, Y_N|M, \theta, I) = p(E_1, E_2, \ldots, E_N|M, \theta, I)$$

$$= \prod_{i=1}^{N} p(E_i|M, \theta, I), \tag{4.46}$$

where $\prod_{i=1}^{N}$ stands for the product of N of these terms. We have already encountered Equation (4.46) in the simple spectral line problem of Section 3.6 (see Equation (4.42)).

---

[5] For more details on the convolution integral and how to evaluate it using the Fast Fourier Transform, see Sections B.4 and B.10.

[6] We deal with the effect of correlated errors in Section 10.2.2.

4.8.2 Probabilistic model

In the second case, our information about the model is uncertain. Here, we will distinguish between three different situations.

1. The model prediction, z_i, includes a statistical noise component ϵ_i.

$$z_i = m(x_i|\theta) + \epsilon_i. \tag{4.47}$$

Equation (4.38) can be rewritten as

$$y_i = z_i + e_i = m(x_i|\theta) + \epsilon_i + e_i. \tag{4.48}$$

In this case, the data, y_i, can differ from the model, $m(x_i|\theta)$, because of a component e_i due to measurement errors, and a component ϵ_i due to a statistical uncertainty in our model. The two error terms are assumed to be uncorrelated. For example, suppose our data consist of a radar return signal from an unidentified aircraft. We could compare the signal to samples of measured radar return signals from a set of known aircraft for different orientations to arrive at the most probable identification. In this case, these sample measurements of known aircraft, which include a noise component, constitute our model, $m(x_i|\theta)$.

Suppose that the probability distribution of ϵ_i is described by a Gaussian with standard deviation σ_{mi}. Then

$$p(Z_i|M,\theta,I) = \frac{1}{\sqrt{2\pi}\sigma_{mi}} \exp\left\{ \frac{-(z_i - m(x_i|\theta))^2}{2\sigma_{mi}^2} \right\}$$

$$= \frac{1}{\sqrt{2\pi}\sigma_{mi}} \exp\left\{ \frac{-\epsilon_i^2}{2\sigma_{mi}^2} \right\} = f_Z(z_i). \tag{4.49}$$

Suppose also that the error term, e_i, in Equation (4.38), has a Gaussian probability distribution with a standard deviation, σ_i, of the form

$$p(E_i|M,\theta,I) = \frac{1}{\sqrt{2\pi}\sigma_i} \exp\left\{ \frac{-e_i^2}{2\sigma_i^2} \right\} = f_E(y_i - z_i). \tag{4.50}$$

Then according to Equation (4.44), $p(Y_i|M,\theta,I)$ is the convolution of the two Gaussian probability distributions. It is easy to show[7] that the result is another Gaussian given by

$$p(Y_i|M,\theta,I) = \frac{1}{\sqrt{2\pi}\sqrt{\sigma_i^2 + \sigma_{mi}^2}} \exp\left\{ \frac{-(y_i - m(x_i|\theta))^2}{2(\sigma_i^2 + \sigma_{mi}^2)} \right\}. \tag{4.51}$$

---

[7] Simply evaluate Equation (4.44) using *Mathematica* with limits on the integral of $\pm\infty$ after substituting for $f_Z(z_i)$ and $f_E(y_i - z_i)$ using Equations (4.49) and (4.50), respectively.

If the Y_i terms are all independent, then

$$p(D|M,\theta,I) = p(Y_1, Y_2, \ldots, Y_N|M,\theta,I)$$

$$= (2\pi)^{-N/2} \left\{ \prod_{i=1}^{N} (\sigma_i^2 + \sigma_{mi}^2)^{-1/2} \right\} \exp\left\{ \sum_{i=1}^{N} \left(\frac{-(y_i - m(x_i|\theta))^2}{2(\sigma_i^2 + \sigma_{mi}^2)} \right) \right\}.$$

$$(4.52)$$

2. In the second situation, our information about the model prediction, z_i, is only uncertain because of uncertainty in the value of the independent variable x_i. For example, we might be interested in fitting a straight line to some data with errors in both coordinates. Let x_{i0} be the nominal value of the independent variable and x_i the true value. Then $\delta x_i = x_i - x_{i0}$, is the uncertainty in x_i. Now suppose the probability distribution of x_i is a Gaussian given by

$$p(X_i|I) = \frac{1}{\sqrt{2\pi}\sigma_{xi}} \exp\left\{ \frac{-(x_i - x_{i0})^2}{2\sigma_{xi}^2} \right\} = f_X(x_i),$$

$$(4.53)$$

where the scale of δx_i is set by σ_{xi}.

Our goal here is to compute an expression for $p(Z_i|M,\theta,I) = f_Z(z_i)$, for use in Equation (4.44). In Section 5.12, we will show how to compute the probability distribution of a function of x_i if we know the probability distribution of x_i. In our case, this function is $z_i = m(x_i|\theta)$. The function $m(x_i|\theta)$ must be a monotonic and differentiable function over the range of x_i of interest. Then there exists an inverse function $x_i = m^{-1}(z_i|\theta)$ which is monotonic and differentiable. Thus, for every interval dx_i there is a corresponding interval dz_i. The result is

$$f_Z(z_i) = f_X(x_i) \left| \frac{dx_i}{dz_i} \right| = f_X(m^{-1}(z_i|\theta)) \left| \frac{dx_i}{dz_i} \right|,$$

$$(4.54)$$

which is valid provided the derivative does not change significantly over a scale of order $2\sigma_{xi}$.

Let's evaluate Equation (4.54) for the straight-line model, $z_i = m(x_i|A, B) = A + Bx_i$. In that case,

$$x_i = m^{-1}(z_i|A, B) = \frac{1}{B}z_i - \frac{A}{B},$$

$$(4.55)$$

so

$$x_i - x_{i0} = \frac{1}{B}(z_i - z_{i0}).$$

$$(4.56)$$

Also, it is apparent that

$$\left|\frac{dx_i}{dz_i}\right| = \frac{1}{|B|}.$$ (4.57)

Combining Equations (4.53), (4.54), (4.56) and (4.57), we obtain

$$f_Z(z_i) = \frac{1}{\sqrt{2\pi}|B|\sigma_{xi}} \exp\left\{\frac{-(z_i - z_{i0})^2}{2B^2\sigma_{xi}^2}\right\}.$$ (4.58)

We now have everything we need to evaluate Equation (4.44). Again, $p(Y_i|M,\theta,I)$ is the convolution of the two Gaussian probability distributions. The result is

$$p(Y_i|M,A,B,I) = \frac{1}{\sqrt{2\pi}\sqrt{\sigma_i^2 + B^2\sigma_{xi}^2}} \exp\left\{\frac{-(y_i - m(x_{i0}|A,B))^2}{2(\sigma_i^2 + B^2\sigma_{xi}^2)}\right\}.$$ (4.59)

If the Y_i terms are all independent, then,

$$p(D|M,A,B,I) = (2\pi)^{-N/2} \left(\prod_{i=1}^{N}(\sigma_i^2 + B^2\sigma_{xi}^2)^{-1/2}\right)$$

$$\times \exp\left\{\sum_{i=1}^{N} \frac{-(y_i - m(x_{i0}|A,B))^2}{2(\sigma_i^2 + B^2\sigma_{xi}^2)}\right\}.$$ (4.60)

The reader is directed to Section 11.7 for a worked problem of this kind.

3. The model prediction, z_i, is uncertain because of statistical uncertainties in both the model and the value of the independent variable(s), x_i. In this case, if we again assume Gaussian distributions for the uncertain quantities, Equation (4.60) becomes

$$p(D|M,A,B,I) = (2\pi)^{-N/2} \left(\prod_{i=1}^{N}(\sigma_i^2 + \sigma_{mi}^2 + B^2\sigma_{xi}^2)^{-1/2}\right)$$

$$\times \exp\left\{\sum_{i=1}^{N} \frac{-(y_i - m(x_{i0}|A,B))^2}{2(\sigma_i^2 + \sigma_{mi}^2 + B^2\sigma_{xi}^2)}\right\}.$$ (4.61)

4.9 Summary

In any Bayesian analysis, the prior information defines the hypothesis space of interest, prior probability distributions and the means for computing $p(D|H_i,I)$, the likelihood function. In this chapter, we have given examples of how to encode prior information into a probability distribution (commonly referred to as the sampling distribution) for use in computing the likelihood term. We saw how the well-known binomial, multinomial, hypergeometric and Poisson distributions correspond to different prior information. In the process, we learned that Bayesian inference is concerned with logical connections between propositions which may or may not correspond to causal physical influences. We introduced the notion of exchangeable distributions and learned how to compute probabilities for situations where the prior information, at first sight, appears very imprecise. In Section 4.7.1, we gained important insight into the fundamental difference between Bayesian and frequentist approaches to inference. Finally, in Section 4.8, we learned how to construct likelihood functions for both deterministic and probabilistic models.

4.10 Problems

1. A bottle contains 50 black balls and 30 red balls. The bottle is first shaken to mix up the balls. What is the probability that blindfolded, you will pick two red balls in three tries?

2. Let $I \equiv$ "A tin is purchased from a company that makes an equal number of two types. Both contain 90 buttons which are identical except that 2/3 of the buttons in one tin are black (the rest are white) and 2/3 of the buttons in the other tin are white (the rest are black). You can't distinguish the tins from their outside."

   Let $D \equiv$ "In a sample of ten buttons drawn from the tin, seven are black."
   Let $B \equiv$ "We are drawing from the black tin."
   Let $W \equiv$ "We are drawing from the white tin."
   Compute the odds $= \dfrac{p(B|D, I)}{p(W|D, I)}$, assuming $p(B|I) = p(W|I)$.

3. A tin contains 17 black buttons and 6 white buttons. The tin is first shaken to mix up the buttons. What is the probability that blindfolded, you will pick a white button on the third pick if you don't know what was picked on the first two picks?

4. A bottle contains three green balls and three red balls. The bottle is first shaken to mix up the balls. What is the probability that blindfolded, you will pick a red ball on the third pick, if you learn that at least one red ball was picked on the first two picks?

5. A spacecraft carrying two female and three male astronauts makes a trip to Mars. The plan calls for a two-person detachable capsule to land at site A on the planet and a second one-person capsule to land at site B. The other two astronauts remain in orbit. Which three will board the two capsules is decided by a lottery, consisting of picking names from a box. What is the probability that a female occupies the one-person capsule if we know that at least one member of the other capsule is female, but we are not told the order in which the astronauts were picked?

6. In a particular water sample, ten bacteria are found, of which three are of type A. Let $Q \equiv$ "the probability that any particular bacterium is of type A is between q and $q + dq$." Plot the posterior $p(Q|D, I)$. What prior probability distribution did you assume and why?

7. In a particular water sample, ten bacteria are found, of which three are of type A. What is the probability of obtaining six type A bacteria, in a second independent water sample containing 12 bacteria in total?

8. In a radio astronomy survey, 41 quasars were detected in a total sample of 90 sources. Let $F \equiv$ "the probability that any particular source is a quasar is between f and $f + df$." Plot the posterior $p(F|D, I)$ assuming a uniform prior for F.

9. In problem 7, what is the probability of obtaining at least three type A bacteria?

10. A certain solution contains three types of bacteria: A, B, and C. Given $p(A|I) = 0.2$, $p(B|I) = 0.3$, and $p(C|I) = 0.5$, what is the probability of obtaining a sample of ten bacteria with three type A, three type B and four type C?

11. A total of five γ-ray photons were detected from a particular star in one hour. What is the probability that three photons will be detected in the next hour of observing?

12. On average, five γ-ray photons are detected from a particular star each hour. What is the probability that three photons were detected in the first hour of a two-hour observation that recorded eight photons in total?

13. In the opinion poll problem of Section 4.2.3, re-plot Figure 4.3 for $n = 55$.

14. In the opinion poll problem of Section 4.2.3, compute the probability that the Liberals will achieve a majority of at least 51%, for $n = 55$ and everything else the same.

15. We want to fit a straight line model of the form $y_i = a + bx_i$ to the list of x, y pairs given below. The data have Gaussian distributed errors in both the x and y coordinates with $\sigma_x = 1$ and $\sigma_y = 2$. Assume uniform priors for a and b, with boundaries that enclose the range of parameter space where there is a significant contribution from the likelihood function. This means that we can treat the prior as a constant, and write $p(a, b|D, M, I) \propto p(D|M, a, b, I)$.
$\{\{-5, -1.22\}, \{-4, -3.28\}, \{-3, -2.52\}, \{-2, 3.74\}, \{-1, 3.01\}, \{0, -1.80\}, \{1, 2.49\}, \{2, 5.48\}, \{3, 0.42\}, \{4, 4.80\}, \{5, 4.22\}\}$

(a) Plot the data with error bars in both coordinates.

(b) Show a contour plot of the joint posterior PDF, $p(a, b|D, I)$.

(c) For what choice of a, b is $p(a, b|D, I)$ a maximum? You can use the *Mathematica* command
FindMaximum[$p(a, b|D, I), \{a, 0.0\}, \{b, 0.5\}$].

(d) Show the best fit line and data with error bars on the same plot.

(e) Compute and plot the marginal distributions $p(a|D, I)$ and $p(b|D, I)$. One way to do this is to compute a table of the joint posterior values for a grid of a, b values and approximate the integrals required for marginalization by a summation over the rows or columns. Make sure to normalize your marginal distributions so $\int p(a|D, I) da = 1 \approx \sum_i p(a_i|D, I) \Delta a$.

5

Frequentist statistical inference

5.1 Overview

We now begin three chapters which are primarily aimed at a discussion of the main concepts of frequentist statistical inference. This is currently the prevailing approach to much of scientific inference, so a student should understand the main ideas to appreciate current literature and understand the strengths and limitations of this approach.

In this chapter, we introduce the concept of a random variable and discuss some general properties of probability distributions before focusing on a selection of important sampling distributions and their relationships. We also introduce the very important Central Limit Theorem in Section 5.9 and examine this from a Bayesian viewpoint in Section 5.10. The chapter concludes with the topic of how to generate pseudo-random numbers of any desired distribution, which plays an important role in Monte Carlo simulations.

In Chapter 6, we address the question of what is a statistic and give some common important examples. We also consider the meaning of a frequentist confidence interval for expressing the uncertainty in parameter values. The reader should be aware that study of different statistics is a very big field which we only touch on in this book. Some other topics normally covered in a statistics course like the fitting of models to data are treated from a Bayesian viewpoint in later chapters.

Finally, Chapter 7 concludes our brief summary of frequentist statistical inference with the important topic of frequentist hypothesis testing and discusses an important limitation known as the *optional stopping problem*.

5.2 The concept of a random variable

Recall from Section 1.1 that conventional "frequentist" statistical inference and Bayesian inference employ fundamentally different definitions of probability. In frequentist statistics, when we write the probability $p(A)$, the argument of the probability is called a *random variable*. It is a quantity that can be considered to take on various values throughout an ensemble or a series of repeated experiments. For example:

1. A measured quantity which contains random errors.
2. Time intervals between successive radioactive decays.

Before proceeding, we need an operational definition of a random variable. From this, we discover that the random variable is not the particular number recorded in one measurement, but rather, it is an abstraction of the measurement operation or observation that gives rise to that number.

Definition: A random variable, X, transforms the possible outcomes of an experiment (measurement operation) to real numbers.

Example: Suppose we are interested in measuring a pollutant's concentration level for each of n time intervals. The observations (procedure for producing a real number) $X_1, X_2, \ldots, X_n$ form a sample of the pollutant's concentration. Before the instrument actually records the concentration level during the ith trial, the observation, X_i, is a random variable. The recorded value, x_i, is not a random variable, but the actual measured value of the observation, X_i.

Question: Why do we need to have n random variables X_i? Why not one random variable X for which $x_1, x_2, \ldots, x_n$ are the realizations of the random variable during the n observations?

Answer: Because we often want to determine the joint probability of getting x_1 on trial 1, x_2 on trial 2, etc. If we think of each observation as a random variable, then we can distinguish between situations corresponding to:

1. Sampling with replacement so that no observation is affected by any other (i.e., independent $X_1, X_2, \ldots, X_n$). In this case, all observations are random variables with identical probability distributions.
2. Sampling without replacement. In this case, the observations are not independent and hence are characterized by different probability distributions. Think of an urn filled with black and white balls. When we don't replace the drawn balls, the probability of say a black on each draw is different.

5.3 Sampling theory

The most important aspect of frequentist statistics is the process of drawing conclusions based on *sample* data drawn from the *population* (which is the collection of all possible samples). The concept of the population assumes that in principle, an infinite number of measurements (under identical conditions) are possible. The use of the term *random variable* conveys the idea of an intrinsic uncertainty in the measurement characterized by an underlying population.

Question: What does the term "random" really mean?

Answer: When we randomize a collection of balls in a bottle by shaking it, this is equivalent to saying that the details of this operation are not understood or too complicated to handle. It is sometimes necessary to assume that certain complicated details, while undeniably relevant, might nevertheless have little numerical effect on the answers to certain questions, such as the probability of drawing r black balls from a bottle in n trials when n is sufficiently small.

According to E. T. Jaynes (2003), the belief that "randomness" is some kind of property existing in nature is a form of Mind Projection Fallacy which says, in effect, "I don't know the detailed causes – therefore Nature is indeterminate." For example, later in this chapter we discuss how to write computer programs which generate seemingly "random" numbers, yet all these programs are completely deterministic. If you did not have a copy of the program, there is almost no chance that you could discover it merely by examining more output from the program. Then the Mind Projection Fallacy might lead to the claim that no rule exists. At scales where quantum mechanics becomes important, the prevailing view is that nature is indeterminate. In spite of the great successes of the theory of quantum mechanics, physicists readily admit that they currently lack a satisfactory understanding of the subject. The Bayesian viewpoint is that the limitation in scientific inference results from incomplete information.

In both Bayesian and frequentist statistical inference, certain sampling distributions (e.g., binomial, Poisson, Gaussian) play a central role. To the frequentist, the sampling distribution is a model of the probability distribution of the underlying population from which the sample was taken. From this point of view, it makes sense to interpret probabilities as long-run relative frequencies.

In a Bayesian analysis, the sampling distribution is a mathematical description of the uncertainty in predicting the data for any particular model because of incomplete information. It enables us to compute the likelihood $p(D|H, I)$.

In Bayesian analysis, any sampling distribution corresponds to a particular state of knowledge. But as soon as we start accumulating data, our state of knowledge changes. The new information necessarily modifies our probabilities in a way that can be incomprehensible to one who tries to interpret probabilities as physical causations or long-run relative frequencies.

5.4 Probability distributions

Now that we have a better understanding of what a random variable is let's restate the frequentist definition of probability more precisely. It is commonly referred to as the *relative frequency definition*.

Relative frequency definition of probability: If an experiment is repeated n times under identical conditions and n_x outcomes yield a value of the random variable $X = x$, the limit of n_x/n, as n becomes very large,[1] is defined as $p(x)$, the probability that $X = x$.

Experimental outcomes can be either discrete or continuous. Associated with each random variable is a probability distribution. A probability distribution may be

---

[1] See Bernoulli's law of large numbers discussed in Section 4.2.1.

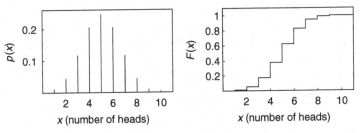

Figure 5.1 The left panel shows the discrete probabilities for the number of heads in ten throws of a fair coin. The right panel shows the corresponding cumulative distribution function.

quantitatively and conveniently described by two functions $p(x)$ and $F(x)$ which are given below for the discrete and continuous cases.

1. Discrete random variables

Probability distribution function: (Also called the *probability mass function*). $p(x_i)$ gives the probability of obtaining the particular value of the random variable $X = x_i$.

(a) $p(x) = p\{X = x\}$
(b) $p(x) \geq 0$ for all x
(c) $\sum_x p(x) = 1$

Cumulative probability function: this gives the probability that the random variable will have a value $\leq x$.

(a) $F(x) = p\{X \leq x\} = \sum_{x_i=0}^{x_i=x} p(x_i)$
(b) $0 \leq F(x) \leq 1$
(c) $F(x_j) > F(x_i)$ if $x_j > x_i$
(d) $F\{X > x\} = 1 - F(x)$

Figure 5.1 shows the discrete probability distribution (binomial) describing the number of heads in ten throws of a fair coin. The right panel shows the corresponding cumulative distribution function.

2. Continuous random variables[2]

Probability density function: $f(x)$

(a) $p\{a \leq X \leq b\} = \int_a^b f(x)\, dx$
(b) $f(x) \geq 0(-\infty < x < \infty)$
(c) $\int_{-\infty}^{+\infty} f(x)dx = 1$

---

[2] *Continuous density function* defined by $f(X = x) = \lim_{\delta x \to 0} [f(x < X < x + \delta x)]/\delta x$.

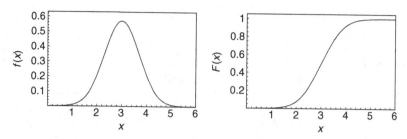

Figure 5.2 The left panel shows a continuous probability density function and the right panel shows the corresponding cumulative probability density function.

Cumulative probability density function:

(a) $F(x) = p\{X \leq x\} = \int_{-\infty}^{x} f(x) dx$
(b) $F(-\infty) = 0; F(+\infty) = 1$
(c) $p\{a < X < b\} = F(b) - F(a)$
(d) $\frac{dF(x)}{dx} = f(x)$

Figure 5.2 shows an example of a continuous probability density function (left panel) and the corresponding cumulative probability density function (right panel).

5.5 Descriptive properties of distributions

The expectation value for a function, $g(X)$, of a random variable, X, is the weighted average of the function over all possible values of x. We will designate the expectation value of $g(X)$ by $\langle g(X) \rangle$, which is given by

$$\langle g(X) \rangle = \begin{cases} \sum_{\text{all } x} g(x) p(x) & \text{(discrete)}, \\ \int_{-\infty}^{+\infty} g(x) f(x) dx & \text{(continuous)}. \end{cases} \tag{5.1}$$

The result, if it exists, is a fixed number (not a function) and a property of the probability distribution of X. The expectation defined above is referred to as the *first moment* of the distribution $g(X)$. The shape of a probability distribution can be rigorously described by the value of its moments:

The rth **moment** of the random variable X about the origin ($x = 0$) is defined by

$$\mu_r' = \langle X^r \rangle = \begin{cases} \sum_x x^r p(x) & \text{(discrete)}, \\ \int_{-\infty}^{+\infty} x^r f(x) dx & \text{(continuous)}. \end{cases} \tag{5.2}$$

Mean $= \mu_1' = \langle X \rangle = \mu =$ first moment about the origin. This is the usual measure of the location of a probability distribution.

The rth **central moment** (origin $=$ mean) of X is defined by

$$\mu_r = \langle (X - \mu)^r \rangle = \begin{cases} \sum_x (x - \mu)^r p(x) & \text{(discrete)}, \\ \int_{-\infty}^{+\infty} (x - \mu)^r f(x) dx & \text{(continuous)}. \end{cases} \tag{5.3}$$

The distinction between μ_r and μ'_r is simply that in the calculation of μ_r the origin is shifted to the mean value of x.

First central moment: $\langle (X - \mu) \rangle = \langle X \rangle - \mu = 0$.

Second central moment: $\text{Var}(X) = \sigma_x^2 = \langle (X - \mu)^2 \rangle$, where $\sigma_x^2 =$ usual measure of dispersion of a probability distribution.

$$\langle (X - \mu)^2 \rangle = \langle (X^2 - 2\mu X + \mu^2) \rangle = \langle X^2 \rangle - 2\mu \langle X \rangle + \mu^2$$

$$= \langle X^2 \rangle - 2\mu^2 + \mu^2 = \langle X^2 \rangle - \mu^2 = \langle X^2 \rangle - \langle X \rangle^2 \qquad (5.4)$$

Therefore, $\sigma^2 = \langle X^2 \rangle - \langle X \rangle^2$.

The *standard deviation*, σ, equal to the square root of the variance, is a useful measure of the width of a probability distribution.

It is frequently desirable to compute an estimate of σ^2 as the data are being acquired. Equation (5.4) tells us how to accomplish this, by subtracting the square of the average of the data from the average of the data values squared. Later, in Section 6.3, we will introduce a more accurate estimate of σ^2 called the sample variance.

Box 5.1

Question: What is the variance of the random variable $Y = aX + b$?
Solution:

$$\text{Var}(Y) = \langle (Y - \mu_y)^2 \rangle = \langle \{(aX + b) - (a\mu_X + b)\}^2 \rangle$$

$$= \langle \{aX - a\mu\}^2 \rangle$$

$$= \langle a^2 X^2 - 2a^2 \mu X + a^2 \mu^2 \rangle = a^2 (\langle X^2 \rangle - \langle X \rangle^2)$$

$$= a^2 \text{Var}(X)$$

Third central moment: $\mu_3 = \langle (X - \mu)^3 \rangle$.

This is a measurement of the asymmetry or skewness of the distribution. For a symmetric distribution, $\mu_3 = 0$ and $\mu_{2n+1} = 0$ for any integer value of n.

Fourth central moment: $\mu_4 = \langle (X - \mu)^4 \rangle$.

μ_4 is called *kurtosis* (another shape factor). It is a measure of how flat-topped a distribution is near its peak. See Figure 5.3 and discussion in the next section for an example.

5.5.1 Relative line shape measures for distributions

The shape of a distribution cannot be entirely judged by the values of μ_3 and μ_4 because they depend on the units of the random variable. It is better to use measures relative to the distribution's dispersion.

$\alpha_3 > 0 \equiv$ positively skewed $\rightarrow$

$\alpha_3 > 0 \equiv$ negatively skewed $\rightarrow$

$\alpha_3 = 0 \equiv$ symmetric $\rightarrow$

$\alpha_4 < 3$ leptokurtic $\equiv$ highly-peaked $\rightarrow$

$\alpha_4 < 3$ platykurtic $\equiv$ flat-topped $\rightarrow$

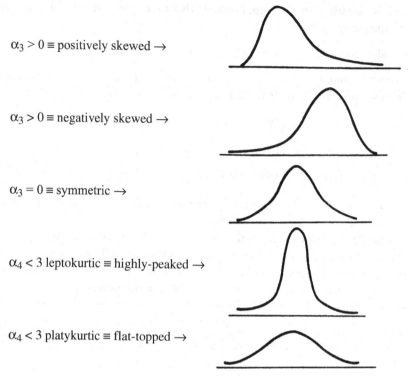

Figure 5.3 Single peak distributions with different coefficients of skewness and kurtosis.

Coefficient of skewness: $\alpha_3 = \dfrac{\mu_3}{(\mu_2)^{3/2}}.$

Coefficient of kurtosis: $\alpha_4 = \dfrac{\mu_4}{(\mu_2)^{2}}.$

Figure 5.3 illustrates a single peaked distribution for different α_3 and α_4 coefficients. Note: $\alpha_4 = 3$ for any Gaussian distribution so distributions with $\alpha_4 > 3$ are more sharply peaked than a Gaussian, while those with $\alpha_4 < 3$ are more flat-topped.

5.5.2 Standard random variable

A random variable X can always be converted to a standard random variable Z using the following definition:

$$Z = \frac{X - \mu}{\sigma_x}. \tag{5.5}$$

Z has a mean $\langle Z \rangle = 0$, and variance $\langle Z^2 \rangle = \sigma_z^2 = 1$.

For any particular value x of X, the quantity $z = (x - \mu)/\sigma_x$ indicates the deviation of x from the expected value of X in terms of standard deviation units. At several points in this chapter we will find it convenient to make use of the standard random variable.

5.5.3 *Other measures of central tendency and dispersion*

Median: The median is a measure of the central tendency in the sense that half the area of the probability distribution lies to the left of the median and half to the right. For any continuous random variable, the median is defined by

$$p(X \le \text{median}) = p(X \ge \text{median}) = 1/2. \qquad (5.6)$$

If a distribution has a strong central peak, so that most of its area is under a single peak, then the median is an estimator of the central peak. It is a more robust estimator than the mean: the median fails as an estimator only if the area in the tail region of the probability distribution is large, while the mean fails if the first moment of the tail is large. It is easy to construct examples where the first moment of the tail is large even though the area is negligible.

Mode: Defined to be a value, x_m of X, that maximizes the probability function (if X is discrete) or probability density (if X is continuous). Note: this is only meaningful if there is a single peak.

If X is continuous, the mode is the solution to

$$\frac{df(x)}{dx} = 0, \quad \text{for} \quad \frac{d^2f(x)}{dx^2} < 0. \qquad (5.7)$$

An example of the mode, median and mean for a particular PDF is shown in Figure 5.4.

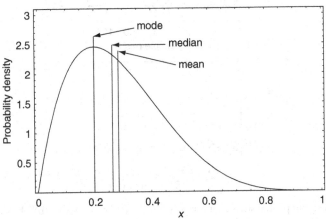

Figure 5.4 The mode, median and mean are three different measures of this probability density function.

5.5.4 *Median baseline subtraction*

Suppose you want to remove the baseline variations in some data without suppressing the signal. Many automated signal detection schemes only work well if these baselines variations are removed first. The upper panel of Figure 5.5 depicts the output from a detector system with a signal profile represented by narrow Gaussian-like features sitting on top of a slowly varying baseline with noise. How do we handle this problem?

Solution: Use running median subtraction.

One way to remove the slowly varying baseline is to subtract a running median. The signal at sample location i is replaced by the original signal at i minus the median of all values within $\pm(N - 1)/2$ samples. N is chosen so it is large compared to the signal profile width and short compared to baseline changes.

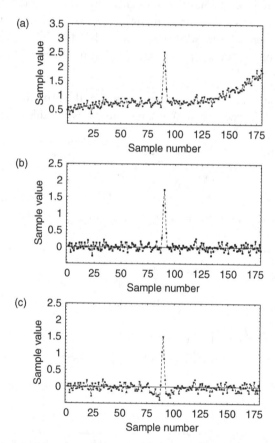

Figure 5.5 (a) A signal profile sitting on top of a slowly varying baseline. (b) The same data with the baseline variations removed by a running median subtraction. (c) The same data with the baseline variations removed by a running mean subtraction; notice the negative bowl in the vicinity of the source profile.

Question: Why is median subtraction more robust than mean subtraction?

Answer: When the N samples include some of the signal points, both the mean value and median will be elevated so that when the running subtraction occurs the signal will sit in a negative bowl as is illustrated in Figure 5.5(c).

With mean subtraction, the size of the bowl will be proportional to the signal strength. With median subtraction, the size of the bowl is smaller and essentially independent of the signal strength for signals greater than noise. To understand why, consider a running median subtraction with $N = 21$ and a signal profile, which for simplicity is assumed to have a width of only 1 sample. First, imagine a histogram of the 21 sample values when no signal is present, i.e., just a Gaussian noise histogram with some median, m_0. Now suppose a signal of strength S is added to sample 11, shifting it in the direction of increasing signal strength. Let T_{11} be the value of sample 11 before the signal was added. There are two cases of interest. (a) If $T_{11} > m_0$ then $T_{11} + S > m_0$ and the addition of the signal produces no change in the median value, i.e., the number of sample values on either side of m_0 is unchanged. (b) If $T_{11} < m_0$, then the addition of S can cause the sample to move to the other side of m_0 thus increasing the median by a small amount to m_1. The size of S required to produce this small shift is $S \sim$ the RMS noise. Once sample 11 has been shifted to the other side, no further increase in the value of S will change the median. Figure 5.5(b) shows the result of a 21-point running median subtraction. The baseline curvature has been nicely removed and there is no noticeable negative bowl in the vicinity of the source.

In the case of a running mean subtraction, the change in the mean of our 21 samples is directly proportional to the signal strength S, which gives rise to the very noticeable negative bowl that can be seen in Figure 5.5(c).

Mean deviation (alternative measure of *dispersion*)

$$\langle |X - \mu| \rangle = \begin{cases} \sum_{\text{all } x} |x - \mu| \, p(x) & \text{(discrete)}, \\ \int_{-\infty}^{+\infty} |x - \mu| f(x) dx & \text{(continuous)}. \end{cases} \tag{5.8}$$

For long-tailed distributions, the effect on the mean deviation of the values in the tail is less than the effect on the standard deviation.

5.6 Moment generating functions

In Section 5.5 we looked at various useful moments of a random variable. It would be convenient if we could describe all moments of a random variable in one function. This function is called the *moment generating function*. We will use it directly to compute moments for a variety of distributions. We will also employ the moment generating function in the derivation of the Central Limit Theorem, in Section 5.9, and in the proof of several theorems in Chapter 6. The moment generating function, $m_x(t)$, of the random variable X is defined by

$$m_x(t) = \langle e^{tX} \rangle = \begin{cases} \sum_x e^{tx} p(x) & \text{(discrete)}, \\ \int_{-\infty}^{+\infty} e^{tx} f(x) dx & \text{(continuous)}, \end{cases} \tag{5.9}$$

where t is a dummy variable. The moment generating function exists if there is a positive constant ϵ such that $m_x(t)$ is finite for $|t| \le \epsilon$. The moments themselves are the coefficients in a Taylor series expansion of the moment generating function (see Equation ((5.12)) below) which converges for $|t| \le \epsilon$.

It can be shown that if a moment generating function exists, then it completely determines the probability distribution of X, i.e., if two random variables have the same moment generating function, they have the same probability distribution.

The rth moment about the origin (see Equation (5.2)) is obtained by taking the rth derivative of $m_x(t)$ with respect to t and then evaluating the derivative at $t = 0$ as shown in Equation (5.10).

$$\left. \frac{d^r m_x(t)}{dt^r} \right|_{t=0} = \left. \frac{d^r}{dt^r} \langle e^{tX} \rangle \right|_{t=0} = \left\langle \frac{d^r e^{tX}}{dt^r} \right\rangle_{t=0}$$

$$= \langle X^r e^{tX} \rangle_{t=0} = \langle X^r \rangle \tag{5.10}$$

$$= \mu'_r.$$

For moments about the mean (central moments), we can use the *central moment generating function*.

$$m_{x-\mu}(t) = \langle \exp\{t(x - \mu)\} \rangle. \tag{5.11}$$

Now we use a Taylor series expansion of the exponential,

$$\langle \exp[t(X - \mu)] \rangle = \left\langle 1 + t(X - \mu) + \frac{t^2(X - \mu)^2}{2!} + \frac{t^3(X - \mu)^3}{3!} \cdots \right\rangle. \tag{5.12}$$

From the expansion, one can see clearly that each successive moment is obtained by taking the next higher derivative with respect to t, each time evaluating the derivative at $t = 0$.

Example:

Let X be a random variable with probability density function

$$f(x) = \begin{cases} \frac{1}{\theta} \exp(-x/\theta), & \text{for } x > 0, \theta > 0 \\ 0, & \text{elsewhere}. \end{cases} \tag{5.13}$$

Determine the moment generating function and variance:

$$m_x(t) = \frac{1}{\theta} \int_0^\infty \exp(tx) \exp(-x/\theta) dx$$

$$= \frac{1}{\theta} \int_0^\infty \exp[-(1-\theta t)x/\theta] dx \qquad (5.14)$$

$$= \frac{\theta}{\theta(1-\theta t)} \exp[-(1-\theta t)x/\theta]|_0^\infty$$

$$= (1-\theta t)^{-1} \quad (\text{for } t < 1/\theta)$$

$$\frac{dm_x(t)}{dt}\Big|_{t=0} = \theta(1-\theta t)^{-2}\big|_{t=0} = \theta = \langle X \rangle \qquad (5.15)$$

$$\frac{d^2 m_x(t)}{dt^2}\Big|_{t=0} = 2\theta^2(1-\theta t)^{-3}\big|_{t=0} = 2\theta^2 = \langle X^2 \rangle. \qquad (5.16)$$

From Equation (5.4), the variance, σ^2, is given by

$$\sigma^2 = \langle X^2 \rangle - \langle X \rangle^2 = 2\theta^2 - \theta^2 = \theta^2. \qquad (5.17)$$

5.7 Some discrete probability distributions

5.7.1 Binomial distribution

The *binomial distribution*[3] is one of the most useful discrete probability distributions and arises in any repetitive experiment whose result is either the occurrence or non-occurrence of an event (only two possible outcomes, like tossing a coin). A large number of experimental measurements contain random errors which can be represented by a limiting form of the binomial distribution called the *normal* or *Gaussian distribution* (Section 5.8.1).

Let X be a random variable representing the number of successes (occurrences) out of n independent trials such that the probability of success for any one trial is p.[4] Then X is said to have a binomial distribution with probability mass function

$$p(x) = p(x|n,p) = \frac{n!}{(n-x)!\,x!} p^x (1-p)^{n-x}, \quad \text{for} \quad x = 0, 1, \dots, n; \; 0 \le p \le 1,$$

$$(5.18)$$

which has two parameters n and p.

---

[3] A Bayesian derivation of the binomial distribution is presented in Section 4.2.
[4] Note: any time the symbol p appears without an argument, it will be taken to be a number representing the probability of a success. $p(x)$ is a probability distribution either discrete or continuous.

Cumulative distribution function:

$$F(x) = \sum_{i=0}^{x} p(i) = \sum_{i=0}^{x} \binom{n}{i} p^i (1-p)^{(n-i)}$$

$\binom{n}{i} = $ short-hand notation for number of combinations of n items taken i at a time.   (5.19)

Box 5.2 *Mathematica* cumulative binomial distribution:

Needs["Statistics `DiscreteDistributions` "]

The probability of at least x successes in n binomial trials is given by

(1 − CDF[BinomialDistribution[*n*, *p*], *x*])

→ answer = 0.623 ($n = 10, p = 0.5, x = 4$)

Moment generating function of a binomial distribution:

We can apply Equation (5.9) to compute the moment generating function of the binomial distribution.

$$m_x(t) = \langle e^{tx} \rangle = \sum_{x=0}^{n} e^{tx} \binom{n}{x} p^x (1-p)^{n-x} \tag{5.20}$$

$$m_x(t) = \sum_{x=0}^{n} \frac{n!}{(n-x)!\, x!} (e^t p)^x (1-p)^{n-x}$$

$$= (1-p)^n + n(1-p)^{n-1}(e^t p) + \cdots + \frac{n!}{(n-k)!}(1-p)^{n-k}(e^t p)^n$$

$$+ \cdots + (e^t p)^n$$

$$= \text{binomial expansion of } [(1-p) + e^t p]^n.$$

Therefore, $m_x(t) = [1 - p + e^t p]^n$.

From the first derivative, we compute the mean, which is given by

$$\text{mean} = \mu_1' = \frac{dm_x(t)}{dt}\Big|_{t=0} = n[1 - p + e^t p]^{n-1} e^t p|_{t=0} = np.$$

The second derivative yields the second moment:

$$\mu_2' = \frac{d^2 m_x(t)}{dt^2}\Big|_{t=0}$$

$$= [n(n-1)(1 - p + e^t p)^{n-2}(e^t p)^2 + n(1 - p + e^t p)^{n-1} e^t p]|_{t=0}$$

$$= n(n-1)p^2 + np.$$

But $\mu'_2 = \langle X^2 \rangle$, and therefore, the variance σ^2 is given by

$$\sigma^2 = \langle (X - \mu)^2 \rangle = \langle X^2 \rangle - \langle X \rangle^2 = \langle X^2 \rangle - \mu^2$$

$$= n(n-1)p^2 + np - (np)^2 \qquad (5.21)$$

$$\sigma^2 = np(1-p) \quad \text{(variance of binomial distribution)}.$$

Box 5.3 *Mathematica* binomial mean and variance

The same results could be obtained in *Mathematica* with the commands:
Mean[BinomialDistribution[n, p]]
Variance[BinomialDistribution[n, p]].

5.7.2 The Poisson distribution

The Poisson distribution was derived by the French mathematician Poisson in 1837, and the first application was to the description of the number of deaths by horse kicking in the Prussian army. The Poisson distribution resembles the binomial distribution if the probability of occurrence of a particular event is very small. Let X be a random variable representing the number of independent random events that occur at a constant average rate in time or space. Then X is said to have a *Poisson distribution* with probability function

$$p(x|\lambda) = \begin{cases} \dfrac{e^{-\lambda} \lambda^x}{x!}, & \text{for } x = 0, 1, 2, \ldots \text{ and } \lambda > 0 \\ 0, & \text{elsewhere.} \end{cases} \qquad (5.22)$$

The parameter of the Poisson distribution is λ, the average number of occurrences of the random event in some time or space interval. $p(x|\lambda)$ is the probability of x occurrences of the event in a specified interval.[5]

The Poisson distribution is a limiting case of the binomial distribution in the limit of large n and small p.

The following calculation illustrates the steps in deriving the Poisson distribution as a limiting case of the binomial distribution.

$$\text{Binomial distribution:} \quad p(x|n, p) = \frac{n!}{(n-x)!x!} p^x (1-p)^{n-x}, \qquad (5.24)$$

---

[5] In Section 4.7, we derived the Poisson distribution by using probability theory as logic, directly from a statement of a particular state of prior information. In that treatment, the Poisson distribution was written as

$$p(n|r, t, I) = \frac{e^{-rt}(rt)^n}{n!}. \qquad (5.23)$$

From a comparison of Equations (5.23) and (5.22), it is clear that the symbol λ, the average number of occurrences in a specified interval, is equal to rt where r is rate of occurrence and t is a specified time interval. Also, in the current chapter, the symbol x will be used in place of n.

where p is the probability of a single occurrence in a sample n in some time interval. Multiply the numerator and denominator of Equation (5.24) by n^x and substitute the following expansion:

$$n!/(n-x)! = n(n-1)(n-2)\ldots(n-x-1). \tag{5.25}$$

With these changes, Equation (5.24) becomes

$$
\begin{aligned}
p(x|n,p) &= \frac{n(n-1)(n-2)\ldots(n-x-1)}{n^x x!}(np)^x(1-p)^{n-x} \\
&= \frac{n(n-1)\ldots(n-x-1)}{n^x}\frac{\lambda^x}{x!}(1-p)^{n-x} \tag{5.26} \\
&= \frac{1(1-\frac{1}{n})(1-\frac{2}{n})\ldots(1-\frac{(x-1)}{n})}{(1-p)^x}\frac{\lambda^x}{x!}(1-p)^n,
\end{aligned}
$$

where λ has replaced the product np.

Now $(1-p)^n \equiv [(1-p)^{-1/p}]^{-np} = [(1-p)^{-1/p}]^{-\lambda}$ and by definition $\lim\limits_{z\to 0}(1+z)^{1/z} = e$. Let $z = -p$, then $\lim\limits_{p\to 0}[(1-p)^{-1/p}]^{-\lambda} = e^{-\lambda}$.

Moreover,

$$\lim_{n\to\infty}(1-1/n)(1-2/n)\ldots(1-(x-1)/n) = 1 \tag{5.27}$$

and,

$$\lim_{p\to 0}(1-p)^x = 1. \tag{5.28}$$

Therefore,

$$\lim_{n\to\infty, p\to 0} p(x|n,p) = \frac{e^{-\lambda}\lambda^x}{x!}, \quad x = 0,1,2,\ldots \tag{5.29}$$

Thus, the Poisson distribution is a limiting case of the binomial distribution in the limit of large n and small p. To make use of the binomial distribution we need to know both n and p. In some instances the only information we have is their product, i.e., the mean number of occurrences, λ. For example, traffic accidents are rare events and the number of accidents per unit of time is well described by the Poisson distribution. The number of traffic accidents that occur each day is usually recorded by the police department, but not the number of cars that are not involved in an accident.

Mean of a Poisson distribution:

$$\mu = \langle X\rangle = \sum_{x=0}^{\infty} x\frac{\lambda^x e^{-\lambda}}{x!} = \lambda e^{-\lambda}\sum_{x=0}^{\infty}\frac{\lambda^{x-1}}{(x-1)!}. \tag{5.30}$$

Let $y = x - 1$

$$\mu = \lambda e^{-\lambda} \sum_{y=0}^{\infty} \frac{\lambda^y}{y!} = \lambda e^{-\lambda} e^{\lambda} = \lambda. \tag{5.31}$$

The mean of a Poisson distribution $= \lambda$. (For a binomial distribution $\mu = np$)

Cumulative distribution:

$$F(x|\lambda) = \sum_{x_i=0}^{x} \frac{\lambda^{x_i} e^{-\lambda}}{x_i!}. \tag{5.32}$$

Poisson variance:

$$\sigma^2(X) = \langle X^2 \rangle - \langle X \rangle^2 \tag{5.33}$$

$$\langle X^2 \rangle = \langle X(X-1) + X \rangle = \langle X(X-1) \rangle + \langle X \rangle. \tag{5.34}$$

Then

$$\langle X(X-1) \rangle = \sum_{x=0}^{\infty} \frac{x(x-1)\lambda^x e^{-\lambda}}{x!}$$

$$= e^{-\lambda}\lambda^2 \sum_{x=2}^{\infty} \frac{\lambda^{x-2}}{(x-2)!} \tag{5.35}$$

$$= \lambda^2 e^{-\lambda} e^{+\lambda}$$

$$= \lambda^2 = \mu^2.$$

Then

$$\langle X^2 \rangle = \mu^2 + \mu$$

$$\sigma^2(X) = \mu^2 + \mu - \langle X \rangle^2 = \mu$$

$$\rightarrow \sigma(X) = \sqrt{\mu}.$$

Note: for a binomial distribution, $\sigma^2 = np(1-p) \rightarrow np = \mu$ as $p \rightarrow 0$.

Figure 5.6 illustrates how the shape of the Poisson distribution varies with λ. As λ increases, the shape of the Poisson distribution asymptotically approaches a Gaussian distribution. The dashed curve in the $\lambda = 40$ panel is a Gaussian distribution with a mean $= \lambda$ and a standard deviation $= \sqrt{\lambda}$.

Examples of situations described by a Poisson distribution:

- Number of telephone calls on a line in a given interval.
- Number of shoppers entering a store in a given interval.
- Number of failures of a product in a given interval.
- Number of photons detected from a distant quasar in a given time interval.
- Number of meteorites to fall per unit area of land.

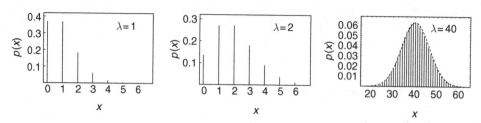

Figure 5.6 As λ increases, the shape of the Poisson distribution becomes more symmetric.

5.7.3 Negative binomial distribution

Imagine a binomial scenario involving a sequence of independent trials where the probability of success of each trial is p. Instead of fixing the number of trials, n, suppose we continue the trials until exactly k successes have occurred.[6] Here, the random variable is n, the number of trials necessary for exactly k successes. If the independent trials continue until the kth success, then the last trial must have been a success. Prior to the last trial, there must have been $k-1$ successes in $n-1$ trials. The number of distinct ways $k-1$ successes can be observed in $n-1$ trials is $\binom{n-1}{k-1}$. Therefore, the probability of k successes in the n trials with the last being a success is

$$p(n|k,p) = \binom{n-1}{k-1}p^{k-1}(1-p)^{n-k} \times p^1 = \binom{n-1}{k-1}p^k(1-p)^{n-k}. \tag{5.36}$$

Equation (5.36) is called the *negative binomial distribution.*

Let the number of trials required to achieve k successes $= X + k$. Then random variable X is the number of failures before k successes, which is given by

$$p(x|k,p) = \begin{cases} \binom{k+x-1}{k-1}p^k(1-p)^x; & x = 0,1,2,\ldots \\ & k = 1,2,\ldots \\ & 0 \leq p \leq 1 \\ 0; & \text{elsewhere.} \end{cases}$$

For the special case of one success $k=1$, the above distribution is known as the *geometric distribution.*

$$p(x|p) = p(1-p)^x. \tag{5.37}$$

The geometric random variable represents the number of failures before the first success.

---

[6] For example, an astronomer could plan to continue taking spectra of candidate stars until exactly 50 white dwarfs have been detected.

5.8 Continuous probability distributions

5.8.1 Normal distribution

The *normal (Gaussian) distribution* is the most important and widely used probability distribution. One of the reasons why the normal distribution is so useful is because of the *Central Limit Theorem*. This theorem will be discussed in detail later, but briefly, it says the following: suppose you have a radioactive source and you measure the average number of decays in one hundred 10-second intervals. (We know that the individual counts obey a Poisson distribution). If you repeated the experiment many times and hence determined a large number of averages then, according to the Central Limit Theorem, the averages will be normally distributed.

The distribution of the sample means (from populations with a finite mean and variance) approaches a normal distribution as the number of terms in the mean approaches infinity. It can be shown to be the limit of a binomial distribution as $n \to \infty$ and $np \gg 1$.

Corollary: Whenever a random variable can be assumed to be the result of a large number of small effects, the distribution is approximately normal.

Gaussian probability density function:

$$f_X(x) = f(x|\mu, \sigma) = \frac{1}{\sqrt{2\pi}\sigma} \exp\left\{ -\frac{(x-\mu)^2}{2\sigma^2} \right\} \tag{5.38}$$

$$\text{for} -\infty < x < \infty; \ -\infty < \mu < \infty; \ 0 < \sigma^2 < \infty.$$

Box 5.4 *Mathematica* **evaluation of a Gaussian or normal distribution:**

Needs["Statistics`ContinuousDistributions`"]

The line above loads a package containing a wide range of continuous distributions of importance to statistics, and the following line computes the probability density function at x for a normal distribution with mean μ and standard deviation σ.

PDF[NormalDistribution[μ, σ],x]
$\to$ answer = 0.45662 ($\mu = 2.0$, $\sigma = 0.4$, $x = 1.5$)

The mean and standard deviation of the distribution are given by
Mean[NormalDistribution[μ, σ]]
$\to$ answer = μ
StandardDeviation[NormalDistribution[μ, σ]]
$\to$ answer = σ

Central moment generating function:

$$m_{X-\mu}(t) = \langle \exp\{t(X-\mu)\}\rangle$$

$$= \frac{1}{\sqrt{2\pi}\sigma} \int_{-\infty}^{+\infty} \exp\{t(x-\mu)\} \exp\left\{-\frac{(x-\mu)^2}{2\sigma^2}\right\} dx$$

$$= \frac{1}{\sqrt{2\pi}\sigma} \int_{-\infty}^{+\infty} \exp\left(-\frac{1}{2\sigma^2}\{(x-\mu)^2 - 2\sigma^2 t(x-\mu)\}\right) dx.$$

Adding and subtracting $\sigma^4 t^2$ in the term in the curly braces:

$$\{(x-\mu)^2 - 2\sigma^2(x-\mu)t + \sigma^4 t^2 - \sigma^4 t^2\} = \{[x-\mu-\sigma^2 t]^2 - \sigma^4 t^2\}$$

$$\rightarrow m_{X-\mu}(t) = \exp\left(\frac{\sigma^2 t^2}{2}\right) \frac{1}{\sqrt{2\pi}\sigma} \int_{-\infty}^{+\infty} \exp\left\{-\frac{(x-\mu-\sigma^2 t^2)^2}{2\sigma^2}\right\} dx$$

$$= \exp\left(\frac{\sigma^2 t^2}{2}\right) = 1 + \frac{\sigma^2 t^2}{2} + \frac{\sigma^4 t^4}{4 \times 2!} + \frac{\sigma^6 t^6}{8 \times 3!} \cdots$$

$$\mathrm{Var}(X) = \frac{d^2 m_{X-\mu}(t)}{dt^2}\bigg|_{t=0} = \sigma^2$$

$\mu_3 = 0$ and $\mu_4 = \dfrac{4!\sigma^4}{4 \times 2!} = 3\sigma^4$

$\alpha_4 = \text{coefficient of kurtosis} = \dfrac{\mu_4}{(\mu_2)^2} = 3.$

Note: for a Poisson distribution, $\alpha_4 = 3 + \dfrac{1}{\mu} \rightarrow 3$ as $\mu \rightarrow \infty$.

Also, for a binomial distribution $\alpha_4 = 3 + \dfrac{[1 - 6p(1-p)]}{np(1-p)} \rightarrow 3$ as $n \rightarrow \infty$.

Convention:

If the random variable X is known to follow a normal distribution with mean μ and variance σ^2, then it is common to abbreviate this by

$$X \approx N(\mu, \sigma^2).$$

For convenience, the following transformation to the standard random variable is often made:

$$Z = \frac{X - \mu}{\sigma} \approx N(0, 1).$$

In terms of Z the normal distribution becomes

$$f(Z) = \frac{1}{\sqrt{2\pi}} \exp -\frac{z^2}{2}.$$

Cumulative distribution function:

$$F(x) \equiv F(x|\mu, \sigma) = p(X \le x) = \frac{1}{\sqrt{2\pi}\sigma} \int_{-\infty}^{x} \exp\left\{-\frac{(t-\mu)^2}{2\sigma^2}\right\} dt.$$

This integral cannot be integrated in closed form. $F(x|\mu, \sigma)$ can be tabulated as a function of μ and σ, which requires a separate table for each pair of values. Since there are an infinite number of values for μ and σ, this task is not practical.

Instead, it is common to calculate the cumulative distribution function of the standard random variable Z. Then:

$$p(X \leq x) = p\left[Z \leq \frac{x - \mu}{\sigma}\right] = \frac{1}{\sqrt{2\pi}} \int_{-\infty}^{(z = \frac{x-\mu}{\sigma})} \exp\left(-\frac{z'^2}{2}\right) dz' = F(z).$$

Usually, $F(z)$ is expressed in terms of the *error function*, erf(z).

$$\text{erf}(z) = \frac{2}{\sqrt{\pi}} \int_0^z \exp\left(-u^2\right) du$$

$$\text{erf}(-z) = -\text{erf}(z).$$

(5.39)

Then $F(z) = \frac{1}{2} + \frac{1}{2}\text{erf}(z/\sqrt{2})$. The error function is in many computer libraries.

Box 5.5

In *Mathematica*, it can be evaluated with the command **Erf[z]**.

However, it is simpler to compute the cumulative probability with the *Mathematica* command: **CDF[NormalDistribution[μ, σ], x]**.

For any normally distributed random variable:

$$p(\mu - \sigma \leq X \leq \mu + \sigma) = 0.683$$
$$p(\mu - 2\sigma \leq X \leq \mu + 2\sigma) = 0.954$$
$$p(\mu - 3\sigma \leq X \leq \mu + 3\sigma) = 0.997$$
$$p(\mu - 4\sigma \leq X \leq \mu + 4\sigma) = 0.999\,937$$
$$p(\mu - 5\sigma \leq X \leq \mu + 5\sigma) = 0.999\,999\,43.$$

Figure 5.7 shows graphs of the normal distribution (left) and the cumulative normal distribution (right) for three different values of σ.

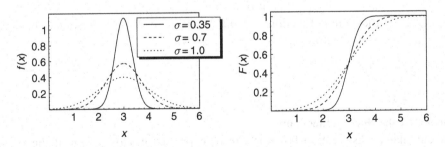

Figure 5.7 Graphs of the normal distribution (left) and the cumulative normal distribution (right).

5.8.2 Uniform distribution

Examples of a uniform distribution include round-off errors and quantization of noise in linear analog-to-digital conversion.

A random variable is said to be uniformly distributed over the interval (a, b) if

$$f(x|a, b) = \begin{cases} 1/(b-a), & \text{for } a \le x \le b \\ 0, & \text{elsewhere} \end{cases} \tag{5.40}$$

$$\text{mean} = (a+b)/2; \qquad \alpha_3 = 0 \cdot$$
$$\text{variance} = (b-a)^2/12; \quad \alpha_4 = 9/5$$
$$\text{no mode}; \qquad \text{median} = \text{mean}.$$

The special case of $a = 0$ and $b = 1$ plays a key role in the computer simulation of values of a random variable with a specified distribution, which will be discussed in Section 5.13.

Cumulative distribution function:

$$F(x|a, b) = \begin{cases} 0, & \text{for } x < a \\ (x-a)/(b-a), & \text{for } a \le x \le b \\ 1, & \text{for } x > b. \end{cases} \tag{5.41}$$

5.8.3 Gamma distribution

The gamma distribution is used extensively in several diverse areas. For example, it is used to represent the random time until the occurrence of some event which occurs only if exactly α independent sub-events occur where the sub-events occur at an average rate $\lambda = 1/\theta$ per unit of time.

$$f(x) = f(x|\alpha, \theta) = \begin{cases} \dfrac{1}{\Gamma(\alpha)\theta^\alpha} x^{\alpha-1} \exp(-x/\theta), & \text{for } x > 0, \alpha, \theta > 0 \\ 0, & \text{elsewhere.} \end{cases} \tag{5.42}$$

$$\text{mean} = \alpha\theta$$
$$\text{variance} = \alpha\theta^2$$
$$\alpha_3 = 2/\sqrt{\alpha}$$
$$\alpha_4 = 3\left(1 + \tfrac{2}{\alpha}\right)$$

Note: $\Gamma(n)$, the gamma function $= \int_0^\infty u^{n-1} \exp(-u)du$ for $n > 0$. Some properties of the gamma function are:

1. $\Gamma(n+1) = n!$ (for n an integer)
2. $\Gamma(n+1) = n\Gamma(n)$
3. $\Gamma(1/2) = \sqrt{\pi}$

Cumulative distribution function:
The cumulative distribution function can be expressed in closed form if the shape parameter α is a positive integer.

$$F(x|\alpha, \theta) = 1 - \left[1 + \frac{x}{\theta} + \frac{1}{2!}\left(\frac{x}{\theta}\right)^2 + \cdots + \frac{1}{(\alpha - 1)!}\left(\frac{x}{\theta}\right)^{\alpha-1}\right] \exp\left(-x/\theta\right). \quad (5.43)$$

Example:

Suppose a metal specimen will break after exactly two stress cycles. If stress occurs independently and at an average rate of 2 per 100 hours, determine the probability that the length of time until failure is within one standard deviation of the average time.

Solution: Let X be a random variable representing the length of time until the second stress cycle. X is gamma-distributed with $\alpha = 2$ and $\theta = 50$.

$$\mu = \text{mean} = \alpha\theta = 2 \times 50 = 100$$

$$\text{standard deviation} = \sqrt{\alpha\theta^2} = \sqrt{2 \times 50^2} = 70.71$$

$$p(\mu - \sigma < X < \mu + \sigma) = p(29.29 < X < 170.71)$$

$$= F(170.71|2, 50) - F(29.28|2, 50).$$

Equation (5.43) for the cumulative distribution function reduces to:

$$F(x|\alpha, \theta) = 1 - (1 + x/50) \exp\left(-x/50\right); \qquad x > 0$$

$$\rightarrow p(\mu - \sigma < X < \mu + \sigma) = 0.7376.$$

When α is an integer, the gamma distribution is known as the *Erlang probability model* after the Danish scientist who used it to study telephone traffic problems.

5.8.4 Beta distribution

While we are on the subject of sampling distribution, here is one that plays a useful role in Bayesian inference. The family of *beta distributions* allows for a wide variety of shapes.

$$f(x) \equiv f(x|\alpha, \beta) = \begin{cases} \dfrac{\Gamma(\alpha + \beta)}{\Gamma(\alpha)\Gamma(\beta)} x^{\alpha-1}(1 - x)^{\beta-1}; & 0 < x < 1 \\ & \alpha, \beta > 0 \qquad (5.44) \\ 0; & \text{elsewhere.} \end{cases}$$

$$\text{mean} = \alpha/(\alpha + \beta); \text{variance} = \frac{\alpha\beta}{(1 + \beta)^2(\alpha + \beta + 1)}.$$

Note: the α appearing in the beta distribution has no connection with the α used in the previously mentioned gamma distribution.

Some examples of the beta distribution are illustrated in Figure 5.8. Any smooth unimodal distribution in the interval $x = 0$ to 1 is likely to be reasonably well approximated by a beta distribution, so it is often possible to approximate a Bayesian prior distribution in this way. If the likelihood function is a binomial distribution, then the Bayesian posterior will have a simple analytic form; namely, another beta distribution. More generally, when both the prior and posterior belong to the same distribution family (in this case the beta distribution), then the prior and

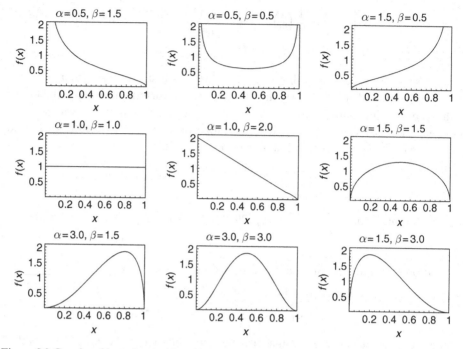

Figure 5.8 Graphs of beta density function for various values of α, β.

likelihood are called *conjugate distributions*. This can greatly simplify any calculations that involve the posterior distribution. The beta distribution is often referred to as a *conjugate prior* for the binomial likelihood.[7] Other well-known examples of conjugate priors are the Gaussian (when dealing with a Gaussian likelihood) and the gamma distribution (when dealing with a Poisson likelihood). In each case, the posterior and prior are members of the same family of distributions.

5.8.5 Negative exponential distribution

The negative exponential distribution is a special case of the gamma distribution for $\alpha = 1$.

---

[7] For example, if the prior is a beta distribution, $\mathrm{Be}(\alpha, \beta)$, then we can write

$$p(x|I) \propto x^{\alpha-1}(1-x)^{\beta-1} \quad (0 \le x \le 1). \tag{5.45}$$

Suppose the likelihood is a binomial, with $p(y|n, x) =$ the probability of obtaining y successes in n trials where the probability of success in any trial is x. Then we can write the likelihood as

$$p(D|x, I) \propto x^y(1-x)^{n-y}. \tag{5.46}$$

The posterior is proportional to the product of Equations (5.45) and (5.46), and given by

$$p(x|D, I) \propto x^{\alpha+y-1}(1-x)^{\beta+n-y-1} \tag{5.47}$$

which is another beta distribution, $\mathrm{Be}(\alpha + y, \beta + n - y)$.

$$f(x) \equiv f(x|\theta) = \begin{cases} \dfrac{1}{\theta}\exp(-x/\theta), & \text{for } x > 0, \theta > 0 \\ 0, & \text{elsewhere.} \end{cases} \tag{5.48}$$

The random variable X is the waiting time until the occurrence of the first Poisson event. That is, the negative exponential distribution can model the length of time between successive Poisson events. It has been used extensively as a time-to-failure model in reliability problems and in waiting-line problems. θ is the mean time between Poisson events.

The *cumulative negative exponential distribution function* is given by

$$F(x|\theta) = 1 - \exp(-x/\theta).$$

5.9 Central Limit Theorem

Let $X_1, X_2, X_3, \ldots, X_n$ be n independent and identically distributed (IID) random variables with unspecified probability distributions, and having a finite mean, μ, and variance, σ^2. The sample average, $\overline{X} = (X_1 + X_2 + X_3 + \cdots + X_n)/n$ has a distribution with mean μ and variance σ^2/n that tends to a normal (Gaussian) distribution as $n \to \infty$. In other words, the standard random variable,

$$\frac{(\overline{X} - \mu)}{\sigma/\sqrt{n}} \to \text{standard normal distribution.}$$

Proof:

As a proof, we will show that the moment generating function of $(\overline{X} - \mu)\sqrt{n}/\sigma$ tends to that of a standard normal as $n \to \infty$.

Let $\quad z_i = (X_i - \mu)/\sigma, \quad i = 1, n$

$$\left.\begin{aligned} \langle z_i \rangle &= 0 \\ \langle z_i^2 \rangle &= 1 \end{aligned}\right\} \text{by definition.}$$

Let $Y = \dfrac{(\overline{X} - \mu)}{\sigma/\sqrt{n}}$.

Now $\displaystyle\sum_{i=1}^{n} z_i = \frac{1}{\sigma}\sum(X_i - \mu) = \frac{n}{\sigma}(\overline{X} - \mu) = \sqrt{n}\,Y$ or $Y = \frac{1}{\sqrt{n}}\displaystyle\sum_{i=1}^{n} z_i$.

Then the moment generating function $m_Y(t)$ is given by

$$m_Y(t) = \langle \exp(tY) \rangle = \left\langle \exp\left(t\sum_{i=1}^{n} \frac{z_i}{\sqrt{n}} \right) \right\rangle$$

$$= \left\langle \exp\left(t\frac{z_i}{\sqrt{n}} \right) \right\rangle^n \qquad \text{since the } z_i\text{'s are IID.}$$

Now
$$\exp\left(t\frac{z_i}{\sqrt{n}}\right) = 1 + \frac{tz_i}{\sqrt{n}} + \frac{t^2 z_i^2}{2!n} + \frac{t^3 z_i^3}{3!n^{3/2}} + \cdots$$

and since $\langle z_i \rangle = 0$; $\langle z_i^2 \rangle = 1$ for all i.

$$\left\langle \exp\left(\frac{tz_i}{\sqrt{n}}\right) \right\rangle = 1 + \frac{t^2}{2n} + \frac{t^3 \langle z_i^3 \rangle}{3!n^{3/2}} + \cdots$$

$$m_Y(t) = \left[1 + \frac{t^2}{2n} + \frac{t^3 \langle z_i^3 \rangle}{3!n^{3/2}} + \cdots\right]^n$$

$$= \left[1 + \frac{1}{n}\left\{\frac{t^2}{2} + \frac{t^3 \langle z_i^3 \rangle}{3!\sqrt{n}} + \cdots\right\}\right]^n$$

$$= \left[1 + \frac{u}{n}\right]^n, \quad \text{where } u = \left\{\frac{t^2}{2} + \frac{t^3 \langle z_i^3 \rangle}{3!\sqrt{n}} + \cdots\right\}$$

$$\lim_{n\to\infty}\left[1 + \frac{u}{n}\right]^n = e^u.$$

In the limit as $n \to \infty$, all terms in $u \to 0$ except the first, $t^2/2$, since all other terms have an n in the denominator.

$$\lim_{n\to\infty} m_Y(t) = \lim_{n\to\infty} e^u = \exp\frac{t^2}{2}$$
$$= \text{moment generating function of a standard normal.}$$

This completes the proof.

5.10 Bayesian demonstration of the Central Limit Theorem

The proof of the CLT given in Section 5.9 does little to develop the reader's intuition on how it works and what are its limitations. To help on both counts, we give the following demonstration of the CLT which is adapted from the work of M. Tribus (1969). In data analysis, it is common practice to compute the average of a repeated measurement and perform subsequent analysis using the average value.

The probability density function (PDF) of the average is simply related to the PDF of the sum which is evaluated below using probability theory as logic. In this demonstration, we will be concerned with measurements of the length of a widget[8] which is composed of many identical components that have all been manufactured on the same assembly line. Because of variations in the manufacturing process, the widgets do not all end up with exactly the same length. The components are analogs of the data points and the widget is the analog of the sum of a set of data points.

---

[8] A widget is some unspecified gadget or device.

$I \equiv$ "a widget is composed of two components. Length of widget = sum of component lengths."

$Y \equiv$ "Length of widget lies between y and $y + dy$."

Note: Y is a logical proposition which appears in the probability function and y is an ordinary algebraic variable.

$X_1 \equiv$ "Length of component 1 lies between x_1 and $x_1 + dx_1$."

$X_2 \equiv$ "Length of component 2 lies between x_2 and $x_2 + dx_2$."

We are given that

$$p(X_1|I) = f_1(x_1)$$

$$p(X_2|I) = f_2(x_2).$$

Problem: Find $p(Y|I)$.

Now Y depends on propositions X_1 and X_2. To evaluate $p(Y|I)$, we first extend the conversation (Tribus, 1969) to include these propositions by writing down the joint probability distribution $p(Y, X_1, X_2|I)$. We can then solve for $p(Y|I)$ by using the marginalizing operation as follows:

$$p(Y|I) = \int\int dX_1 dX_2\, p(Y, X_1, X_2|I)$$

$$= \int\int dX_1 dX_2\, p(X_1|I) p(X_2|I) p(Y|X_1, X_2, I),$$

where we assume X_1 and X_2 are independent.

Since $y = x_1 + x_2$,

$$p(Y|X_1, X_2, I) = \delta(y - x_1 - x_2)$$

$$\rightarrow p(Y|I) = \int dx_1\, f_1(x_1) \int dx_2\, f_2(x_2) \delta(y - x_1 - x_2).$$

The presence of the delta function in the second integral serves to pick out the value of the integrand at $x_2 = y - x_1$, so we have

$$p(Y|I) = \int dx_1\, f_1(x_1) f_2(y - x_1). \tag{5.49}$$

The right hand side of this equation is the *convolution integral*.[9] The convolution operation is demonstrated in Figure 5.9 for the case where both $f_1(x)$ and $f_2(x)$ are uniform PDFs of the same width. The result is a triangular distribution.

---

[9] For more details on the convolution integral and how to evaluate it using the Fast Fourier Transform, see Sections B.4 and B.10.

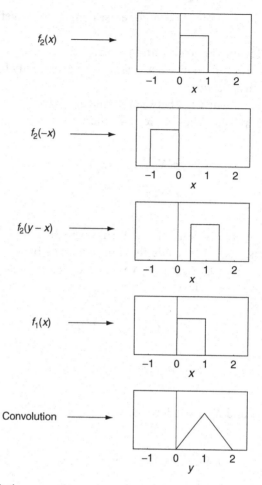

Figure 5.9 The convolution operation.

What if the widget is composed of three components?
Let $Z \equiv$ "Length of widget is between z and $z + dz$."

$$p(Z|I) = \int \int \int dX_1 \, dX_2 \, dX_3 \, p(Z, X_1, X_2, X_3|I)$$

$$= \int \int dY \, dX_3 \, p(Z, X_3, Y|I)$$

$$= \int dY \, p(Y|I) \int dX_3 \, p(X_3|I) \, p(Z|X_3, Y, I)$$

where $p(Z|X_3, Y, I) = \delta(z - y - x_3)$ and $p(Y|I)$ is the solution to the two-component case.

$$p(Z|I) = \int dy\, p(Y|I) f_3(z - y)$$

$$= \int dy\, f(y) f_3(z - y).$$

Another convolution!

Shown below, the probability density function (PDF) of the average is simply related to the PDF of the sum (which we have just evaluated).
Let $x_A = (x_1 + x_2 + x_3)/3 = z/3$ or $z = 3x_A$.

$$f(x_A)dx_A = f(z)dz = 3f(z)dx_A$$
$$\rightarrow p(X_A|I) = 3p(Z|I).$$

Figure 5.10 compares the PDF of the average for the case of $n = 1, 2, 4$ and 8 components. According to the Central Limit Theorem, $p(X_A|I)$ tends to a Gaussian distribution as the number of data being averaged becomes larger. After averaging only four components, the PDF has already taken on the appearance of a Gaussian. If instead of a uniform distribution, our starting PDF had two peaks (bimodal), then a larger number of components would have been required before the PDF of the average was a reasonable approximation to a Gaussian. On the basis of this analysis, we come to the following generalization of the Central Limit Theorem: "Any quantity that stems from a large number of sub-processes is expected to have a Gaussian distribution." The Central Limit Theorem (CLT) is both remarkable and of great practical value in data analysis. In frequentist statistics, we are often uncertain of the form of the sampling distribution the data are drawn from. The equivalent problem in a Bayesian analysis is the choice of likelihood function to use. By working with the averages of data points (frequently as few as five points), we can appeal to the CLT and make use of a Gaussian distribution for the sampling distribution or likelihood function. The CLT also provides a deep understanding for why measurement uncertainties frequently have a Gaussian distribution. This is because the

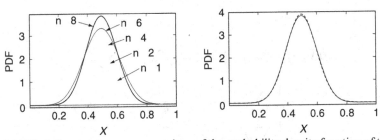

Figure 5.10 The left panel shows a comparison of the probability density function of the average for the case of $n = 1, 2, 4$ and 8 components. The right panel compares the $n = 8$ case to a Gaussian (dashed curve) with the same mean and variance. The two curves are so similar it is difficult to separate them.

measured quantity is often the result of a large number of effects, i.e., is some kind of averaged resultant of these effects (random variables in frequentist language). Since the distribution of the average of a collection of random variables tends to a Gaussian, this is often what we observe.

Two exceptions to the Central Limit Theorem exist. They are:

1. One of the $p(X_i|I)$ is much broader than all of the others. It is apparent from the above demonstration that convolving a very wide uniform distribution with a narrow uniform distribution will give a result that is essentially the same as the original wide uniform distribution.
2. The variances of one or more of the individual $p(X_i|I)$ distributions are infinite. A Cauchy or Lorentzian distribution is an example of such a distribution:

$$p(x|\alpha, \beta, I) = \frac{\beta}{\pi[\beta^2 + (x - \alpha)^2]}.$$

Its very wide wings lead to an infinite second moment, i.e., the variance of $\overline{X}$ is infinite and the sample mean is not a useful quantity. One example of this is the natural line shape of a spectral line.

5.11 Distribution of the sample mean

It is apparent from the previous section that the PDF of the sample mean (average) rapidly approaches a Gaussian in shape and the width of this Gaussian becomes narrower as the number of samples in the average increases. In this section, we want to quantify this latter effect using the frequentist approach.

Let a random sample $X_1, X_2, \ldots, X_n$ consist of n IID random variables such that

$$\langle X_i \rangle = \mu \quad \text{and} \quad \text{Var}(X_i) = \sigma^2.$$

Then
$$\langle \overline{X} \rangle = \langle \tfrac{1}{n} \sum X_i \rangle = \tfrac{1}{n} \sum \langle X_i \rangle = \tfrac{1}{n} \sum \mu$$

$$\rightarrow \langle \overline{X} \rangle = \mu,$$

and

$$\text{Var}(\overline{X}) = \left\langle (\overline{X} - \mu)^2 \right\rangle = \left\langle \left(\frac{1}{n} \sum X_i - \frac{1}{n} \sum \mu \right)^2 \right\rangle$$

$$= \frac{1}{n^2} \left\langle \sum (X_i - \mu)^2 \right\rangle = \frac{1}{n^2} \sum \left\langle (X_i - \mu)^2 \right\rangle$$

$$= \frac{1}{n^2} \sum \sigma^2 = \frac{1}{n^2} n\sigma^2$$

$$\text{Var}(\overline{X}) = \frac{\sigma^2}{n}.$$

(5.50)

The following is true for any distribution with a finite variance:

$$\langle \overline{X} \rangle = \mu$$

$$\mathrm{Var}(\overline{X}) = \frac{\sigma^2}{n}$$

Conclusion: The distribution of a sample average $\overline{X}$ sharpens around μ as the sample size n increases.[10] Signal averaging is based on this principle.

5.11.1 Signal averaging example

Every second a spectrometer output consisting of 64 voltage levels corresponding to 64 frequencies, is sampled by the computer. These voltages are added to a memory buffer containing the results of all previous one-second spectrometer readings. The accumulated spectra are shown in Figure 5.11 at different stages. Although no signal is evident above the noise level in the first spectrum, the signal is clearly evident (near channel 27) after eight spectra have been summed.

Let $S_i \propto n\overline{X}_i = $ the signal in the ith channel of n summed spectra, and

$N_i \propto n\sqrt{\mathrm{Var}(\overline{X}_i)} = $ the noise in the ith channel. Then

$$\frac{S_i}{N_i} = \frac{n\overline{X}_i}{n\sigma/\sqrt{n}} = \sqrt{n}\frac{\overline{X}_i}{\sigma}.$$

In radio astronomy, we are often trying to detect signals which are 10^{-3} of the noise. For a $S/N \approx 5$, $n \approx 2.5 \times 10^7$. However, in these cases, the time required for one independent sample is often much less than one microsecond, and is determined by the radio astronomy receiver bandwidth and detector time constant.

5.12 Transformation of a random variable

In Section 5.13, we will want to generate random variables having a variety of probability distributions for use in simulating experimental data. To do that, we must first learn about the probability distribution of a transformed random variable.

---

[10] What happens to the average of samples drawn from a distribution which has an infinite variance? In this case, the error bar for the sample mean does not decrease with increasing n. Even though the sample mean is not a good estimator of the distribution mean μ, we can still employ Bayes' theorem to compute the posterior PDF of μ from the available samples. The PDF continues to sharpen about μ as the number of samples increases. For a good numerical demonstration of this point, see the lighthouse problem discussed by Sivia (1996) and Gull (1988a).

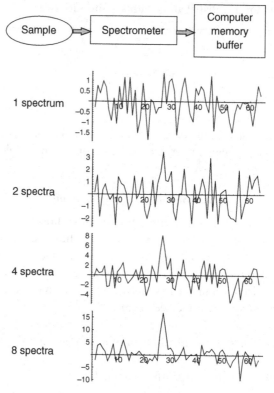

Figure 5.11 Signal averaging. Every second a spectrometer output consisting of 64 voltage levels, corresponding to 64 frequencies, is added to a computer memory buffer. The summed spectra are shown at different stages. Although no signal is evident above the noise level in the first spectrum, the signal is clearly evident (near channel 27) after eight spectra have been summed.

Problem: How do we obtain the probability density function, $f_Y(y)$, of the transformed random variable y, where $y = g(x)$, from knowledge of the probability density function, $f_X(x)$, of the original random variable X?

The function $y = g(x)$ must be a monotonic (increasing or decreasing), differentiable function of x. Then there exists an inverse function $x = g^{-1}(y)$ which is also monotonic and differentiable and for every interval dx there is a corresponding interval dy. Then the probability that $y \leq Y \leq y + dy$ must equal the probability that $x \leq X \leq x + dx$, or

$$|f_Y(y)dy| = |f_X(x)dx|. \tag{5.51}$$

Since probabilities are always positive, we can write

$$f_Y(y) = f_X(x)\left|\frac{dx}{dy}\right|. \tag{5.52}$$

Example:

Find $f_Z(z)$, where Z is the **standard random variable** defined by $Z = (X - \mu)/\sigma$, where $\mu = $ mean and $\sigma = $ standard deviation.

$$z = g(x) = \frac{x - \mu}{\sigma}. \tag{5.53}$$

$$x = g^{-1}(z) = \sigma z + \mu \text{ and } \frac{dx}{dz} = \sigma. \tag{5.54}$$

Then from Equation (5.52), we can write $f_Z(z) = f_X(x)\sigma$.

Suppose $f_X(x) = \frac{1}{\sqrt{2\pi}\sigma} \exp -\frac{(x - \mu)^2}{2\sigma^2}$ (normal distribution).

Then from Equations (5.52) and (5.53), we obtain

$$f_Z(z) = \frac{1}{\sqrt{2\pi}} \exp -\frac{z^2}{2}. \tag{5.55}$$

5.13 Random and pseudo-random numbers

Computer simulations have become an extremely useful tool for testing data analysis algorithms and analyzing complex systems, which are often comprised of many interdependent components. Some examples of their use are given below.

- To simulate experimental data in the design of a complex detector system in many branches of science.
- To test the effectiveness or completeness of some complex analysis program.
- To compute the uncertainties in the parameter estimates derived from nonlinear model fitting.
- To calculate the solution to a statistical mechanics problem which is not amenable to analytical solution.
- To make unpredictable data for use in cryptography, to deal with a variety of authentication and confidentiality problems.

What is usually done is to assume an appropriate probability distribution for each distinct component and to generate a sequence of random or pseudo-random values for each. There are many procedures, known by the generic name of *Monte Carlo*,[11] that follow these lines and use the commodity called random numbers, which have to be manufactured somehow. Typically, the sequences of random numbers are generated by numerical algorithms that can be repeated exactly; such sequences are not truly random. However, they exhibit enough random properties to be sufficient for most applications. We consider below, possible ways of generating random values from some discrete and continuous probability distributions.

---

[11] In Chapter 12, we discuss the important topic of Markov chain Monte Carlo (MCMC) methods, which are dramatically increasing our ability to evaluate the integrals required in a Bayesian analysis of very complicated problems.

The uniform distribution on the interval (0, 1) plays a key role in the generation of random values. From it, we can generate random numbers for any other distribution using the following theorem:

Theorem:

For any continuous random variable X, the cumulative distribution function $F(x|\theta)$ with parameter θ may be represented by a random variable u, which is uniformly distributed on the unit interval.

Proof:

By definition, $F(x|\theta) = \int_{-\infty}^{x} f(t|\theta)dt$. For each value of x, there is a corresponding value of $F(x|\theta)$ which is necessarily in the interval (0,1). Also, $F(x|\theta)$ is a random variable by virtue of the randomness of X. For each value u of the random variable u, the function $u = F(x|\theta)$ defines a one-to-one correspondence between U and X having an inverse relationship $x = F^{-1}(u)$.

Recall that it was shown earlier how to obtain the PDF $f(y)$ of the transformed random variable $Y = g(X)$ from the knowledge of the PDF of X. The result was

$$f_y(y) = f_X(x|\theta)\left|\frac{dx}{dy}\right| = f_X[g^{-1}(y)|\theta]\left|\frac{dg^{-1}(y)}{dy}\right|. \tag{5.56}$$

In the present case, this means

$$f_U(u) = f_X[F^{-1}(u)|\theta]\left|\frac{dx}{du}\right|. \tag{5.57}$$

Since

$$u = F(x|\theta) \rightarrow \frac{du}{dx} = \frac{dF(x|\theta)}{dx} = f_X(x|\theta)$$

$$\rightarrow \frac{dx}{du} = \{f_X(x|\theta)\}^{-1}. \tag{5.58}$$

But $x = F^{-1}(u)$. Substituting for x in Equation (5.58), we obtain

$$\frac{dx}{du} = \{f_X[F^{-1}(u)|\theta]\}^{-1}. \tag{5.59}$$

Substituting Equation (5.59) into Equation (5.57) yields

$$f_U(u) = \frac{f_X[F^{-1}(u)|\theta]}{f_X[F^{-1}(u)|\theta]} \tag{5.60}$$

$$f_U(u) = 1, \quad 0 \le u \le 1.$$

The essence of the theorem is that in many instances, we are able to determine the value of x corresponding to a value of u such that $F(x|\theta) = u$. For this reason, practically all computer systems have a built-in capability of generating random values for a uniform distribution on the unit interval.

Therefore, to generate random variables for any continuous distribution, we need only generate a random number, u, from a uniform distribution and then solve

$$\int_{-\infty}^{x} f_X(x|\theta)dx = u \text{ for } x,$$

where $f_X(x|\theta)$ is the **PDF** of distribution of interest.

The procedure is illustrated in Figure 5.12. Suppose we want to generate random numbers with a PDF represented by panel (a). Construct the cumulative distribution function, $F(x)$, as shown in panel (b). Generate a sequence of random numbers which have a uniform distribution in the interval 0 to 1. Locate each of these on the y-axis of panel (c) and draw a line parallel to the x-axis to intersect $F(x)$. Drop a perpendicular to the horizontal axis and read off the x value. The distribution of random x values

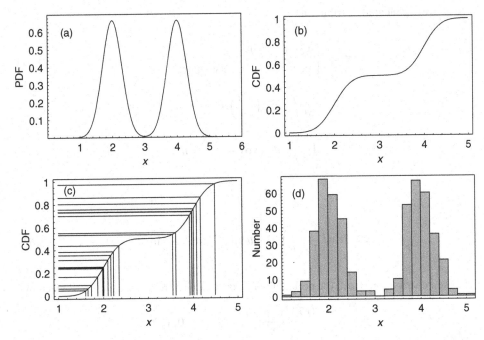

Figure 5.12 This figure illustrates the construction of random numbers which have a distribution corresponding to that shown in panel (a). It makes use of the cumulative distribution function (CDF) shown in panel (b), and a sequence of random numbers that have a uniform distribution in the interval 0 to 1. The construction is illustrated in panel (c) for 20 random numbers. Panel (d) shows a histogram of 500 random numbers generated by this process. See the text for details.

derived in this way will have the desired probability distribution. Panel (c) illustrates this construction for 20 random numbers. A histogram of 500 of these random x values is shown in panel (d).

Examples:

1. Uniform distribution on the interval (a, b):

$$f(x|a, b) = \frac{1}{(b - a)}, (a \leq x \leq b).$$

First, generate a random value of u in the interval $(0, 1)$, equate it to the cumulative distribution function, integrate and solve:

$$(b - a)^{-1} \int_a^x dt = u$$

$$x = u(b - a) + a, \ (a \leq x \leq b).$$

2. Negative exponential distribution:

$$f(x) = \begin{cases} \frac{1}{\theta}\exp(-x/\theta), & \text{for } x > 0, \ \theta > 0 \\ 0, & \text{elsewhere.} \end{cases} \tag{5.61}$$

$$\frac{1}{\theta} \int_0^x \exp\left(-\frac{t}{\theta}\right) dt = u$$

$$\left(\frac{1}{\theta}\right)(-\theta) \exp\left(-\frac{t}{\theta}\right)\Big|_0^x = u$$

$$1 - \exp\left(-\frac{x}{\theta}\right) = u$$

$$x = \theta \ln\left(\frac{1}{1 - u}\right) = -\theta \ln(1 - u) = -\theta \ln(u). \tag{5.62}$$

3. Poisson distribution: recall the probability of exactly x occurrences in a time T is given by

$$p(x|T) = \frac{(rT)^x \exp(-rT)}{x!}, \quad x = 0, 1, 2, \ldots$$

where $r =$ average rate of occurrences and $\lambda = rT$ is the average number of occurrences in time T. Since the time difference between independent Poisson occurrences has a negative exponential distribution, one can generate a random Poisson value by generating successive negative exponential random values using $t = -\theta \ln(u)$. The process continues until the sum of $x + 1$ values of t exceeds the prescribed length T. The Poisson random value, therefore, is x. Recall θ = mean time between Poisson events = $1/r$.

5.13.1 Pseudo-random number generators

Considerable effort has been focused on finding methods for generating uniform distributions of numbers in the range [0,1] (Knuth, 1981; Press *et al.*, 1992). These numbers can then be transformed into other ranges and other types of distributions (e.g., Poisson, normal) as we have seen.

The procedure below illustrates one approach to pseudo-random number generation called *Linear Congruent Generators* (LCG) which generate a sequence of integers $I_1, I_2, I_3, \ldots$ each between 0 and $(m - 1)/m$, where m is a large number, by the following operations:

Step 1: $I_{i+1} = aI_i + c$ where a and c are integers. This generates an upward-going sequence from a seed I_0.
Step 2: Modulus$[I, m] = I - \text{IntegerPart}[I/m] \times m$
e.g., Modulus$[5, 3] = 5 - \text{Integer Part}[5/3] \times 3 = 2$
This reduces the above sequence to a random one with values in the range 0 to $m - 1$ (actually a distribution of round-off errors).
Also written as $I_{i+1} = aI_i + c(\text{Mod } m)$.
Step 3: $U = \frac{1}{m} \text{Modulus}[I, m]$
This gives the desired sequence U between 0 and $(m - 1)/m$.
Notice the smallest difference between terms is $1/m$, which means the numbers a LCG produces comprise a set of m equally spaced rational fractions in the range $0 \le x \le (m - 1)/m$.

Problems:

1. The sequence repeats itself with some period which is $\le m$.
2. For certain choices of parameters, some generators skip many of the possible numbers and give an incomplete set. A series that generates all the m distinct integers $(0 < n < m - 1)$ during each period is called a *full period*.
3. Contains subtle serial correlations. See *Numerical Recipes* (Press *et al.*, 1992) for more details.

Established rules for choosing parameters that give a long and full period are given by Knuth (1981) and by Park and Miller (1988). One way to reduce all of the above problems is to use a compound LCG or shuffling generator which works as follows:

1. Use two LCGs.
2. Use first LCG to generate N lists of random numbers.
3. Use second LCG to calculate a number l between 1 and N, then select top number from lth list (return this number back to the bottom of that list).
4. Period of compound LCG $\approx$ product of periods of individual LCGs.

Box 5.6 *Mathematica* pseudo-random numbers:

We can use *Mathematica* to generate pseudo-random numbers with a wide range of probability distributions. The following command will yield a list of 10 000 uniform random numbers in the interval $0 \rightarrow 1$:

Table[Random[ ],{10000}]

Random uses the Wolfram rule 30 cellular automaton generator for integers (Wolfram, 2002).

To obtain a table of pseudo-random numbers with a Gaussian distribution use the following commands. The first line loads a package containing a wide range of continuous distributions of interest for statistics.

Needs["Statistics `ContinuousDistributions` "]

Table [Random[NormalDistribution[μ, σ]], {10000}]

Mathematica uses the time of day as a seed for random number generation. To ensure you always get the same sequence of pseudo-random numbers, you need to provide a specific seed (e.g., 99) with the command:
SeedRandom[99]

5.13.2 Tests for randomness

Most computers have lurking in their library routines a random number generator typically with the name RAN. $X = \text{RAN(ISEED)}$ is a typical calling sequence. ISEED is some arbitrary initialization value. Any random number generator needs testing before use as the example discussed below illustrates. Four common approaches to testing are:

- Random walk.
- Compare the actual distribution of the pseudo-random numbers to a uniform distribution using a statistical test. Two commonly used frequentist tests are the Kolmogorov–Smirnov test and the χ^2 goodness-of-fit test. The latter is discussed in Section 7.3.1.
- Examine the Fourier spectrum.
- Test for correlations between neighboring random numbers.

Examples of the latter two tests are given in this section.

Panel (a) of Figure 5.13 shows the power spectral density of 262 144 pseudo-random numbers generated using *Mathematica*. The frequency axis is the number of cycles in the 262 144 steps. There do not appear to be any significant peaks indicative of a periodicity. Note: the uniformly distributed pseudo-random numbers in the interval

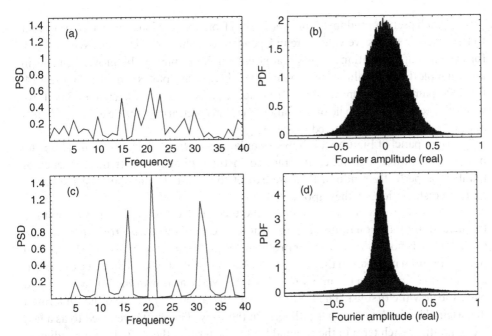

Figure 5.13 Panel (a) shows the power spectral density of 262 144 pseudo-random numbers generated using *Mathematica*. The frequency axis is the number of cycles in the 262 144 steps. Panel (b) shows a histogram of the real part of the Fourier amplitudes. For comparison, panels (c) and (d) demonstrate how sensitive the PSD and Fourier amplitude histogram are to a repeating sequence of random numbers (5.24 cycles of 50 000 random numbers).

$0 \rightarrow 1$ were transformed to be uniform in the interval $-0.5 \rightarrow 0.5$ by subtracting 0.5 from each, so there is no DC (zero frequency) component in the spectrum. Panel (b) shows a histogram of the real part of the Fourier amplitudes which has a Gaussian shape. From the Central Limit Theorem, we expect the histogram to be a Gaussian, since each amplitude corresponds to a weighted sum (weighted by a sine wave) of a very large number of random values.

Panels (c) and (d) demonstrate how sensitive the power spectral density (PSD) and Fourier amplitude histogram are to a repeating sequence of random numbers. Panel (c) shows the PSD for a sequence of 262 144 random numbers consisting of 5.24 cycles of 50 000 random numbers generated with *Mathematica*. Again the frequency axis is the number of cycles in the 262 144 steps. This time, one can clearly see peaks in the PSD at multiples of 5.24, and the histogram has become much narrower. Note: when the sequence is an exact multiple of the repeat period, the vast majority of Fourier amplitudes are zero and the histogram takes on the appearance of a sharp spike or delta function, sitting on a broad plateau. A program to carry out the above calculations can be found in the section of the *Mathematica* tutorial entitled, "Fourier test of random numbers."

Since each pseudo-random number is derived from the previous value, it is important to test whether successive values are independent or exhibit correlations. We can look for evidence of correlations between adjacent random numbers by grouping them in pairs and plotting the value of one against the other. Such a plot is shown in Figure 5.14, for 3000 pairs of random numbers. If adjacent pairs were completely correlated, we would expect the points to lie on a straight line. This is clearly not the case as the points appear to be randomly scattered over the figure.

The right panel of Figure 5.14 shows a similar correlation test involving neighboring points, taken three at a time, and plotted in three dimensions. If the sequence of numbers is perfectly random, then we expect the points to have an approximately uniform distribution, as they appear to do.

It is possible to extend and quantify these correlation tests. The most common frequentist tool for quantifying correlation is called the *autocorrelation function* (ACF). Here is how it works for our problem: let $\{x_i\}$ be a list of uniformly distributed pseudo-random numbers in the interval 0 to 1. Now subtract the mean value of the list, $\bar{x}$, to obtain a new list in the interval -0.5 to 0.5. Make a copy of the list $\{x_i - \bar{x}\}$ and place it below the first list. Then shift the copy to the left by j terms so the ith term in the original list is above the $(i+j)$th term in the copy. This shift is referred to as a lag. Next, multiply each term in the original list by the term in the shifted list immediately below, and compute the average of these products (for all terms that overlap in the two lists), which we designate $\rho(j)$. We can repeat this process and compute $\rho(j)$ for a wide range of lags, ranging from $j = 0$ to some large value. If the numbers in the list are

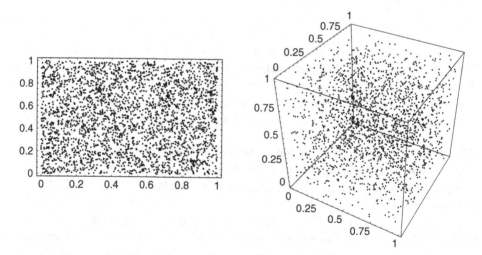

Figure 5.14 The left panel shows a correlation test between successive pairs of random numbers, obtained from the *Mathematica* pseudo-random number generator. The coordinates of each point are given by the pair of random numbers. The right panel shows a three-dimensional correlation test involving successive random numbers taken three at a time. The right panel was plotted with the *Mathematica* command **ScatterPlot3D**.

truly random then any term in the original list will be completely independent of any term in the shifted copy. So each multiplication is equally likely to be a positive or negative quantity. The average of a large number of random positive and negative quantities tends to zero. Of course for $j = 0$ (no shift), the two terms are identical so the products are all positive quantities and there is no cancellation. Thus, for a list of completely random numbers, a plot of the ACF, $\rho(j)$, will look like a spike at $j = 0$ and be close to zero for all $j \geq 1$. If the terms are not completely independent, then we expect the plot of $\rho(j)$ to decay gradually towards zero over a range of j values.

The formula for $\rho(j)$ given below differs slightly from the operation just described, in that instead of computing the average, we sum the product terms for each j and then normalize by dividing by

$$\sqrt{\sum_{\text{overlap}} (x_i - \bar{x})^2} \times \sqrt{\sum_{\text{overlap}} (x_{i+j} - \bar{x})^2}.$$

With this normalization, the maximum value of the ACF is 1.0 and it allows the ACF to handle a wider variety of correlation problems than the particular one we are interested in here:

$$\rho(j) = \frac{\sum_{\text{overlap}}[(x_i - \bar{x})(x_{i+j} - \bar{x})]}{\sqrt{\sum_{\text{overlap}}(x_i - \bar{x})^2} \times \sqrt{\sum_{\text{overlap}}(x_{i+j} - \bar{x})^2}}, \tag{5.63}$$

where the summation is carried out over the subset of samples that overlap. Figure 5.15 shows a plot of the ACF for a sequence of 10 000 uniformly distributed

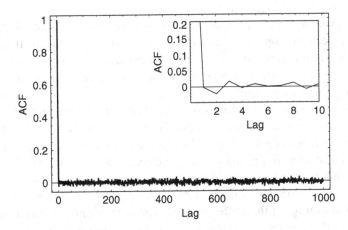

Figure 5.15 The autocorrelation function (ACF) for a sequence of 10 000 uniformly distributed pseudo-random numbers generated by the *Mathematica* **Random** command. The larger plot spans a range of 1000 lags, while the blow-up in the corner shows the first ten lags. Clearly, for lags ≥ 1, the ACF is essentially zero indicating no detectable correlation.

pseudo-random numbers generated by the *Mathematica* **Random** command. The larger plot spans a range of 1000 lags, while the blow-up in the corner shows the first ten lags. Clearly, for lags ≥ 1, the ACF is essentially zero indicating no detectable correlation. The noise-like fluctuations for $j \geq 1$ arise because of incomplete cancellation of the product terms for a finite list of random numbers.

For quite a few years, RAN3 in *Numerical Recipes* (not an LCG but based on a subtractive method) was considered a reliable portable random number generator, but even this has been called into question (see Barber *et al.*, 1985; Vattulainen *et al.*, 1994; Maddox, 1994; and Fernandez and Rivero, 1996). However, what is random enough for one application may not be random enough for another. In the near future, random number generators based upon a physical process, like Johnson noise from a resistor or a reverse-biased zener diode, will be incorporated into every computer. Intel already supplies such devices on some chipsets for PC-type computers. One can anticipate that users of these hardware-derived random numbers will again be concerned with just how random these numbers are.

5.14 Summary

The most important aspect of frequentist statistics is the process of drawing conclusions based on sample data drawn from the population (which is the collection of all possible samples). The use of the term random variable conveys the idea of an intrinsic uncertainty in the measurement characterized by an underlying population. A random variable is not the particular number recorded in one measurement, but rather, it is an abstraction of the measurement operation or observation that gives rise to that number, e.g., X may represent the random variable and x the realization of the random variable in one measurement.

To the frequentist, the sampling distribution is a model of the probability distribution of the underlying population from which the sample was taken. In a Bayesian analysis, the sampling distribution is a mathematical description of the uncertainty in predicting the data for any particular model because of incomplete information. It enables us to compute the likelihood $p(D|H, I)$.

We considered various descriptive properties of probability distributions: moments, moment generating functions (useful in proofs of important theorems) and measures of the central tendency of a distribution (mode, median and mean). This was followed by a discussion of some important discrete and continuous probability distributions.

The most important probability distribution is the Gaussian or normal distribution. This is because a measured quantity is often the result of a large number of effects, i.e., is some kind of average of these effects. According to the Central Limit Theorem, the distribution of the average of a collection of random variables tends to a Gaussian. We also learned that in most circumstances, the distribution of the sample average sharpens around the sample mean as the sample size increases, which is the basis of signal averaging that plays such an important role in experimental work.

Finally, we learned how to generate pseudo-random samples from an arbitrary probability distribution, a topic that is of great importance in experimental simulations and Monte Carlo techniques.

5.15 Problems

1. Write a small program to reproduce Figure 5.5 starting from the raw data "badbase.dat" (supplied with the *Mathematica* tutorial) and using 21 points for the running median and mean. For the first and last ten points of the data, just subtract the average of the ten from each point.
2. When sampling from a normal distribution with mean $\mu = 2$ and $\sigma = 1$, compute $P(\mu - 2.7\sigma \leq X \leq \mu + 2.7\sigma)$.
3. As one test of your pseudo-random number generator, generate a sample of 50 000 random numbers in the interval 0 to 1 with a uniform distribution. Compare the mean and variance of the sequence to that expected for a uniform distribution. Does the sample mean agree with the expected mean to within one standard error of the sample mean?
4. Generate 10 000 pseudo-random numbers with a beta distribution with $\alpha = 2$ and $\beta = 4$. See the *Mathematica* example in Section 5.13.1 of the book and use **BetaDistribution** instead of **NormalDistribution**. Plot a histogram of your random numbers, and on the same plot, overlay a beta distribution for comparison. Compute the mean and median of your simulated data set. See the **BinCounts**, **Mean**, and **Median** commands in *Mathematica*.
5. Let $X_1, X_2, X_3, \ldots, X_n$ be n independent and identically distributed (IID) random variables with a beta PDF given by

$$f(x) \equiv f(x|\alpha, \beta) = \begin{cases} \dfrac{\Gamma(\alpha + \beta)}{\Gamma(\alpha)\Gamma(\beta)} x^{\alpha-1}(1 - x)^{\beta-1}, & \begin{array}{l} \text{for } 0 < x < 1 \\ \alpha, \beta > 0, \end{array} \\ 0, & \text{elsewhere,} \end{cases} \tag{5.64}$$

where $\alpha = 2$ and $\beta = 4$.
What is the probability density function (PDF) of $\overline{X}$, the average of n measurements? As an alternative to averaging large numbers of samples (simplest approach) you could make use of the convolution theorem and the Fast Fourier Transform (FFT) (remember to zero pad). Note: if you are not familiar with using the FFT and zero padding, you will find this approach much more challenging. Note: the Discrete Fourier Transform, the FFT and zero padding are discussed in Appendix B.

   a) By way of an answer, plot the PDFs for $n = 1, 3, 5, 8$ and display all four distributions on the same plot. Be careful to normalize each distribution for unit area.

b) Compute the mean and variance of the four PDFs (do not simply quote an expected theoretical value) for each value of n.

c) Compare your result for the $n = 5$ case to a Gaussian with the same mean and variance drawn on the same graph. Repeat for $n = 8$. What conclusions do you draw?

6

What is a statistic?

6.1 Introduction

In this chapter, we address the question "What is a statistic"? In particular, we look at what role statistics play in scientific inference and give some common useful examples. We will examine their role in the two basic inference problems: hypothesis testing (the frequentist equivalent of model selection) and parameter estimation, with emphasis on the latter. Hypothesis testing will be dealt with in Chapter 7.

Recall that an important aspect of frequentist statistical inference is the process of drawing conclusions based on sample data drawn from the population (which is the collection of all possible samples). The concept of the population assumes that in principle, an infinite number of measurements (under identical conditions) are possible. Suppose $X_1, X_2, \ldots, X_n$ are n independent and identically distributed (IID) random variables that constitute a random sample from the population for which $x_1, x_2, \ldots, x_n$ is one realization. The population is assumed to have an intrinsic probability distribution (or density function) which, if known, would allow us to predict the likelihood of the sample $x_1, x_2, \ldots, x_n$.

For example, suppose the random variable we are measuring is the time interval between successive decays of a radioactive sample. In this case, the population probability density function is a negative exponential (see Section 5.8.5), given by $f(x|\theta) = [\exp(-x/\theta)]/\theta$. The likelihood is given by

$$\mathcal{L}(x_1, x_2, \ldots, x_n|\theta) = \prod_{i=1}^{n} f(x_i|\theta). \tag{6.1}$$

This particular population probability density function is characterized by a single parameter θ. Another population probability distribution that arises in many problems is the normal (Gaussian) distribution which has two parameters, μ and σ^2.

In most problems, the parameters of the underlying population probability distribution are not known. Without knowledge of their values, it is impossible to compute the desired probabilities. However, a population parameter can be estimated from a *statistic*, which is determined from the information contained in a random sample. It is for this reason that the notion of a statistic and its sampling distribution is so important in statistical inference.

Definition: A *statistic* is any function of the observed random variables in a sample such that the function does not contain any unknown quantities.

One important statistic is the sample mean $\overline{X}$ given by

$$\overline{X} = (X_1 + X_2 + \cdots + X_n)/n = \frac{1}{n}\sum_{i=1}^{n} X_i. \tag{6.2}$$

Note: we are using a capital $\overline{X}$ which implies we are talking about a random variable. All statistics are random variables and to be useful, we need to be able to specify their sampling distribution.

For example, we might be interested in the mean redshift[1] of a population of cosmic gamma-ray burst (GRB) sources. This would provide information about the distances of these objects and their mean energy. GRBs are the most powerful type of explosion known in the universe. The parameter of interest is the mean redshift which we designate μ. A parameter of a population is always regarded as a fixed and usually unknown constant. Let Z be a random variable representing GRB redshifts. Suppose the redshifts, $\{z_1, z_2, \ldots, z_7\}$, of a sample of seven GRB sources are obtained after a great deal of effort. What can we conclude about the population mean redshift μ from our sample, i.e., how accurately can we determine μ from our sample?

This can be a fairly difficult question to answer using the individual measurements, z_i, because we don't know the form of the sampling distribution for GRB source redshifts. Happily, in this case, we can proceed with our objective by exploiting the Central Limit Theorem (CLT) which predicts the sampling distribution of the *sample mean statistic*. The way to think about this is as follows: consider a thought experiment in which we are able to obtain redshifts for a very large number of samples (hypothetical reference set) of GRB redshifts. Each sample consists of seven redshift measurements. The means of all these samples will have a distribution. According to the CLT, the distribution of sample means tends to a Gaussian as the number n of observations tends to infinity. In practice, a Gaussian sampling distribution is often employed when $n \geq 5$. Of course, we don't have the results from this hypothetical reference set, only the results from our one sample, but at least we know that the shape of the sampling distribution characterizing our sample mean statistic is approximately a Gaussian. This allows us to make a definite statement about the uncertainty in the population mean redshift μ which we derive from our one sample of seven redshift measurements. Just how we do this is discussed in detail in Section 6.6.2. In the course of answering that question, we will encounter the sample variance statistic, S^2, and develop the notion of a sampling distribution of a statistic.

---

[1] Redshift is a measure of the wavelength shift produced by the Doppler effect. In 1929, Edwin Hubble showed that we live in an expanding universe in which the velocity of recession of a galaxy is proportional to its distance. A recession velocity shifts the observed wavelength of a spectral line to longer wavelengths, i.e., to the red end of the optical spectrum.

6.2 The χ^2 distribution

The *sampling distribution* of any particular statistic is the probability distribution of that statistic that would be determined from an infinite number of independent samples, each of size n, from an underlying population. We start with a treatment of the χ^2 sampling distribution.[2] We will prove in Section 6.3 that the χ^2 distribution describes the distribution of the variances of samples taken from a normal distribution. The χ^2 distribution is a special case of the gamma distribution:

$$f(x|\alpha, \theta) = \frac{1}{\Gamma(\alpha)\theta^\alpha} x^{\alpha-1} \exp\left(\frac{-x}{\theta}\right) \tag{6.3}$$

with $\theta = 2$ and $\alpha = \nu/2$, where ν is called the *degree of freedom*.

The χ^2 distribution has the following properties:

$$f(x|\nu) = \frac{1}{\Gamma\left(\frac{\nu}{2}\right)2^{\frac{\nu}{2}}} x^{\frac{\nu}{2}-1} \exp\left(-\frac{x}{2}\right) \tag{6.4}$$

$$\langle x \rangle = \nu; \qquad \text{Var}[x] = 2\nu. \tag{6.5}$$

The coefficients of skewness (α_3) and kurtosis (α_4) are given by

$$\alpha_3 = \frac{4}{\sqrt{2\nu}}; \qquad \alpha_4 = 3\left(1 + \frac{4}{\nu}\right). \tag{6.6}$$

Finally, the *moment generating function* of χ_ν^2 with ν degrees of freedom is given by

$$m_{\chi_\nu^2}(t) = (1 - 2t)^{-\frac{\nu}{2}}. \tag{6.7}$$

We now prove two useful theorems pertaining to the χ^2 distribution.

Theorem 1:
Let $\{X_i\} = X_1, X_2, \ldots, X_n$ be an IID sample from a normal distribution $N(\mu, \sigma)$. Let $Y = \sum_{i=1}^{n}(X_i - \mu)^2/\sigma^2 = \sum_{i=1}^{n} Z_i^2$, where Z_i are standard random variables. Then Y has a chi-squared (χ_n^2) distribution with n degrees of freedom.

Proof:
Let $m_Y(t) = $ the moment generating function (recall Section 5.6) of Y. From Equation (5.9), we can write

$$\begin{aligned} m_Y(t) &= \langle e^{tY} \rangle = \langle e^{t\sum_i Z_i^2} \rangle \\ &= \langle e^{tZ_1^2} \times e^{tZ_2^2} \times \cdots \times e^{tZ_n^2} \rangle. \end{aligned} \tag{6.8}$$

---

[2] The χ^2 statistic plays an important role in fitting models to data using the least-squares method, which is discussed in great detail in Chapters 10 and 11.

Since the random variable Z is IID then,

$$m_Y(t) = \langle e^{tZ_1^2} \rangle \times \langle e^{tZ_2^2} \rangle \times \cdots \times \langle e^{tZ_n^2} \rangle$$
$$= m_{Z_1^2}(t) \times m_{Z_2^2}(t) \times \cdots \times m_{Z_n^2}(t). \tag{6.9}$$

The moment generating function for each Z_i is given by

$$m_{Z^2}(t) = \int_{-\infty}^{+\infty} f(z) \exp(tZ^2) dZ$$
$$= \frac{1}{\sqrt{2\pi}} \int \exp(tZ^2) \exp\left(\frac{-Z^2}{2}\right) dZ \tag{6.10}$$
$$= \frac{1}{\sqrt{2\pi}} \int \exp\left[\frac{-Z^2}{2}(1 - 2t)\right] dZ,$$

where we have made use of the fact that $f(z)$ is also a normal distribution, i.e., a Gaussian.

Multiplying and dividing Equation (6.10) by $(1 - 2t)^{-\frac{1}{2}}$ we get

$$m_{Z^2}(t) = (1 - 2t)^{-\frac{1}{2}} \underbrace{\int_{-\infty}^{+\infty} \frac{1}{\sqrt{2\pi}(1 - 2t)^{-\frac{1}{2}}} \exp\left[\frac{-Z^2}{2(1 - 2t)^{-1}}\right] dZ}_{\text{Integral of normal distribution} = 1} \tag{6.11}$$

$$\Rightarrow m_{Z^2}(t) = (1 - 2t)^{-\frac{1}{2}}.$$

Therefore,

$$m_Y(t) = (1 - 2t)^{-\frac{n}{2}}. \tag{6.12}$$

Comparison of Equations (6.12) and (6.7) shows that Y has a χ^2 distribution, with n degrees of freedom, which we designate by χ_n^2. Figure 6.1 illustrates the χ^2 distribution for three different choices of the number of degrees of freedom.

Example:
In Section 5.9, we showed that for any IID sampling distribution with a finite variance, $(\overline{X} - \mu)\sqrt{n}/\sigma$ tends to $N(0, 1)$ as $n \to \infty$, and therefore $[(\overline{X} - \mu)\sqrt{n}/\sigma]^2$ is approximately χ_1^2 with one degree of freedom.[3]

---

[3] When sampling from a normal distribution, the distribution of $(\overline{X} - \mu)\sqrt{n}/\sigma$ is always $N(0, 1)$ regardless of the value of n.

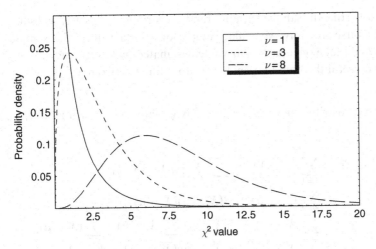

Figure 6.1 The χ^2 distribution for three different choices of the number of degrees of freedom.

Theorem 2:
If X_1 and X_2 are two independent χ^2-distributed random variables with ν_1 and ν_2 degrees of freedom, then $Y = X_1 + X_2$ is also χ^2-distributed with $\nu_1 + \nu_2$ degrees of freedom.

Proof:
Since X_1 and X_2 are independent, the moment generating function of Y is given by

$$m_y(t) = m_{X_1}(t) \times m_{X_2}(t) = (1 - 2t)^{-\frac{\nu_1}{2}} \times (1 - 2t)^{-\frac{\nu_2}{2}} \qquad (6.13)$$

$$= (1 - 2t)^{-\frac{(\nu_1 + \nu_2)}{2}}, \qquad (6.14)$$

which equals the moment generating function of a χ^2 random variable with $\nu_1 + \nu_2$ degrees of freedom.

6.3 Sample variance S^2

We often want to estimate the variance (σ^2) of a population from an IID sample taken from a normal distribution. We usually don't know the mean (μ) of the population so we use the sample mean $(\overline{X})$ as an estimate. To estimate σ^2 we use another random variable called the *sample variance* (S^2), defined as follows:

$$S^2 = \sum_{i=1}^{n} \frac{(X_i - \overline{X})^2}{n - 1}. \qquad (6.15)$$

Just why we define the sample variance random variable in this way will soon be made clear. Of course, for any particular sample of n data values, the sample random variable would take on a particular value designated by lower case s^2.

Here is a useful theorem that enables us to estimate σ from S:

Theorem 3:

The sampling distribution of $(n-1)S^2/\sigma^2$ is χ^2 with $(n-1)$ degrees of freedom.

Proof:

$$\begin{aligned}
(n-1)S^2 &= \sum_{i=1}^{n}(X_i - \overline{X})^2 = \sum[(X_i - \mu) - (\overline{X} - \mu)]^2 \\
&= \sum[(X_i - \mu)^2 - 2(X_i - \mu)(\overline{X} - \mu) + (\overline{X} - \mu)^2] \\
&= \sum(X_i - \mu)^2 - 2(\overline{X} - \mu)\sum(X_i - \mu) + \sum(\overline{X} - \mu)^2 \qquad (6.16) \\
&= \sum(X_i - \mu)^2 - 2(\overline{X} - \mu)n(\overline{X} - \mu) + n(\overline{X} - \mu)^2 \\
&= \sum[(X_i - \mu)^2] - n(\overline{X} - \mu)^2.
\end{aligned}$$

Therefore,

$$\frac{(n-1)S^2}{\sigma^2} + \underbrace{\frac{(\overline{X} - \mu)^2}{\sigma^2/n}}_{\chi_1^2} = \underbrace{\sum_{i=1}^{n}\frac{(X_i - \mu)^2}{\sigma^2}}_{\chi_n^2}. \qquad (6.17)$$

From Theorem 2, $(n-1)S^2/\sigma^2$ is χ_{n-1}^2 with $(n-1)$ degrees of freedom.

The expectation value of a quantity that has a χ^2 distribution with $(n-1)$ degrees of freedom is equal to the number of degrees of freedom (see Equation (6.5)). Therefore,

$$\left\langle \frac{(n-1)S^2}{\sigma^2} \right\rangle = n - 1. \qquad (6.18)$$

But,

$$\left\langle \frac{(n-1)S^2}{\sigma^2} \right\rangle = \langle S^2 \rangle \frac{(n-1)}{\sigma^2} = n - 1. \qquad (6.19)$$

Therefore,

$$\langle S^2 \rangle = \sigma^2. \qquad (6.20)$$

This provides justification for our definition of S^2 – its expectation value is the population variance. Note: this does not mean that S^2 will equal σ^2 for any particular sample.

Note 1: We have just established Equation (6.20) when sampling for a normal distribution. We now show that Equation (6.20) is valid for IID sampling from any arbitrary distribution with finite variance. From Equation (6.16), we can write

$$\langle S^2 \rangle = \frac{\sum \langle (X_i - \mu)^2 \rangle - n \langle (\overline{X} - \mu)^2 \rangle}{n - 1}. \tag{6.21}$$

But $\langle (X_i - \mu)^2 \rangle = \text{Var}(X_i) = \sigma^2$ by definition, and $\langle (\overline{X} - \mu)^2 \rangle = \text{Var}(\overline{X}) = \sigma^2/n$ from Equation (5.50). It follows that

$$\langle S^2 \rangle = \frac{n\sigma^2 - \sigma^2}{n - 1} = \sigma^2. \tag{6.22}$$

Thus, Equation (6.22) is valid for IID sampling from an arbitrary distribution with finite variance. In the language of frequentist statistics, we say that S^2, as defined in Equation (6.15), is an unbiased estimator of σ^2.

Standard error of the sample mean: We often want to quote a typical error for the mean of a population based on our sample. According to Equation (5.50), $\text{Var}(\overline{X}) = \sigma^2/n$ for any distribution with finite variance. Since we do not normally know σ^2, the variance of the population, we use the sample variance as an estimate.

The *standard error of the sample mean* is defined as $\dfrac{S}{\sqrt{n}}$. $\tag{6.23}$

In Section 6.6.2 we will use a Student's t distribution to be more precise about specifying the uncertainty in our estimate of the population mean from the sample mean.

Note 2: In a situation where we know population μ but not σ^2, define S^2:

$$S^2 = \sum_{i=1}^{n} \frac{(X_i - \mu)^2}{n}. \tag{6.24}$$

It is easily shown that with this definition, nS^2/σ^2 is χ_n^2 with n degrees of freedom. We lose one degree of freedom when we estimate μ from $\overline{X}$.

Example:
A random sample of size $n = 16$ (IID sample) is drawn from a population with a normal distribution of unknown mean (μ) and variance (σ^2). We compute the sample variance, S^2, and want to determine

$$p(\sigma^2 < 0.49 S^2). \tag{6.25}$$

Solution: Equation (6.25) is equivalent to

$$p\left(\frac{S^2}{\sigma^2} > 2.041 \right). \tag{6.26}$$

We know that the random variable $X = (n-1)S^2/\sigma^2$ has a χ^2 distribution with $(n-1)$ degrees of freedom. In this case, $(n-1) = 15 = \nu$ degrees of freedom. Therefore,

$$p\left(\frac{S^2}{\sigma^2} > 2.041\right) = p\left((n-1)\frac{S^2}{\sigma^2} > 30.61\right). \tag{6.27}$$

Let

$$\alpha = p((n-1)S^2/\sigma^2 > 30.61).$$

Then

$$1 - \alpha = p((n-1)S^2/\sigma^2 \le 30.61),$$

or more generally, $1 - \alpha = p(X \le x_{1-\alpha})$ where $x_{1-\alpha}$ is the particular value of the random variable X for which the cumulative distribution $p(X \le x_{1-\alpha}) = 1 - \alpha$. $x_{1-\alpha}$ is called the $(1 - \alpha)$ *quantile value* of the distribution, and $p(X \le x_{1-\alpha}|\nu)$ is given by

$$p(X \le x_{1-\alpha}|\nu) = \frac{1}{\Gamma(\frac{\nu}{2})2^{\frac{\nu}{2}}} \int_0^{x_{1-\alpha}} t^{\frac{\nu}{2}-1} \exp\left(-\frac{t}{2}\right) dt = 1 - \alpha. \tag{6.28}$$

For $\nu = 15$ degrees of freedom, 30.61 corresponds to $\alpha = 0.01$ or $x_{0.990}$. Thus, the probability that $\sigma^2 < 0.49S^2 = 1\%$. Figure 6.2 shows the χ^2 distribution for $\nu = 15$ degrees of freedom and the $1 - \alpha = 0.99$ quantile value.

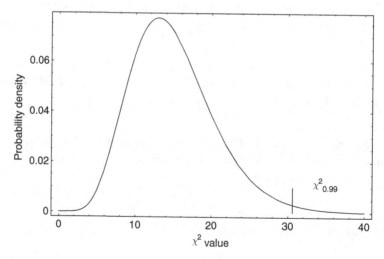

Figure 6.2 The χ^2 distribution for $\nu = 15$ degrees of freedom. The vertical line marks the $1 - \alpha = 0.99$ quantile value. The area to the left of this line corresponds to $1 - \alpha$.

We can evaluate Equation (6.28) with the following *Mathematica* command:

Box 6.1 *Mathematica* χ^2 significance

Needs ["Statistics `ContinuousDistributions` "]
The line above loads a package containing a wide range of continuous distributions of importance to statistics, and the following line computes α, the area in the tail of the χ^2 distribution to the right of $\chi^2 = 30.61$, for $\nu = 15$ degrees of freedom.

$$\textbf{GammaRegularized}\left[\frac{\nu}{2}, \frac{\chi^2}{2}\right] = \textbf{GammaRegularized}\left[\frac{15}{2}, \frac{30.61}{2}\right] = 0.01$$

In statistical hypothesis testing (to be discussed in the next chapter), α is referred to as the *significance* or the *one-sided P-value* of a statistical test.

6.4 The Student's *t* distribution

Recall, when sampling from a normal distribution with known standard deviation, σ, the distribution of the standard random variable $Z = (\overline{X} - \mu)\sqrt{n}/\sigma$ is $N(0, 1)$. In practice, σ is usually not known. The logical thing to do is to replace σ by the sample standard deviation S. The usual inference desired is that there is a specified probability that $\overline{X}$ lies within $\pm S$ of the true mean μ.

Unfortunately, the distribution of $(\overline{X} - \mu)\sqrt{n}/S$ is not $N(0, 1)$. However, it is possible to determine the exact sampling distribution of $(\overline{X} - \mu)\sqrt{n}/S$ when sampling from $N(\mu, \sigma)$ with both μ and σ^2 unknown. To this end, we examine the *Student's t distribution*.[4] The following useful theorem pertaining to the Student's *t* distribution is given without proof.

Theorem 4:
Let Z be a standard normal random variable and let X be a χ^2 random variable with ν degrees of freedom. If Z and X are independent, then the random variable

$$T = \frac{Z}{\sqrt{X/\nu}} \tag{6.29}$$

has a Student's *t* distribution with ν degrees of freedom and a probability density given by

$$f(t|\nu) = \frac{\Gamma[\frac{\nu+1}{2}]}{\sqrt{\pi\nu}\,\Gamma(\frac{\nu}{2})}\left[1 + \left(\frac{t^2}{\nu}\right)\right]^{-\frac{(\nu+1)}{2}}; \quad (-\infty < t < +\infty), \nu > 0. \tag{6.30}$$

---

[4] The *t* distribution is named for its discoverer, William Gosset, who wrote a number of statistical papers under the pseudonym "Student." He worked as a brewer for the Guinness brewery in Dublin in 1899. He developed the *t* distribution in the course of analyzing the variability of various materials used in the brewing process.

The Student's t distribution has the following properties:

$$\langle T \rangle = 0 \quad \text{and} \quad \text{Var}(T) = \frac{\nu}{(\nu - 2)} \quad \nu > 2. \tag{6.31}$$

When sampling $N(\mu, \sigma)$ we know that $(\overline{X} - \mu)\sqrt{n}/\sigma$ is $N(0, 1)$. We also know that $(n - 1)S^2/\sigma^2$ is χ^2 with $(n - 1)$ degrees of freedom. Therefore, we can identify Z with $(\overline{X} - \mu)\sqrt{n}/\sigma$ and X with $(n - 1)S^2/\sigma^2$. Therefore,

$$T = \frac{\frac{(\overline{X} - \mu)}{(\sigma/\sqrt{n})}}{\sqrt{\frac{(n-1)S^2/\sigma^2}{n-1}}} = \frac{(\overline{X} - \mu)}{(\sigma/\sqrt{n})} \frac{\sigma}{S} = \frac{(\overline{X} - \mu)}{(S/\sqrt{n})}. \tag{6.32}$$

Therefore, $(\overline{X} - \mu)\sqrt{n}/S$ is a random variable with a Student's t distribution with $n - 1$ degrees of freedom. Figure 6.3 shows a comparison of a Student's t distribution for three degrees of freedom, and a standard normal. The broader wings of the Student's t distribution are clearly evident.

The $(1 - \alpha)$ quantile value for ν degrees of freedom, $t_{1-\alpha,\nu}$, is given by

$$p(T \leq t_{1-\alpha,\nu}) = \frac{\Gamma[\frac{(\nu+1)}{2}]}{\sqrt{\pi\nu}\,\Gamma(\frac{\nu}{2})} \int_{-\infty}^{t_{1-\alpha,\nu}} \left[1 + \left(\frac{t^2}{\nu}\right)\right]^{-\frac{(\nu+1)}{2}} dt = 1 - \alpha. \tag{6.33}$$

Example:

Suppose a cigarette manufacturer claims that one of their brands has an average nicotine content of 0.6 mg per cigarette. An independent testing organization

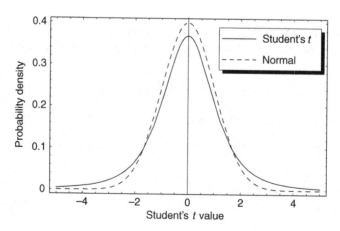

Figure 6.3 Comparison of a standard normal distribution and a Student's t distribution for 3 degrees of freedom.

measures the nicotine content of 16 such cigarettes and has determined the sample average and the sample standard deviation to be 0.75 and 0.197 mg, respectively. If we assume the amount of nicotine is a normal random variable, how likely is the sample result given the manufacturer's claim?

$T = (\overline{X} - \mu)\sqrt{n}/S$ has a Student's t distribution.
$\overline{x} = 0.75$ mg, $s = 0.197$ mg, and $n = 16$,
so the number of degrees of freedom $\nu = 15$.

Manufacturer claims $\mu = 0.6$ mg corresponds to a

$$t = \frac{(0.75 - 0.6)}{0.197/\sqrt{16}} = 3.045. \qquad (6.34)$$

The Student's t distribution is a continuous distribution, and thus we cannot calculate the probability of any specific t value since there is no area under a point. The question of how likely the t value is, given the manufacturer's claim, is usually interpreted as what is the probability by chance that $T \geq 3.045$. The area of the distribution beyond the sample t value gives us a measure of how far out in the tail of the distribution the sample value resides.

Box 6.2   *Mathematica* solution:

We can solve the above problem with the following commands:

Needs["Statistics `ContinuousDistributions` "]

The following line computes the area in the tail of the
T distribution beyond $T = 3.045$.
(1 − CDF[StudentTDistribution[ν], 3.045]) → answer $= 0.004$ ($\nu = 15$)

where **CDF[StudentTDistribution [ν], 3.045]** stands for the *cumulative density function* of the T distribution from $T = -\infty \rightarrow 3.045$.

Therefore, $p(T > 3.045) = \alpha = 0.004$ or 0.4%, i.e., the manufacturer's claim is very improbable. The way to think of this is to imagine we could repeatedly obtain samples of 16 cigarettes and compute the value of t for each sample. The fraction of these t values that we would expect to fall in the tail area beyond $t > 3.045$ is only 0.4%. If the manufacturer's claim were reasonable, we would expect that the t value of our actual sample would not fall so far out in the tail of the distribution. If you are still puzzled by this reasoning, we will have a lot more to say about it in Chapter 7. We will revisit this example in Section 7.2.3.

Note: although $(\overline{x} - \mu)/s = 0.15/0.197 < 1$, s is not a meaningful uncertainty for $\overline{x}$ – only for x_i. The usual measure of the uncertainty in $\overline{x}$ is $s/\sqrt{n} = 0.049$. The quantity $s/\sqrt{n}$ is called the *standard error of the sample mean*.

6.5 F distribution (F-test)

The F distribution is used to find out if two data sets have significantly different variances. For example, we might be interested in the effect of a new catalyst in the brewing of beer so we compare some measurable property of a sample brewed with the catalyst to a sample from the control batch made without the catalyst. What effect has the catalyst had on the variance of this property?

Here, we develop the appropriate random variable for use in making inferences about the variances of two independent normal distributions based on a random sample from each. Recall that inferences about σ^2, when sampling from a normal distribution, are based on the random variable $(n-1)S^2/\sigma^2$, which has a χ^2_{n-1} distribution.

Theorem 5:

Let X and Y be two independent χ^2 random variables with ν_1 and ν_2 degrees of freedom. Then the random variable

$$F = \frac{X/\nu_1}{Y/\nu_2} \tag{6.35}$$

has an F distribution with a probability density function

$$p(f\,|\nu_1,\nu_2) = \begin{cases} \frac{\Gamma[(\nu_1+\nu_2)/2]}{\Gamma(\nu_1/2)\Gamma(\nu_2/2)} \left(\frac{\nu_1}{\nu_2}\right)^{\frac{\nu_1}{2}} \frac{f^{\frac{1}{2}(\nu_1-2)}}{(1+f\nu_1/\nu_2)^{\frac{1}{2}(\nu_1+\nu_2)}}, & (f>0) \\ 0, & \text{elsewhere.} \end{cases} \tag{6.36}$$

An F distribution has the following properties:

$$\langle F \rangle = \frac{\nu_2}{\nu_2-2}, \quad (\nu_2 > 2). \tag{6.37}$$

(Surprisingly, $\langle F \rangle$ depends only on ν_2 and not on ν_1.)

$$\text{Var}(F) = \frac{\nu_2^2(2\nu_2+2\nu_1-4)}{\nu_1(\nu_2-1)^2(\nu_2-4)}, \quad (\nu_2 > 4) \tag{6.38}$$

$$\text{Mode} = \frac{\nu_2(\nu_1-2)}{\nu_1(\nu_2+2)}. \tag{6.39}$$

Let $X = (n_1-1)S_1^2/\sigma_1^2$ and $Y = (n_2-1)S_2^2/\sigma_2^2$. Then,

$$F_{12} = \frac{X/\nu_1}{Y/\nu_2} = \frac{X/(n_1-1)}{Y/(n_2-1)} = \frac{S_1^2/\sigma_1^2}{S_2^2/\sigma_2^2}. \tag{6.40}$$

Box 6.3 *Mathematica* **example:**

The sample variance is $s_1^2 = 16.65$ for $n_1 = 6$ IID samples from a normal distribution with a population variance σ_1^2, and $s_2^2 = 5.0$ for $n_2 = 11$ IID samples from a second independent normal distribution with a population variance σ_2^2. If we assume that $\sigma_1^2 = \sigma_2^2$, then from Equation (6.40), we obtain $f = 3.33$ for $\nu_1 = n_1 - 1 = 5$ and $\nu_2 = n_2 - 1 = 10$ degrees of freedom. What is the probability of getting an f value ≥ 3.33 by chance if $\sigma_1^2 = \sigma_2^2$?

Needs["Statistics `ContinuousDistributions` "]

The following line computes the area in the tail of the F distribution beyond $f = 3.33$.

(1 − CDF[FRatioDistribution[ν_1, ν_2], 3.33]) → answer = 0.05

where **CDF[FRatioDistribution[ν_1, ν_2], 3.33]** stands for the cumulative density function of the F distribution from $f = 0 \to 3.33$. Another way to compute this tail area is with
FRatioPValue[fratio, ν_1, ν_2]

The F distribution for this example is shown in Figure 6.4.

What if we had labeled our two measurements of s the other way around so $\nu_1 = 10, \nu_2 = 5$ and $s_1^2/s_2^2 = 1/3.33$? The equivalent question is: what is the probability that $f \leq 1/3.33$ which we can evaluate by

CDF[FRatioDistribution [ν_1, ν_2], 1/3.33]? Answer : 0.05

Not surprisingly, we obtain the same probability.

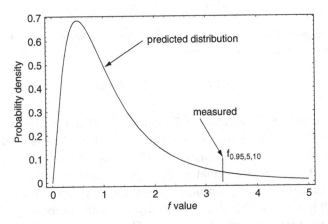

Figure 6.4 The F distribution for $\nu_1 = 5, \nu_2 = 10$ degrees of freedom. The measured value of 3.33, indicated by the line, corresponds to $f_{0.95,5,10}$, the 0.95 quantile value.

6.6 Confidence intervals

In this section, we consider how to specify the uncertainty of our estimate of any particular parameter of the population, based on the results of our sample. We start by considering the uncertainty in the population mean μ when it is known that we are sampling from a population with a normal distribution. There are two cases of interest. In the first, we will assume that we know the variance σ^2 of the underlying population we are sampling from. More commonly, we don't know the variance and must estimate it from the sample. This is the second case.

6.6.1 Variance σ^2 known

Let $\{X_i\}$ be an IID $N(\mu, \sigma^2)$ random sample of $n = 10$ measurements from a population with unknown μ but known $\sigma = 1$. Let $\overline{X}$ be the sample mean random variable which will have a sample mean standard deviation, $\sigma_m = \sigma/\sqrt{n} = 1/\sqrt{10} = 0.32$, to two decimal places. The probability that $\overline{X}$ will be within one $\sigma_m = 0.32$ of μ is approximately 0.68 (from Section 5.8.1). We can write this as

$$p(\mu - 0.32 < \overline{X} < \mu + 0.32) = 0.68. \tag{6.41}$$

Since we are interested in making inferences about μ from our sample, we rearrange Equation (6.41) as follows:

$$
\begin{aligned}
p(\mu - 0.32 < \overline{X} < \mu + 0.32) &= p(-0.32 < \overline{X} - \mu < 0.32) \\
&= p(0.32 > \mu - \overline{X} > -0.32) \\
&= p(\overline{X} + 0.32 > \mu > \overline{X} - 0.32) = 0.68,
\end{aligned}
$$

or,

$$p(\overline{X} - 0.32 < \mu < \overline{X} + 0.32) = 0.68. \tag{6.42}$$

Suppose the measured sample mean is $\overline{x} = 5.40$. Can we simply substitute this value into Equation (6.42), which would yield

$$p(5.08 < \mu < 5.72) = 0.68? \tag{6.43}$$

We need to be careful how we interpret Equations (6.42) and (6.43).

Equation (6.42) says that if we repeatedly draw samples of the same size from this population, and each time compute specific values for the random interval $(\overline{X} - 0.32, \overline{X} + 0.32)$, then we would expect 68% of them to contain the unknown mean μ. In frequentist theory, a probability represents the percentage of time that something will happen. It says nothing directly about the probability that any one realization of a random interval will contain μ. The specific interval $(5.08, 5.72)$ is but one realization of the random interval $(\overline{X} - 0.32, \overline{X} + 0.32)$ based on the data of a single sample. Since the probability of 0.68 is with reference to the random interval $(\overline{X} - 0.32, \overline{X} + 0.32)$, it would be incorrect to say that the probability of μ being contained in the interval (5.08, 5.72) is 0.68.

However, the 0.68 probability of the *random interval* does suggest that our confidence in the interval (5.08, 5.72) for containing the unknown mean μ is high and we refer to it as a *confidence interval*. It is only in this sense that we are willing to assign a degree of confidence in the statement $5.02 < \mu < 5.72$.

Meaning of a confidence interval: When we write $p(5.08 < \mu < 5.72)$, we are not making a probability statement in a classical sense but rather are expressing a degree of confidence. In general, we write $p(5.08 < \mu < 5.72) = 1 - \alpha$ where $1 - \alpha$ is called the *confidence coefficient*. It is important to remember that the "68% confidence" refers to the probability of the *test*, not to the *parameter*.

If you listen closely to the results of a political poll, you will hear something like the following: "In a recent poll, 55% of a sample of 800 voters indicated they would vote for the Liberals. These results are reliable within ±3.5%, 19 times out of 20." What this means is that if you repeated the poll using the same methodology, then 95% (19 out of 20) of the time you would get the same result within 3.5%. In this case, the 95% confidence interval is 51.5 to 58.5%. A Bayesian analysis of the same polling data was given in Section 4.2.3.

Figure 6.5 shows 68% confidence intervals for the means of 20 samples of a random normal distribution with a $\mu = 5.0$ and $\sigma = 1.0$. Each sample consists of ten measurements. Notice that 13 out of 20 intervals contain the true mean of 5. The number expected for 68% confidence intervals is 13.6.

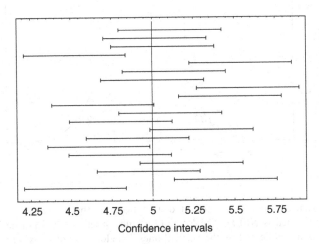

Confidence intervals

Figure 6.5 68% confidence intervals for the means of 20 samples of a random normal distribution with a $\mu = 5.0$ and $\sigma = 1.0$. Each sample consists of ten measurements.

A general procedure for finding confidence intervals:

If we wish to find a general procedure for finding confidence intervals, we must first return to Equation (6.42).

$$p(-0.32 < \overline{X} - \mu < 0.32) = 0.68. \tag{6.44}$$

More generally, we can write

$$p(L_1 < \overline{X} - \mu < L_2) = 1 - \alpha, \tag{6.45}$$

where L_1 and L_2 stand for the lower and upper limits of our confidence interval. We need to develop expressions for L_1 and L_2. The limits are obtained from our sampling distribution, which in this particular case is the sampling distribution for the sample mean random variable. Recall that $\overline{X}$ is $N(\mu, \sigma^2/n)$, so the distribution of the standard random variable $Z = (\overline{X} - \mu)\sqrt{n}/\sigma$ is $N(0,1)$. Figure 6.6 shows the distribution of Z. In terms of Z, we can write Equation (6.45) as

$$p\left(\frac{L_1}{\sigma/\sqrt{n}} < Z < \frac{L_2}{\sigma/\sqrt{n}}\right) = 1 - \alpha. \tag{6.46}$$

The desired sampling distribution is

$$f(z) = \frac{1}{\sqrt{2\pi}} \exp\left\{-\frac{z^2}{2}\right\}. \tag{6.47}$$

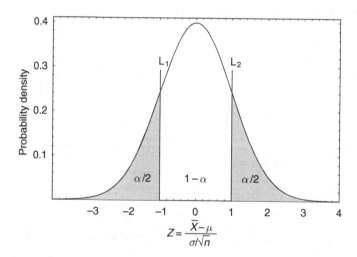

Figure 6.6 The figure shows the expected distribution for the standard random variable $Z = (\overline{X} - \mu)\sqrt{n}/\sigma$ which is N(0,1). The lower (L_1) and upper (L_2) boundaries of the $1 - \alpha$ confidence interval are indicated by the vertical lines. The location of L_1 is set by the requirement that the shaded area to the left of L_1 is equal to $\alpha/2$. Similarly, the shaded area to the right of L_2 is equal to $\alpha/2$.

The limits L_1 and L_2 are evaluated from the following two equations:

$$\int_{-\infty}^{\frac{L_1}{\sigma/\sqrt{n}}} f(z)dz = \frac{\alpha}{2}, \tag{6.48}$$

and

$$\int_{\frac{L_2}{\sigma/\sqrt{n}}}^{+\infty} f(z)dz = \frac{\alpha}{2}. \tag{6.49}$$

Let $Z_{\frac{\alpha}{2}} = L_1\sqrt{n}/\sigma$ and $Z_{1-\frac{\alpha}{2}} = L_2\sqrt{n}/\sigma$. Then

$$p(Z_{\frac{\alpha}{2}} < Z < Z_{1-\frac{\alpha}{2}}) = 1 - \alpha. \tag{6.50}$$

It follows that

$$L_1 = Z_{\frac{\alpha}{2}}\frac{\sigma}{\sqrt{n}} \quad \text{and} \quad L_2 = Z_{1-\frac{\alpha}{2}}\frac{\sigma}{\sqrt{n}}.$$

But for a standard normal, $Z_{\frac{\alpha}{2}} = -Z_{1-\frac{\alpha}{2}}$; therefore,

$$L_1 = -L_2 = -Z_{1-\frac{\alpha}{2}}\frac{\sigma}{\sqrt{n}}. \tag{6.51}$$

We can now generalize Equations (6.41) and (6.42) and write

$$p\left(\mu - \frac{\sigma}{\sqrt{n}}Z_{1-\frac{\alpha}{2}} < \overline{X} < \mu + \frac{\sigma}{\sqrt{n}}Z_{1-\frac{\alpha}{2}}\right) = 1 - \alpha, \tag{6.52}$$

and

$$p\left(\overline{X} - \frac{\sigma}{\sqrt{n}}Z_{1-\frac{\alpha}{2}} < \mu < \overline{X} + \frac{\sigma}{\sqrt{n}}Z_{1-\frac{\alpha}{2}}\right) = 1 - \alpha. \tag{6.53}$$

Therefore, the $100(1 - \alpha)\%$ confidence interval for μ is

$$\bar{x} \pm \frac{\sigma}{\sqrt{n}}Z_{1-\frac{\alpha}{2}}. \tag{6.54}$$

Clearly, the larger the sample size, the smaller the width of the interval.

Box 6.4   *Mathematica* example:

We can compute the 68% confidence interval for the mean of a population, with
known variance $= 0.1$, from a list of data values with the following commands:

Needs["Statistics `ConfidenceIntervals` "]

The line above loads the confidence intervals package and the line below computes
the confidence interval for a normal distribution.

MeanCI[data, KnownVariance → 0.1, ConfidenceLevel → 0.68]

Where **data** is a list of the sample data values. If the variance is unknown, leave out
the **KnownVariance → 0.1** option and then the confidence interval will be based on
a Student's t distribution.

6.6.2 Confidence intervals for μ, unknown variance

Again, we know that the distribution of $\overline{X}$ is $N(\mu, \sigma^2/n)$, but since we do not know σ
we are unable to use this distribution to compute the desired confidence interval.
Fortunately, we can obtain the confidence interval using the Student's t statistic which
makes use of the sample variance which we can compute. Recall, that the random
variable $T = (\overline{X} - \mu)\sqrt{n}/S$ has a Student's t distribution with $(n-1)$ degrees of
freedom. Figure 6.7 shows a Student's t distribution for $\nu = n - 1 = 9$ degrees of

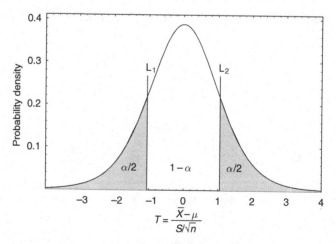

Figure 6.7 The figure shows the Student's t distribution for the $T = (\overline{X} - \mu)\sqrt{n}/S$ statistic, for
$\nu = n - 1 = 9$ degrees of freedom. The lower (L_1) and upper (L_2) boundaries of the $1 - \alpha$ confidence interval are indicated by the vertical lines. The location of L_1 is set by the requirement that the
shaded area to the left of L_1 is equal to $\alpha/2$. Similarly, the shaded area to the right of L_2 is equal to $\alpha/2$.

freedom. For a Student's t distribution, the $t_{1-\frac{\alpha}{2}, n-1}$ quantile value is defined by the equation

$$p(-t_{1-\frac{\alpha}{2}, n-1} < T < t_{1-\frac{\alpha}{2}, n-1}) = 1 - \alpha, \tag{6.55}$$

which we can rewrite as

$$p\left(-\frac{s}{\sqrt{n}} t_{1-\frac{\alpha}{2}, n-1} < \overline{X} - \mu < \frac{s}{\sqrt{n}} t_{1-\frac{\alpha}{2}, n-1}\right) = 1 - \alpha. \tag{6.56}$$

We can obtain values for L_1 and L_2 by comparing Equation (6.56) to Equation (6.45). This yields $L_1 = -L_2 = -(s/\sqrt{n}) t_{1-\frac{\alpha}{2}, n-1}$. The final $1 - \alpha$ confidence interval is

$$\overline{x} \pm \frac{s}{\sqrt{n}} t_{1-\frac{\alpha}{2}, n-1}. \tag{6.57}$$

Box 6.5   *Mathematica* example:

In the introduction to this chapter, we posed a problem concerning the mean redshift of a population of cosmic gamma-ray bursts (GRB), based on a sample of seven measured GRB redshifts (the number known at the time of writing). The redshifts are: 1.61, 0.0083, 1.619, 0.835, 3.420, 1.096, 0.966. We now want to determine the 68% confidence interval for the mean redshift for the population GRB sources. We neglect the uncertainties in the individual measured redshifts as they are much smaller than the spread of the seven values. Although we do not know the probability density function for the population, we know from the CLT that the distribution of the sample mean random variable $(\overline{X})$ will be approximately normal for $n = 7$. Furthermore, $<\overline{X}> = \mu$, the mean of the population, and $\mathrm{Var}(\overline{X}) = \sigma^2/n$, where $\sigma^2 = $ the variance of the population. Since we do not know σ^2, we use the measured sample variance,

$$s^2 = \sum_{i=1}^{7} \frac{(x_i - \overline{x})^2}{n - 1}.$$

We can thus evaluate the Student's t value for our sample and use this to arrive at our 68% confidence interval for the population mean redshift (recall Equation (6.56)).

data $= \{1.61, 0.0083, 1.619, 0.835, 3.420, 1.096, 0.966\}$

MeanCI[data, ConfidenceLevel $\rightarrow$ **0.68]** $= \{0.93, 1.80\}$

In this case, we have left out the **KnownVariance** option to the confidence interval command, **MeanCI**, because the population variance is unknown. **MeanCI** now returns a confidence interval based on the Student's t distribution.

6.6.3 Confidence intervals: difference of two means

One of the most fundamental problems that occurs in experimental science, is that of analyzing two independent measurements of the same physical quantity, one "control" and one "trial," taken under slightly different experimental conditions, e.g., drug testing. Here, we are interested in computing confidence intervals for the difference in the means of the control population and the trial population, when sampling from two independent normal distributions.

1) If μ_x and μ_y are unknown and σ_x and σ_y are known, then the random variable

$$Z = \frac{\overline{X} - \overline{Y} - (\mu_x - \mu_y)}{\sqrt{\sigma_x^2/n_x + \sigma_y^2/n_y}} \tag{6.58}$$

has a normal distribution $N(0,1)$. The $100(1-\alpha)\%$ confidence interval for $(\mu_x - \mu_y)$ is

$$\overline{x} - \overline{y} \pm Z_{1-\frac{\alpha}{2}} \sqrt{\frac{\sigma_x^2}{n_x} + \frac{\sigma_y^2}{n_y}}, \tag{6.59}$$

where the quantile $Z_{1-\frac{\alpha}{2}}$ is such that

$$p(Z \le Z_{1-\frac{\alpha}{2}}) = 1 - \frac{\alpha}{2}. \tag{6.60}$$

At this point you may be asking yourself what value of α to use in presenting your results. There are really two types of questions we might be interested in. First, do the data indicate that the means are significantly different? This type of question is addressed in Chapter 7, which deals with hypothesis testing. The other type of question, which is being addressed here, concerns estimating the difference of the two means. For this question, it is common practice to use an $\alpha = 0.32$, corresponding to a 68% confidence interval. We will look at this issue again in more detail in Section 7.2.1.

2) If σ_x and σ_y are unknown but assumed equal, the random variable

$$T = \frac{\overline{X} - \overline{Y} - (\mu_x - \mu_y)}{S_D\sqrt{1/n_x + 1/n_y}}$$

has a Student's t distribution with $(n_x + n_y - 2)$ degrees of freedom, where

$$S_D^2 = \frac{[(n_x - 1)S_x^2 + (n_y - 1)S_y^2]}{(n_x + n_y - 2)}. \tag{6.61}$$

The $100(1-\alpha)\%$ confidence interval is

$$\overline{x} - \overline{y} \pm S_D t_{1-\frac{\alpha}{2},\,(n_x+n_y-2)}. \tag{6.62}$$

3) If σ_x and σ_y are unknown and assumed to be unequal, the random variable

$$T = \frac{\overline{X} - \overline{Y} - (\mu_x - \mu_y)}{S_p}$$

is distributed approximately as Student's t with ν degrees of freedom (Press *et al.*, 1992), where

$$S_p^2 = \frac{S_x^2}{n_x} + \frac{S_y^2}{n_y}, \tag{6.63}$$

and,

$$\nu = \frac{\left[\frac{S_x^2}{n_x} + \frac{S_y^2}{n_y}\right]^2}{\frac{[S_x^2/n_x]^2}{n_x-1} + \frac{[S_y^2/n_y]^2}{n_y-1}}. \tag{6.64}$$

See Section 7.2.2 for a worked example of the use of the Student's t test for these conditions.

6.6.4 Confidence intervals for σ^2

Here, we are interested in computing confidence intervals for σ^2 when sampling from a normal distribution with unknown mean. Recall that $(n-1)S^2/\sigma^2$ is χ^2_{n-1}. Then it follows that the $100(1-\alpha)\%$ interval for σ^2 is

$$\left[\frac{(n-1)s^2}{\chi^2_{1-\frac{\alpha}{2},n-1}}, \frac{(n-1)s^2}{\chi^2_{\frac{\alpha}{2},n-1}}\right]. \tag{6.65}$$

Box 6.6 *Mathematica* **example:**

We can compute the 68% confidence interval for the variance of a population with unknown mean, from a list of data values designated by data, with the following command:

VarianceCI[data,ConfidenceLevel → 0.68]

6.6.5 Confidence intervals: ratio of two variances

In this section, we want to determine confidence intervals for the ratio of two variances when sampling from two independent normal distributions. In this case, the $100(1-\alpha)\%$ confidence interval for σ_y^2/σ_x^2 is

$$\left[\frac{s_y^2}{s_x^2} \frac{1}{f_{1-\frac{\alpha}{2}, n_y-1, n_x-1}}, \frac{s_y^2}{s_x^2} f_{1-\frac{\alpha}{2}, n_x-1, n_y-1} \right]. \tag{6.66}$$

Box 6.7   *Mathematica* example:

We can compute the 68% confidence interval for the ratio of the population variance of *data1* to the population variance of *data2* with the following command:

VarianceRatioCI[data1,data2,ConfidenceLevel $\rightarrow$ 0.68]

6.7 Summary

In this chapter, we introduced the $\overline{X}$, χ^2, S^2, Student's t and F statistics and showed how they are useful in making inferences about the mean (μ) and variance (σ^2) of an underlying population from a random sample. In the frequentist camp, the usefulness of any particular statistic stems from our ability to predict its distribution for a very large number of hypothetical repeats of the experiment. In this chapter we have assumed that each of the data sets, either real or hypothetical, is an IID sample from a normal distribution, in some cases appealing to the Central Limit Theorem to satisfy this requirement. For IID normal samples, we found that:

1.

$$S^2 = \sum_{i=1}^{n} \frac{(X_i - \overline{X})^2}{n - 1}$$

is an unbiased estimator of the sample variance, σ^2.

2. The sampling distribution of

$$(n - 1)\frac{S^2}{\sigma^2} = \sum_{i=1}^{n} \frac{(X_i - \overline{X})^2}{\sigma^2}$$

is χ^2 with $(n - 1)$ degrees of freedom. This statistic has a wide range of uses that we will learn in subsequent chapters (e.g., Method of Least Squares) and is clearly useful in specifying confidence intervals for estimates of σ^2 based on the sample variance S^2.

3. One familiar form of the Student's t statistic is

$$T = \frac{(\overline{X} - \mu)}{(S/\sqrt{n})}.$$

The statistic is particularly useful for computing confidence intervals for the population mean, μ, based on the sample $\overline{X}$ and S^2. In Chapter 9, we will see the Student's t distribution reappear in a Bayesian context characterizing the posterior PDF for a particular state of information.

4. The F statistic is given by

$$F_{12} = \frac{S_1^2/\sigma_1^2}{S_2^2/\sigma_2^2}.$$

We saw how this can be used to compare the ratio of the variances of two populations from a measurement of the sample variance of each population.

Once the distribution of the statistic is specified, it is possible to interpret the significance of the particular value of the statistic corresponding to our one actual measured sample. For example, we were able to test a cigarette manufacturer's claim about the mean nicotine content by comparing the value of the Student's t statistic for a measured sample to the distribution for a hypothetical reference set. In case you didn't fully comprehend this line of reasoning, we will go into it in more detail in the next chapter, which deals with frequentist hypothesis testing.

Throughout this chapter, the expression "number of degrees of freedom" has cropped up in connection with each choice of statistic. Its precise meaning is defined in the definition of the χ^2, Student's t, and F distributions. Roughly, what it translates to in practice is the number of data points (or data bins if the data are binned) in the sample used to compute the statistic, minus the number of additional parameters (like S^2 in the Student's t statistic) that have to be estimated from the same sample.

A major part of inferring the values of the μ and σ^2 of the population concerns their estimated uncertainties. In the frequentist case, this amounts to estimating a confidence interval, e.g., 68% confidence interval. Keep in mind that a frequentist confidence interval says nothing directly about the probability that a single confidence interval, derived from your one actual measured sample, will contain the true population value. Then what does the 68% confidence mean? It means that if you repeated the measurement a large number of times, each time computing a 68% confidence interval from the new sample, then 68% of these intervals will contain the true value.

6.8 Problems

1. Suppose you are given the IID normal data sample $\{0.753, 3.795, 4.827, 2.025\}$. Compute the sample variance and standard deviation. What is the standard error of the sample mean?
2. What is the 95% confidence interval for the mean of the IID normal sample $\{0.753, 3.795, 4.827, 2.025\}$?
3. Compute the area in the tail of a χ^2 distribution to the right of $\chi^2 = 30.61$, for $\nu = 10$ degrees of freedom.
4. The sample variance is $s_1^2 = 16.65$ for $n_1 = 6$ IID samples from a normal distribution with a population variance σ_1^2. Also, $s_2^2 = 5.0$ for $n_2 = 11$ IID samples from a second independent normal distribution with a population variance σ_2^2. If we assume that $\sigma_1^2 = 2\sigma_2^2$, then from Equation (6.40), we obtain $f = 1.665$ for $\nu_1 = n_1 - 1 = 5$ and $\nu_2 = n_2 - 1 = 10$ degrees of freedom. What is the probability of getting an f value ≥ 3.33 by chance if $\sigma_1^2 = 2\sigma_2^2$?

7

Frequentist hypothesis testing

7.1 Overview

One of the main objectives in science is that of inferring the truth of one or more hypotheses about how some aspect of nature works. Because we are always in a state of incomplete information, we can never prove any hypothesis (theory) is true. In Bayesian inference, we can compute the probabilities of two or more competing hypotheses directly for our given state of knowledge.

In this chapter, we will explore the frequentist approach to hypothesis testing which is considerably less direct. It involves considering each hypothesis individually and deciding whether to (a) reject the hypothesis, or (b) fail to reject the hypothesis, on the basis of the computed value of a suitable choice of statistic. This is a very big subject and we will give only a limited selection of examples in an attempt to convey the main ideas. The decision on whether to reject a hypothesis is commonly based on a quantity called a P-value. At the end of the chapter we discuss a serious problem with frequentist hypothesis testing, called the "optional stopping problem."

7.2 Basic idea

In hypothesis testing we are interested in making inferences about the truth of some hypothesis. Two examples of hypotheses which we analyze below are:

- The radio emission from a particular galaxy is constant.
- The mean concentration of a particular toxin in river sediment is the same at two locations.

Recall that in frequentist statistics, the argument of a probability is restricted to a random variable. Since a hypothesis is either true or false, it cannot be considered a random variable and therefore we must indirectly infer the truth of the hypothesis (in contrast to the direct Bayesian).

In the river toxin example, we proceed by assuming that the mean concentration is the same for both locations and call this the *null hypothesis*. We then choose a statistic, such as the sample mean, that can be computed from our one actual data set. The value of the statistic can also be computed in principle for a very large number of hypothetical repeated measurements of the river sediment under identical conditions. Our choice of statistic must be one whose distribution is predictable for this reference set of hypothetical repeats, assuming the truth of our null hypothesis. We then compare the actual value of the statistic, computed from our one actual data set, to the predicted reference distribution. If it falls in a very unlikely spot (i.e., way out in the tail of the predicted distribution) we choose to reject the null hypothesis at some confidence level on the basis of the measured data set.

If the statistic falls in a reasonable part of the distribution, this does not mean that we accept the hypothesis; only that we fail to reject it. At best, we can substantiate a particular hypothesis by failing to reject it and rejecting every other competing hypothesis that has been proposed. It is an argument by contradiction designed to show that the null hypothesis will lead to an absurd conclusion and should therefore be rejected on the basis of the measured data set. It is not even logically correct to say we have disproved the hypothesis, because, for any one data set it is still possible by chance that the statistic will fall in a very unlikely spot far out in the tail of the predicted distribution. Instead, we choose to reject the hypothesis because we consider it more fruitful to consider others.

7.2.1 Hypothesis testing with the χ^2 statistic

Figure 7.1 shows radio flux density measurements of a radio galaxy over a span of 6100 days made at irregular intervals of time. The observations were obtained as part of a project to study the variability of galaxies at radio wavelengths. The individual radio flux density measurements are given in Table 7.1.

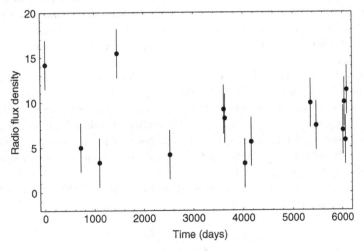

Figure 7.1 Radio astronomy measurements of a galaxy over time.

Table 7.1 *Radio astronomy flux density measurements for a galaxy.*

Day Number	Flux Density (mJy)
0.0	14.2
718.0	5.0
1097.0	3.3
1457.1	15.5
2524.1	4.2
3607.7	9.2
3630.1	8.2
4033.1	3.2
4161.3	5.6
5355.9	9.9
5469.1	7.4
6012.4	6.9
6038.3	10.0
6063.2	5.8
6089.3	11.4

Below we outline the steps involved in the current hypothesis test:

1. Choose as our null hypothesis that the galaxy has an unknown but constant flux density. If we can demonstrate that this hypothesis is absurd at say the 95% confidence level, then this provides indirect evidence that the radio emission is variable. Previous experience with the measurement apparatus indicates that the measurement errors are independently normal with a $\sigma = 2.7$.

2. Select a suitable statistic that (a) can be computed from the measurements, and (b) has a predictable distribution. More precisely, (b) means that we can predict the distribution of values of the statistic that we would expect to obtain from an infinite number of repeats of the above set of radio measurements under identical conditions. We will refer to these as our hypothetical reference set. More specifically, we are predicting a probability distribution for this reference set.

    To refute the null hypothesis, we will need to show that scatter of the individual measurements about the mean is larger than would be expected from measurement errors alone. A useful measure of the scatter in the measurements is the sample variance. We know from Section 6.3, that the random variable $(n-1)S^2/\sigma^2$ (usually called the χ^2 statistic) has a χ^2 distribution with $(n-1)$ degrees of freedom when the measurement errors are known to be independently normal. From Equation (6.4), it is clear that the distribution depends only on the number of degrees of freedom.

3. Evaluate the χ^2 statistic from the measured data. Let's start with the expression for the χ^2 statistic for our data set:

$$\chi^2 = \sum_{i=1}^{n} \frac{(x_i - \bar{x})^2}{\sigma^2}, \tag{7.1}$$

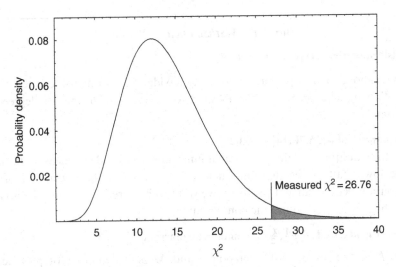

Figure 7.2 The χ^2 distribution predicted on the basis of our null hypothesis with 14 degrees of freedom. The value computed from the measurements, $\chi^2 = 26.76$, is indicated by the vertical bar.

where x_i represents the ith flux density value, $\bar{x} = 7.98$ mJy is the average of our sample values, and $\sigma = 2.7$, as given above. The number of degrees of freedom $\nu = n - 1 = 15 - 1 = 14$, where n is the number of flux density measurements.[1] Equation (7.1) becomes

$$\chi^2 = \sum_{i=1}^{n} \frac{(x_i - \bar{x})^2}{\sigma^2} = \sum_{i=1}^{n} \frac{(x_i - 7.98)^2}{2.7^2} = 26.76. \tag{7.3}$$

4. Plot the computed value of $\chi^2 = 26.76$ on the χ^2 distribution predicted for 14 degrees of freedom. This is shown in Figure 7.2. The χ^2 computed from our one actual data set is shown by the vertical line. The question of how unlikely is this value of χ^2 is by convention interpreted in terms of the area in the tail of the χ^2 distribution beyond this line, which is called the *P-value* or *significance*. We need to specify an area because there is no probability associated with a point. We can evaluate this from

$$P\text{-value} = 1 - F(\chi^2) = 1 - \int_0^{\chi^2} \frac{1}{\Gamma(\frac{\nu}{2})2^{\frac{\nu}{2}}} x^{\frac{\nu}{2}-1} \exp\left(-\frac{x}{2}\right) dx, \tag{7.4}$$

where $F(\chi^2)$ is the *cumulative χ^2 distribution*. Alternatively, we can evaluate the P-value with the following *Mathematica* command.

---

[1] The null hypothesis did not specify the assumed value for μ, the constant flux density, so we estimated it from the mean of the data. Whenever we estimate a model parameter from the data, we lose one degree of freedom. If the null hypothesis had specified μ, then we would have used the following expression for the χ^2 statistic:

$$\chi^2 = \sum_{i=1}^{n} \frac{(x_i - \mu)^2}{\sigma^2}, \tag{7.2}$$

which has a χ^2 distribution with n degrees of freedom.

Box 7.1   *Mathematica* χ^2 P-value

Needs["Statistics `HypothesisTests` "]

The line above loads a package containing a wide range of hypothesis tests, and the following line computes the P-value for a $\chi^2 = 26.76$ and 14 degrees of freedom:

ChiSquarePValue[26.76,14] $\to$ **0.02**

Note: the ChiSquarePValue has a maximum value of 0.5 and will measure the area in the lower tail of the distribution if $\chi^2_{measured}$ falls in the lower half of the distribution. In the current problem, we want to be sure we measure the area to the right of $\chi^2_{measured}$, so use the command:

GammaRegularized $\left[\frac{N-M}{2}, \frac{\chi^2}{2}\right]$ = **GammaRegularized** $\left[\frac{14}{2}, \frac{26.76}{2}\right]$

where N is the number of data points and M is the number of parameters estimated from the data.

Note: in some problems, it is relevant to use a *two-sided test* (see Section 7.2.3) using the χ^2 statistic, e.g., testing that the population variance is equal to a particular value.

---

5. Finally, compute our confidence in rejecting the null hypothesis which is equal to the area of the χ^2 distribution to the left of $\chi^2 = 26.76$. This area is equal to $(1 - \text{P-value}) = 0.98$ or 98%.

While the above recipe is easy to compute, it undoubtedly contains many perplexing features. Most among them is the strangely convoluted definition of the key determinant of falsification, the P-value, also known as the significance α.

What precisely does the P-value mean? It means that if the flux density of this galaxy is really constant, and we repeatedly obtained sets of 15 measurements under the same conditions, then only 2% of the χ^2 values derived from these sets would be expected to be greater than our one actual measured value of 26.76. At this point, you may be asking yourself why we should care about a probability involving results never actually obtained, or how we choose a P-value to reject the null hypothesis. In some areas of science, a P-value threshold of 0.05 (confidence of 95%) is used; in other areas, the accepted threshold for rejection is 0.01,[2] i.e., it depends on the scientific culture you are working in.

Unfortunately, P-values are often incorrectly viewed as the probability that the hypothesis is true. There is no objective means for deciding the latter without specifying an alternative hypothesis, H_1, to the null hypothesis. The point is that any

---

[2] Note: because experimental errors are frequently underestimated, and hence χ^2 values overestimated, it is not uncommon to require a P-value < 0.001 before rejecting a hypothesis.

particular P-value might arise even if the alternative hypothesis is true.[3] The concept of an alternative hypothesis is introduced in Section 7.2.3. In Section 9.3, we will consider a Bayesian analysis of the galaxy variability problem.

There is another useful way of expressing a statistical conclusion like that of the above hypothesis test. Instead of the P-value, we can measure how far out in the tail the statistic falls in units of the $\sigma = \sqrt{\text{variance}}$ of the reference distribution. For example, the variance of a χ^2 distribution $= 2\nu = 28$ and the expectation value of $\chi^2 = \langle \chi^2 \rangle = \nu = 14$. Therefore:

$$\frac{\chi^2_{\text{obs}} - \langle \chi^2 \rangle}{\sigma} = \frac{\chi^2_{\text{obs}} - \langle \chi^2 \rangle}{\sqrt{2\nu}} = \frac{26.76 - 14}{\sqrt{28}} = 2.4 \equiv 2.4\sigma.$$

In many branches of science, a minimum of a 3σ effect is required for a claim to be taken with any degree of seriousness. More often, referees of scientific journals will require a 5σ result to recommend publication. It depends somewhat on how difficult or expensive it is to get more data.

Now suppose we were studying a sample of 50 galaxies for evidence of variable radio emission. If all 50 galaxies were actually constant, then for a confidence threshold of 98%, we would expect to detect only one false variable in the sample by chance. If we found that ten galaxies had χ^2 values that exceeded the 98% quantile value, then we would expect nine of them were not constant. If, on the other hand, we were studying a sample of 10^4 galaxies, we would expect to detect approximately 200 false variables.

It is easy to see how to extend the use of the χ^2 test described above to other more complex situations. Suppose we had reason to believe that the radio flux density was decreasing linearly with time at a known rate, m, with respect to some reference time, t_0. In that case,

$$\chi^2 = \sum_{i=1}^{n} \frac{(x_i - [m(t_i - t_0)])^2}{\sigma^2}, \tag{7.5}$$

where t_i is the time of the ith sample. We will have much more use for the χ^2 statistic in later chapters dealing with linear and nonlinear model fitting (see Section 10.8).

7.2.2 Hypothesis test on the difference of two means

Table 7.2 gives measurements of a certain toxic substance in the river sediment at two locations in units of parts per million (ppm). In this example, we want to test the hypothesis that the mean concentration of this toxin is the same at the two locations. How do we proceed? Sample 1 consists of 12 measurements taken from location 1, and

---

[3] The difficulty in interpreting P-values has been highlighted in many papers (e.g., Berger and Sellke, 1987; Delampady and Berger, 1990; Sellke *et al.*, 2001). The focus of these works is that P-values are commonly considered to imply considerably greater evidence against the null hypothesis H_0 than is actually warranted.

Table 7.2 *River sediment toxin concentration measurements at two locations.*

Location 1 (ppm)	Location 2 (ppm)
13.2	8.9
13.8	9.1
8.7	8.3
9.0	6.0
8.6	7.7
9.9	9.9
14.2	9.9
9.7	8.9
10.7	
8.3	
8.5	
9.2	

sample 2 consists of 8 measurements from location 2. For each location, we can compute the sample mean. From the frequentist viewpoint, we can compare sample 1 to an infinite set of hypothetical data sets that could have been realized from location 1. For each of the hypothetical data sets, we could compute the mean of the 12 values. According to the Central Limit Theorem, we expect that the means for the hypothetical data sets will have an approximately normal distribution. Let $\overline{X}_1$ and $\overline{X}_2$ be random variables representing means for locations 1 and 2, respectively. It is convenient to work with the standard normal distributions given by

$$Z_1 = \frac{\overline{X}_1 - \mu_1}{\sigma_1/\sqrt{n_1}}, \tag{7.6}$$

and

$$Z_2 = \frac{\overline{X}_2 - \mu_2}{\sigma_2/\sqrt{n_2}}, \tag{7.7}$$

where μ_1 and μ_2 are the population means and σ_1 and σ_2 the population standard deviations. Similarly, we expect

$$Z = \frac{\overline{X}_1 - \overline{X}_2 - (\mu_1 - \mu_2)}{\sqrt{\sigma_1^2/n_1 + \sigma_2^2/n_2}} \tag{7.8}$$

to be approximately normal as well (see Section 6.6.3). In the present problem, the null hypothesis represented by H_0 corresponds to

$$H_0 \equiv \mu_1 - \mu_2 = 0. \tag{7.9}$$

Since we do not know the population standard deviation for Z, we need to estimate it from our measured sample. Then according to Section 6.6.3, the random variable T given by

$$T = \frac{\overline{X}_1 - \overline{X}_2 - (\mu_1 - \mu_2)}{S_p} \tag{7.10}$$

has a Student's t distribution. All that remains is to specify the value of S_p and the number of degrees of freedom, ν, for the Student's t distribution.

In the present problem, we cannot assume $\sigma_1 = \sigma_2$ so we use Equations (6.63) and (6.64):

$$S_p^2 = \frac{S_1^2}{n_1} + \frac{S_2^2}{n_2}, \tag{7.11}$$

and

$$\nu = \frac{\left[\frac{S_1^2}{n_1} + \frac{S_2^2}{n_2}\right]^2}{\frac{[S_1^2/n_1]^2}{n_1-1} + \frac{[S_2^2/n_2]^2}{n_2-1}}. \tag{7.12}$$

Note: Equation (6.30) for the Student's t probability density is valid even if ν is not an integer.

Of course, we could test the hypothesis that the standard deviations are the same but we leave that for a problem at the end of the chapter. Inserting the data we find the t statistic $= 2.23$ and $\nu = 17.83$ degrees of freedom. We are now ready to test the null hypothesis by comparing the measured value of the t statistic to the distribution of t values expected for a reference set of hypothetical data sets for 17.83 degrees of freedom. Figure 7.3 shows the reference t distribution and the measured value of 2.23. We can compute the area to the right of $t = 2.23$ which is called the *one-sided P-value*. We can evaluate this with the following *Mathematica* command:

Needs["Statistics `HypothesisTests` "]

StudentTPValue[2.23,17.83,OneSided − > True] → 0.0193

This P-value is the fraction of hypothetical repeats of the experiment that are expected to have t values ≥ 2.23 if the null hypothesis is true. In this problem, we would expect an equal number of hypothetical repeats to fall in the same area in the opposite tail region by chance.

What we are really interested in is the fraction of hypothetical repeats that would be extreme enough to fall in either of these tail regions (shaded regions of Figure 7.3) or what is called the two-sided P-value. We can evaluate this with the following *Mathematica* command:

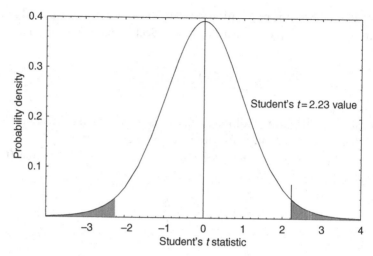

Figure 7.3 Reference Student's t distribution with 17.83 degrees of freedom. The measured t statistic is indicated by a line and the shaded areas correspond to upper and lower tail areas in a two-sided hypothesis test.

Needs["Statistics `HypothesisTests` "]

StudentTPValue[2.23,17.83,TwoSided − > True] → 0.0386

Mathematica provides an easier way of computing this P-value (i.e., computes the t value and degrees of freedom for you) with

MeanDifferenceTest[data1,data2,diff,TwoSided − > True, FullReport − > True]

where **data1** and **data2** are the two data lists.
If the variances of the two data sets are known to be the same then use

MeanDifferenceTest[data1,data2,diff,EqualVariances − > True, TwoSided − > True,FullReport − > True]

Our confidence in rejecting the null hypothesis is equal to the area outside of these extreme tail regions which equals 1 − (two-sided P-value) = 0.961 or 96.1%. If we use a typical threshold for rejection of say 95%, then in this case, we just reject the null hypothesis.

7.2.3 One-sided and two-sided hypothesis tests

In the galaxy example, where we used the χ^2 statistic, we computed the confidence in rejecting the null hypothesis using a one-sided tail region. Why didn't we use a two-sided tail region as in the river toxin problem? Here, we introduce the concept of the

Table 7.3 *Type I and type II errors in hypothesis testing.*

Possible decisions	Possible consequences	Errors
Reject H_0	when in fact H_0 true when in fact H_0 false	← Type *I* error (conviction)
Fail to reject H_0	when in fact H_0 true when in fact H_0 false	← Type *II* error (acquittal)

alternative hypothesis, i.e., alternative to the null hypothesis. In the galaxy problem, the alternative hypothesis is that the radio emission is variable. If the alternative hypothesis were true, then examination of Equation (7.1) indicates that for a given value of the measurement error, σ, we expect the value of χ^2 to be greater[4] when the source is variable than when it is constant. In this case, we would expect to measure χ^2 values in the upper tail but not the lower, which is why we used a one-sided test. In the river toxin problem, the alternative hypothesis is that the mean toxin levels at the two locations are different. In this case, if the alternative were true, we would expect t values in either tail region, which is why we used the two-sided P-value test.

The rules of the game are that the null hypothesis is regarded as true unless sufficient evidence to the contrary is presented. If this seems a strange way of proceeding, it might prove useful to consider the following courtroom analogy. In the courtroom, the null hypothesis stands for "the accused is presumed innocent until proven otherwise." Table 7.3 illustrates the possible types of errors that can arise in a hypothesis test.

In hypothesis testing, a type *I* error is considered more serious (i.e., the possibility of convicting an innocent party is considered worse than the possibility of acquitting a guilty party). A type *I* error is only possible if we reject H_0, the null hypothesis. It is not possible to minimize both the type *I* and type *II* errors. The normal procedure is to select the maximum size of the type *I* error we can tolerate and construct a test procedure that minimizes the type *II* error.[5] This means choosing a threshold value for the statistic which if exceeded will lead us to reject H_0. For example, suppose we are dealing with a one-sided upper tail region test and we are willing to accept a maximum type *I* error of 5%. This means a threshold value of the test statistic anywhere in the upper 5% tail area satisfies our type *I* error requirement. The size of the type *II* error is a minimum at the lower boundary of this region, i.e., the larger the value of the test statistic, the more likely it is we will acquit a possibly guilty party.

Suppose we had used a two-tail test in the radio galaxy problem of Section 7.2.1 rather than the upper tail test. Recall that the alternative hypothesis is only expected to

---

[4] The only way for the variability to reduce χ^2 is if the fluctuations in the galaxies' output canceled measurement errors.
[5] The test that minimizes the type *II* error is often referred to as having maximum "power" in rejecting the null hypothesis when it really is false.

give rise to larger values of χ^2 than those expected on the basis of the null hypothesis. In a two-tail test we would divide the rejection area equally between the lower and upper tails, i.e., for a 98% confidence threshold, that would mean the upper and lower tail areas would each have an area of 1%. The χ^2 value required for rejecting the null hypothesis in the upper tail region would be larger in a two-tail test than for a one-tail test. Thus, for a given confidence level, the two-tail test would increase the chance of a type *II* error, because we would have squandered rejection area in the lower tail region, a region of χ^2 that would not be accessed under the assumption of our alternative hypothesis.

In the river toxin example of Section 7.2.2, the alternative hypothesis can give rise to values of the Student's t statistic in either tail region. In this case, we will want to reject H_0 if the t value falls far enough out in either tail region. In this case, divide the area corresponding to the maximum acceptable type *I* error equally between the two tail regions. To minimize the type *II* area, choose threshold values for the test statistic that are at the inner boundaries of these two tails.

In practice, the role of the alternative hypothesis is mainly to help decide whether to use an upper tail region, a lower tail region or both tails in our statistical test. The choice depends on what is physically meaningful and minimizes the size of the type *II* error. In Section 6.4, we used the Student's t statistic in an analysis of a cigarette manufacturer's claim regarding nicotine content. Since we would reject the claim if the t value fell sufficiently far out in either tail region, we should use a two-sided test in this case.

Typically, in frequentist hypothesis testing involving the use of P-values, a specific value for the type *II* error is not normally computed. Instead it is used as an argument to decide where in the tail region to locate the decision value of the test statistic, as outlined above.

7.3 Are two distributions the same?

We have previously considered tests to compare the means and variances of two samples. Now generalize the questions and ask the simple question: "Can we reject the null hypothesis that the two samples are drawn from the same population?" Rejecting the null hypothesis in effect implies that the two data sets are from different distributions. Failing to reject the null hypothesis only shows that the data sets can be consistent with a single distribution.

Deciding whether two distributions are different is a problem that occurs in many research areas.

Example 1:
Are stars uniformly distributed in the sky? That is, is the distribution of stars as a function of latitude the same as the distribution of the sky area with latitude? In this case, the data set (location of stars) and comparison distribution (sky area) are continuous.

Example 2:
Are the educational patterns in Vancouver and Toronto the same? Is the distribution of people as a function of "last grade attended" the same? Here, both data sets are discrete or binned.

Example 3:
Are the distribution of grades in a particular physics course normally distributed? Here, the grades are discrete or binned and the distribution we are comparing to is continuous. In this latter case, we might be comparing with a normal distribution with a given μ and σ^2 or alternatively we might not know μ and σ^2 and be interested only in whether the shape is normal.

One can always turn continuous data into binned data by grouping the events into specified ranges of the continuous variable(s). Binning involves a loss of information, however, and there is considerable arbitrariness as to how the bins should be chosen.

The accepted test for differences between binned distributions is the *Pearson χ^2 test*. For continuous data as a function of a simple variable, the most generally accepted test is the *Kolmogorov–Smirnov test*.

7.3.1 Pearson χ^2 goodness-of-fit test

Let a random sample of size n from the distribution of a random variable X be divided into k mutually exclusive and exhaustive classes (or bins) and let N_i $(i = 1, \ldots, k)$ be the number of observations in each class (or bin). We want to test the simple null hypothesis

$$H_0 \equiv p(x) = p_0(x)$$

where the claimed probability model $p_0(x)$ is completely specified with regard to all parameters. Since $p_0(x)$ is completely specified, we can determine the probability p_i of obtaining an observation in the ith class under H_0, where by necessity $\sum_{i=1}^{k} p_i = 1$.

Let $n_i = np_i =$ expected number in each class according to the null hypothesis, H_0. Usually, H_0 does not predict n, and this is obtained by setting

$$n = \sum_{i=1}^{k} N_i.$$

This has the effect of reducing the number of degrees of freedom in the χ^2 test by one.[6] Note: N_i is an integer while the n_i's may not be.

---

[6] Note on the number of degrees of freedom: If H_0 does predict the n_i's and there is no *a priori* constraint on any of the N_i's, then $\nu =$ number of bins, k.

More commonly, the n_i's are normalized after the fact so that their sum equals the sum of the N_i's, the total number of events measured. In this case, $\nu = k - 1$.

If the model that gives the n_i's had additional free parameters that were adjusted after the fact to agree with the data, then each of these additional "fitted" parameters reduces the ν by one. The number of these additional fitted parameters (not including the normalization of the n_i's) is commonly called the "number of constraints" so the number of degrees of freedom is $\nu = k - 1$ when there are "zero constraints."

Question: What is the form of $p_0(x)$?

Answer: Since there are k mutually exclusive categories with probabilities $p_1, p_2, \ldots, p_k$, then under the null hypothesis, the probability of the grouped sample is the same as the probability of a *multinomial distribution* discussed in Section 4.3.

In what follows, we will deduce the appropriate test statistic for H_0 which is known as the *Pearson chi-square goodness-of-fit test*. Start with the simple case where $k = 2$; thus, $p_0(x)$ is a binomial distribution.

$$p(x|n,p) = \frac{n!}{(n-x)!x!} p^x (1-p)^{n-x}, \quad x = 0, 1, \ldots, n. \tag{7.13}$$

In this case, $x = n_1$, $p = p_1$, $n - x = n_2$ and $(1 - p) = p_2$. Recall that $\sigma = \sqrt{np(1-p)}$ for the binomial distribution. Consider the standardized random variable

$$Y = \frac{N_1 - np_1}{\sqrt{np_1(1-p_1)}}. \tag{7.14}$$

For $np_1 \gg 1$, the distribution of Y is approximately the standard normal. Recall that the square of a standard normal variable,

$$\sum_{i=1}^{n} \frac{(x_i - \mu)^2}{\sigma^2},$$

is χ^2-distributed with n degrees of freedom. Thus we expect the statistic

$$\frac{(N_1 - np_1)^2}{np_1(1-p_1)} \tag{7.15}$$

to be approximately χ^2-distributed with one degree of freedom.

Note:

$$\left(\frac{1}{np_1} + \frac{1}{np_2}\right) = \frac{n(p_1 + p_2)}{n^2 p_1 p_2} = \frac{1}{np_1(1-p_1)}.$$

$$\frac{(N_1 - np_1)^2}{np_1(1-p_1)} = \frac{(N_1 - np_1)^2}{np_1} + \frac{[(n - N_2) - n(1-p_2)]^2}{np_2}$$

$$= \frac{(N_1 - np_1)^2}{np_1} + \frac{(N_2 - np_2)^2}{np_2} \tag{7.16}$$

$$= \sum_{i=1}^{2} \frac{(N_i - np_i)^2}{np_i}.$$

Following this reasoning, it can be shown that for $k \geq 2$ the statistic,

$$\sum_{i=1}^{k} \frac{(N_i - np_i)^2}{np_i} \tag{7.17}$$

is approximately χ^2-distributed with $\nu = k - 1$ degrees of freedom. N_i is the observed frequency of the ith class and np_i is the corresponding expected frequency under the null hypothesis. Any term in Equation (7.17) with $N_i = np_i = 0$ should be omitted from the sum. A term with $p_i = 0$ and $N_i \neq 0$ gives an infinite χ^2, as it should, since in this case, the N_i's cannot possibly be drawn from the p_i's.

Strictly speaking, the χ^2 P-value is the probability that the sum of squares of ν standard normal random variables will be greater than χ^2. In general, the terms in the sum of Equation (7.17) will not be individually normal. However, if either the number of bins is large ($\gg 1$), or the number of events in each bin is large ($\gg 1$), then the χ^2 P-value is a good approximation for computing the significance of χ^2 values given by Equation (7.17). Its use to estimate the significance of the Pearson χ^2 goodness-of-fit test is standard.

Example 1:
In this first example, we apply the Pearson χ^2 goodness-of-fit test to a $k = 2$ bin problem and compare with the exact result expected using a test based on the binomial distribution. Suppose a certain theory predicts that the fraction of radio sources that are expected to be quasars at an observing wavelength of 20 cm is 70%. A sample of 90 radio sources is selected and each source is optically identified. Only 54 turn out to be quasars. At what confidence level can we reject the above theory? Thus, our null hypothesis is that the theory is true.

	Quasars	Other
Predicted	63	27
Observed	54	36

Number of degrees of freedom = number of bins $- 1 = 1$.

$$\chi^2_\nu = \sum_{i=1}^{2} \frac{(N_i - np_i)^2}{np_i} = \frac{(54 - 63)^2}{63} + \frac{(36 - 27)^2}{27} = 4.29.$$

Our alternative hypothesis in this case is that the theory is not true. Based on the alternative hypothesis, we would choose an upper tail test for the Pearson χ^2 statistic. The observed χ^2 of 4.29 corresponds to a P-value of 3.8%.

What is the corresponding P-value predicted by the binomial distribution? Recall that the binomial distribution (Equation (7.13)) predicts the probability of x successes in n trials when the probability of a success in any one trial is p. In this case, a success means the source is a quasar. According to the theory, $p = 0.7$. First we calculate the area in the tail region extending from $n = 0$ to $n = 54$ which equals 2.7%. The true P-value is double this, or 5.4%, because we need to use a two-tailed test since we would reject the null hypothesis if the observed number fell far enough out in either tail of the binomial distribution. Examination of Equation (7.15) demonstrates that both of these binomial tails contribute to a single χ^2 tail. Thus, using the χ^2 test, we would reject the null hypothesis at the 95% confidence level but just fail to reject the hypothesis using the more accurate test based on the binomial distribution. Comparison of the results, P-values = 3.8% (χ^2) and = 5.4% (binomial) indicates the approximate level of agreement to be expected from the Pearson χ^2 test in this simple two-bin test and an $n \approx 100$.

Now repeat the above test; only this time, we will use a sample of 2000 radio sources of which 1360 prove to be quasars.

$$\chi^2_{\nu-1} = \sum_{i=1}^{2} \frac{(N_i - np_i)^2}{np_i} = \frac{(1360 - 1400)^2}{1400} + \frac{(640 - 600)^2}{600} = 3.81.$$

In this case, the observed χ^2 of 3.81 corresponds to a P-value (significance α) of 5.1%. The more accurate P-value based on the binomial distribution is equal to 5.4%. As expected, as we increase the sample size, the agreement between the two tests becomes much closer. In this case, both tests just fail to reject the null hypothesis at the 95% level.

Example 2:

Now we consider an example involving a large number of bins or classes. Table 7.4 compares the total number of goals scored per game in four seasons of World Cup soccer matches (years 1990, 1994, 1998, and 2002), with the expected number if the number of goals is Poisson distributed. Only the goals scored in the 90 minutes regulation time are considered. This leaves out goals scored in extra time or in penalty shoot-outs. Based on the information provided, is there reason to believe at the 95% confidence level that the number of goals is not a Poisson random variable?

The Poisson distribution is given by

$$p(x) = \frac{(\lambda)^x e^{-\lambda}}{x!},$$

where λ = average number of goals per game (a parameter that must be estimated from the data). Each parameter estimated from the data decreases the number of degrees of freedom by one. From the data of Table 7.4, we compute $\lambda = 2.4785$.

The probability of exactly zero goals under the null hypothesis of a Poisson distribution is

$$p(0) = \frac{(2.4785)^0 e^{-2.4785}}{0!} = 0.0839.$$

For $n = 232$ games, the expected number of games with zero goals is $232 \times 0.0839 = 19.46$. Even though the Poisson distribution makes non-zero predictions for seven or more goals, the expected number rapidly falls below the resolution of our data set. There is no requirement that the bins be of equal size so our last bin is for ≥ 7 goals.

For $k = 8$ classes, the number of degrees of freedom = 6. We lose one degree of freedom from the normalizing $n = \sum_{i=1}^{k} N_i$, and another from estimating λ from the data. The value of $\chi^2 = 2.66$, which is derived from the data in Table 7.4,

Table 7.4 *World Cup goal statistics.*

Number of goals	Actual number of games	Expected number of games	$\dfrac{[N_i - np_i(\lambda)]^2}{np_i(\lambda)}$
0	19	19.46	0.0108
1	49	48.23	0.0124
2	60	59.76	0.0009
3	47	49.37	0.1142
4	32	30.59	0.0647
5	18	15.16	0.5302
6	3	6.26	1.7009
≥ 7	4	3.15	0.2267
Totals	232	232	2.6607

corresponds to a P-value (significance α) of 0.85. This corresponds to a confidence in rejecting the null hypothesis of 15%, which is much less than the 95% usually required. Thus, we fail to reject the null hypothesis that the number of goals scored is Poisson distributed. For more statistical analysis of the World Cup soccer data, see Chu (2003).

7.3.2 Comparison of two-binned data sets

In this case,

$$\chi^2 = \sum_i \frac{(R_i - S_i)^2}{R_i + S_i}, \tag{7.18}$$

where R_i and S_i are the number of events in bin i for the first and second data set, respectively. It is instructive to compare Equation (7.18) with Equation (7.17). The term in the denominator of Equation (7.17), the predicted number of counts in the ith bin, is a measure of the expected variance in the counts. The variance of the difference of two random variables is the sum of their variances, which explains why the denominator in Equation (7.18) is $R_i + S_i$.

If the data were collected in such a way that $\sum R_i = \sum S_i$, then $\nu = k - 1$. If this requirement were absent, the number of degrees of freedom would be k.

7.4 Problem with frequentist hypothesis testing

We now consider a serious problem with frequentist hypothesis testing referred to as the optional stopping problem (e.g., Loredo, 1990; Berger and Berry, 1988). The optional stopping problem is best illustrated by an example. Consider the following astronomical fable motivated by a tutorial given by Tom Loredo at a Maximum Entropy and Bayesian Methods meeting:

An Astronomical Fable

Theorist: I predict the fraction of nearby stars that are like the sun (G spectral class) is $f = 0.1$.

Observer: I count five G stars out of $N = 102$ total stars observed. This gives me a P-value = 4.3%. Your theory is rejected at the 95% level.

Theorist: Let me check that: I can use the binomial distribution to compute the probability of observing five or fewer G stars out of a total of 102 stars observed for a predicted probability $f = 0.1$.

$$\text{P-value} = 2 \times \sum_{n=0}^{5} p(n \,|\, N, f), \tag{7.19}$$

where,

$$p(n \,|\, N, f) = \frac{N!}{n!(N-n)!} f^n (1-f)^{N-n}. \tag{7.20}$$

The factor of 2 in Equation (7.19) is because a two-tailed test is required here. My hypothesis could be rejected if either too few or too many G stars were counted.

I get a P-value = 10% so my theory is still alive. You have failed to reject my theory at the 95% level.

Observer: Never trust a theorist with your data! I planned my observations by deciding beforehand that I would observe until I saw $n_G = 5$ G stars, and then stop. The random quantity your theory predicts is thus N, not n_G. The correct reference distribution is the negative binomial. Thus,

$$\text{P-value} = 2 \times \sum_{N=102}^{\infty} p(N \,|\, n_G, f), \tag{7.21}$$

where,

$$p(N \,|\, n_G, f) = \binom{N-1}{n_G - 1} f^{n_G} (1-f)^{N-n_G}. \tag{7.22}$$

I get a P-value = 4.3% as I claimed.

Theorist: What if bad weather ended your observations before you saw five G stars?

Observer: I'd either throw out the data, or include the probability of bad weather.

Theorist: But then you should include it in the analysis now, because the weather *could* have been bad.

MORAL: Never trust a frequentist with your data!

The problem with the frequentist approach is that we need to specify a reference set of hypothetical samples that could have been observed, but were not, in order to compute the P-value of our observed sample. Thus, the decision on whether to reject the null hypothesis based on the P-value depends on the thoughts of the investigator about data that might have been observed but were not. Clearly, the theorist and observer had different thoughts about what was the appropriate reference set and thus arrived at quite different conclusions. To avoid this problem, experiments must therefore be carefully planned beforehand (e.g., the stopping rule specified before the experiment commences) to be amenable to frequentist analysis and if the plan is altered during execution for any reason (for example, if the experimenter runs out of funds), the data are worthless and cannot be analyzed.

The fact that P-value hypothesis testing depends on considerations like the intentions of the investigator and unobserved data indicates a potentially serious flaw in the logic behind the use of P-values. Surely if our plan for an experiment has to be altered (e.g., astronomical observations cut short due to bad weather), we should still be able to analyze the resulting data provided we are fully aware of the physical details of the experiment. Clearly, our state of information has changed. Fortunately in Bayesian inference, the stopping rule plays no role in the analysis. There is no ambiguity over which quantity is to be considered a "random variable," because the notion of a random variable and consequent need for a reference set of hypothetical data is absent from the theory. All that is required is a specification of the state of knowledge that allows us to compute the likelihood function.

7.4.1 Bayesian resolution to optional stopping problem

In the Bayesian approach, where the probability assignments describe the state of knowledge defined in the problem, such paradoxes disappear. Here, we are interested in the posterior probability of f, the fraction of all nearby stars that are G stars.

$$p(f \mid D, I) = \frac{p(f \mid I)p(D \mid f, I)}{p(D \mid I)}. \tag{7.23}$$

The Bayesian calculation focuses on the functional dependence of the likelihood on the hypotheses corresponding to different choices of f. Both the binomial and negative binomial distributions depend on f in the same way, so Bayesian calculations by the theorist and the observer lead to the same conclusion, as we now demonstrate.

1. Binomial case:

$$p(D|f, I) = p(n_G|N, f) = \frac{N!}{(N - n_G)!(n_G)!} f^{n_G}(1 - f)^{N - n_G}$$

$$p(D|I) = \int df \, p(f|I) \, p(D|f, I) \tag{7.24}$$

$$p(f|D, I) = \frac{p(f|I)f^{n_G}(1 - f)^{N - n_G}}{\int df \, p(f|I)f^{n_G}(1 - f)^{N - n_G}},$$

where the factorial terms cancel out because they appear in both the numerator and denominator.

2. Negative binomial case:

$$p(D|f, I) = p(N|n_G, f) = \binom{N - 1}{n_G - 1} f^{n_G}(1 - f)^{N - n_G}$$

$$p(f|D, I) = \frac{p(f|I)f^{n_G}(1 - f)^{N - n_G}}{\int df \, p(f|I)f^{n_G}(1 - f)^{N - n_G}}. \tag{7.25}$$

Again the factorial terms cancel out because they appear in both the numerator and denominator. Equations (7.24) and (7.25) are identical so the conclusions are the same; theorist and observer agree. Figure 7.4 shows the Bayesian posterior PDF for the fraction of G stars assuming a uniform prior for $p(f|I)$.

The frequentist calculations, on the other hand, focus on the dependence of the sampling distribution on the data N and n_G. Since the binomial and negative binomial distributions depend on N and n_G in different ways, one would be led to different conclusions depending on the distribution chosen. Variations of weather and

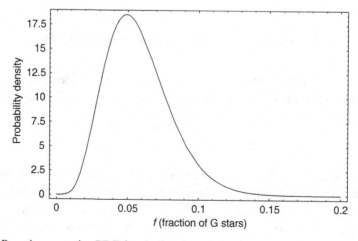

Figure 7.4 Bayesian posterior PDF for the fraction of G stars.

equipment can affect N and n_G but not f, and thus only the Bayesian conclusion is consistently the same.

In the frequentist hypothesis test, we were attempting to reject the null hypothesis that the theorist's prediction ($f = 0.1$) is correct. Recall that in a Bayesian analysis, we cannot compute the probability of a single hypothesis in isolation but only in comparison to one or more alternative hypotheses. The posterior PDF shown in Figure 7.4 allows us to compare the probability density at $f = 0.1$ to the probability density at any other value of f. Assuming a uniform prior, the PDF is a maximum close to $f \approx 0.05$.

7.5 Problems

1. In Section 7.2.2, we tested the hypothesis that the river sediment toxin concentrations at the two locations are the same. Using the same data, test whether the variances of the data are the same for the two locations. Should you use a one-sided or a two-sided hypothesis test in this case? Some choices of *Mathematica* commands to use to answer this question are given in the following box:

Needs["Statistics `HypothesisTests` "]

VarianceRatioTest[data1,data2,ratio,FullReport − > True]

or

VarianceRatioTest[data1,data2,ratio,TwoSided − > True, FullReport − > True]

Note: OneSided − > True, is the default.

Both are based on the FRatioPValue[fratio,numdef,dendef] calculation of Section 6.5.

---

2. Table 7.5 gives measurements of a certain river sediment toxic substance at two locations in units of parts per million (ppm). The sampling is assumed to be from two independent normal populations.

   a) Determine the 95% confidence intervals for the means and variances of the two data sets.
   b) At what confidence level (express as a %) can you reject the hypothesis that the two samples are from populations with the same variance? Explain why you chose to use a one-sided or two-sided hypothesis test.
   c) At what confidence level can you reject the hypothesis that the two samples are from populations with the same mean? Assume the population variances are unknown but equal and use a two-sided hypothesis test.
   d) At what confidence level can you reject the hypothesis that the two samples are from populations with the same mean? Assume the population variances are unknown and unequal, and use a two-sided hypothesis test.

Table 7.5 *Measurements of the concentration of a river sediment toxin in ppm at two locations.*

Location 1	Location 2
17.1	7.0
11.1	12.0
12.6	6.8
12.1	9.3
5.9	8.9
7.7	9.4
10.5	9.6
15.3	7.6
10.5	
10.5	

Table 7.6 *The distribution of a sample of 100 radiation measurements.*

Count Obtained	Number of Occurrences
0	1
1	6
2	18
3	17
4	23
5	10
6	15
7	4
8	4
9	1
10	0
11	0
12	1

Tips: The following *Mathematica* commands may prove useful.

StudentTPValue FRatioPValue VarianceRatioTest MeanDifferenceTest MeanCI VarianceCI

3. In Example 1 of Section 7.3.1, suppose 41 quasars were detected in a total sample of 90 radio sources. With what confidence could you reject the hypothesis that 70% of radio sources are quasars?

4. Generate a list of 50 000 random numbers with a uniform distribution in the interval 0 to 1. Divide this interval into 500 bins of equal size and count the number

of random numbers in each bin (see *Mathematica* command **BinCounts**). Use the Pearson χ^2 goodness-of-fit test to see if you can reject the hypothesis that the counts have a uniform distribution at a 95% confidence level.

5. A distribution of background radiation measurements in a radioactively contaminated site are given in Table 7.6 based on a sample of 100 measurements. Use the Pearson χ^2 goodness-of-fit test to see if you can reject the hypothesis that the counts have a Poisson distribution at a 95% confidence level. Include a plot of the data.

8

Maximum entropy probabilities

8.1 Overview

This chapter can be thought of as an extension of the material covered in Chapter 4 which was concerned with how to encode a given state of knowledge into a probability distribution suitable for use in Bayes' theorem. However, sometimes the information is of a form that does not simply enable us to evaluate a unique probability distribution $p(Y|I)$. For example, suppose our prior information expresses the following constraint:

$I \equiv$ "the mean value of $\cos y = 0.6$."

This information alone does not determine a unique $p(Y|I)$, but we can use I to test whether any proposed probability distribution is acceptable. For this reason, we call this type of constraint information *testable information*. In contrast, consider the following prior information:

$I_1 \equiv$ "the mean value of $\cos y$ is probably > 0.6."

This latter information, although clearly relevant to inference about Y, is too vague to be testable because of the qualifier "probably."

Jaynes (1957) demonstrated how to combine testable information with Claude Shannon's entropy measure of the uncertainty of a probability distribution to arrive at a unique probability distribution. This principle has become known as the *maximum entropy principle* or simply MaxEnt.

We will first investigate how to measure the uncertainty of a probability distribution and then find how it is related to the entropy of the distribution. We will then examine three simple constraint problems and derive their corresponding probability distributions. In the course of this examination, we gain further insight into the special properties of a Gaussian distribution. We also explore the application of MaxEnt to situations where the constraints are uncertain and consider an application to image restoration/reconstruction.[1] The last section deals with a promising Bayesian image reconstruction/compression technique called the Pixon[TM] method.

8.2 The maximum entropy principle

The major use of Bayes' theorem is to update the probability of a hypothesis when new data become available. However, for certain types of constraint information, it is not always obvious how to use it directly in Bayes' theorem. This is because the information does not easily enable us to evaluate a prior probability distribution or evaluate the likelihood function.

As an example, consider the following problem involving a six-sided die: each side has a unique number of dots on it, ranging in number from one to six. Suppose the die is thrown a very large number of times and on each throw, the number of dots appearing on the top face is recorded. The book containing the results of the individual throws is then unfortunately lost. The only information remaining is the average number of dots on the repeated throws. Using only this prior information, how can we arrive at a unique assignment for the probability that the top face will have *n* dots on any one throw (i.e., we want to obtain a prior probability for each side of the die)?

Principle: Out of all the possible probability distributions which agree with the given constraint information, select the one that is maximally non-committal with regard to missing information.

Question: How do we accomplish the goal of being maximally non-committal about missing information?

Answer: The greater the missing information, the more uncertain the estimate. Therefore, make estimates that maximize the uncertainty in the probability distribution, while still being maximally constrained by the given information.

What is uncertainty and how do we measure it?

Jaynes argued that the best measure of uncertainty to maximize is the *entropy* of the probability distribution, an idea which was first introduced by Claude Shannon in his pioneering work on information theory.

We start by developing our intuition about uncertainty:

Example 1:

Consider an experiment with only two possible outcomes. For which of the three probability distributions listed below is the outcome most uncertain?

---

[1] Image restoration, the recovery of images from image-like data, usually means removing the effects of point-spread-function blurring and noise. Image reconstruction means the construction of images from more complexly encoded data (e.g., magnetic resonance imaging data or from the Fourier data measured in radio astronomy aperture synthesis). In the remainder of the chapter, we will use the term *image reconstruction* to refer to both.

(1) $p_1 = p_2 = \frac{1}{2}$ ←The outcome here most uncertain
(2) $p_1 = \frac{1}{4}$, $p_2 = \frac{3}{4}$
(3) $p_1 = \frac{1}{100}$, $p_2 = \frac{99}{100}$

Example 2:

Consider an experiment with different numbers of outcomes

(1) $p_1 = p_2 = \frac{1}{2}$
(2) $p_1 = p_2 = p_3 = p_4 = \frac{1}{4}$
(3) $p_1 = p_2 = \cdots = p_8 = \frac{1}{8}$ ← Most uncertain

i.e., If there are n equally probable outcomes, the uncertainty $\propto n$.

8.3 Shannon's theorem

In 1948, Claude Shannon published a landmark paper on information theory in which he developed a measure of the uncertainty of a probability distribution which he labeled 'entropy.' He demonstrated that the expression for entropy has a meaning quite independent of thermodynamics. Shannon showed that the uncertainty, $S(p_1, p_2, \ldots, p_n)$, of a discrete probability distribution p_i is given by the entropy of the distribution, which is

$$S(p_1, p_2, \ldots, p_n) = -\sum_{i=1}^{n} p_i \ln(p_i) = \text{entropy}. \qquad (8.1)$$

The theorem is based on the following assumptions:

(1) Some real numbered measure of the uncertainty of the probability distribution $(p_1, p_2, \ldots, p_n)$ exists, which we designate by

$$S(p_1, p_2, \ldots, p_n).$$

(2) S is a continuous function of the p_i. Otherwise an arbitrary small change in the probability distribution could lead to the same big change in the amount of uncertainty as a big change in the probability distribution.
(3) $S(p_1, p_2, \ldots, p_n)$ should correspond to common sense in that when there are many possibilities, we are more uncertain than when there are few. This condition implies that in the case where the p_i are all equal (i.e., $p_i = 1/n$),

$$S\left(\frac{1}{n}, \ldots, \frac{1}{n}\right) = nf\left(\frac{1}{n}\right)$$

shall be a monotonic increasing function of n.
(4) $S(p_1, p_2, \ldots, p_n)$ is a consistent measure. If there is more than one way of working out its value, we must get the same answer for every possible way.

8.4 Alternative justification of MaxEnt

Here, we consider how we might go about assigning a probability distribution for the sides of a weighted die given only constraint information about the die. Let p_i = the probability of the ith side occurring in any toss where i = number of dots on that side. We now impose the constraint that

$$\text{mean number of dots} = \sum_{i=1}^{6} i \, p(i) = 4. \qquad (8.2)$$

Note: the mean value for a fair die is 3.5. Our job is to come up with a unique set of p_i values consistent with this constraint.

As a start, let's consider what we can infer about the probabilities of the six sides from prior information consisting of the mean number of dots from ten throws of the die. Suppose $I \equiv$ "in ten tosses of a die, the mean number of dots was four." We will solve this problem and then consider what happens as the number of tosses becomes very large.

For a finite number of tosses, there are a finite number of possible outcomes. Let $h_1 \rightarrow h_n$ be the set of hypotheses representing these different outcomes. Some example outcomes are given in Table 8.1. Which hypothesis is the most probable?

In the die problem just discussed, we can use our information to reject all hypotheses which predict a mean $\neq 4$. This still leaves us a large number of possible hypotheses.

Our intuition tells us that in the absence of any additional information, certain hypotheses are more likely than others (e.g., h_1 is less likely than h_2 or h_4). Let's try and refine our intuition.

If we knew the individual p_i's, we could calculate the probability of each h_i. It is given by the multinomial distribution

Table 8.1 *Some hypotheses about the possible outcomes of tossing a die ten times.*

# of dots	h_1	h_2	h_3	h_4	$\cdots$	h_n
1	0/10	1/10	1/10	1/10		2/10
2	0/10	1/10	2/10	1/10		1/10
3	0/10	1/10	2/10	1/10		3/10
4	10/10	1/10	2/10	2/10		2/10
5	0/10	6/10	2/10	4/10		1/10
6	0/10	0/10	1/10	1/10		1/10
mean	4.0	4.0	3.5	4.0	$\cdots$	3.2

$$p(n_1, n_2, \ldots, n_6 | N, p_1, p_2, \ldots, p_6) = \left\{ \underbrace{\frac{N!}{n_1! n_2! \ldots n_6!}}_{W} \underbrace{p_1^{n_1} \times p_2^{n_2} \times \cdots \times p_6^{n_6}}_{P} \right. \qquad (8.3)$$

where $N = \Sigma_i n_i = $ the total number of throws of the die. W is the number of different ways that h_i can occur in N trials, usually called the multiplicity. P is the probability of any particular realization of hypothesis h_i.

Without knowing the p_i's we have no way of computing the P term.[2] In what follows, we will ignore this term and investigate the consequences of the W term alone.

Let's evaluate the multiplicity, W, for h_1 and h_4.

$$W_{h_1} = \frac{10!}{0!0!0!10!0!0!} = 1,$$

$$W_{h_4} = \frac{10!}{1!1!1!2!4!1!} = 75\,600.$$

It is obvious that h_1 can only occur in one way, but to our surprise we find that h_4 can occur in 75 600 different ways. In the absence of additional information, if we were to carry out a large number of repeats of ten tosses, we would expect h_4 to occur 75 600 times more often than h_1. Thus, amongst the hypotheses that satisfy our constraint, the one with the largest multiplicity is the one we would consider most probable. Call this outcome W_{max}.

From W_{max} we can derive the frequency of occurrence of any particular side of the die and use this as an estimate of the probability of that side. A problem arises because for a small number of throws like ten, the frequency determined for one or more sides might be zero. To set the probability of these sides to zero would be unwarranted since what it really means is that $p_i < 1/10$.

The general concept of using the multiplicity to select from the h_i satisfying the constraint is good, but we need to refine it further. Suppose we were to use the average of the ten throws as the best estimate of the average of a much larger number of throws $N \gg 10$. For a much larger N, there would be a correspondingly larger number of hypotheses about probability distributions which could satisfy the average constraint. In this case, the smallest increment in p_i will be $1/N$ instead of $1/10$. The one with the largest multiplicity (W_{max}) will be a smoother version of what we got earlier with only ten throws, and if N is large enough, it will be very unlikely for any of the p_i to be exactly zero, unless of course the average was either one or six dots.

We like the smoother version of the probability distribution that comes about from using a larger value of N; however, this gives rise to a new difficulty. We will see in Equation (8.7) that as N increases, W_{max} increases at such a rate that there are

---

[2] Clearly, if the mean number of dots is significantly different from 3.5, the value for a fair die, then the constraint information is telling us that some sides are more probable than others. A challenge for the future is to see how this information can be used to constrain the P term.

(essentially) infinitely more ways W_{max} can be realized than other not-too-different probability distributions. Since we started with an average constraint pertaining to only ten throws, this degree of discrimination against acceptable competing h_i is unwarranted. Happily, if we use Stirling's approximation for $\ln N!$, we can factor $\ln W$ into two terms as we now show. *Stirling's approximation* for large N is

$$\ln N! = N \ln N - N. \tag{8.4}$$

Writing $n_i = Np_i$, the multiplicity becomes

$$
\begin{aligned}
\ln W &= N \ln N - N - \sum_{i=1}^{6} Np_i \ln Np_i + \sum_{i=1}^{6} Np_i \\
&= N \ln N - N - \sum Np_i \ln(Np_i) + \sum Np_i \\
&= N \ln N - N - N\left(\sum p_i \ln p_i + \ln N\right) + N \\
&= -N \sum_{i=1}^{6} p_i \ln p_i
\end{aligned}
$$

$$\ln W = -N \sum_{i=1}^{6} p_i \ln p_i = N \times \text{entropy} = NS, \tag{8.5}$$

where

$$S = -\sum_{i=1}^{6} p_i \ln p_i. \tag{8.6}$$

Equation (8.5) factors $\ln W$ into two terms: the number of throws, N, and the entropy term, which depends only on the desired p_i's. Maximizing entropy achieves the desired smooth probability distribution.

Since the multiplicity $W = \exp(NS)$, it follows that

$$\frac{W_{max}}{W} = \exp[N(S_{max} - S)] = \exp(N\Delta S), \tag{8.7}$$

where W_{max} is the multiplicity of the probability distribution with maximum entropy, S_{max}, and W is the multiplicity of a distribution with entropy S. The actual relative probabilities of two different probability distributions is proportional to the ratio of their multiplicities or $\propto \exp(N \times \Delta\text{entropy})$. Clearly, the degree of discrimination depends strongly on N. For N large, Equation (8.7) tells us that there are (essentially) infinitely more ways the outcome corresponding to maximum entropy (MaxEnt) can be realized than any outcome having a lower entropy.

Jaynes (1982) showed that the quantity $2N\Delta S$ has a χ^2 distribution with $M - k - 1$ degrees of freedom, where M is the number of possible outcomes and k is the number of constraints. In our problem of the die, $M = 6$ and $k = 1$. This allows us to compute explicitly the range of S about S_{max} corresponding to any confidence level.

8.5 Generalizing MaxEnt

8.5.1 Incorporating a prior

In Equation (8.3), we argued that without knowing the p_i's we have no way of computing term P. Using the principle of indifference, we assigned the same value for P for each of the acceptable hypotheses and concluded that the relative probability of acceptable hypotheses is proportional to the multiplicity term. In the present generalization, we allow for the possibility of prior information about the $\{p_i\}$. For example, suppose that the index i enumerates the individual pixels in an image of a very faint galaxy taken with the Hubble Space Telescope. Our constraint information in this case is the set of measured image pixel values. However, because of noise, these constraints are uncertain. In Section 8.8.2 we will learn how to make use of MaxEnt with uncertain constraints. In general, to find the MaxEnt image requires an iterative procedure which starts from an assumed prior image which is often taken to be flat, i.e., all p_i equal. However, if we already have another lower resolution image of the same galaxy taken with a ground-based telescope, then this would be a better prior image to start from. In this way, we can have a prior estimate of the p_i values in Equation (8.3).

For the moment we will return to the case where our constraints are certain and we will let $\{m_i\}$ be our prior estimate of $\{p_i\}$. For example, maybe we know that two sides of the die have two dots and that the other four sides have 3, 4, 5, and 6 dots, respectively. Substituting into Equation (8.3), and generalizing the discussion to a discrete probability distribution where i varies from 1 to M (instead of $i = 1$ to 6 for the die), we obtain

$$p(n_1, n_2, \ldots, n_M | N, p_1, p_2, \ldots, p_M) = \frac{N!}{n_1!\, n_2! \ldots n_M!} m_1^{n_1} \times m_2^{n_2} \times \cdots \times m_M^{n_M}. \quad (8.8)$$

Taking the natural logarithm of both sides yields

$$
\begin{aligned}
\ln[p(n_1, n_2, \ldots, n_M | N, p_1, p_2, \ldots, p_M)] &= \sum_{i=1}^{M} n_i \ln[m_i] + \ln[N!] - \sum_{i=1}^{M} \ln[n_i!] \\
&= \sum_{i=1}^{M} n_i \ln[m_i] - N \sum_{i=1}^{M} p_i \ln[p_i]
\end{aligned}
\quad (8.9)
$$

where we have used Stirling's approximation (Equation (8.4)) in the last line. Substituting for $n_i = Np_i$, we obtain

$$
\begin{aligned}
\frac{1}{N} \ln[p(n_1, n_2, \ldots, n_M | N, p_1, p_2, \ldots, p_M)] &= \sum_{i=1}^{M} p_i \ln[m_i] - \sum_{i=1}^{M} p_i \ln[p_i] \\
&= -\sum_{i=1}^{M} p_i \ln[p_i/m_i] = S.
\end{aligned}
\quad (8.10)
$$

This generalized entropy is known by various names including the *Shannon–Jaynes entropy* and the *Kullback entropy*. It is also sometimes written with the opposite sign (so it has to be minimized) and referred to as the *cross-entropy*.

8.5.2 *Continuous probability distributions*

The correct measure of uncertainty in the continuous case (Jaynes, 1968; Shore and Johnson, 1980) is:

$$S_c = - \int p(y) \ln \left[\frac{p(y)}{m(y)} \right] dy. \tag{8.11}$$

The quantity $m(y)$, called the *Lebesgue measure* (Sivia, 1996), ensures that the entropy expression is invariant under a change of variables, $y \to y' = f(y)$, because both $p(y)$ and $m(y)$ transform in the same way. Essentially, the measure takes into account how the (uniform) bin-widths in y-space translate to a corresponding set of (variable) bin-widths in the alternative y'-space. If $m(y)$ is a constant, this equation reduces to

$$S_c = - \int p(y) \ln p(y) dy + \ln m(y) \int p(y) dy$$
$$= - \int p(y) \ln p(y) dy + \text{constant}. \tag{8.12}$$

To find the maximum entropy solution, we are interested in derivatives of Equation (8.12), and for this, a constant prior has no effect.

8.6 How to apply the MaxEnt principle

In this section, we will demonstrate how to use the MaxEnt principle to encode some testable information into a probability distribution. We will need to use the Lagrange multipliers of variational calculus in which MaxEnt plays the role of a variational principle,[3] so we first briefly review that topic.

8.6.1 *Lagrange multipliers of variational calculus*

Suppose there are M distinct possibilities $\{y_i\}$ to be considered where $i = 1$ to M. We want to compute $p(y_i|I)$ (abbreviated by p_i) subject to a testable constraint. If S represents the entropy of $p(y_i|I)$, then the condition for maximum entropy is given by

$$dS = \frac{\partial S}{\partial p_1} dp_1 + \cdots + \frac{\partial S}{\partial p_M} dp_M = 0.$$

Without employing a constraint, the dp_i's are independent and the only solution is if all the coefficients are individually equal to 0. Suppose we are given the constraint

---

[3] One desirable feature of a variational principle is that it does not introduce correlations between the p_i values unless information about these correlations is contained in the constraints. In Section 8.8.1, we show that the MaxEnt variational principle satisfies this condition.

$\sum p_i^2 = R$, where R is a constant. Rewrite the constraint[4] as $C = \sum p_i^2 - R = 0$. With this constraint, any permissible dp_i's must satisfy

$$
\begin{aligned}
dC = 0 &= \frac{\partial C}{\partial p_1} dp_1 + \cdots + \frac{\partial C}{\partial p_M} dp_M \\
&= 2p_1 dp_1 + \cdots + 2p_M dp_M.
\end{aligned}
\tag{8.13}
$$

We can combine dS and dC in the form

$$
dS - \lambda dC = 0,
\tag{8.14}
$$

where λ is an undetermined multiplier.

$$
\left(\frac{\partial S}{\partial p_1} - \lambda 2 p_1 \right) dp_1 + \cdots + \left(\frac{\partial S}{\partial p_M} - \lambda 2 p_M \right) dp_M = 0.
$$

Now if λ is chosen so $(\partial S/\partial p_1 - \lambda 2 p_1) = 0$, then the equation reduces to

$$
\left(\frac{\partial S}{\partial p_2} - \lambda 2 p_2 \right) dp_2 + \cdots + \left(\frac{\partial S}{\partial p_M} - \lambda 2 p_M \right) dp_M = 0.
\tag{8.15}
$$

But the remaining $M - 1$ variables dp_i can be considered independent so their coefficients must also equal zero to satisfy Equation (8.15). This set of M equations, together with the constraint equation, can be solved for the $\{p_i\}$ and the Lagrange multiplier λ. It can be shown that this procedure does lead to a global maximum in S (e.g., Tribus, 1969).

8.7 MaxEnt distributions

Before deriving MaxEnt probability distributions for some common forms of testable information, we first examine some general properties of MaxEnt distributions.

8.7.1 General properties

Suppose we are given the following constraints:

$$
\sum_{i=1}^{M} p_i = 1
$$
$$
\sum_{i=1}^{M} f_1(y_i) p_i = \langle f_1 \rangle = \overline{f_1}
$$
$$
\vdots
$$
$$
\sum_{i=1}^{M} f_r(y_i) p_i = \langle f_r \rangle = \overline{f_r}
$$

where M is the number of discrete probabilities. For example, suppose we have the constraint information $\langle \cos(y) \rangle = \overline{f_1}$. In this case, $f_1(y_i) = \cos(y_i)$. Equation (8.14) can be written with the help of Equation (8.10) as

---

[4] In principle, R might be a $f(p_1, \ldots, p_M)$ and thus lead to a term in the differential.

$$d\left[-\sum_{i=1}^{M}p_i\ln p_i + \sum_{i=1}^{M}p_i\ln m_i - \lambda\left(\sum_{i=1}^{M}p_i - 1\right) - \lambda_1\left(\sum_{i=1}^{M}f_1(y_i)p_i - \overline{f_1}\right) - \right.$$
$$\left. \cdots - \lambda_r\left(\sum_{i=1}^{M}f_r(y_i)p_i - \overline{f_r}\right)\right] = 0. \tag{8.16}$$

Assuming m_i is a constant, then $\sum_{i=1}^{M}p_i\ln m_i = \ln m_i \sum_{i=1}^{M}p_i = \ln m_i$ and $d\ln m_i = 0$. In this case, the above equation simplifies to

$$\sum_{i=1}^{M}\left[-\ln p_i - p_i\frac{\partial\ln p_i}{\partial p_i} - \lambda - \lambda_1 f_1(y_i) - \cdots - \lambda_r f_r(y_i)\right]dp_i$$
$$= \sum_{i=1}^{M}[-\ln p_i - (1+\lambda) - \lambda_1 f_1(y_i) - \cdots - \lambda_r f_r(y_i)]dp_i = 0. \tag{8.17}$$

For each i, we can solve for p_i.

$$p_i = \exp[-(1+\lambda)]\exp[-\lambda_1 f_1(y_i) - \cdots - \lambda_r f_r(y_i)]$$
$$= \exp[-\lambda_0]\exp\left[-\sum_{j=1}^{r}\lambda_j f_j(y_i)\right], \tag{8.18}$$

where $\lambda_0 = 1 + \lambda$. Using the first constraint, we obtain

$$\sum_{i=1}^{M}p_i = \exp[-\lambda_0]\sum_{i=1}^{M}\exp\left[-\sum_{j=1}^{r}\lambda_j f_j(y_i)\right] = 1, \tag{8.19}$$

which can be rewritten as

$$\exp[+\lambda_0] = \sum_{i=1}^{M}\exp\left[-\sum_{j=1}^{r}\lambda_j f_j(y_i)\right]. \tag{8.20}$$

Now differentiate Equation (8.20) with respect to λ_k, and multiply through by $-\exp[-\lambda_0]$ to obtain

$$-\frac{\partial\lambda_0}{\partial\lambda_k} = \sum_{i=1}^{M}\exp[-\lambda_0]\exp\left[-\sum_{j=1}^{r}\lambda_j f_j(y_i)\right]f_k(y_i)$$
$$= \sum_{i=1}^{M}p_i f_k(y_i) = \langle f_k\rangle. \tag{8.21}$$

This leads to the following useful result, that we make use of in Section 8.7.4.

$$-\frac{\partial\lambda_0}{\partial\lambda_k} = \langle f_k\rangle = \overline{f_k}. \tag{8.22}$$

From Equation (5.4), we can write the variance of f_k as

$$\text{Var}(f_k) = \langle f_k^2 \rangle - (\langle f_k \rangle)^2. \tag{8.23}$$

We obtain $\langle f_k^2 \rangle$ from a second derivative of Equation (8.21). Substituting that into Equation (8.23) yields

$$\text{Var}(f_k) = \frac{\partial^2 \lambda_0}{\partial \lambda_k^2} - \left(\frac{\partial \lambda_0}{\partial \lambda_k}\right)^2. \tag{8.24}$$

8.7.2 Uniform distribution

Suppose the only known constraint is the minimal constraint possible for a probability distribution $\sum_{i=1}^{M} p_i = 1$. Following Equation (8.14), we can write

$$d\left[\underbrace{-\sum_{i=1}^{M} p_i \ln[p_i/m_i]}_{\text{entropy}} \underbrace{-\lambda\left(\sum_{i=1}^{M} p_i - 1\right)}_{\text{constraint}}\right] = 0 \tag{8.25}$$

$$d\left[-\sum_{i=1}^{M} p_i \ln p_i + \sum_{i=1}^{M} p_i \ln m_i - \lambda\left(\sum_{i=1}^{M} p_i - 1\right)\right] = 0$$

$$\sum_{i=1}^{M}\left(-\ln p_i - p_i\frac{\partial \ln p_i}{\partial p_i} + \ln m_i - \lambda\frac{\partial p_i}{\partial p_i}\right)dp_i = 0$$

$$\sum_{i=1}^{M}(-\ln[p_i/m_i] - 1 - \lambda)dp_i = 0.$$

The addition of the Lagrange undetermined multiplier makes

$$(-\ln[p_i/m_i] - 1 - \lambda = 0)$$

for one p_i and the remaining $(M-1)$ of the dp_i's independent. So for all p_i, we require

$$-\ln[p_i/m_i] - 1 - \lambda = 0, \tag{8.26}$$

or,

$$p_i = m_i e^{-(1+\lambda)}. \tag{8.27}$$

Since $\sum p_i = 1$,

$$\sum_{i=1}^{M} m_i e^{-(1+\lambda)} = 1 = e^{-(1+\lambda)} \sum_{i=1}^{M} m_i. \tag{8.28}$$

Since $\sum_{i=1}^{M} m_i = 1$, then $\lambda = -1$ and thus

$$p_i = m_i. \tag{8.29}$$

Suppose our prior information leads us to assume $m_i =$ a constant $= 1/M$. Then p_i describes a uniform distribution. In the continuum limit, we would write Equation (8.29) as

$$p(y|I) = m(y). \tag{8.30}$$

Thus, for $m(y) =$ a constant and the minimal constraint, $\int p(y) = 1$, the uniform distribution has maximum entropy.

8.7.3 Exponential distribution

In this case, we assume an additional constraint that the average value of y_i is known and equal to μ, so we have two constraints.

(1) $\sum_i^M p_i = 1$ (constraint 1)
(2) $\sum_i^M y_i p_i = \mu$ (constraint 2: known mean)

For example, in the die problem of Section 8.4, we could be told the average number of dots on a very large number of throws of the die but not be given the results of the individual throws. In this case, we use two Lagrange multipliers, λ and λ_1. Following Equation (8.14), we can write

$$d\left[\underbrace{-\sum_{i=1}^M p_i \ln[p_i/m_i]}_{\text{entropy}} - \lambda \underbrace{\left(\sum_{i=1}^M p_i - 1\right)}_{\text{constraint 1}} - \lambda_1 \underbrace{\left(\sum_{i=1}^M y_i p_i - \mu\right)}_{\text{constraint 2}}\right] = 0 \tag{8.31}$$

$$\sum_{i=1}^M \left(-\ln[p_i/m_i] - p_i \frac{\partial \ln p_i}{\partial p_i} - \lambda \frac{\partial p_i}{\partial p_i} - y_i \lambda_1 \frac{\partial p_i}{\partial p_i}\right) dp_i = 0 \tag{8.32}$$

$$\sum_{i=1}^M (-\ln[p_i/m_i] - 1 - \lambda - y_i \lambda_1) dp_i = 0.$$

Again, the addition of the Lagrange undetermined multipliers makes $(-\ln[p_i/m_i] - 1 - \lambda - y_i \lambda_1 = 0)$ for one p_i and the remaining $(M-1)$ of the dp_i's independent. So for all p_i, we require

$$-\ln[p_i/m_i] - 1 - \lambda - y_i \lambda_1 = 0, \tag{8.33}$$

or,

$$p_i = m_i \, e^{-(1+\lambda)} e^{-\lambda_1 y_i}. \tag{8.34}$$

We can now apply our two constraints to determine the Lagrange multipliers.

$$\sum_{i=1}^M p_i = 1 = e^{-(1+\lambda)} \sum_{i=1}^M m_i \, e^{-\lambda_1 y_i}. \tag{8.35}$$

Therefore

$$e^{-(1+\lambda)} = \frac{1}{\sum_{i=1}^{M} m_i e^{-\lambda_1 y_i}}. \tag{8.36}$$

From the second constraint, we have

$$\sum_{i=1}^{M} y_i p_i = \mu = \frac{\sum_{i=1}^{M} y_i m_i e^{-\lambda_1 y_i}}{\sum_{i=1}^{M} m_i e^{-\lambda_1 y_i}}, \tag{8.37}$$

or

$$\sum_{i=1}^{M} y_i m_i e^{-\lambda_1 y_i} - \mu \sum_{i=1}^{M} m_i e^{-\lambda_1 y_i} = 0. \tag{8.38}$$

For any particular value of μ, the above equation can be solved numerically for λ_1. If we set $m_i = 1/6$ and $y_i = i$, then Equation (8.38) can be used to solve for the probability of the six sides of the die problem discussed in Section 8.4. We illustrate this with the following *Mathematica* commands assuming $\mu = 2.2$ and the result is shown in Figure 8.1.

Box 8.1   *Mathematica* commands: MaxExt die problem

First, we define the function $q[\mu\_]$ for an arbitrary value of μ:

$$q[\mu\_]: = \sum_{i=1}^{6} i \; \mathbf{Exp}[-i \; \lambda 1] - \mu \sum_{i=1}^{6} \mathbf{Exp}[-i \; \lambda 1];$$

The next line solves for $\lambda 1$ with $\mu = 2.2$:

sol = Solve[q[2.2] == 0, λ1]

For each $\mu, \lambda 1$ has multiple complex solutions and one real solution $\lambda 1$real. Pick out the real solution using **Cases**.

λ1real = Cases[λ1/.sol,_Real][[1]]

Next, evaluate the expression for the probability of the *i*th side, **probi**

$$\mathbf{probi} = \frac{\mathbf{Exp}[-i \; \lambda 1 \mathbf{real}]}{\sum_{j=1}^{6} \mathbf{Exp}[-j \; \lambda 1 \mathbf{real}]}$$

Finally create a table of the probabilities of the 6 sides.

prob = Table[{i, probi}, {i, 6}]

{{1, 0.421273}, {2, 0.251917}, {3, 0.150644},
{4, 0.0900838}, {5, 0.0538692}, {6, 0.0322133}}

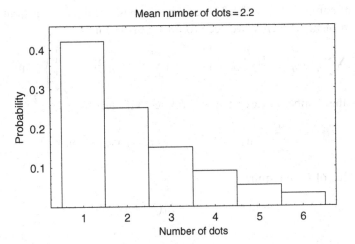

Figure 8.1 The figure shows the probability of the six sides of a die given the constraint that the average number of dots $\mu = 2.2$.

We can simply generalize Equation (8.34) for p_i to the continuous case $p(y|I)$:

$$p(y|I) = m(y)e^{-(1+\lambda)}e^{-\lambda_1 y}. \tag{8.39}$$

If we assume $m(y)$ is a constant, then we have

$$p(y|I) \propto e^{-\lambda_1 y}. \tag{8.40}$$

The normalization and λ_1 can easily be evaluated if the limits of integration extend from 0 to ∞. The result is

$$p(y|\mu) = \frac{1}{\mu}e^{-y/\mu} \quad \text{for } y \geq 0. \tag{8.41}$$

8.7.4 Normal and truncated Gaussian distributions

In this section, we assume $m(y)$, the prior for $p(y)$, has the following form:

$$m(y) = \begin{cases} 1/(y_H - y_L), & \text{if } y_L \leq y \leq y_H \\ 0, & \text{if } y_L > y \text{ or } y > y_H. \end{cases} \tag{8.42}$$

In this case, we assume an additional constraint that the variance of y is equal to σ^2. The two constraints in this case are:

(1) $\int_{y_L}^{y_H} p(y)dy = 1$

(2) $\int_{y_L}^{y_H} (y - \mu)^2 p(y)dy = \sigma^2$

Because $m(y)$ is a constant, we solve for $p(y)$ which maximizes

$$-\int p(y) \ln p(y)dy$$

subject to the constraints 1 and 2. This optimization is best done as the limiting case of a discrete problem; explicitly, we need to find the solution to

$$d\left\{-\sum_{i=1}^{M} p_i \ln p_i - \lambda\left[\sum_{i=1}^{M} p_i - 1\right] - \lambda_1\left[\sum_{i=1}^{M}(y_i - \mu)^2 p_i - \sigma^2\right]\right\} = 0, \qquad (8.43)$$

where M is the number of discrete probabilities. This leads to

$$\sum_{i=1}^{M}[-\ln p_i - 1 - \lambda - \lambda_1(y_i - \mu)^2]dp_i = 0. \qquad (8.44)$$

For each value of i, we require

$$-\ln p_i - 1 - \lambda - \lambda_1(y_i - \mu)^2 = 0, \qquad (8.45)$$

or,

$$p_i = e^{-(1+\lambda)}e^{-\lambda_1(y_i-\mu)^2}$$
$$= e^{-\lambda_0}e^{-\lambda_1(y_i-\mu)^2}, \qquad (8.46)$$

where $\lambda_0 = 1 + \lambda$. This generalizes to the continuum assignment

$$p(y) = e^{-\lambda_0}e^{-\lambda_1(y-\mu)^2}. \qquad (8.47)$$

We can solve for λ_1 and λ_0 from our two constraints. From the first constraint,

$$\int_{y_L}^{y_H} p(y)dy = 1 = e^{-\lambda_0}\int_{y_L}^{y_H} e^{-\lambda_1(y-\mu)^2}dy. \qquad (8.48)$$

Compare this equation to the equation for the *error function*[5] erf(z) (see Equation (5.39)) given by

$$\text{erf}(z) = \frac{2}{\sqrt{\pi}}\int_0^z \exp(-u^2)\,du. \qquad (8.49)$$

The solution for λ_0 in Equation (8.48), in terms of the error function, is

$$\lambda_0 = \ln\left[\frac{\sqrt{\pi}}{2\sqrt{\lambda_1}}\right] + \ln\left[\text{erf}\left\{\sqrt{\lambda_1}(y_H - \mu)\right\} - \text{erf}\left\{\sqrt{\lambda_1}(y_L - \mu)\right\}\right]. \qquad (8.50)$$

We will consider two cases which depend on the limits of integration (y_L, y_H).

Case I (Normal Gaussian)

Suppose the limits of integration satisfy the condition[6]

$$\sqrt{\lambda_1}(y_H - \mu) \gg 1 \text{ and } \sqrt{\lambda_1}(y_L - \mu) \ll -1. \qquad (8.51)$$

---

[5] The error function has the following properties: erf$(\infty) = 1$, erf$(-\infty) = -1$ and erf$(-z) = -$erf(z).
[6] Note: erf$(1) = 0.843$, erf$(\sqrt{2}) = 0.955$, erf$(2) = 0.995$, and erf$(3) = 0.999978$.

In this case,

$$\mathrm{erf}\left\{\sqrt{\lambda_1}(y_H - \mu)\right\} \approx 1 \text{ and } \mathrm{erf}\left\{\sqrt{\lambda_1}(y_L - \mu)\right\} \approx -1,$$

and Equation (8.50) simplifies to

$$\lambda_0 \approx \ln\left[\frac{\sqrt{\pi}}{2\sqrt{\lambda_1}}\right] + \ln[2] = \ln\left[\sqrt{\frac{\pi}{\lambda_1}}\right]. \tag{8.52}$$

We now make use of Equation (8.22) to obtain an equation for λ_1:

$$-\frac{\partial \lambda_0}{\partial \lambda_1} = -\frac{\partial \ln\left[\sqrt{\pi/\lambda_1}\right]}{\partial \lambda_1} = \frac{1}{2\lambda_1} = \sigma^2. \tag{8.53}$$

Combining Equations (8.53) and (8.52), we obtain

$$\lambda_1 = \frac{1}{2\sigma^2}; \quad e^{-\lambda_0} = \frac{1}{\sqrt{2\pi}\sigma}.$$

The result,

$$p(y) = \frac{1}{\sqrt{2\pi}\sigma} e^{-(y-\mu)^2/2\sigma^2}, \tag{8.54}$$

is a Gaussian. Thus, for a given σ^2 and a uniform prior that satisfies Equation (8.51), a Gaussian distribution has the greatest uncertainty (maximum entropy). Now that we have evaluated λ_0 and λ_1, we can rewrite Equation (8.51) in the more useful form

$$\frac{(y_H - \mu)}{\sqrt{2}\sigma} \gg 1 \quad \text{and} \quad \frac{(\mu - y_L)}{\sqrt{2}\sigma} \gg 1. \tag{8.55}$$

We frequently deal with problems where the qth data value, y_q, is described by an equation of the form

$$y_q = y_{pq} + e_q \quad \Rightarrow \quad y_q - y_{pq} = e_q,$$

where y_{pq} is the model prediction for the qth data value and e_q is an error term. In Section 4.8.1, we showed that for a deterministic model, M_j, the probability of the data, $p(Y_q|M_j, I)$, is equal to $p(E_q|M_j, I)$, the probability of the errors. If we interpret the μ in Equation (8.54) as y_{pq}, the model prediction, then this equation becomes the MaxEnt sampling distribution for the qth error term in the likelihood function.

This is a very important result. It says that unless we have some additional prior information which justifies the use of some other sampling distribution, then use a Gaussian sampling distribution. It makes the fewest assumptions about the information you don't have and will lead to the most conservative estimates (i.e., greater uncertainty than you would get from choosing a more appropriate distribution based on more information).

In a situation where we do not know the appropriate sampling distribution, we will also, in general, not know the actual variance (σ^2) of the distribution. In that case, we can treat the σ of the Gaussian sampling distribution as an unknown nuisance parameter with a range specified by our prior information. The one restriction to this argument is that the prior upper bound on σ must satisfy Equation (8.55). If possible data values, represented by the variable y, are unrestricted, then this condition simply requires that the upper bound on σ be finite. In some experiments, the range of possible data values is limited, e.g., positive values only. In that case, the MaxEnt distribution may become a truncated Gaussian, as discussed in Case II below. We now consider a simple example that exploits a MaxEnt Gaussian sampling distribution with unknown σ. We will make considerable use of this approach in later chapters starting with Section 9.2.3.

Example:
Suppose we want to estimate the location of the start of a stratum in a core sample taken by a Martian rover. The rover transmits a low resolution scan, which allows the experimenter to refine the region of interest for analysis by a higher resolution instrument aboard the rover. Unfortunately, the rover made a rough landing and ceases operation after only two high resolution measurements have been completed. In this example, we simulate a sample of two measurements made with the high resolution instrument for a stratum starting position of 20 units along the core sample. For the simulation, we assume a bimodal distribution of measurement errors as shown in panel (a) of Figure 8.2. We further suppose that the distribution of measurement errors is unknown by the scientist, named Jean, who will perform the analysis.

Jean needs to choose a sampling distribution for use in evaluating the likelihood function in a Bayesian analysis of the high resolution core sample. In the absence of additional information, she picks a Gaussian sampling distribution, because from the above argument, the Gaussian will lead to the most conservative estimates (i.e., greater uncertainty than you would get from choosing a more appropriate distribution based on more information). Based on the low resolution core sample measurements, she assumes a uniform prior for the mean location extending from 15 to 25 units. She assumes a Jeffreys prior for σ and estimates a conservative upper limit to σ of 4 units. She estimates the lower limit, $\sigma = 0.4$ units, by setting it equal to the digital read out accuracy.

To see how well the parameterized Gaussian sampling distribution performs, we simulated five independent samples, each consisting of two measurements. Panels (b), (c), (d), (e), and (f) of Figure 8.2 show a comparison of the posterior PDFs for the stratum start location computed using: (1) the true sampling distribution (solid curve), and (2) a Gaussian with an unknown σ (dotted curve). The actual measurements are indicated by the arrows in the top of each panel.

It is quite often the case that we don't know the true likelihood function. In some cases, we have a sufficient number of repeated measurements (say five or more) that we can appeal to the CLT (see Section 5.11) and work with the average value, whose distribution will be closely Gaussian with a σ given by Equation (5.50). However, in

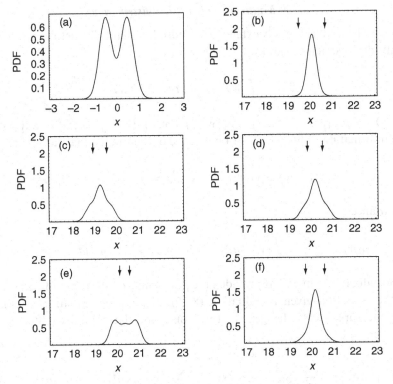

Figure 8.2 Panel (a) shows the bimodal distribution of instrumental measurement errors. Panels (b), (c), (d), (e), and (f) show a comparison of the posterior PDFs for the stratum start location derived from five simulated data samples. Each sample consists of the two data points indicated by the two arrows at the top of each panel. The solid curve shows the result obtained using the true sampling distribution. The dotted curve shows the result using a Gaussian with unknown σ.

this example, we had only two measurements. Instead, we appealed to the MaxEnt principle and used a Gaussian likelihood function, marginalizing over the unknown variance. From Figure 8.2, it is apparent that the Gaussian likelihood function performed quite well compared to the true bimodal likelihood function. The conservative nature of the Gaussian assumption is apparent from the much broader tails. Further details of this type of analysis are given in Section 9.2.3.

Case II (Truncated Gaussian)

When the condition specified by Equation (8.55) is not satisfied, it is still possible to compute a MaxEnt sampling distribution that we refer to as a *truncated Gaussian*, but there is no simple analytic solution for the Lagrange multipliers. The λ's need to be solved for numerically.[7]

---

[7] See Tribus (1969) for a more detailed discussion of this case.

8.7.5 *Multivariate Gaussian distribution*

The MaxEnt procedure is easily extended to multiple variables by defining the entropy as a multi-dimensional integral:

$$S = -\int p(Y)\ \ln[p(Y)/m(Y)]dY, \qquad (8.56)$$

where $p(Y) = p(y_1, y_2, \ldots, y_N|I) = p(\{y_i\}|I)$ and $\int dY = \int \cdots \int d^N y$. Suppose the testable information only consists of knowledge of their individual variances,

$$\langle (y_i - \mu_i)^2 \rangle = \int (y_i - \mu_i)^2 p(y_1, y_2, \ldots, y_N) d^N y = \sigma_{ii} = \sigma_i^2 \quad (i = 1 \text{ to } N), \qquad (8.57)$$

and covariances,

$$\langle (y_i - \mu_i)(y_j - \mu_j) \rangle = \int (y_i - \mu_i)(y_j - \mu_j)\ p(y_1, y_2, \ldots, y_N) d^N y = \sigma_{ij}. \qquad (8.58)$$

In Appendix E, we show that provided the prior limits on the range of each variable satisfy the condition given in Equation (8.55), maximizing Equation (8.56), with uniform measure, yields the general form of a correlated multivariate Gaussian distribution:

$$p(Y|\{\mu_i, \sigma_{ij}\}) = \frac{1}{(2\pi)^{N/2}\sqrt{\det \mathbf{E}}} \exp\left[-\frac{1}{2}\sum_{ij}(y_i - \mu_i)[\mathbf{E}^{-1}]_{ij}(y_j - \mu_j)\right], \qquad (8.59)$$

where

$$\sum_{ij} = \sum_{i=1}^{N}\sum_{j=1}^{N},$$

and

$$\mathbf{E} = \begin{pmatrix} \sigma_{11} & \sigma_{12} & \sigma_{13} & \cdots & \sigma_{1N} \\ \sigma_{21} & \sigma_{22} & \sigma_{23} & \cdots & \sigma_{2N} \\ \cdot & \cdot & \cdot & \cdot & \cdot \\ \sigma_{N1} & \sigma_{N2} & \sigma_{N3} & \cdots & \sigma_{NN} \end{pmatrix}. \qquad (8.60)$$

In most applications that we will encounter, the y_i variable will represent possible values of a datum and be labeled d_i. Equation (8.59) can then be rewritten as a likelihood:

$$p(D|M, I) = \frac{1}{(2\pi)^{N/2}\sqrt{\det \mathbf{E}}} \exp\left[-\frac{1}{2}\sum_{ij}(d_i - f_i)[\mathbf{E}^{-1}]_{ij}(d_j - f_j)\right], \qquad (8.61)$$

where f_i is the model prediction for the ith datum.

If the variables are all independent, i.e., the covariance terms are all zero, then Equation (8.61) reduces to

$$p(D|M,I) = \prod_{i=1}^{N} \frac{1}{\sqrt{2\pi}\sigma_i} \exp\left\{-\frac{(d_i - f_i)^2}{2\sigma_i^2}\right\}$$

$$= \left(\prod_{i=1}^{N} \frac{1}{\sqrt{2\pi}\sigma_i}\right) \exp\left\{-\sum_{r=1}^{R} \frac{(d_i - f_i)^2}{2\sigma_i^2}\right\}.$$

(8.62)

In Chapter 10, we discuss the concepts of covariance and correlation in more detail and make use of a multivariate Gaussian in least-squares analysis.

8.8 MaxEnt image reconstruction

It is convenient to think of a probability distribution as a special case of a PAD (positive, additive distribution). Another example of a PAD is the intensity or power, $f(x, y)$, of incoherent light as a function of position (x, y), in an optical image. This is positive and additive because the integral $\iint f(x, y)dxdy$ represents the signal energy recorded by the image. By contrast, the amplitude of incoherent light, though positive, is not additive. A probability distribution is a PAD which is normalized so

$$\int_{-\infty}^{+\infty} p(Y)dY = 1.$$

Question: What form of entropy expression should we maximize in image reconstruction to best represent the PAD $f(x, y)$ values?
Answer: Derived by Skilling (1989).

$$S(f, m) = -\int\int \left[f(x, y) - m(x, y) - f(x, y)\ln\left(\frac{f(x, y)}{m(x, y)}\right)\right]dxdy,$$

where $m(x, y)$ is the prior estimate of $f(x, y)$. If $f(x, y)$ and $m(x, y)$ are normalized, this reduces to the simpler form:

$$-\int\int f(x, y)\ln\left[\frac{f(x, y)}{m(x, y)}\right]dxdy.$$

8.8.1 The kangaroo justification

The following simple argument (Gull and Skilling, 1984) gives additional insight into the use of entropy for a PAD. Imagine that we are given the following information:

a) One third of kangaroos have blue eyes.
b) One third of kangaroos are left-handed.

How can we estimate the proportion of kangaroos that are both blue-eyed (BE) and left-handed (LH) using only the above information? The joint proportions of LH and BE can be represented by a 2×2 probability table which is shown in Table 8.2(a). The probabilities p_1, p_2, p_3 and p_4 must satisfy the given constraints:

a) $p_1 + p_2 = \frac{1}{3}$ (1/3 of kangaroos have blue eyes)
b) $p_1 + p_3 = \frac{1}{3}$ (1/3 of kangaroos are left-handed)
c) $p_1 + p_2 + p_3 + p_4 = 1$.

Feasible solutions have one remaining degree of freedom which we parameterize by the variable z. Table 8.2(b) shows the parameterized joint probability table. The parameter z is constrained by the above three constraints to $0 \leq z \leq \frac{1}{3}$. Below we consider three feasible solutions:

1. The first corresponds to the independent case

$$p(BE, LH) = p(BE)p(LH) = \left(\frac{1}{3}\right)\left(\frac{1}{3}\right) = \frac{1}{9}$$

$$z = \frac{1}{9}$$

which leads to the contingency table shown in Figure 8.3(a).
2. Case of maximum positive correlation.

$$p(BE, LH) = p(BE)p(LH|BE) = \left(\frac{1}{3}\right)(1) = \frac{1}{3}$$

$$z = \frac{1}{3}.$$

3. Case of maximum negative correlation.

$$p(BE, LH) = p(BE)p(LH|BE) = \left(\frac{1}{3}\right)(0) = 0$$

$$z = 0.$$

Table 8.2 *Panel (a) is the joint probability table for the kangaroo problem. In panel (b), the table is parameterized in terms of the remaining one degree of freedom represented by the variable z*

Blue eyes	Left-Handed	
	True	False
True	p_1	p_2
False	p_3	p_4

(a)

Blue eyes	Left-Handed	
	True	False
True	$0 \leq z \leq \frac{1}{3}$	$\frac{1}{3} - z$
False	$\frac{1}{3} - z$	$\frac{1}{3} + z$

(b)

(a) Independent

	Left-Handed True	Left-Handed False
Blue Eyes True	$\frac{1}{9}$	$\frac{2}{9}$
Blue Eyes False	$\frac{2}{9}$	$\frac{4}{9}$

(b) Positive correlation

	Left-Handed True	Left-Handed False
Blue Eyes True	$\frac{1}{3}$	0
Blue Eyes False	0	$\frac{2}{3}$

(c) Negative correlation

	Left-Handed True	Left-Handed False
Blue Eyes True	0	$\frac{1}{3}$
Blue Eyes False	$\frac{1}{3}$	$\frac{1}{3}$

Figure 8.3 The three panels give the joint probabilities for (a) the independent case, (b) maximum positive correlation, and (c) maximum negative correlation.

Suppose we must choose one answer – which is the best?

The answer we select cannot be thought of as being any more likely than any other choice, because there may be some degree of genetic correlation between eye color and handedness. However, it is nonsensical to select either positive or negative correlations without having any relevant prior information. Therefore, based on the available information, the independent choice $p_1 = p(BE, LH) = 1/9$ is preferred.

Question: Is there some function of the p_i which, when maximized subject to the known constraints, yields the same preferred solution? If so, then it would be a good candidate for a general variational principle which could be used in situations that were too complicated for our common sense. Skilling (1988) showed that the only functions with the desired property, $p(BE, LH) = 1/9$, are those related monotonically to the entropy:

$$S = -\sum_{i=1}^{4} p_i \ln p_i$$
$$= -z \ln z - 2\left(\frac{1}{3} - z\right) \ln\left(\frac{1}{3} - z\right) - \left(\frac{1}{3} + z\right) \ln\left(\frac{1}{3} + z\right).$$

Three proposed alternatives are listed in Table 8.3. Only one of the four $(-\sum p_i \ln p_i)$ gives the preferred uncorrelated result.

Table 8.3 *Solutions to the kangaroo problem obtained by maximizing four different functions, subject to the constraints.*

Variation Function	Optimal z	Implied Correlation
$-\sum p_i \ln p_i$	$1/9 = 0.1111$	uncorrelated
$-\sum p_i^2$	$1/12 = 0.0833$	negative
$\sum \ln p_i$	0.1303	positive
$\sum p_i^{1/2}$	0.1218	positive

But what have kangaroos got to do with image restoration/reconstruction? Consider the following restatement of the problem.

a) One third of the flux comes from the top half of the image.
b) One third of the flux comes from the left half of the image.

What proportion of the flux comes from the top left quarter? All the advertised functionals except $(-\sum p_i \ln p_i)$ imply either a positive or negative correlation in the distribution of the flux in the four quadrants based on the given information. Thus, these functionals fail to be consistent with our prior information on even the simplest non-trivial image problem. Inconsistencies are not expected to disappear just because practical data are more complicated.

8.8.2 *MaxEnt for uncertain constraints*

Example:
In image reconstruction, we want the most probable image when the data are incomplete and noisy. In this example,

$B \equiv$ "proposition representing prior information"
$I_i \equiv$ "proposition representing a particular image."

Apply Bayes' theorem:

$$p(I_i|D, B) \propto p(I_i|B)p(D|I_i, B). \tag{8.63}$$

Suppose the image consists of M pixels ($j = 1 \rightarrow M$)
Let $d_j =$ measured value for pixel j
   $I_{ij} =$ predicted value for pixel j based on image hypothesis I_i
   $e_j = d_j - I_{ij} =$ error due to noise which is assumed to be IID Gaussian.

In this situation, the measured d_j values are the constraints, which are uncertain because of noise. Thus,

$$p(d_j|I_{ij}, B) = p(e_j|I_{ij}, B) \propto \exp\left[-\frac{e_j^2}{2\sigma_j^2}\right] = \exp\left[-\frac{1}{2}\left(\frac{d_j - I_{ij}}{\sigma_j}\right)^2\right] \tag{8.64}$$

and

$$p(D|I_i, B) \propto \prod_{j=1}^{m} \exp\left[-\frac{1}{2}\left(\frac{d_j - I_{ij}}{\sigma_j}\right)^2\right]$$

$$= \exp\left[-\frac{1}{2}\sum_{j=1}^{m}\left(\frac{d_j - I_{ij}}{\sigma_j}\right)^2\right] = \exp\left[-\frac{\chi^2}{2}\right]. \tag{8.65}$$

Determination of $p(I_i|B)$:
Suppose we made trial images, I_i, by taking N quanta and randomly throwing them into the M image pixels. Then $p(I_i|B)$ is given by a multinomial distribution,

$$p(I_i|B) = \frac{N!}{n_1! \ldots n_M!} \frac{1}{M^N} = \frac{W}{M^N}, \qquad (8.66)$$

where W is the multiplicity.
   Recall for large N,

$$\ln W \to -N \sum_j p_j \ln p_j = N \times \text{entropy} = NS,$$

where as $N \to \infty$, $p_j \to n_j/N = \text{constant}$.
   Therefore,

$$p(I_i|B) = \frac{1}{M^N} \exp(NS). \qquad (8.67)$$

In general, we don't know the number of discrete quanta in the image, so we write

$$p(I_i|B) = \exp(\alpha S). \qquad (8.68)$$

Substituting Equations (8.68) and (8.65) into Equation (8.63), we obtain

$$p(I_i|D, B) = \exp\left(\alpha S - \frac{\chi^2}{2}\right). \qquad (8.69)$$

We want to maximize $p(I_i|D, B)$ or $(\alpha S - \chi^2/2)$.
   In "classic" MaxEnt, the α parameter is set so the misfit statistic χ^2 is equal to the number of data points N. This in effect overestimates χ^2, since some effective number λ_1 parameters are being "fitted" in doing the image reconstruction.
   The full Bayesian approach treats α as a parameter of the hypothesis space which can be estimated by marginalizing over the image hypothesis space. Improved images can also be obtained by introducing prior information about the correlations between image pixels, enforcing smoothness. More details on MaxEnt image reconstruction can be found in Buck (1991), Gull and Skilling (1984), Skilling (1989), Gull (1989a), and Sivia (1996).
   Two examples that illustrate some of the capabilities of MaxEnt image reconstruction are shown in Figures 8.4 and 8.5. Figure 8.4 illustrates how the maximum entropy method is capable of increasing the contrast of an image, and can also increase its sharpness if the measurements are sufficiently accurate. Figure 8.5 illustrates how the maximum entropy method automatically allows for missing data (Skilling and Gull, 1985). These and other examples, along with information on commercial software products, are available from Maximum Entropy Data Consultants, Ltd. (http://www.maxent.co.uk/).

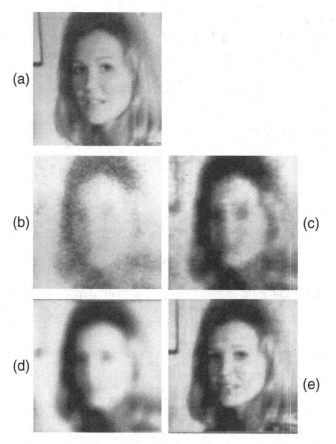

Figure 8.4 The original high-resolution low-noise image is shown in panel (a). Panel (b) shows the blurred original with high added noise. The MaxEnt reconstruction of the blurred noisy image is shown in (c). This demonstrates how the maximum entropy method suppresses noise, yielding a higher contrast image. Panel (d) shows the blurred original with low added noise. The MaxEnt reconstructed image, shown in (e), demonstrates how well maximum entropy de-blurs if the data are accurate enough. (Courtesy S. F. Gull, Maximum Entropy Data Consultants.)

8.9 Pixon multiresolution image reconstruction

Piña and Puetter (1993) and Puetter (1995) describe another very promising Bayesian approach to image reconstruction, which they refer to as the Pixon^{TM} method. Instead of representing the image with pixels of a constant size, they introduce an image model where the size of the pixel varies locally according to the structure in the image. Their generalized pixels are called *pixons*. A map of the pixon sizes is called an *image model*. The Pixon method seeks to find the best joint image and image model that is consistent with the data based on a χ^2 goodness-of-fit criterion, and that can represent the structure in the image by the smallest number of pixons. For example, suppose we have a 1024 by 1024 image of the sky containing a galaxy which occupies the inner 100

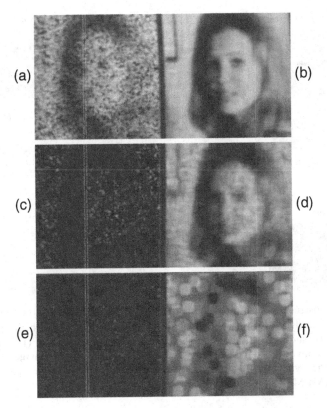

Figure 8.5 This figure demonstrates how the maximum entropy method automatically allows for missing data. Panel (a) shows the original image when 50% of the pixels, selected at random, have been removed. Panel (b) shows the corresponding MaxEnt reconstructed image. Panel (c) shows the original image when 95% of the pixels have been removed. Panel (d) shows the corresponding MaxEnt reconstructed image. Panel (e) shows the original image when 99% of the pixels have been removed. Panel (f) shows the corresponding MaxEnt reconstructed image. (Courtesy S. F. Gull, Maximum Entropy Data Consultants.)

by 100 pixels. In principle, we need many numbers to encode the significant structure in the galaxy region, but only one number to encode information in the featureless remainder of the image. Because the Pixon method constructs a model that represents the significant structure by the smallest number of parameters (pixons), it has the smallest Occam penalty.

Figure 8.6 shows an image reconstructions of a mock data set. The original image is shown on the far left along with a surface plot (center row). The original image is convolved (blurred) with the point-spread-function (PSF) shown at the bottom of the first column. Then noise (see bottom of second column) is added to the smoothed (PSF-convolved) data to produce the input (surface plot in middle panel) to the image reconstruction algorithm. To the right are a Pixon method reconstruction and a maximum entropy reconstruction. The algorithms used are the MEMSYS 5

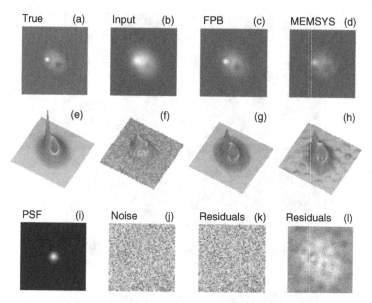

Figure 8.6 Reconstruction of a mock data set. The original image is shown on the far left (a) along with a surface plot (e). This image is convolved (blurred) with the point-spread-function (PSF) (i) shown at the bottom of the first column. Then noise (j) is added to the smoothed (PSF-convolved) data to produce the input (b), (f) to the image reconstruction algorithm. To the right are a Pixon method reconstruction (FPB) and a maximum entropy reconstruction (MEMSYS). (Courtesy Pixon LLC.)

algorithms, a powerful set of commercial maximum entropy (ME) algorithms available from Maximum Entropy Data Consultants, Ltd. The ME reconstructions were performed by Nick Weir, a recognized ME and MEMSYS expert. The reconstructions were supplemented by Nick Weir's multi-correlation channel approach.

The Pixon method reconstructions use the *Fractal–Pixon Basis* (FPB) approach (Piña and Puetter, 1993; Puetter, 1995). The "Fractal" nomenclature has since been dropped, so the term FPB simply refers to the "standard" Pixon method. It can be seen that the FPB reconstruction has no signal correlated residuals and is effectively artifact (false-source) free, whereas these problems are obvious in the MEMSYS reconstruction. The absence of signal correlated residuals and artifacts can be understood from the underlying theory of the Pixon method (Puetter, 1995).

Figure 8.7 shows the Pixon method applied to X-ray mammography, taken from the Pixon[TM] homepage located at http://www.pixon.com, or alternatively, http://casswww.ucsd.edu/personal/puetter/pixonpage.html. The raw X-ray image appears to the left. In this example, a breast phantom is used (material with X-ray absorption properties similar to the human breast). A small fiber (400 micrometer diameter) is present in the phantom. The signature of the fiber is rather faint in the direct X-ray image. The Pixon method reconstruction is seen to the right. Here, the signature of the fiber is obvious. Such image enhancement is of clear benefit to the discovery of weak

Mammogram (standard phantom, American College of Radiology)

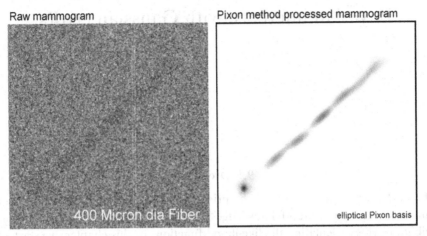

Figure 8.7 An example of the Pixon method applied to X-ray mammography. The raw X-ray image appears to the left. In this example, a breast phantom is used (material with X-ray absorption properties similar to the human breast). In this case, a small fiber (400 micrometer diameter) is present in the phantom. The Pixon method reconstruction is seen to the right. (Courtesy Pixon LLC.)

X-ray signatures. As can be seen, the X-ray signature of the fiber is very close to the noise level. This is evidenced by the break-up of the continuous fiber into pieces in the Pixon image. The Pixon method recognized that in certain locations, the X-ray signal present is not statistically significant. In these locations, the fiber vanished in the reconstructed image.

8.10 Problems

1. Use the maximum entropy method to compute and plot the probability of each side of a six-sided loaded die given that exhaustive tests have determined that the expectation value of the number of dots on the uppermost face = 4.6.
2. Use the maximum entropy method to compute and plot the probability of each side of a six-sided loaded die given that exhaustive tests have determined that the expectation value of the number of dots on the uppermost face = α, for $\alpha = 1.1$ to 5.9 in steps of 0.1. Plot the probability of each side versus α. Plot the probability of all six sides on one plot versus α.
3. Evaluate a unique probability distribution for $p(Y|I)$ (the question posed in Section 8.1) using the MaxEnt principle together with the constraint: "the mean value of $\cos y = 0.6$." Our prior information also tells us that $m(Y|I)$, the prior estimate of $p(Y|I)$, is a constant in the range 0 to 2π. In working out the solution, you will encounter the modified Bessel functions of the first kind, designated by **BesselI[n, z]** in *Mathematica*. You may also find the command **FindRoot[]** useful.

9

Bayesian inference with Gaussian errors

9.1 Overview

In the next three chapters, we will be primarily concerned with estimating model parameters when our state of knowledge leads us to assign a Gaussian sampling distribution when calculating the likelihood function. In this chapter, we start with a simple problem of computing the posterior probability of the mean of a data set. Initially, we assume the variance of the sampling distribution is known and then consider the case where the variance is unknown. We next look at the question of how to determine whether the signal present in the data is constant or variable. In the final section, we consider a Bayesian treatment of a fundamental problem that occurs in experimental science – that of analyzing two independent measurements of the same physical quantity, one "control" and one "trial."

9.2 Bayesian estimate of a mean

Here we suppose that we have collected a set of N data values $\{d_1, \ldots, d_N\}$ and we are assuming the following model is true:

$$d_i = \mu + e_i,$$

where e_i represents the noise component of the ith data value. For this one data set, and any prior information, we want to obtain the Bayesian estimate of μ. We will investigate three interesting cases. In all three cases, our prior information about e_i leads us to adopt an independent Gaussian sampling distribution.[1] In Section 9.2.1, we analyze the case where the noise σ is the same for all e_i. In Section 9.2.2, we treat the more general situation where the σ_i are unequal. Section 9.2.3 considers the case where the σ_i are assumed equal but the value is unknown.

---

[1] Note: if we had prior evidence of dependence, i.e., correlation, it is a simple computational detail to take this into account as shown in Section 10.2.2.

9.2.1 Mean: known noise σ

In this situation, we will assume that the variance of the noise is already known. We might, for example, know this from earlier measurements with the same apparatus in similar conditions. We also assume the prior information gives us lower and upper limits on μ but no preference for μ in that range.

The problem is to solve for $p(\mu|D,I)$. The first step is to write down Bayes' theorem:

$$p(\mu|D,I) = \frac{p(\mu|I)\,p(D|\mu,I)}{p(D|I)},$$ (9.1)

where the likelihood $p(D|\mu,I)$ is sometimes written as $\mathcal{L}(\mu)$. Our assumed prior for μ is given by

$$p(\mu|I) = K(\text{constant}), \quad \mu_{\mathrm{L}} \leq \mu \leq \mu_{\mathrm{H}}$$
$$= 0, \quad \text{otherwise.}$$

Evaluate K from

$$\int_{\mu_{\mathrm{L}}}^{\mu_{\mathrm{H}}} p(\mu|I)d\mu = \int_{\mu_{\mathrm{L}}}^{\mu_{\mathrm{H}}} K d\mu = 1.$$

Therefore,

$$K = \frac{1}{\mu_{\mathrm{H}} - \mu_{\mathrm{L}}} = \frac{1}{R_{\mu}},$$

where $R_{\mu} \equiv$ range of μ. This gives the normalized prior,

$$p(\mu|I) = \frac{1}{R_{\mu}}.$$ (9.2)

The likelihood is given by

$$p(D|\mu,I) = \prod_{i=1}^{N} \frac{1}{\sigma\sqrt{2\pi}} \exp\left\{ -\frac{(d_i - \mu)^2}{2\sigma^2} \right\}$$

$$= \sigma^{-N}(2\pi)^{-\frac{N}{2}} \exp\left\{ -\frac{\sum_{i=1}^{N}(d_i - \mu)^2}{2\sigma^2} \right\}$$ (9.3)

$$= \sigma^{-N}(2\pi)^{-\frac{N}{2}} \exp\left\{ -\frac{Q}{2\sigma^2} \right\},$$

where we have abbreviated $\sum_{i=1}^{N}(d_i - \mu)^2$ by Q.

Expanding Q, we obtain

$$Q = \sum_{i=1}^{N}(d_i - \mu)^2 = \sum d_i^2 + \sum \mu^2 - 2\sum d_i\mu$$

$$= \sum d_i^2 + N\mu^2 - 2N\mu\bar{d} \quad \{\bar{d} \equiv \frac{1}{N}\sum d_i\}$$

$$= N(\mu^2 - 2\mu\bar{d} + \bar{d}^2) + \sum d_i^2 - N\bar{d}^2$$

$$= N(\mu - \bar{d})^2 + \sum d_i^2 - N\bar{d}^2 \tag{9.4}$$

$$= N(\mu - \bar{d})^2 + \sum d_i^2 - 2N\bar{d}^2 + N\bar{d}^2$$

$$= N(\mu - \bar{d})^2 + \sum d_i^2 - 2\bar{d}\sum d_i + \sum \bar{d}^2$$

$$= N(\mu - \bar{d})^2 + \sum (d_i - \bar{d})^2$$

$$= N(\mu - \bar{d})^2 + Nr^2,$$

where $r^2 = \frac{1}{N}\sum(d_i - \bar{d})^2$ is the mean square deviation from $\bar{d}$. Now substitute Equation (9.4) into Equation (9.3):

$$p(D|\mu, I) = \sigma^{-N}(2\pi)^{-\frac{N}{2}}\exp\left\{-\frac{Nr^2}{2\sigma^2}\right\}\exp\left\{-\frac{N(\mu - \bar{d})^2}{2\sigma^2}\right\}. \tag{9.5}$$

We can express $p(D|I)$ as

$$p(D|I) = \int_{\mu_L}^{\mu_H} d\mu\, p(\mu|I)p(D|\mu, I). \tag{9.6}$$

Substitution of Equations (9.2), (9.5) and (9.6) into Equation (9.1) yields the desired posterior:

$$p(\mu|D, I) = \frac{\frac{1}{R_\mu}\sigma^{-N}(2\pi)^{-\frac{N}{2}}\exp\left\{-\frac{Nr^2}{2\sigma^2}\right\}\exp\left\{-\frac{N(\mu-\bar{d})^2}{2\sigma^2}\right\}}{\frac{1}{R_\mu}\sigma^{-N}(2\pi)^{-\frac{N}{2}}\exp\left\{-\frac{Nr^2}{2\sigma^2}\right\}\int_{\mu_L}^{\mu_H} d\mu\,\exp\left\{-\frac{N(\mu-\bar{d})^2}{2\sigma^2}\right\}}. \tag{9.7}$$

Equation (9.7) simplifies to

$$p(\mu|D, I) = \frac{\exp\left\{-\frac{(\mu-\bar{d})^2}{2\sigma^2/N}\right\}}{\int_{\mu_L}^{\mu_H}\exp\left\{-\frac{(\mu-\bar{d})^2}{2\sigma^2/N}\right\}d\mu} = \frac{\text{NUM}}{\text{DEN}}. \tag{9.8}$$

Therefore,

$$p(\mu|D, I) = \frac{1}{\text{DEN}}\exp\left\{-\frac{(\mu - \bar{d})^2}{2\sigma^2/N}\right\}. \tag{9.9}$$

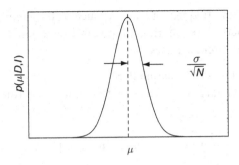

Figure 9.1 The posterior probability density function for $p(\mu|D, I)$.

Since the denominator (DEN) evaluates to a constant (see Equation (9.11), the posterior, within the range μ_L to μ_H, is simply a Gaussian with variance equal to σ^2/N. Thus, the uncertainty in the mean is inversely proportional to the square root of the sample size, which is the basis of signal averaging as discussed in Section 5.11.1. Figure 9.1 shows the resulting posterior probability density function for $p(\mu|D, I)$ in the limit of $\mu_H = +\infty$ and $\mu_L = -\infty$.

It is interesting to compare this Bayesian result to the frequentist confidence intervals for the mean when sampling from a normal distribution as discussed in Section 6.6. In the frequentist approach, we were not able to make any probability statement in connection with a single confidence interval derived from one data set $\{d_i\}$. For example, the interpretation of the 68% confidence interval was: if we repeatedly draw samples of the same size from a population, and each time compute specific values for the 68% confidence interval, $\bar{d} \pm \sigma/\sqrt{N}$, then we expect 68% of these confidence intervals to contain the unknown mean μ. In the frequentist case, the problem was to find the mean of a hypothetical population of possible measurements for which our sample was but one realization.

In the Bayesian case, we are making a probability statement about the value of a model parameter. From our posterior probability density function, we can always compute the probability that the model parameter μ lies within $\pm\sigma/\sqrt{N}$ of the sample mean $\bar{d}$. It turns out that when the prior bounds for μ are so wide that they are far outside the range indicated by the data, the value of this probability is 68%. In this particular instance, the boundaries of the Bayesian 68% credible region are the same as the frequentist 68% confidence interval. However, if we decrease the range of the prior bounds for μ, the probability contained within the frequentist 68% confidence boundary increases and reaches 100% when the prior boundaries coincide with $\pm\sigma/\sqrt{N}$.

The differences in conclusions drawn between a Bayesian and a frequentist analysis of the same data are a consequence of the different definitions of probability used in the two approaches. Recall that in the frequentist case, the argument of a probability must be a random variable. Because a parameter is not a random variable, the frequentist approach does not permit the probability density of a parameter to be calculated directly or allow for the inclusion of a prior probability for the parameter.

The interpretation of any frequentist statistic, such as the sample mean, is always in relation to a hypothetical population of possible samples that could have been obtained under similar circumstances.

Detail: Calculation of DEN in Equation (9.8)

To evaluate DEN, compare with the error function $\mathrm{erf}(x)$ (the *Mathematica* command is Erf[x]).

$$\mathrm{erf}(x) = \frac{2}{\sqrt{\pi}} \int_0^x \exp(-u^2)du, \quad \text{note: } \mathrm{erf}(-x) = -\mathrm{erf}(x) \tag{9.10}$$

let

$$\frac{N(\mu - \overline{d})^2}{2\sigma^2} = u^2 \;\Rightarrow\; u = (\mu - \overline{d})(\frac{2\sigma^2}{N})^{-\frac{1}{2}},$$

therefore,

$$du = \left(\frac{2\sigma^2}{N}\right)^{-\frac{1}{2}} d\mu \quad \text{or} \quad d\mu = \left(\frac{2\sigma^2}{N}\right)^{\frac{1}{2}} du,$$

and

$$\mathrm{DEN} = \left(\frac{2\sigma^2}{N}\right)^{\frac{1}{2}} \frac{\sqrt{\pi}}{2} \left[\frac{2}{\sqrt{\pi}} \int_{u_L}^{u_H} \exp(-u^2)du\right].$$

We can rewrite the integral limits in DEN as follows:

$$\int_{u_L}^{u_H} = \int_{-\infty}^{u_H} - \int_{-\infty}^{u_L}$$

$$= \int_{-\infty}^{0} + \int_{0}^{u_H} - \int_{-\infty}^{0} - \int_{0}^{u_L}$$

$$= \int_{0}^{u_H} - \int_{0}^{u_L}$$

therefore,

$$\mathrm{DEN} = \left(\frac{2\sigma^2}{N}\right)^{\frac{1}{2}} \frac{\sqrt{\pi}}{2} \underbrace{[\mathrm{erf}(u_H) - \mathrm{erf}(u_L)]}, \tag{9.11}$$

$$(\approx 2 \text{ if } u_H \gg 1 \text{ and } u_L \ll -1)$$

where

$$u_H = \left(\frac{2\sigma^2}{N}\right)^{-\frac{1}{2}} (\mu_H - \overline{d}),$$

$$u_L = \left(\frac{2\sigma^2}{N}\right)^{-\frac{1}{2}} (\mu_L - \overline{d}).$$

9.2.2 Mean: known noise, unequal σ

In this situation, we again assume that the noise variance is known but that it can differ for each data point. The likelihood is given by

$$
\begin{aligned}
p(D|\mu, I) &= \prod_{i=1}^{N} \frac{1}{\sigma_i \sqrt{2\pi}} \exp\left\{ -\frac{(d_i - \mu)^2}{2\sigma_i^2} \right\} \\
&= \left[\prod_{i=1}^{N} \sigma_i^{-1} \right] (2\pi)^{-\frac{N}{2}} \exp\left\{ -\sum_{i=1}^{N} \frac{(d_i - \mu)^2}{2\sigma_i^2} \right\} \\
&= \left[\prod_{i=1}^{N} \sigma_i^{-1} \right] (2\pi)^{-\frac{N}{2}} \exp\left\{ -\sum_{i=1}^{N} \frac{w_i(d_i - \mu)^2}{2} \right\} \\
&= \left[\prod_{i=1}^{N} \sigma_i^{-1} \right] (2\pi)^{-\frac{N}{2}} \exp\left\{ -\frac{Q}{2} \right\},
\end{aligned}
\tag{9.12}
$$

where $w_i = 1/\sigma_i^2$ is called the weight of data value d_i. In this case, Q is given by

$$
\begin{aligned}
Q &= \sum_{i=1}^{N} w_i(d_i - \mu)^2 = \sum w_i d_i^2 + \mu^2 \sum w_i - 2\mu \sum w_i d_i \\
&= \left(\sum w_i \right) \left\{ \mu^2 - 2\mu \frac{\sum w_i d_i}{\sum w_i} + \frac{\sum w_i d_i^2}{\sum w_i} \right\} \\
&= \left(\sum w_i \right) \left\{ \left(\mu^2 - 2\mu \frac{\sum w_i d_i}{\sum w_i} + \frac{(\sum w_i d_i)^2}{(\sum w_i)^2} \right) - \frac{(\sum w_i d_i)^2}{(\sum w_i)^2} + \frac{\sum w_i d_i^2}{\sum w_i} \right\} \\
&= \left(\sum w_i \right) \left\{ \left(\mu - \frac{\sum w_i d_i}{\sum w_i} \right)^2 - \frac{(\sum w_i d_i)^2}{(\sum w_i)^2} + \frac{\sum w_i d_i^2}{\sum w_i} \right\} \\
&= \left(\sum w_i \right) \left(\mu - \frac{\sum w_i d_i}{\sum w_i} \right)^2 - \frac{(\sum w_i d_i)^2}{\sum w_i} + \sum w_i d_i^2.
\end{aligned}
\tag{9.13}
$$

Only the first term, which contains the unknown mean μ, will appear in our final equation for the posterior probability of the mean, as we see below. Although the second and third terms do not appear in the final result, they can be shown to equal the weighted mean square residual (r_w^2) times the sum of the weights $(\sum w_i)$. The weighted mean square residual, r_w^2, is given by

$$
r_w^2 = \frac{1}{\sum w_i} \left\{ \sum w_i \left(d_i - \frac{\sum w_i d_i}{\sum w_i} \right)^2 \right\},
\tag{9.14}
$$

where $\sum w_i d_i / (\sum w_i) = \overline{d_w}$ is the weighted mean of the data values. Thus

$$
r_w^2 = \frac{1}{\sum w_i} \left\{ \sum w_i (d_i - \overline{d_w})^2 \right\},
\tag{9.15}
$$

and

$$Q = \frac{(\mu - \overline{d_w})^2}{1/\sum w_i} + r_w^2 \sum w_i. \tag{9.16}$$

Now substitute Equation (9.16) into (9.12).

$$p(D|\mu, I) = \left[\prod_{i=1}^N \sigma_i^{-1}\right](2\pi)^{-\frac{N}{2}} \exp\left(-\frac{r_w^2 \sum w_i}{2}\right) \exp\left(-\frac{(\mu - \overline{d_w})^2}{2/\sum w_i}\right)$$

$$= \left[\prod_{i=1}^N \sigma_i^{-1}\right](2\pi)^{-\frac{N}{2}} \exp\left(-\frac{r_w^2 \sum w_i}{2}\right) \exp\left(-\frac{(\mu - \overline{d_w})^2}{2\sigma_w^2}\right), \tag{9.17}$$

where $\sigma_w^2 = 1/(\sum w_i)$. Substitution of Equations (9.2), (9.17) and (9.6) into Equation (9.1) yields the desired posterior:

$$p(\mu|D, I) = \frac{\frac{1}{R_\mu}[\prod_i \sigma_i^{-1}](2\pi)^{-\frac{N}{2}} \exp(-\frac{r_w^2 \sum w_i}{2}) \exp\left(-\frac{(\mu - \overline{d_w})^2}{2\sigma_w^2}\right)}{\frac{1}{R_\mu}[\prod_i \sigma_i^{-1}](2\pi)^{-\frac{N}{2}} \exp(-\frac{r_w^2 \sum w_i}{2}) \int_{\mu_L}^{\mu_H} d\mu \exp\left(-\frac{(\mu - \overline{d_w})^2}{2\sigma_w^2}\right)}. \tag{9.18}$$

Therefore,

$$p(\mu|D, I) = \frac{\exp\left\{-\frac{(\mu - \overline{d_w})^2}{2\sigma_w^2}\right\}}{\int_{\mu_L}^{\mu_H} \exp\left\{-\frac{(\mu - \overline{d_w})^2}{2\sigma_w^2}\right\} d\mu}. \tag{9.19}$$

Since the denominator evaluates to a constant, the posterior, within the range μ_L to μ_H, is simply a Gaussian with variance $\sigma_w^2 = 1/(\sum w_i)$. The most probable value of μ is the weighted mean $\overline{d_w} = \sum w_i d_i/(\sum w_i)$.

9.2.3 Mean: unknown noise σ

In this section, we assume that the variance, σ^2, of the noise is unknown but is assumed to be the same for each measurement,[2] d_i. As in the previous case, we proceed by writing down the assumed model:

$$d_i = \mu + e_i.$$

In general, e_i consists of the random measurement errors plus any real signal in the data that cannot be explained by the model. For example, suppose that unknown to us, the data contained a periodic signal superposed on the mean. In this connection, the periodic signal would act like an additional unknown noise term. It is often the case, that nature is more complex than our current model. In the absence of a detailed knowledge

---

[2] See Section 12.9 on extrasolar planets for a discussion of the case where the variance is not constant.

of the effective noise distribution, we could appeal to the Central Limit Theorem and argue that if the effective noise stems from a large number of sub-processes then it is expected to have a Gaussian distribution. Alternatively, the MaxEnt principle tells us that a Gaussian distribution would be the most conservative choice (i.e., maximally non-committal about the information we don't have). For a justification of this argument, see Section 8.7.4. The only requirement is that the noise variance be finite. In what follows, we will assume the effective noise has a Gaussian distribution with unknown σ.

Now we have two unknowns in our model, μ and σ. The joint posterior probability $p(\mu, \sigma | D, I)$ is given by Bayes' theorem:

$$p(\mu, \sigma | D, I) = \frac{p(\mu, \sigma | I) p(D | \mu, \sigma, I)}{p(D | I)}. \tag{9.20}$$

We are interested in $p(\mu | D, I)$ regardless of what the true value of σ is. In this problem, σ is a nuisance parameter so we marginalize over σ:

$$p(\mu | D, I) = \int p(\mu, \sigma | D, I) d\sigma. \tag{9.21}$$

From the product rule: $p(\mu, \sigma | I) = p(\mu | I) p(\sigma | \mu, I)$.
Assuming the prior for σ is independent of the prior for μ, then

$$p(\mu, \sigma | I) = p(\mu | I) p(\sigma | I). \tag{9.22}$$

Combining Equations (9.20), (9.21), and (9.22),

$$p(\mu | D, I) = \frac{p(\mu | I) \int p(\sigma | I) p(D | \mu, \sigma, I) d\sigma}{p(D | I)}, \tag{9.23}$$

where

$$p(D | I) = \int p(\mu | I) \int p(\sigma | I) p(D | \mu, \sigma, I) d\sigma d\mu. \tag{9.24}$$

As before we assume a flat prior for μ in the range μ_L to μ_H. Therefore

$$p(\mu | I) = \frac{1}{R_\mu}, \quad (R_\mu = \mu_H - \mu_L). \tag{9.25}$$

σ is a scale parameter, so it can only take on positive values $0 \rightarrow \infty$. Realistic limits do not go all the way to zero and infinity. For example, we always know that σ cannot be less than a value determined by the digitizing accuracy with which we record the data; nor so great that the noise power would melt the apparatus. Let σ_L and σ_H be our prior limits on σ.

We will assume a Jeffreys prior for the scale parameter σ:

$$p(\sigma | I) = \begin{cases} K/\sigma, & \sigma_L \leq \sigma \leq \sigma_H \\ 0, & \text{otherwise.} \end{cases}$$

The constant K is determined from the condition

$$\int_{\sigma_L}^{\sigma_H} p(\sigma|I)d\sigma = 1 \quad \Rightarrow \quad K = \frac{1}{\ln(\sigma_H/\sigma_L)}$$

$$p(\sigma|I) = \frac{1}{\sigma \ln \sigma_H/\sigma_L}.$$

(9.26)

Question: Why did we choose σ instead of the variance $v = \sigma^2$ as our second parameter or does it matter? To the extent that both v and σ are "equally natural" parameterizations of the width of a Gaussian, it is desirable that investigators using either parameter reach the same conclusions.

Answer: A feature of the Jeffreys prior is that it is invariant to such a reparameterization as we now demonstrate. We start with the requirement that

$$p(v|I)dv = p(\sigma|I)d\sigma.$$

(9.27)

The Jeffreys prior for σ can be written as

$$p(\sigma|I)d\sigma = \frac{K}{\sigma}d\sigma,$$

(9.28)

where K is a constant that depends on the prior upper and lower bounds on σ. Since $\sigma = v^{1/2}$, $d\sigma = (1/2)v^{-1/2}dv$. Upon substitution into Equation (9.28), we obtain

$$p(v|I)dv = \frac{K}{2v}dv = \frac{K'}{v}dv.$$

(9.29)

Thus, choosing a Jeffreys prior for σ is equivalent to assuming a Jeffreys prior for v. It is easy to show that this would not be the case for a uniform prior.

Another example of "equally natural" parameters is the choice of whether to use the frequency or period of an unknown periodic signal. Again, it is easy to show that the choice of a Jeffreys prior for frequency is equivalent to assuming a Jeffreys prior for the period.

Calculation of the likelihood function:

$$\mathcal{L}(\mu, \sigma) = p(D|\mu, \sigma, I) = \prod_{i=1}^{N} \frac{1}{\sigma\sqrt{2\pi}} e^{-[(d_i-\mu)^2]/2\sigma^2}$$

$$= \sigma^{-N}(2\pi)^{-N/2}e^{-Q/2\sigma^2},$$

(9.30)

where Q depends on μ and is given by Equation (9.4).

$$Q = N(\mu - \overline{d})^2 + Nr^2.$$

(9.31)

Substituting Equations (9.25), (9.26), and (9.30) into Equation (9.23), we obtain

$$p(\mu|D, I) = \frac{(2\pi)^{-\frac{N}{2}} \frac{1}{R_\mu \ln\frac{\sigma_H}{\sigma_L}} \int_{\sigma_L}^{\sigma_H} \sigma^{-(N+1)} e^{-\frac{Q}{2\sigma^2}} d\sigma}{(2\pi)^{-\frac{N}{2}} \frac{1}{R_\mu \ln\frac{\sigma_H}{\sigma_L}} \int_{\mu_L}^{\mu_H} \int_{\sigma_L}^{\sigma_H} \sigma^{-(N+1)} e^{-\frac{Q}{2\sigma^2}} d\sigma d\mu}$$

$$= \frac{\int_{\sigma_L}^{\sigma_H} \sigma^{-(N+1)} e^{-\frac{Q}{2\sigma^2}} d\sigma}{\int_{\mu_L}^{\mu_H} \int_{\sigma_L}^{\sigma_H} \sigma^{-(N+1)} e^{-\frac{Q}{2\sigma^2}} d\sigma d\mu}.$$

(9.32)

Now we change variables. Let $\tau = Q/2\sigma^2$; therefore,

$$\sigma = \sqrt{\frac{Q}{2\tau}} \quad \text{and} \quad d\sigma = -\frac{1}{2}\tau^{-\frac{3}{2}}\sqrt{\frac{Q}{2}} d\tau$$

$$\sigma^{-(N+1)} = \left(\frac{2\tau}{Q}\right)^{\frac{N+1}{2}} = 2^{\frac{N+1}{2}} \tau^{\frac{N+1}{2}} Q^{-\left(\frac{N+1}{2}\right)}$$

(9.33)

$$\sigma^{-(N+1)} d\sigma = -2^{\frac{N}{2}-1}\tau^{\frac{N}{2}-1} Q^{-\left(\frac{N}{2}\right)} d\tau$$

and therefore,

$$p(\mu|D, I) = \frac{\int_{\tau_L}^{\tau_H} Q^{-\left(\frac{N}{2}\right)} \tau^{\frac{N}{2}-1} e^{-\tau} d\tau}{\int_{\mu_L}^{\mu_H} d\mu Q^{-\left(\frac{N}{2}\right)} \int_{\tau_H}^{\tau_L} \tau^{\frac{N}{2}-1} e^{-\tau} d\tau}$$

$$\approx \frac{Q^{-\left(\frac{N}{2}\right)}}{\int_{\mu_L}^{\mu_H} d\mu Q^{-\left(\frac{N}{2}\right)}},$$

(9.34)

where $\tau_L = Q/(2\sigma_H^2)$ and $\tau_H = Q/(2\sigma_L^2)$.

The integral with respect to τ in equation (9.34) can be evaluated in terms of the incomplete gamma function (see equation (C.15) of Appendix C). The τ integral clearly depends on μ, but provided $\sigma_L \ll r$ and $\sigma_H \gg r$, where $r =$ the RMS residual of the most probable model fit, the integral is effectively constant. As an example, for $N = 10, \sigma_L = 0.5r$ and $\sigma_H = 5r$, the τ integral deviates by $\leq 1\%$ for values of $|\mu - \bar{d}| \leq 2.3r$. However, at $|\mu - \bar{d}| = 2.3r$, the term $Q^{-N/2}$ in equation (9.34) has reached a value of 10^{-4} of its value at $|\mu - \bar{d}| = 0$. For larger values of $|\mu - \bar{d}|$, the τ integral decreases monotonically. At $|\mu - \bar{d}| = 3r$ the integral is only down by 5%, but now $Q^{-N/2}$ is down by a factor of 10^5.

Use Equation (9.31) to substitute for Q:

$$p(\mu|D, I) \approx \frac{[Nr^2 + N(\mu - \bar{d})^2]^{-\frac{N}{2}}}{\int_{\mu_L}^{\mu_H} d\mu [Nr^2 + N(\mu - \bar{d})^2]^{-\frac{N}{2}}}$$

(9.35)

$$p(\mu|D, I) \approx \frac{\left[1 + \frac{(\mu - \bar{d})^2}{r^2}\right]^{-\frac{N}{2}}}{\int_{\mu_L}^{\mu_H} d\mu \left[1 + \frac{(\mu - \bar{d})^2}{r^2}\right]^{-\frac{N}{2}}},$$

where the quantity $Nr^2 = \sum(d_i - \bar{d})^2$, which is independent of μ, has been factored out of the numerator and denominator and canceled. Now compare

$$\left[1 + \frac{(\mu - \bar{d})^2}{r^2}\right]^{-\frac{N}{2}} \tag{9.36}$$

with the Student's t distribution which was discussed in Section 6.4.

$$f(t|\nu) = \frac{\Gamma\left[\frac{(\nu+1)}{2}\right]}{\sqrt{\pi\nu}\,\Gamma\left(\frac{\nu}{2}\right)}\left[1 + \frac{t^2}{\nu}\right]^{-\frac{(\nu+1)}{2}}. \tag{9.37}$$

If we set

$$\frac{t^2}{\nu} = \frac{(\mu - \bar{d})^2}{r^2}, \tag{9.38}$$

and the number of degrees of freedom $\nu = N - 1$, then Equation (9.36) has the same form as the Student's t distribution.[3]

From this comparison, it is clear that the posterior probability for μ when σ is unknown is a Student's t distribution. If $\mu_{\mathrm{L}} = -\infty$ and $\mu_{\mathrm{H}} = +\infty$, then

$$p(\mu|D, I) \approx \frac{\Gamma\left(\frac{N}{2}\right)}{\sqrt{\pi}\,\Gamma\left(\frac{N-1}{2}\right)}\frac{1}{r}\left[1 + \frac{(\mu - \bar{d})^2}{r^2}\right]^{-\frac{N}{2}}. \tag{9.39}$$

If the limits on μ do not extend to $\pm\infty$, then the constant outside the square brackets will be different but computable from a Student's t distribution and the known prior limits on μ. In practice, if μ_{L} and μ_{H} are well outside some measure of the range of μ argued for by the likelihood function, then the result is the same as setting the prior limits of μ to $\pm\infty$.

We can easily generalize the results of this section to more complicated models than one that predicts the mean. Suppose the data were described by the following model:

$$d_i = m_i(\theta) + e_i,$$

where θ represents a set of model parameters with a prior $p(\theta|I)$. Then from Equation (9.34), we can write

$$p(\theta|D, I) \approx \frac{p(\theta|I)Q^{-\left(\frac{N}{2}\right)}}{\int d\theta\, p(\theta|I)Q^{-\left(\frac{N}{2}\right)}}, \tag{9.40}$$

where Q is given by

$$Q = \sum_{i=1}^{N}(d_i - m_i(\theta))^2. \tag{9.41}$$

---

[3] In Equation (9.36), $r^2 = \frac{1}{N}\sum(d_i - \bar{d})^2 = \frac{N-1}{N}S^2$, where S^2 is the frequentist sample variance as defined in Equation (6.15).

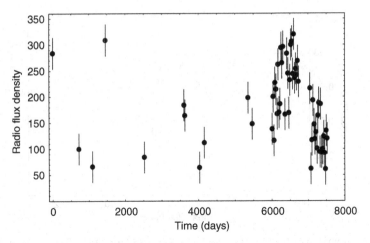

Figure 9.2 Plot of the radio source measurements and 1σ measurement errors.

Example:

Often we encounter situations in which our model plus known instrumental errors fail to adequately describe the full range of variability in the data. We illustrate this with an example from radio astronomy. In this case, we are interested in inferring the mean flux density of a celestial radio source from repeated measurements with a radio telescope with well-known noise properties.

Figure 9.2 shows 56 measurements of the radio flux density of a galaxy. The individual measurement errors are known to have a Gaussian distribution with a $\sigma_1 = 30$ units of radio flux density. It is obvious from the scatter in the measurements compared to the error bars that there is some additional source of uncertainty or the signal strength is variable. For example, additional fluctuations might arise from propagation effects in the interstellar medium between the source and observer. In the absence of prior information about the distribution of the additional scatter, both the Central Limit Theorem and the MaxEnt principle (Section 8.7.4) lead us to adopt a Gaussian distribution because it is the most conservative choice. Let $\sigma_2 =$ the standard deviation of this Gaussian.[4] The resulting likelihood function is the convolution of the Gaussian model of the additional scatter and the Gaussian measurement error distribution (see Section 4.8.2). The result is another Gaussian with a variance, $\sigma^2 = \sigma_1^2 + \sigma_2^2$.

---

[4] Since σ_2 is an unknown nuisance parameter we will need to marginalize over some prior range. For values of σ_2 close to the upper bounds of this range, the lower tail of the Gaussian distribution may extend into negative values of source strength. If the scatter arises from variations in the source strength then this situation is non-physical. Thus, it would be more exact to adopt a truncated Gaussian but the mathematics is greatly complicated. The current analysis must therefore be viewed as approximate.

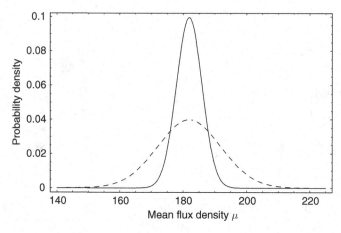

Figure 9.3 Comparison of the computed results for the posterior PDF for the mean radio flux density assuming σ known (solid curve), and marginalizing over an unknown σ (dashed curve).

We have computed the posterior probability of the mean flux density $p(\mu|D, I)$ in two ways. First, assuming the known measurement errors and Equation (9.8), the result is shown as the solid curve in Figure 9.3. Next we assumed σ was unknown and plotted the result after marginalizing over values of σ in the range 30 to 400 units, using Equation (9.39). This results in the much broader dashed curve shown in Figure 9.3. In the latter analysis, where we marginalize over σ, we are in effect estimating σ from the data and any variability which is not described by the model is assumed to be noise (the following section provides a justification of this statement). This approach leads to a broader posterior probability distribution which reflects the larger effective noise when using a model that assumes the source flux density is constant. Note: if the effective noise had been equal to the measurement error ($\sigma = 30$) then the result for $p(\mu|D, I)$ would have been the same as if we had used a fixed noise of 30. A justification of this claim is given in the following section.

9.2.4 Bayesian estimate of σ

In the previous section, we computed the Bayesian estimate of the mean of a data set when the σ of the Gaussian sampling distribution is unknown. It is also of interest to see what the data have to say about σ. This can be answered by computing $p(\sigma|D, I)$, the posterior marginal for σ. Following Equation (9.23), we write

$$p(\sigma|D, I) = \frac{p(\sigma|I) \int p(\mu|I)p(D|\mu, \sigma, I)d\mu}{p(D|I)}. \tag{9.42}$$

Substituting Equations (9.30) and (9.31) into Equation (9.42), we obtain

$$
p(\sigma|D,I) = \frac{(2\pi)^{-\frac{N}{2}} \frac{1}{R_\mu \ln\frac{\sigma_H}{\sigma_L}} \sigma^{-(N+1)} e^{-\left(\frac{Nr^2}{2\sigma^2}\right)} \int_{\mu_L}^{\mu_H} e^{-\frac{N(\mu-\bar{d})^2}{2\sigma^2}} d\mu}{(2\pi)^{-\frac{N}{2}} \frac{1}{R_\mu \ln\frac{\sigma_H}{\sigma_L}} \int_{\sigma_L}^{\sigma_H} \sigma^{-(N+1)} e^{-\left(\frac{Nr^2}{2\sigma^2}\right)} \int_{\mu_L}^{\mu_H} e^{-\frac{N(\mu-\bar{d})^2}{2\sigma^2}} d\mu d\sigma}
$$

$$
p(\sigma|D,I) = \frac{\sigma^{-(N+1)} e^{-\left(\frac{Nr^2}{2\sigma^2}\right)} \int_{\mu_L}^{\mu_H} e^{-\frac{N(\mu-\bar{d})^2}{2\sigma^2}} d\mu}{\int_{\sigma_L}^{\sigma_H} \sigma^{-(N+1)} e^{-\left(\frac{Nr^2}{2\sigma^2}\right)} \int_{\mu_L}^{\mu_H} e^{-\frac{N(\mu-\bar{d})^2}{2\sigma^2}} d\mu d\sigma} \tag{9.43}
$$

$$
= \frac{\sigma^{-(N+1)} e^{-\left(\frac{Nr^2}{2\sigma^2}\right)} \sqrt{2\pi} \frac{\sigma}{\sqrt{N}}}{\int_{\sigma_L}^{\sigma_H} \sigma^{-(N+1)} e^{-\left(\frac{Nr^2}{2\sigma^2}\right)} \sqrt{2\pi} \frac{\sigma}{\sqrt{N}} d\sigma}
$$

$$
= C\sigma^{-N} e^{-\left(\frac{Nr^2}{2\sigma^2}\right)}.
$$

In the above equation, we have made use of the fact that the integral of a normalized Gaussian is equal to 1, i.e.,

$$
\frac{1}{\sqrt{2\pi} \frac{\sigma}{\sqrt{N}}} \int_{\mu_L}^{\mu_H} e^{-\left(\frac{(\mu-\bar{d})^2}{2\sigma^2/N}\right)} d\mu = 1; \tag{9.44}
$$

therefore,

$$
\int_{\mu_L}^{\mu_H} e^{-\left(\frac{N(\mu-\bar{d})^2}{2\sigma^2}\right)} d\mu = \sqrt{2\pi} \frac{\sigma}{\sqrt{N}}. \tag{9.45}
$$

The most probable value (mode) of Equation (9.43) is the solution of

$$
\frac{\partial p}{\partial \sigma} = [-N\hat{\sigma}^{-N-1} + Nr^2 \hat{\sigma}^{-N-3}]Ce^{-Nr^2/2\hat{\sigma}^2} = 0. \tag{9.46}
$$

The solution is

$$
\hat{\sigma} = r. \tag{9.47}
$$

Thus, $p(\sigma|D,I)$ has a maximum at $\sigma = r$, the RMS deviation from $\bar{d}$.

Since $p(\sigma|D,I)$ is not a simple Gaussian, it is of interest to compare the mode to $\langle\sigma\rangle$ and $\langle\sigma^2\rangle$, the expectation values of σ and σ^2, respectively. They are given by

$$
\langle\sigma\rangle = \int_0^\infty \sigma p(\sigma|D,I)d\sigma, \tag{9.48}
$$

$$
\langle\sigma^2\rangle = \int_0^\infty \sigma^2 p(\sigma|D,I)d\sigma. \tag{9.49}
$$

These equations can be evaluated using an inverse gamma integral and a change of variables. The results are

$$\langle\sigma\rangle = \frac{\sqrt{N}r}{\sqrt{2}}\frac{\Gamma[(N-2)/2]}{\Gamma[(N-1)/2]},\tag{9.50}$$

and

$$\langle\sigma^2\rangle = \frac{Nr^2}{N-1} = \frac{1}{N-1}\sum_{i=1}^{N}(d_i - \overline{d})^2.\tag{9.51}$$

These three summaries $(\hat{\sigma}, \langle\sigma\rangle, \sqrt{\langle\sigma^2\rangle})$ of $p(\sigma|D, I)$ are all different; the distribution is not symmetric like a Gaussian. Figure 9.4 illustrates the three summaries assuming a value of $r = 2$. For $N = 3$, $\langle\sigma\rangle$ can differ from $\hat{\sigma}$ by as much as a factor ≈ 2, but this difference drops to 15% by $N = 10$. As N increases, the summaries asymptotically approach r, the RMS residual. Of the three, $\sqrt{\langle\sigma^2\rangle}$ is the most representative, lying between the other two.

The reader should recognize the expression for $\langle\sigma^2\rangle$ is identical to the equation for the frequentist sample variance, $S^2 = \sum(d_i - \overline{d})^2/(N-1)$, that is used to estimate the population variance for an IID sample taken from a normal distribution (see Section 6.3). Of course in Bayesian analysis, the concept of a population of hypothetical samples plays no role.

The main message of this section is that in problems where σ is unknown, the effect of marginalizing over σ is roughly equivalent to setting $\sigma =$ RMS residual of the most probable model. Thus, anything in the data that can't be explained by the model is treated as noise, leading to the most conservative estimates of model parameters. It is a very safe thing to do.

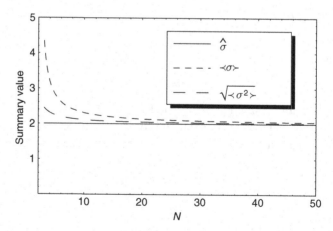

Figure 9.4 A comparison of the three summaries for the marginal probability density function for $p(\sigma|D, I)$ assuming an RMS residual $r = 2$.

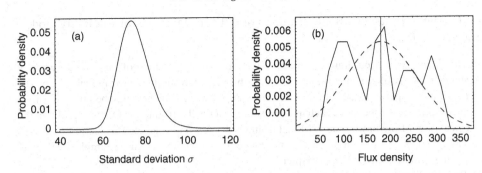

Figure 9.5 Panel (a) shows the marginal probability density function for $p(\sigma|D, I)$ for the radio galaxy data of Figure 9.2. Panel (b) compares the effective Gaussian sampling distribution employed in the analysis with a normalized histogram of the actual data values.

Returning to the radio source example (Figure 9.2) of the previous section, we now use Equation (9.43) to estimate the posterior marginal $p(\sigma|D, I)$ for these data, which are shown in panel (a) of Figure 9.5. The three summaries in this case are $\hat{\sigma} = 73.8$, $\langle\sigma\rangle = 75.6$, $\sqrt{\langle\sigma^2\rangle} = 74.5$. Recall that in the absence of prior information on the sampling distribution for the radio source measurements, we adopted a Gaussian with unknown σ. Panel (b) compares the effective Gaussian sampling distribution (based on $\sqrt{\langle\sigma^2\rangle} = 74.5$)) employed in the analysis with a normalized histogram of the actual data values. The Gaussian is centered at $\mu = 182$, the posterior maximum.

So far in this chapter, we have been concerned with fitting a simple linear model with one parameter, the mean. In Chapter 10, we will be concerned with linear models with M parameters. We will also have occasion to marginalize over an unknown noise σ. Again, we can compute the posterior marginal $p(\sigma|D, I)$, after marginalizing over the M model parameters. Assuming a Jeffreys prior for σ with prior boundaries well outside the region of the posterior peak, it can be shown that the value of $\langle\sigma^2\rangle$ is given by

$$\langle\sigma^2\rangle = \frac{1}{(N - M)}\sum_{i=1}^{N}(d_i - \overline{d})^2. \tag{9.52}$$

9.3 Is the signal variable?

In Section 7.2.1, we used a frequentist hypothesis test to decide whether we could reject, at the 95% confidence level, the null hypothesis that the radio signal from a galaxy is constant. If we can reject this hypothesis, then it provides indirect evidence that the signal is variable. A Bayesian analysis of the same data allows one to directly compare the probabilities of two hypotheses: $H_c \equiv$ "the signal is constant," and $H_v \equiv$ "the signal is variable." To compute $p(H_v|D, I)$, it is first necessary to specify a model for the signal variability. Some examples of different categories of variability models are given below.

1. The signal varies according to some specific non-periodic function of time, $f(t|\theta)$, where θ stands for a set of model parameters, e.g., slope and intercept in a linear model. The model might make specific predictions concerning the parameters or they may be unknown nuisance parameters. Of course, each nuisance parameter will introduce an Occam penalty in the calculation of the Bayes factor. Model fitting is discussed in Chapters 10, 11, 12.
2. The signal varies according to some specific periodic function of time. Examples of this are discussed in Section 12.9 and Chapter 13.
3. The signal varies according to some unknown periodic function of time (Gregory and Loredo, 1992; Loredo, 1992; Gregory, 1999). In this case, it is possible to proceed if we assume a model, or family of models, that is capable of describing an arbitrary shape of periodic variability with the minimum number of parameters. An example of this will be discussed in Section 13.4.
4. The signal varies according to some unknown non-periodic function of time. Again, it is possible to proceed if we assume a model, or family of models, that is capable of describing an arbitrary shape of variability with the minimum number of parameters. An example of this is given by Gregory and Loredo (1993).
5. The model only provides information about the statistical properties of the signal variability, i.e., specifies a probability distribution of the signal fluctuations. When combined with a model of the measurement errors (see Section 4.8.2), it can be used as a sampling distribution to compute the likelihood of the data set.
6. Finally, we may only have certain constraints on a model of the signal variability. We can always exploit the MaxEnt principle to arrive at a form of the signal variability distribution that reflects our current state of knowledge. Again, when combined with a model of the measurement errors (see Section 4.8.2), it can be used as a sampling distribution to compute the likelihood of the data set.

9.4 Comparison of two independent samples

The decisions on whether a particular drug is effective, or some human activity is proving harmful to the environment, are important topics to which Bayesian analysis can make a significant contribution. The issue typically boils down to comparing two independent samples referred to as the *trial sample* and the *control sample*. In this section, we will demonstrate a Bayesian approach to comparing two samples based on the treatment given by Bretthorst (1993), which is an extension of earlier work by Dayal (1972), and Dayal and Dickey (1976). His derivation is a generalization of the Behrens–Fisher and two-sample problems,[5] using the traditional F and Student's t distributions.

---

[5] In the frequentist statistical literature, estimating the difference in means assuming the same but unknown standard deviation is referred to as the two-sample problem. Estimating the difference in means assuming different unknown standard deviations is known as the Behrens–Fisher problem.

To start, we need to specify the prior information, I, which includes a statement of the problem, the hypothesis space of interest, and the sampling distribution to be used in calculating the likelihood function. To illustrate the methodology, we will re-visit a problem that was considered using frequentist statistical tools in Section 7.2.2, and in problem 2 at the end of Chapter 7. The problem is to compare the concentrations of a particular toxin in river sediment samples taken from two locations and tabulated in Table 7.2. The location 1 sample was taken upriver (control sample) from a processing plant and the location 2 sample taken downstream (trial sample) from the plant. In Section 7.2.2 we considered whether we could reject the null hypothesis that the mean toxin concentrations are the same at the two locations assuming the standard deviations of the two samples were different. Using the frequentist approach, we were just able to reject the null hypothesis at the 95% confidence level.

The current analysis assumes the two samples can differ in only two ways, the mean toxin concentrations and/or the sample standard deviations. Let $d_{1,i}$ represent the ith measurement in the first sample consisting of N_1 measurements in total. The symbol D_1 will represent the set of measurements $\{d_{1,i}\}$ that constitute sample 1. We will model $d_{1,i}$ by the equation

$$d_{1,i} = c_1 + e_{1,i}, \tag{9.53}$$

where, as usual, $e_{1,i}$ represents an unknown error component in the measurement. We assume that our knowledge of the source of the errors leads us to assume a Gaussian distribution for $e_{1,i}$, with a standard deviation of σ_1. To be more precise, σ_1 is a continuous hypothesis asserting that the noise standard deviation in D_1 is between σ_1 and $\sigma_1 + d\sigma_1$. In some cases, we assume a Gaussian distribution because, in the absence of knowledge of the true sampling distribution, employing a Gaussian distribution is the most conservative choice for the reasons given in Section 8.7.4. We also assume that individual measurements that constitute the sample are independent, i.e., the $e_{1,i}$ are independent.

In the absence of the error component, the model predicts $d_{1,i} = c_1$. Although c_1 will be referred to as the *mean* of D_1, c_1 is more precisely, a continuous hypothesis asserting that the constant signal component in D_1 is between c_1 and $c_1 + dc_1$. We can write a similar equation for the ith measurement of sample 2, which consists of N_2 measurements.

$$d_{2,i} = c_2 + e_{2,i}. \tag{9.54}$$

Again, we will let σ_2 represent the standard deviation of the Gaussian error term. The hypothesis space of interest for our Bayesian analysis is given in Table 9.1.

We will be concerned with answering the following hierarchy of questions:

1. Do the samples differ, i.e., do the mean concentrations and/or the standard deviations differ?
2. If so, how do they differ; in the mean, standard deviation or both?

Table 9.1 *The hypotheses addressed. The symbol in the right hand column is used as an abbreviation for the hypothesis.*

Hypothesis	In words	Symbol
$c_1 = c_2$	Same means	C
$c_1 \neq c_2$	Means differ	$\overline{C}$
$\sigma_1 = \sigma_2$	Same standard deviations	S
$\sigma_1 \neq \sigma_2$	Standard deviations differ	$\overline{S}$
$c_1 = c_2$ and $\sigma_1 = \sigma_2$	Same means and standard deviations	C, S
$c_1 = c_2$ and $\sigma_1 \neq \sigma_2$	Same means, standard deviations differ	$C, \overline{S}$
$c_1 \neq c_2$ and $\sigma_1 = \sigma_2$	Means differ, same standard deviations	$\overline{C}, S$
$c_1 \neq c_2$ and $\sigma_1 \neq \sigma_2$	Means and standard deviations differ	$\overline{C}, \overline{S}$
$c_1 \neq c_2$ and/or $\sigma_1 \neq \sigma_2$	Means and/or standard deviations differ	$\overline{C} + \overline{S}$
$c_1 - c_2 = \delta$	Difference in means $= \delta$	δ
$\sigma_1/\sigma_2 = r$	Ratio of standard deviations $= r$	r

3. If the means differ, what is their difference δ?
4. If the standard deviations differ, what is their ratio r?

To answer the above questions, we need to compute the probabilities of the hypotheses listed in Table 9.1. The discrete hypotheses are represented by the capitalized symbols and the continuous hypotheses by the lower case symbols, δ and r. For example, the symbol C stands for the hypothesis that the means are the same, and $\overline{C}$ stands for the hypothesis that they differ.

9.4.1 Do the samples differ?

The answer to question (1) can be obtained by computing $p(\overline{C} + \overline{S}|D_1, D_2, I)$, the probability that the means and/or the standard deviations are different given the sample data (D_1 and D_2) and the prior information I. Apart from the priors for the parameters $c_1, c_2, \sigma_1, \sigma_2$, we have already specified the prior information I above. To compute the probability that the means and/or the standard deviations differ, we note that from Equation (2.1), we can write

$$p(\overline{C} + \overline{S}|D_1, D_2, I) = p(\overline{C, S}|D_1, D_2, I) = 1 - p(C, S|D_1, D_2, I). \qquad (9.55)$$

Equation (9.55) demonstrates that it is sufficient to compute the probability that the means and the standard deviations are the same. From that, one can compute the probability that the means and/or the standard deviations differ. The hypothesis C, S assumes the means and the standard deviations are the same, so only two parameters (a constant c_1, and a standard deviation σ_1) have to be removed by marginalization:

$$p(C,S|D_1,D_2,I) = \int dc_1 d\sigma_1 p(C,S,c_1,\sigma_1|D_1,D_2,I), \qquad (9.56)$$

where $c_2 = c_1$ and $\sigma_2 = \sigma_1$. The right hand side of this equation may be factored using Bayes' theorem to obtain

$$p(C,S|D_1,D_2,I) = K \int dc_1 d\sigma_1 p(C,S,c_1,\sigma_1|I)p(D_1,D_2|C,S,c_1,\sigma_1,I), \qquad (9.57)$$

where,

$$K = \frac{1}{p(D_1,D_2|I)}. \qquad (9.58)$$

We need to evaluate the probabilities of four basic alternative hypotheses. They are $(C,S), (C,\overline{S}), (\overline{C},S)$ and $(\overline{C},\overline{S})$. Equation (9.57) gives the posterior for (C,S). We could similarly write out the posterior for the other three. For hypothesis $(C,\overline{S})$, which assumes $\sigma_1 \neq \sigma_2$, the result is

$$p(C,\overline{S}|D_1,D_2,I) = K \int dc_1 d\sigma_1 d\sigma_2 p(C,\overline{S},c_1,\sigma_1,\sigma_2|I)$$
$$\times p(D_1,D_2|C,\overline{S},c_1,\sigma_1,\sigma_2,I). \qquad (9.59)$$

Each of the four posteriors has a different numerator on the right hand side but a common denominator, $p(D_1,D_2|I)$. Recall that the denominator in Bayes' theorem, $p(D_1,D_2|I)$, ensures that the posterior is normalized over this hypothesis space. In terms of these basic hypotheses, $p(D_1,D_2|I)$ is the sum of the four numerators and is given by

$$p(D_1,D_2|I) = \int dc_1 d\sigma_1 p(C,S,c_1,\sigma_1|I)p(D_1,D_2|C,S,c_1,\sigma_1,I)$$

$$+ \int dc_1 d\sigma_1 d\sigma_2 p(C,\overline{S},c_1,\sigma_1,\sigma_2|I)p(D_1,D_2|C,\overline{S},c_1,\sigma_1,\sigma_2,I)$$

$$+ \int dc_1 dc_2 d\sigma_1 p(\overline{C},S,c_1,c_2,\sigma_1|I)p(D_1,D_2|\overline{C},S,c_1,c_2,\sigma_1,I) \qquad (9.60)$$

$$+ \int dc_1 dc_2 d\sigma_1 d\sigma_2 p(\overline{C},\overline{S},c_1,c_2,\sigma_1,\sigma_2|I)$$

$$\times p(D_1,D_2|\overline{C},\overline{S},c_1,c_2,\sigma_1,\sigma_2,I).$$

Assuming logical independence of the parameters and the data, Equation (9.57) may be further simplified to obtain

$$p(C, S|D_1, D_2, I) = K \int dc_1 d\sigma_1 p(C, S|I) p(c_1|I) p(\sigma_1|I)$$
$$\times p(D_1|C, S, c_1, \sigma_1, I) p(D_2|C, S, c_1, \sigma_1, I), \tag{9.61}$$

where $p(C, S|I)$ is the prior probability that the means and the standard deviations are the same, $p(c_1|I)$ is the prior probability for the mean, $p(\sigma_1|I)$ is the prior probability for the standard deviation, and $p(D_1|C, S, c_1, \sigma_1, I)$ and $p(D_2|C, S, c_1, \sigma_1, I)$ are the likelihoods of the two data sets.

Assignment of priors

In this calculation we will adopt bounded uniform priors for the location parameters, c_1 and c_2, and Jeffreys priors for the scale parameters, σ_1 and σ_2. Thus, for the mean, c_1, we write

$$p(c_1|I) = \begin{cases} 1/R_c, & \text{if } L \leq c_1 \leq H \\ 0, & \text{otherwise} \end{cases} \tag{9.62}$$

where $R_c \equiv H - L$, and H and L are the limits on the constant c_1 and are assumed known. The same prior will be used for the c_2 constant.

The prior for the standard deviation, σ_1, of the noise component in D_1, is given by

$$p(\sigma_1|I) = \begin{cases} 1/\sigma_1 \log(R_\sigma), & \text{if } \sigma_L \leq \sigma_1 \leq \sigma_H \\ 0, & \text{otherwise} \end{cases} \tag{9.63}$$

where R_σ is the ratio σ_H/σ_L, and σ_H and σ_L are the limits on the standard deviation σ_1 and are also assumed known. Again, the same prior will be assumed for σ_2.

We now come to the difficult issue of choosing prior ranges for the mean toxin concentrations (the means c_1 and c_2 in Equations (9.53) and (9.54)), and the standard deviations (σ_1 and σ_2). Recall from Section 3.5 that in a Bayesian model selection problem, marginalizing over parameters introduces Occam penalties, one for each parameter. Here, the models all contain the same types of parameters, constants and standard deviations, but they contain differing numbers of these parameters. Consequently, the prior ranges are important and will affect model selection conclusions. We also saw in Section 3.8.1 that for a uniform prior, the results are quite sensitive to the prior boundaries.

In general, any scientific enquiry is motivated from a particular prior state of knowledge on which we base our selection of prior boundaries. In the current instance, the motivation is to illustrate some methodology so we will investigate the dependence of the results on four different choices of prior boundaries as given in Table 9.2.

Finally, we need to assign a prior probability for each of the four fundamental hypotheses: $(C, S), (C, \overline{S}), (\overline{C}, S),$ and $(\overline{C}, \overline{S})$. Since the given information, I, indicates no preference, we assign a probability of $1/4$ to each.

Table 9.2 *Different choices for lower and upper bounds on the priors for the mean and standard deviation of the river sediment toxin concentrations.*

Case	Mean lower (ppm)	Mean upper (ppm)	σ_L lower (ppm)	σ_H upper (ppm)
1	2	18	0.4	10
2	7	12	0.4	10
3	2	18	1	4
4	7	12	1	4

There is a danger that the reader will get lost in the forest of calculations required to evaluate the probabilities of the four fundamental hypotheses so we have moved them to Appendix C. If you are planning on applying Bayesian analysis to a non-trivial problem in your own research field, it often helps to see worked examples of other non-trivial problems. Consider Appendix C as such a worked example. After we have evaluated the four basic hypotheses, Equation (9.61) can be used to determine if the data sets are the same. Equation (9.55) can be used to determine the probability that the means and/or the standard deviations differ, and thus, answers the first question of interest, "Do the samples differ?"

9.4.2 How do the samples differ?

We now address the second question: assuming that the two samples differ, how do they differ? There are only three possibilities: the means differ, the standard deviations differ, or both differ. To determine if the means differ, one computes $p(C|D_1, D_2, I)$. Similarly, to determine if the standard deviations differ, one computes $p(S|D_1, D_2, I)$. Using the sum rule, these probabilities may be written

$$p(C|D_1, D_2, I) = p(C, S|D_1, D_2, I) + p(C, \overline{S}|D_1, D_2, I) \qquad (9.64)$$

and

$$p(S|D_1, D_2, I) = p(C, S|D_1, D_2, I) + p(\overline{C}, S|D_1, D_2, I) \qquad (9.65)$$

where $p(C|D_1, D_2, I)$ is computed independent of whether or not the standard deviations are the same, while $p(S|D_1, D_2, I)$ is independent of whether or not the means are the same.

9.4.3 Results

We now have expressions for computing the probability for the first nine hypotheses appearing in Table 9.1. These calculations have been implemented in a special section in the *Mathematica* tutorial entitled, "Bayesian analysis of two independent samples."

This analysis program produces three different types of output: (1) the probability for the four fundamental compound hypotheses; (2) the probability that the means are different, the probability that the variances are different, and the probability that one or both are different; and finally (3) the probability for the difference in means and the ratio of the standard deviations. Table 9.3 illustrates the output for the prior boundaries corresponding to case 4 in Table 9.2, i.e., $7.0 \leq c_1, c_2 \leq 12\,\mathrm{ppm}$ and for the standard deviations $1 \leq \sigma_1, \sigma_2 \leq 4$. The last line gives an odds ratio of 9.25 in favor of the means and/or standard deviations being different.

Recall that the posterior probability is proportional to the product of the prior probability and the likelihood. Following Bretthorst's analysis, we assumed equal

Table 9.3 *Output from Mathematica program: "Bayesian analysis of two independent samples," for the river sediment toxin measurements.*

Data Summary			
No.	Standard Deviation	Average	Data Set
12	2.1771	10.3167	river B.1
8	1.2800	8.5875	river B.2
20	2.0256	9.6251	Combined

Prior mean lower bound	7.0
Prior mean upper bound	12.0
Prior standard deviation lower bound	1.0
Prior standard deviation upper bound	4.0
Number of steps for plotting $p(\delta|D_1, D_2, I)$	200
Number of steps for plotting $p(r|D_1, D_2, I)$	300

Hypothesis	Probability
$C, S \equiv$ same means, same standard deviations	0.0975
$\overline{C}, S \equiv$ different means, same standard deviation	0.2892
$C, \overline{S} \equiv$ same mean, different standard deviations	0.1443
$\overline{C}, \overline{S} \equiv$ different means, different standard deviations	0.4690
$C \equiv$ means are the same	0.2419
$\overline{C} \equiv$ means are different	0.7581
The odds ratio in favor of different means	odds = 3.13
$S \equiv$ standard deviations are the same	0.3867
$\overline{S} \equiv$ standard deviations are different	0.6133
The odds ratio in favor of different standard deviations	odds = 1.59
$C, S \equiv$ same means, same standard deviations	0.0975
$\overline{C} + \overline{S} \equiv$ one or both are different	0.9025
The odds ratio in favor of a difference	odds = 9.25

prior probabilities for the four compound hypotheses: $p(C, S|I)$, $p(C, \overline{S}|I)$, $p(\overline{C}, S|I)$, and $p(\overline{C}, \overline{S}|I)$. With this assumption, the prior odds favoring different means and/or different standard deviations $(\overline{C} + \overline{S})$ is 3.0 to 1. The data acting through the likelihood term are responsible for increasing this from 3 to 9.25 in this case. If instead we had taken as our prior that $p(C, S|I) = p(\overline{C} + \overline{S}|I)$, then the posterior odds ratio would be reduced to 3.08. It is important to remember that all Bayesian probabilities are conditional probabilities, conditional on the truth of the data and prior information. It is thus important in any Bayesian analysis to specify the prior used in the analysis.

Table 9.4 illustrates the dependence of the probabilities of the different hypotheses on different choices of prior boundaries. It is clear from the table that as we increase our prior uncertainty, the hypotheses with more parameters to marginalize over suffer larger Occam penalties and hence their probability is reduced compared to the simpler hypothesis of no change. This might be a good time to review the material on the Occam factor in Section 3.5. In all cases, the odds ratio, Odds$_{\text{diff}}$, favoring different means and/or standard deviations, exceeds 1. In this analysis, we have purposely considered four choices of prior boundaries to see what effect the different boundaries have on the final results. It often requires some careful thought to translate the available background information into an appropriate choice of prior parameter boundaries. Otherwise one might make these boundaries artificially large and as a consequence, the probability of a possibly correct complex model will decrease in relation to simpler models.

How do the present results compare to our earlier frequentist test of the null hypothesis that the mean toxin concentrations are the same (see Section 7.2.2)? On the basis of that analysis, we obtained a P-value $= 0.04$ and thus rejected the null hypothesis at the 96% confidence level. The frequentist P-value is often incorrectly viewed as the probability that the null hypothesis is true. The Bayesian conclusion regarding the question of whether the means are the same is given by $p(C) = 1 - p(\overline{C})$. Although this depends on our prior uncertainty in the means and standard deviations of two samples, the minimum value for $p(C)$ according to Table 9.4 is $\approx 1 - 0.76 = 0.24$. The difficulty in interpreting P-values and confidence

Table 9.4 *Dependence of the probabilities of the hypotheses of interest on the different prior boundaries given in Table 9.2. See Table 9.1 for the meaning of the different hypotheses.*

#	$p(C, S)$	$p(\overline{C}, S)$	$p(C, \overline{S})$	$p(\overline{C}, \overline{S})$	$p(\overline{C})$	$p(\overline{S})$	$p(\overline{C} + \overline{S})$	Odds$_{\text{diff}}$
1	0.293	0.276	0.209	0.222	0.498	0.431	0.707	2.41
2	0.141	0.420	0.101	0.338	0.757	0.439	0.858	6.06
3	0.202	0.190	0.299	0.308	0.499	0.607	0.798	3.95
4	0.098	0.289	0.144	0.469	0.758	0.613	0.902	9.25

levels has been highlighted in many papers (e.g., Berger and Sellke, 1987; Delampady and Berger, 1990; Sellke *et al.*, 2001). The focus of these works is that P-values are commonly considered to imply considerably greater evidence against the null hypothesis than is actually warranted.

Now that one knows that the means and/or standard deviations are not the same, or at the very least are probably not the same, one would like to know what is different between the control and the trial. Are the means different? Are the standard deviations different? Examination of Table 9.4 indicates that the probability the means differ is 0.758, for the choice of prior boundaries corresponding to case 4. The probability that the standard deviations differ is 0.613.

Using the calculations presented so far, one can determine if something is different, and then determine what is different. But after determining what is different, again one's interest in the problem changes. The next step is to estimate the magnitude of the changes.

9.4.4 The difference in means

To estimate the difference in means, one must first introduce this difference into the problem. Defining δ and γ to be the difference and sum, respectively, of the constants c_1 and c_2, one has

$$\delta = c_1 - c_2, \qquad \gamma = c_1 + c_2. \tag{9.66}$$

The two constants, c_1 and c_2, are then given by

$$c_1 = \frac{\gamma + \delta}{2}, \qquad c_2 = \frac{\gamma - \delta}{2}. \tag{9.67}$$

The probability for the difference, δ, is then given by

$$\begin{aligned} p(\delta|D_1, D_2, I) &= p(\delta, S|D_1, D_2, I) + p(\delta, \overline{S}|D_1, D_2, I) \\ &= p(S|D_1, D_2, I)p(\delta|S, D_1, D_2, I) \\ &\quad + p(\overline{S}|D_1, D_2, I)p(\delta|\overline{S}, D_1, D_2, I). \end{aligned} \tag{9.68}$$

This is a weighted average of the probability for the difference in means given that the standard deviations are the same (the two-sample problem) and the probability for the difference in means given that the standard deviations are different (the Behrens–Fisher problem). The weights are just the probabilities that the standard deviations are the same or different. Two of these four probabilities, $p(S|D_1, D_2, I)$ and $p(\overline{S}|D_1, D_2, I) = 1 - p(S|D_1, D_2, I)$, have already been computed in Equation (9.65). The other two probabilities, $p(\delta|S, D_1, D_2, I)$ and $p(\delta|\overline{S}, D_1, D_2, I)$, are derived in Appendix C.3. Figure 9.6 shows the probability density function for the difference in means for the prior boundaries corresponding to case 4 of Table 9.2. Three curves are shown: the probability for the difference in means given that the standard

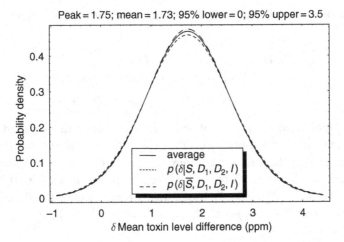

Figure 9.6 Three probability density functions are shown: the probability for the difference in means given that the standard deviations are the same (dotted line); the probability for the difference in means given that the the standard deviations are different (dashed line); and the probability for the difference in means independent of whether or not the standard deviation are the same (solid line).

deviations are the same (dotted line); the probability for the difference in means given that the the standard deviations are different (dashed line); and the probability for the difference in means independent of whether or not the standard deviation are the same (solid line).

All three curves are very similar but it is possible to notice small differences especially near the peak. The parameters listed along the top border of the figure are the peak and mean of the distribution, and the lower and upper boundaries of the 95% credible region. These apply to the weighted average distribution (solid curve).

9.4.5 *Ratio of the standard deviations*

To estimate the ratio of the standard deviations, this ratio must first be introduced into the problem. Defining r and σ to be

$$r = \frac{\sigma_1}{\sigma_2}, \qquad \sigma = \sigma_2 \tag{9.69}$$

and substituting these into the model, Equations (9.53) and (9.54), one obtains

$$d_{1i} = c_1 + \text{noise of standard deviation } r\sigma, \tag{9.70}$$

and

$$d_{2i} = c_2 + \text{noise of standard deviation } \sigma. \tag{9.71}$$

The probability for the ratio of the standard deviations, $p(r|D_1, D_2, I)$, is then given by

$$
\begin{aligned}
p(r|D_1, D_2, I) &= p(r, C|D_1, D_2, I) + p(r, \overline{C}|D_1, D_2, I) \\
&= p(C|D_1, D_2, I)p(r|C, D_1, D_2, I) \\
&\quad + p(\overline{C}|D_1, D_2, I)p(r|\overline{C}, D_1, D_2, I).
\end{aligned}
\tag{9.72}
$$

This is a weighted average of the probability for the ratio of the standard deviations given that the means are the same plus the probability for the ratio of the standard deviations given that the means are different. The weights are just the probabilities that the means are the same or not. Two of the four probabilities, $p(C|D_1, D_2, I)$ and $p(\overline{C}|D_1, D_2, I) = 1 - p(C|D_1, D_2, I)$, have already been computed in Equation (9.64). The other two probabilities, $p(r|C, D_1, D_2, I)$ and $p(r|\overline{C}, D_1, D_2, I)$, are derived in Appendix C.4.

In case 4 of Table 9.4, there is significant evidence in favor of the means being different. Thus, we might expect that the probability for the ratio of the standard deviations, assuming the same means, will differ from the probability for the ratio of the standard deviations assuming that the means are different.

These two distributions, as well as the weighted average, are shown in Figure 9.7. The probability for the ratio of the standard deviations, assuming that the means are the same, is shown as the dotted line. This model does not fit the data as well (the pooled data have a larger standard deviation than either data set separately). Consequently, the uncertainty in this probability distribution is larger compared to

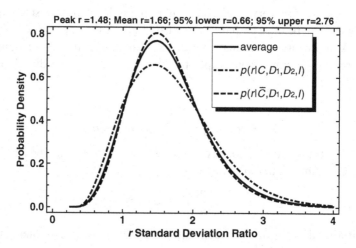

Figure 9.7 Probability density for the ratio of the standard deviations. Three probability density functions are shown: (1) the probability for the ratio of standard deviations given that the means are the same (dotted line); (2) the probability for the ratio of standard deviations given that the means are different (dashed line); (3) the probability for the ratio of standard deviations independent of whether or not the means are the same (solid line).

the other models and the distribution is more spread out. The probability for the ratio of standard deviations assuming different means is shown as the dashed line. This model fits the data better, and results in a more strongly peaked probability distribution. But probability theory tells one to take a weighted average of these two distributions, the solid line. The weights are just the probabilities that the means are the same or different. Here those probabilities are 0.242 and 0.758, respectively. As expected, the weighted average agrees more closely with $p(r|\overline{C}, D_1, D_2, I)$. The parameters listed along the top border of the figure apply to the weighted average distribution (solid curve).

9.4.6 Effect of the prior ranges

We have already discussed the effect of different prior ranges on the model selection conclusions (see Table 9.4). Here we look at their effect on the two-parameter estimation problems. Figure 9.8 shows the weighted average estimate of the difference in the means of the two data sets as given by Equation (9.68), for the four choices of prior boundaries. The results for all four cases are essentially identical. Provided the prior ranges are outside the parameter range selected by the likelihood function, we don't expect much of an effect. This is because in a parameter estimation problem we are comparing a continuum of hypotheses all with the same number of parameters so the Occam factors cancel. However, since we are plotting the weighted average of the two-sample calculation and the Behrens–Fisher calculation, in principle, the prior ranges can affect the weights differently and lead to a small effect. This is particularly

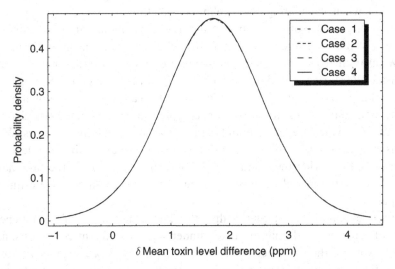

Figure 9.8 Posterior probability of the differences in the mean river sediment toxin concentration for the four different choices of prior boundaries given in Table 9.2. The effects of different choices of prior boundaries are barely discernible near the peak.

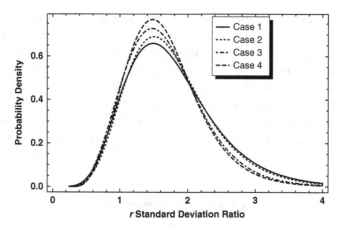

Figure 9.9 Posterior probability for the ratio of standard deviations of the river sediment toxin concentration for the four different choices of prior boundaries given in Table 9.2.

noticeable in the case of the estimation of the standard deviation ratio shown in Figure 9.9, which is a weighted average (Equation (9.72)) of the result assuming no difference in the means and the result which assumes the means are different.

9.5 Summary

This whole chapter has been concerned with Bayesian analysis of problems when our prior information (state of knowledge) leads us to assign a Gaussian sampling distribution when calculating the likelihood function. In some cases, we do this because, in the absence of knowledge of the true sampling distribution, employing a Gaussian distribution is the most conservative choice for the reasons given in Section 8.7.4. We examined a simple model parameter estimation problem – namely, estimating a mean. We started with data for which the σ of the Gaussian sampling distribution was a known constant. This was extended to the case where σ is known but is not the same for all data. Often, nature is more complex than the assumed model and gives rise to residuals which are larger than the instrumental measurement errors. We dealt with this by treating σ as a nuisance parameter which we marginalize over, leading to a Student's t PDF. This has the desirable effect of treating anything in the data that can't be explained by the model as noise, leading to the most conservative estimates of model parameters.

The final section of this chapter dealt with a Bayesian analysis of two independent samples of some physical quantity taken under slightly different conditions, and we wanted to know if there has been a change. The numerical example deals with toxin concentrations in river sediment which are taken at two different locations. One location might be upriver from a power plant and the other just downstream. What other examples can you think of where this type of analysis might be useful? The

Bayesian analysis allows the experimenter to investigate the problem in ways never before possible. The details of this non-trivial problem are presented in Appendix C, and *Mathematica* software to solve the problem is included in the accompanying *Mathematica* tutorial.

9.6 Problems

1. V_i is a set of ten voltage measurements with known but unequal independent Gaussian errors σ_i.

$$V_i = \{4.36, 4.00, 4.87, 5.64, 6.14, 5.92, 3.93, 6.58, 3.78, 5.84\}$$

$$\sigma_i = \{2.24, 1.94, 1.39, 2.55, 1.69, 1.38, 1.00, 1.60, 1.00, 1.00\}$$

a) Compute the weighted mean value of the voltages.
b) Compute and plot the Bayesian posterior probability density for the mean voltage assuming a uniform prior for the mean in the range 3 to 7.
c) Find the 68.3% credible region for the mean and compare the upper and lower boundaries to

$$\mu + \sigma_{\text{mean}}$$

$$\mu - \sigma_{\text{mean}}$$

where

$$\mu = \frac{\sum_i^N w_i d_i}{\sum_i^N w_i}, \qquad (9.73)$$

and

$$\sigma^2_{\text{mean}} = \frac{1}{\sum_i^N w_i}; \quad w_i = 1/\sigma_i^2. \qquad (9.74)$$

d) Compute and plot the Bayesian posterior probability density for the mean voltage assuming a uniform prior for the mean in the range 4.6 to 5.4. Be sure your probability density is normalized to an area of 1 in this prior range. Plot the posterior over the mean range 3 to 7. (Hint: For plotting you may find it useful to examine the item "Define a function which has a different meaning in different ranges of x" in the section "Functions and Plotting" of the *Mathematica* tutorial.)
e) Find the new 68% credible region for the mean based on (d) and compare with that found in (c).

2. In Section 9.2.3, we derived the Bayesian estimate of the mean assuming the noise σ is unknown. The desired quantity, $p(\mu|D, I)$, was obtained from the joint posterior

$p(\mu, \sigma | D, I)$ by marginalizing over the nuisance parameter σ, leading to Equation (9.34). Would we have arrived at the same conclusion if we had started from the joint posterior $p(\mu, \sigma^2 | D, I)$ and marginalized over the variance, σ^2? To answer this question, re-derive Equation (9.34) for this case.

3. Table 7.5 gives measurements of the concentration of a toxic substance at two locations in units of parts per million (ppm). The sampling is assumed to be from two independent normal populations. Assume a uniform prior for the unknown mean concentrations, and a Jeffreys prior for the unknown σs. Use the material discussed in Section 9.4 to evaluate the items (a) to (f) listed below, for two different choices of prior ranges for the means and standard deviations. These two choices are:

1) mean $(1,18)$, $\sigma(0.1, 12)$
2) mean $(7,13)$, $\sigma(1.0, 4.0)$

Note: the prior ranges for the mean and standard deviation are assumed to be the same at both locations.

a) The probabilities of the four models:

    i. two data sets have same mean and same standard deviation,
    ii. have different means and same standard deviation,
    iii. have same mean and different standard deviations,
    iv. have different means and different standard deviations.

b) The odds ratio in favor of different means.
c) The odds ratio in favor of different standard deviations.
d) The odds ratio in favor of different means and/or different standard deviations.
e) Plot a graph of the probability of the difference in means assuming the standard deviations are (*i*) the same, (*ii*) different, and (*iii*) regardless of whether or not the standard deviations are the same. Plot the result for all three on the same graph.
f) Plot a graph of the probability of the ratio of standard deviations assuming the means are (*i*) the same, (*ii*) different, and (*iii*) regardless of whether or not the means are the same. Plot all three on the same graph.
g) Explain the changes in the probabilities of the four models that occur as a result of a change in the prior ranges of the parameters, in terms of the Occam's penalty.

These calculations have been implemented in a special section in the *Mathematica* tutorial accompanying this book. The section is entitled, "Bayesian analysis of two independent samples."

10

Linear model fitting (Gaussian errors)

10.1 Overview

An important part of the life of any physical scientist is comparing theoretical models to data. We now begin three chapters devoted to the nuts and bolts of model fitting. In this chapter, we focus on *linear models*.[1] By a linear model, we mean a model that is linear with respect to the model parameters, not (necessarily) with respect to the indicator variables labeling the data. We will encounter the method of *linear least-squares*, which is so familiar to most undergraduate science students, but we will see it as a special case in a more general Bayesian treatment.

Examples of linear models:

1. $f_i = A_0 + A_1 x_i + A_2 x_i^2 + \cdots$
   where $A_0, A_1, \ldots$ are the linear model parameters,
   and x_i is the independent (indicator) variable.
2. $f_{i,j} = A_0 + A_1 x_i + A_2 y_j + A_3 x_i y_j$
   where $A_0, A_1, \ldots$ are the linear model parameters,
   and x_i, y_j are a pair of independent variables.
3. $f_i = A_1 \cos \omega t_i + A_2 \sin \omega t_i$
   where A_1, A_2 are the linear model parameters,
   and ω is a known constant.
4. $T_i = T f_i$,
   where T is the linear parameter,
   and f_i is a Gaussian line shape of the form $f_i = \exp\left\{\frac{-(\nu_i - \nu_o)^2}{2\sigma_L^2}\right\}$,
   and ν_o and σ_L are known constants.
   Such a model was considered previously in Section 3.6.

It is important to distinguish between linear and nonlinear models. In the fourth example, T_i is called a linear model because T_i is linearly dependent on T. On the other hand, if the center frequency ν_o and/or line width σ_L were unknown, then T_i would be

---

[1] About 10 years ago, Tom Loredo circulated some very useful notes on this topic. Those notes formed the starting point for my own treatment, which is presented in this chapter.

a nonlinear model. Nonlinear parameter estimation will be considered in the following chapter.

In Section 10.2, we first derive the posterior distribution for the amplitudes (i.e., parameters) of a linear model for a signal contaminated with Gaussian noise. A remarkable feature of linear models is that the joint posterior probability distribution $p(A_1, \ldots, A_n | D, I)$ of the parameters is a multivariate (multi-dimensional) Gaussian if we assume a flat prior.[2] That means that there is a single peak in the joint posterior. We derive the most probable amplitudes (which are the same as those found in linear least-squares) and their errors. The errors are given by an entity called the covariance matrix of the parameters, which we will introduce in Section 10.5.1. We also revisit the use of the χ^2 statistic to assess whether we can reject the model in a frequentist hypothesis test. In Section 10.3, we briefly consider the relationship between least-squares model fitting and regression analysis.

In most of this chapter, we assume that the data errors are independent and identically distributed (IID). In Section 10.2.2, we show how to generalize the results to allow for data errors with standard deviations that are not equal and also not independent.

We will also show how to find the boundaries of the full joint credible regions using the χ^2 distribution. We then consider how to calculate the marginal probability distribution for individual model parameters, or for a subset of amplitudes of particular interest. A useful property of Gaussian joint posterior distributions allows us to calculate any marginal distribution by maximizing with respect to the uninteresting amplitudes, instead of integrating them in a marginalization operation which is in general more difficult.

Finally, we derive some results for Bayesian model comparison with Gaussian posteriors and consider some other schemes to decide on the optimum model complexity.

10.2 Parameter estimation

Our task is to infer the parameters of some model function, f, that we sample in the presence of noise. We assume that we have N data values, d_i, that are related to N values of the function f_i, according to

$$d_i = f_i + e_i, \tag{10.1}$$

where e_i represents an unknown "error" component in the measurement of f_i. We assume that our knowledge (or lack thereof!) of the source of the errors is described by a Gaussian distribution for the e_i.[3] For now, we assume the distribution for each e_i to

---

[2] Though only a linear model leads to an exactly Gaussian posterior, nonlinear models may be approximately Gaussian close to the peak of their posterior probability.

[3] If the only knowledge we have about the noise is that it has a finite variance, then the MaxEnt principle of Chapter 8 tells us to assume a Gaussian. This is because it makes the fewest assumptions about the information we don't have and will lead to the most conservative estimates, i.e., greater uncertainty than we would get from choosing a more appropriate distribution based on more information.

be independent of the values of the other errors, and that all of the error distributions have a common standard deviation, σ. We will later generalize the results to remove the restriction of equal and independent data errors.

By a linear model, we mean that f_i can be written as a linear superposition of M functions, $g_{i\alpha}$, where $g_{i\alpha}$ is the value of the αth known function for the ith datum. The M functions are each completely specified (they have no parameters); it is their relative amplitudes that are unknown and to be inferred. Denoting the coefficients of the known functions by A_α, we thus have

$$f_i(A) = \sum_{\alpha=1}^{M} A_\alpha g_{i\alpha}. \tag{10.2}$$

Our task is to infer $\{A_\alpha\}$, which we will sometimes denote collectively with an unadorned A, as we have here. For example, if

$$f_i = A_1 + A_2 x_i + A_3 x_i^2 + A_4 x_i^3 + \cdots = \sum_{\alpha=1}^{M} A_\alpha g_{i\alpha}, \tag{10.3}$$

then $g_i = \{1, x_i, x_i^2, \ldots\}$.

Note: to avoid confusing the various indices that will arise in this calculation, we are using Roman indices to label data values and Greek indices to label model basis functions. Thus, Roman indices can take on values from 1 to N, and Greek indices can take on values from 1 to M. When limits in a sum are unspecified, the sum should be taken over the full range appropriate to its index.

Our goal is to compute the joint posterior probability distribution of the parameters, $p(A_1, \ldots, A_M | D, I)$. According to Bayes' theorem, this will require us to specify priors for the parameters and to evaluate the likelihood function. The likelihood function is the joint probability for all the data values, which we denote collectively by D, given all of the parameters specifying the model function, f_i. This is just the probability that the difference between the data and the specified function values is made up by the noise. With identical, independent Gaussians for the errors, the likelihood function is simply the product of N Gaussians, one for each of the $e_i = d_i - f_i$. Figure 10.1 illustrates graphically the basis for the calculation of the likelihood function $p(D|\{A_\alpha\}I)$ for a model f_i of the form $f_i = A_1 + A_2 x_i + A_3 x_i^2$. The smooth curve is the model prediction for a specific choice of the parameters, namely $A_1 = 0.5$, $A_2 = 0.8$, $A_3 = -0.06$. The predicted values of f_i for each choice of the independent variable x_i are marked by a dashed line. The actual measured value of d_i (represented by a cross) is located at the same value of x_i but above or below f_i as a result of the uncertainty e_i. Since the distribution of e_i values is assumed to be Gaussian, at the location of each f_i value, we have constructed a Gaussian probability density function (which we call a tent) for e_i along the line of fixed x_i, with probability plotted in the z-coordinate.

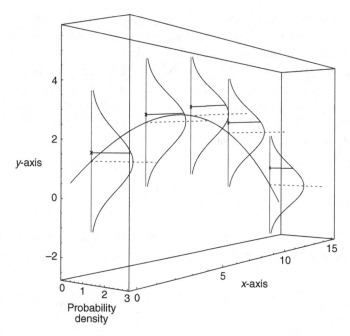

Figure 10.1 This figure illustrates graphically the basis for the calculation of the likelihood function $p(D|\{A_\alpha\}I)$ for a model f_i of the form $f_i = A_1 + A_2 x_i + A_3 x_i^2$. The smooth curve is the model prediction for a specific choice of the parameters. The predicted values of f_i for each choice of the independent variable x_i are marked by a dashed line. The actual measured value of d_i (represented by a cross) is located at the same value of x_i but above or below f_i as a result of the uncertainty e_i. At the location of each f_i value we have constructed a Gaussian probability density function (tent) for e_i along the line of fixed x_i, with probability plotted in the z-coordinate. For the assumed choice of model parameters, the probability of any d_i is proportional to the height of the Gaussian curve directly above the data point which is shown by the solid line.

We have assumed that the width of each Gaussian curve, determined by σ_i, is the same for all f_i but in principle they can all be different. For the assumed choice of model parameters, the probability of any d_i is proportional to the height of the Gaussian curve directly above the data point which is shown by the solid line.

The probability of the data set D is proportional to the product of these Gaussian heights. As we vary the choice of model parameters, the locations of the f_i points and Gaussian tents move up and down while the measured data points stay fixed. For some choice of model parameters, the likelihood will be a maximum. It should be clear that the particular choice of model parameters illustrated in the figure is far from optimal, since the data values are systematically above the model. A better choice of parameters would have the data values distributed about the model. Of course, if our prior information indicated that the probability density function for e_i had a different non-Gaussian shape, then we only need to change the shape of the probability tents.

The product of these N IID Gaussians is given by

$$p(D|\{A_\alpha\}, I) = \frac{1}{\sigma^N (2\pi)^{N/2}} \exp\left[-\frac{1}{2\sigma^2} \sum_{i=1}^{N} (d_i - f_i)^2\right]$$

$$= \frac{1}{\sigma^N (2\pi)^{N/2}} e^{-Q/2\sigma^2}.$$

(10.4)

In Equation (10.4), the quadratic form $Q(\{A_\alpha\})$ is

$$Q(\{A_\alpha\}) = \sum_{i=1}^{N} (d_i - f_i)^2$$

$$= \sum_{i=1}^{N} \left(d_i - \sum_{\alpha=1}^{M} A_\alpha g_{i\alpha}\right)^2$$

(10.5)

$$= \sum_{i=1}^{N} d_i^2 + \sum_{i=1}^{N} \sum_{\alpha\beta} A_\alpha A_\beta g_{i\alpha} g_{i\beta} - 2 \sum_{i=1}^{N} d_i \sum_{\alpha=1}^{M} A_\alpha g_{i\alpha}.$$

To get to our destination, we will take advantage of the quadratic nature of Gaussians to simplify our notation. The new notation will not only make things *look* simpler, it will actually simplify the calculations themselves, and their interpretation.

The new notation will eliminate Roman (data) indices by denoting such quantities as vectors. Thus, the N data are written as $\vec{d}$, the N values of the total model are written as $\vec{f}$, the N error values are written as $\vec{e}$, and the M model functions are written as $\vec{g}_\alpha$. In terms of these N-dimensional vectors, the data equation is

$$\vec{d} = \vec{f} + \vec{e},$$

(10.6)

with

$$\vec{f} = \sum_\alpha A_\alpha \vec{g}_\alpha.$$

(10.7)

Note: $\vec{f}$ is the sum of M vectors, where usually $M < N$. Thus, the model spans an M-dimensional subspace of the N-dimensional data space. Actually, if one or more of the $\vec{g}$'s can be written as a linear superposition of the others, the model spans a subspace of dimension less than M. Hereafter, we will refer to the $\vec{g}$'s as the *basis functions* for the model.

The quadratic form is $Q = (\vec{d} - \vec{f})^2 = \vec{e} \cdot \vec{e} = e^2$, the squared magnitude of the error vector extending from $\vec{f}$ to $\vec{d}$. In terms of the basis functions, the quadratic form can be written

$$
\begin{aligned}
Q(\{A_\alpha\}) &= (\vec{d} - \vec{f})^2 \\
&= d^2 + f^2 - 2\vec{d} \cdot \vec{f} \\
&= d^2 + \sum_{\alpha\beta} A_\alpha A_\beta \vec{g}_\alpha \cdot \vec{g}_\beta - 2\sum_\alpha A_\alpha \vec{d} \cdot \vec{g}_\alpha \\
&= d^2 + \sum_\alpha A_\alpha^2 \vec{g}_\alpha \cdot \vec{g}_\alpha + 2\sum_{\alpha \neq \beta} A_\alpha A_\beta \vec{g}_\alpha \cdot \vec{g}_\beta - 2\sum_\alpha A_\alpha \vec{d} \cdot \vec{g}_\alpha,
\end{aligned}
\tag{10.8}
$$

where we follow the usual notation, $\vec{a} \cdot \vec{b} = \sum_i a_i b_i$, and $a^2 = \vec{a} \cdot \vec{a}$. It follows that

$$
\sum_{\alpha\beta} A_\alpha A_\beta \vec{g}_\alpha \cdot \vec{g}_\beta = \sum_{\alpha\beta} A_\alpha A_\beta \sum_i g_{i\alpha} g_{i\beta}, \tag{10.9}
$$

where

$$
\sum_{\alpha\beta} = \sum_{\alpha=1}^M \sum_{\beta=1}^M. \tag{10.10}
$$

$\vec{g}_\alpha \cdot \vec{g}_\beta$ and $\vec{d} \cdot \vec{g}_\alpha$ are easily computable from the data values d_i and the corresponding values of the basis functions $g_{i\alpha}$.

To estimate the amplitudes, we need to assign a prior density to them. We will simply use a uniform prior that is constant over some range ΔA_α for each parameter, so that

$$
p(\{A_\alpha\} \,|\, I) = \frac{1}{\prod_\alpha \Delta A_\alpha}. \tag{10.11}
$$

Then as long as we are inside the prior range, the posterior density for the amplitudes is just proportional to the likelihood function

$$
p(\{A_\alpha\} | D, I) = C e^{-Q(\{A_\alpha\})/2\sigma^2}. \tag{10.12}
$$

Outside the prior range, the posterior vanishes. In this equation, C is a normalization constant[4]

$$
C = \frac{p(\{A_\alpha\} | I)}{p(D|I)}, \tag{10.13}
$$

---

[4] This is only true for the choice of a uniform prior for $p(\{A_\alpha\} | I)$ appropriate for amplitudes which are location parameters. For a scale parameter like a temperature, we should use a Jeffreys prior and then C is no longer a constant. However, for many parameter estimation problems, the likelihood function selects out a very narrow range of the parameter space over which a Jeffreys prior is effectively constant. Thus, the exact choice of prior used is only critical if there are few data or we are dealing with a model selection problem. In the latter case, the choice of prior can have a big effect as we saw in Section 3.8.1.

where the global likelihood in the denominator is

$$p(D|I) = \int_{\Delta A} d^M A_\alpha p(\{A_\alpha\}|I) p(D|A_\alpha, I)$$

$$= \frac{1}{\prod_\alpha \Delta A_\alpha} \frac{1}{\sigma^N (2\pi)^{N/2}} \int_{\Delta A} d^M A_\alpha e^{-Q/2\sigma^2}. \tag{10.14}$$

We can calculate the value of C if needed – indeed, we will do so below when we discuss model comparison – but since it is independent of the A_α, we don't need to know it to address many parameter estimation questions.

The full joint posterior distribution given by Equation (10.12) is the Bayesian answer to the question, "What do the data tell us about the A_α parameters?" It is usually useful to have simple summaries of the posterior, especially if it is of high dimension, in which case it is often difficult to depict it (either mentally or graphically). We will devote the next few subsections to finding point estimates for the amplitudes (most probable and mean values) and credible regions. We'll also discuss how to summarize the implications of the data for a subset of the amplitudes by marginalizing out the uninteresting amplitudes. In fact, since the mean amplitudes require integrating the posterior over the A_α, we'll have to learn how to do such marginalization integrals before we can calculate the mean amplitudes.

10.2.1 Most probable amplitudes

The most probable values for the amplitudes are the values that maximize the posterior (Equation (10.12)), which (because of our uniform prior) are the values that minimize Q and lead to the "normal equations" of the method of least-squares. Denoting the most probable values by $\hat{A}_\alpha$, we can find them by solving the following set of M equations (one for each value of α):

$$\frac{\partial Q}{\partial A_\alpha}\bigg|_{A=\hat{A}} = 2\sum_\beta \hat{A}_\beta \vec{g}_\beta \cdot \vec{g}_\alpha - 2\vec{g}_\alpha \cdot \vec{d} = 0, \tag{10.15}$$

or,

$$\sum_\beta \hat{A}_\beta \vec{g}_\beta \cdot \vec{g}_\alpha = \vec{g}_\alpha \cdot \vec{d}. \tag{10.16}$$

For the $M = 2$ case, Equation (10.16) corresponds to the two equations

$$\hat{A}_1 \vec{g}_1 \cdot \vec{g}_1 + \hat{A}_2 \vec{g}_2 \cdot \vec{g}_1 = \vec{g}_1 \cdot \vec{d} \tag{10.17}$$

and

$$\hat{A}_1 \vec{g}_1 \cdot \vec{g}_2 + \hat{A}_2 \vec{g}_2 \cdot \vec{g}_2 = \vec{g}_2 \cdot \vec{d}. \tag{10.18}$$

Define the most probable model vector, $\hat{\vec{f}}$, by

$$\hat{\vec{f}} = \sum_{\beta} \hat{A}_{\beta} \vec{g}_{\beta}. \tag{10.19}$$

Using $\hat{\vec{f}}$, Equation (10.16) can be written as

$$\hat{\vec{f}} \cdot \vec{g} = \vec{g} \cdot \vec{d}. \tag{10.20}$$

This doesn't help us solve Equation (10.16), but it gives us a bit of insight: the most probable total model function is the one whose projection on each basis function equals the data's projection on each basis function. Crudely, the most probable model vector explains as much of the data as can be spanned by the M-dimensional model basis.

Equations (10.17) and (10.18) can be written in the following matrix form:

$$\begin{pmatrix} \vec{g}_1 \cdot \vec{g}_1 & \vec{g}_2 \cdot \vec{g}_1 \\ \vec{g}_1 \cdot \vec{g}_2 & \vec{g}_2 \cdot \vec{g}_2 \end{pmatrix} \begin{pmatrix} \hat{A}_1 \\ \hat{A}_2 \end{pmatrix} = \begin{pmatrix} \vec{g}_1 \cdot \vec{d} \\ \vec{g}_2 \cdot \vec{d} \end{pmatrix}. \tag{10.21}$$

Problem: Evaluate Equation (10.21) for a straight line model.

Solution: When fitting a straight line, $f = A_1 + A_2 x$, the two basis functions are $g_{i1} = 1$ and $g_{i2} = x_i$. In this case, the matrix elements are given by:

$$\vec{g}_1 \cdot \vec{g}_1 = \sum_{i}^{N} g_{i1}^2 = \sum_{i}^{N} 1 = N, \tag{10.22}$$

$$\vec{g}_1 \cdot \vec{g}_2 = \vec{g}_2 \cdot \vec{g}_1 = \sum_{i}^{N} g_{i1} g_{i2} = \sum_{i}^{N} x_i, \tag{10.23}$$

$$\vec{g}_2 \cdot \vec{g}_2 = \sum_{i}^{N} g_{i2} g_{i2} = \sum_{i}^{N} x_i^2, \tag{10.24}$$

$$\vec{g}_1 \cdot \vec{d} = \sum_{i}^{N} g_{i1} d_i = \sum_{i}^{N} d_i, \tag{10.25}$$

$$\vec{g}_2 \cdot \vec{d} = \sum_{i}^{N} d_i x_i, \tag{10.26}$$

and Equation (10.21) becomes

$$\begin{pmatrix} N & \sum_i x_i \\ \sum_i x_i & \sum_i x_i^2 \end{pmatrix} \begin{pmatrix} \hat{A}_1 \\ \hat{A}_2 \end{pmatrix} = \begin{pmatrix} \sum_i d_i \\ \sum_i d_i x_i \end{pmatrix}. \tag{10.27}$$

It will prove useful to express Equation (10.16) in a more compact matrix form. Let $\mathbf{G}$ be an $N \times M$ matrix where the αth column contains N values of the αth basis function evaluated at each of the N data locations. As an example, consider the $M = 2$ case again. The two basis functions for the ith data value are g_{i1} and g_{i2}, and $\mathbf{G}$ is given by

$$\mathbf{G} \equiv \begin{pmatrix} g_{11} & g_{12} \\ g_{21} & g_{22} \\ \cdot & \cdot \\ \cdot & \cdot \\ \cdot & \cdot \\ g_{N1} & g_{N2} \end{pmatrix}. \tag{10.28}$$

Now take the transpose of $\mathbf{G}$ which is given by

$$\mathbf{G}^T \equiv \begin{pmatrix} g_{11} & g_{21} & \cdot & \cdot & \cdot & g_{N1} \\ g_{12} & g_{22} & \cdot & \cdot & \cdot & g_{N2} \end{pmatrix}. \tag{10.29}$$

Define the matrix $\psi = \mathbf{G}^T \mathbf{G}$, which for $M = 2$ is given by

$$\begin{aligned} \psi &\equiv \begin{pmatrix} \sum_i g_{i1}^2 & \sum_i g_{i1} g_{i2} \\ \sum_i g_{i2} g_{i1} & \sum_i g_{i2}^2 \end{pmatrix} \\ &= \begin{pmatrix} g_1^2 & \vec{g}_1 \cdot \vec{g}_2 \\ \vec{g}_2 \cdot \vec{g}_1 & g_2^2 \end{pmatrix} \\ &= \begin{pmatrix} \psi_{11} & \psi_{12} \\ \psi_{21} & \psi_{22} \end{pmatrix}. \end{aligned} \tag{10.30}$$

Thus, ψ is a symmetric matrix because $\vec{g}_1 \cdot \vec{g}_2 = \vec{g}_2 \cdot \vec{g}_1$. More generally,

$$\psi_{\alpha\beta} = \vec{g}_\alpha \cdot \vec{g}_\beta = \psi_{\beta\alpha}. \tag{10.31}$$

Finally, if D is a column matrix of data values d_i, then for $M = 2$

$$\mathbf{G}^T \mathbf{D} \equiv \begin{pmatrix} \sum_i g_{i1} d_i \\ \sum_i g_{i2} d_i \end{pmatrix} = \begin{pmatrix} \vec{g}_1 \cdot \vec{d} \\ \vec{g}_2 \cdot \vec{d} \end{pmatrix}. \tag{10.32}$$

Equation (10.21) can now be written as

$$\mathbf{G}^T \mathbf{G} \hat{\mathbf{A}} = \mathbf{G}^T \mathbf{D}. \tag{10.33}$$

The solution to this matrix equation is given by

$$\hat{\mathbf{A}} = (\mathbf{G}^T \mathbf{G})^{-1} \mathbf{G}^T \mathbf{D} = \psi^{-1} \mathbf{G}^T \mathbf{D}, \tag{10.34}$$

or in component form becomes

$$\hat{A}_\alpha = \sum_\beta [\psi^{-1}]_{\alpha\beta} \vec{g}_\beta \cdot \vec{d}. \tag{10.35}$$

In the method of least-squares, the set of equations represented by Equation (10.33) are referred to as the *normal equations*.

Again, for the $M = 2$ case, we can write Equation (10.33) in long form:

$$\begin{pmatrix} \psi_{11} & \psi_{12} \\ \psi_{21} & \psi_{22} \end{pmatrix} \begin{pmatrix} \hat{A}_1 \\ \hat{A}_2 \end{pmatrix} = \begin{pmatrix} \vec{g}_1 \cdot \vec{d} \\ \vec{g}_2 \cdot \vec{d} \end{pmatrix}, \tag{10.36}$$

where $\psi_{12} = \psi_{21}$. The solution (Equation (10.34)) is given by

$$\begin{pmatrix} \hat{A}_1 \\ \hat{A}_2 \end{pmatrix} = \frac{1}{\Delta} \begin{pmatrix} \psi_{22} & -\psi_{12} \\ -\psi_{21} & \psi_{11} \end{pmatrix} \begin{pmatrix} \vec{g}_1 \cdot \vec{d} \\ \vec{g}_2 \cdot \vec{d} \end{pmatrix}, \tag{10.37}$$

and where $\psi_{12} = \psi_{21}$ and the denominator, Δ, is the determinant of the ψ matrix given by

$$\Delta = (\psi_{11}\psi_{22} - \psi_{12}^2). \tag{10.38}$$

$$\hat{A}_1 = \frac{\psi_{22}(\vec{g}_1 \cdot \vec{d}) - \psi_{12}(\vec{g}_2 \cdot \vec{d})}{\psi_{11}\psi_{22} - \psi_{12}^2} \tag{10.39}$$

$$\hat{A}_2 = \frac{-\psi_{12}(\vec{g}_1 \cdot \vec{d}) + \psi_{11}(\vec{g}_2 \cdot \vec{d})}{\psi_{11}\psi_{22} - \psi_{12}^2}. \tag{10.40}$$

Note: the matrix must be non-singular – the basis vectors must be linearly independent – for a solution to exist. We henceforth assume that any redundant basis models have been eliminated, so that ψ is non-singular.[5]

Problem: Evaluate Equation (10.34) for the straight line model.

Solution: Comparison of Equations (10.27) and (10.36) allows an evaluation of all the terms needed for Equations (10.39) and (10.40).

$$\hat{A}_1 = \frac{\sum_i x_i^2 \sum_i d_i - \sum_i x_i \sum_i x_i d_i}{N \sum_i x_i^2 - (\sum_i x_i)^2}$$

$$\hat{A}_2 = \frac{-\sum_i x_i \sum_i d_i + N \sum_i x_i d_i}{N \sum_i x_i^2 - (\sum_i x_i)^2}.$$

[5] Sometimes the data do not clearly distinguish between two or more of the basis functions provided, and ψ gets sufficiently close to being singular that the answer becomes extremely sensitive to round-off errors. The solution in this case is to use *singular value decomposition* which is discussed in Appendix A.

10.2.2 More powerful matrix formulation

Everything we have done so far has assumed that the error associated with each datum is independent of the errors for the others, and that the Gaussian describing our knowledge of the magnitude of the error has the same variance for each datum. In general, however, the errors can have different variances, and could be correlated. In that case, we need to replace the likelihood function $p(D|\{A_\alpha\}, I)$, given by Equation (10.4), by the multivariate Gaussian (Equation (8.61)) which we derived using the MaxEnt principle in Section 8.7.5 and Appendix E. The new likelihood is

$$
\begin{aligned}
p(D|\{A_\alpha\}, I) &= \frac{1}{(2\pi)^{N/2}\sqrt{\det \mathbf{E}}}\exp\left[-\frac{1}{2}\sum_{ij}(d_i - f_i)[\mathbf{E}^{-1}]_{ij}(d_j - f_j)\right] \\
&= \frac{1}{(2\pi)^{N/2}\sqrt{\det \mathbf{E}}}e^{-\chi^2/2},
\end{aligned}
\tag{10.41}
$$

where $\mathbf{E}$ is called the *data covariance matrix*[6]

$$
\mathbf{E} = \begin{pmatrix}
\sigma_{11} & \sigma_{12} & \sigma_{13} & \cdots & \sigma_{1N} \\
\sigma_{21} & \sigma_{22} & \sigma_{23} & \cdots & \sigma_{2N} \\
. & . & . & . & . \\
\sigma_{N1} & \sigma_{N2} & \sigma_{N3} & \cdots & \sigma_{NN}
\end{pmatrix},
\tag{10.42}
$$

and

$$
\sum_{ij} = \sum_{i=1}^{N}\sum_{j=1}^{N}.
$$

How does this affect the "normal equations" of the method of least-squares? We simply need to replace Q/σ^2 appearing in the likelihood, Equation (10.4), by

$$
\chi^2 = \sum_{i,j}(d_i - f_i)[\mathbf{E}^{-1}]_{ij}(d_j - f_j).
\tag{10.43}
$$

If the errors are independent, $\mathbf{E}$ is diagonal, with entries equal to σ_i^2, and Equation (10.43) takes the familiar form

$$
\chi^2 = \sum_i \frac{(d_i - f_i)^2}{\sigma_i^2} = \sum_i \frac{e_i^2}{\sigma_i^2}.
\tag{10.44}
$$

The inverse *data covariance matrix*, $\mathbf{E}^{-1}$, plays the role of a metric[7] in the full N-dimensional vector space of the data. Thus, if we set $\sigma = 1$, and understand $\vec{a} \cdot \vec{b}$

---

[6] If we designate each element in the covariance matrix by σ_{ij}, then the diagonal elements are given by $\sigma_{ii} = \sigma_i^2 = \langle e_i^2 \rangle$, where $\langle e_i^2 \rangle$ is the expectation value of e_i^2. The off-diagonal elements are $\sigma_{ij} = \langle e_i e_j \rangle$ $(i \neq j)$.

[7] The metric of a vector space is useful for answering questions having to do with the geometry of the vector space, such as the distance between two points. In our work, we use it to compute the dot product in the vector space.

to stand for $\vec{a}[\mathbf{E}^{-1}]\vec{b}$ everywhere a dot product occurs in the above analysis, then we already have the desired generalization! Thus, we have that

$$\vec{d}\cdot\vec{d} = \sum_{ij} d_i[\mathbf{E}^{-1}]_{ij}d_j \tag{10.45}$$

and

$$\vec{g}_\alpha\cdot\vec{g}_\beta = \sum_{ij} g_{i\alpha}[\mathbf{E}^{-1}]_{ij}g_{j\beta}. \tag{10.46}$$

The new equivalents of Equations (10.33) and (10.34) are

$$\mathbf{G}^T\mathbf{E}^{-1}\mathbf{G}\hat{\mathbf{A}} = \mathbf{G}^T\mathbf{E}^{-1}\mathbf{D} \tag{10.47}$$

$$\hat{\mathbf{A}} = (\mathbf{G}^T\mathbf{E}^{-1}\mathbf{G})^{-1}\mathbf{G}^T\mathbf{E}^{-1}\mathbf{D} = \mathbf{\Psi}^{-1}\mathbf{G}^T\mathbf{E}^{-1}\mathbf{D}. \tag{10.48}$$

To bring out the changes more clearly, we repeat our earlier Equation (10.34) for the model amplitudes.

$$\hat{\mathbf{A}} = (\mathbf{G}^T\mathbf{G})^{-1}\mathbf{G}^T\mathbf{D} = \psi^{-1}\mathbf{G}^T\mathbf{D}. \tag{10.49}$$

Notice that whenever we employ $\mathbf{E}$, we need to replace ψ by its capitalized form $\mathbf{\Psi}$. Recall the matrix ψ did not incorporate information about the data errors while $\mathbf{\Psi}$ does. In the case that the data errors all have the same σ and are independent, then $\mathbf{\Psi} = \psi/\sigma^2$. The following problem employs the $\mathbf{\Psi}$ matrix for the case of independent errors. We consider a problem with correlated errors in Section 10.6 after we have introduced the correlation coefficient.

Problem: Fit a straight line model to the data given in Table 10.1, where $\overline{d}_i$ is the average of n_i data values measured at x_i. The probability of the individual d_i measurements is IID normal with $\sigma^2 = 8.1$, regardless of the x_i value. Recall from the Central Limit Theorem,

Table 10.1 *Data table.*

x_i	$\overline{d}_i$	n_i
10	0.5	14
20	4.67	3
30	6.25	25
40	10.0	2
50	13.5	3
60	13.7	22
70	17.5	5
80	23.0	2

$p(\overline{d_i}|I)$ will tend to a Gaussian with variance $= \sigma^2/n_i$ as n_i increases even if $p(d_i|I)$ is very non-Gaussian.

Solution: The data covariance matrix $\mathbf{E}$ can be written as

$$\mathbf{E} = \sigma^2 \begin{pmatrix} 1/n_1 & 0 & 0 & \cdots & 0 \\ 0 & 1/n_2 & 0 & \cdots & 0 \\ 0 & 0 & 1/n_3 & \cdots & 0 \\ \cdot & \cdot & \cdot & \cdot & \cdot \\ 0 & 0 & \cdots & 0 & 1/n_N \end{pmatrix}. \tag{10.50}$$

The inverse data covariance matrix $\mathbf{E}^{-1}$ is

$$\mathbf{E}^{-1} = \frac{1}{\sigma^2} \begin{pmatrix} n_1 & 0 & 0 & \cdots & 0 \\ 0 & n_2 & 0 & \cdots & 0 \\ 0 & 0 & n_3 & \cdots & 0 \\ \cdot & \cdot & \cdot & \cdot & \cdot \\ 0 & 0 & \cdots & 0 & n_N \end{pmatrix}$$

$$= \frac{1}{\sigma^2} \begin{pmatrix} 14 & 0 & 0 & \cdots & 0 \\ 0 & 3 & 0 & \cdots & 0 \\ 0 & 0 & 25 & \cdots & 0 \\ \cdot & \cdot & \cdot & \cdot & \cdot \\ 0 & 0 & \cdots & 0 & 2 \end{pmatrix}. \tag{10.51}$$

$$\mathbf{\Psi} = \mathbf{G}^T \mathbf{E}^{-1} \mathbf{G}$$

$$= \begin{pmatrix} 1 & 1 & 1 & \cdots & 1 \\ 10 & 20 & 30 & \cdots & 80 \end{pmatrix} \frac{1}{\sigma^2} \begin{pmatrix} 14 & 0 & 0 & \cdots & 0 \\ 0 & 3 & 0 & \cdots & 0 \\ 0 & 0 & 25 & \cdots & 0 \\ \cdot & \cdot & \cdot & \cdot & \cdot \\ 0 & 0 & \cdots & 0 & 2 \end{pmatrix} \begin{pmatrix} 1 & 10 \\ 1 & 20 \\ 1 & 30 \\ \cdot & \cdot \\ 1 & 80 \end{pmatrix} \tag{10.52}$$

$$= \frac{1}{\sigma^2} \begin{pmatrix} 76 & 3010 \\ 3010 & 152\,300 \end{pmatrix}.$$

$$\mathbf{\Psi}^{-1} = \sigma^2 \begin{pmatrix} 0.061 & -0.0012 \\ -0.0012 & 0.000030 \end{pmatrix}. \tag{10.53}$$

Let $\mathbf{R} = \mathbf{G}^T \mathbf{E}^{-1} \mathbf{D}$

$$= \begin{pmatrix} 1 & 1 & 1 & \cdots & 1 \\ 10 & 20 & 30 & \cdots & 80 \end{pmatrix} \frac{1}{\sigma^2} \begin{pmatrix} 14 & 0 & 0 & \cdots & 0 \\ 0 & 3 & 0 & \cdots & 0 \\ 0 & 0 & 25 & \cdots & 0 \\ \cdot & \cdot & \cdot & \cdot & \cdot \\ 0 & 0 & \cdots & 0 & 2 \end{pmatrix} \begin{pmatrix} 0.5 \\ 4.67 \\ 6.25 \\ \cdot \\ 23.0 \end{pmatrix}. \tag{10.54}$$

Then,

$$\hat{A} = \Psi^{-1}R. \tag{10.55}$$

$$\begin{pmatrix} \hat{A}_1 \\ \hat{A}_2 \end{pmatrix} = \begin{pmatrix} -2.054 \\ 0.275 \end{pmatrix}. \tag{10.56}$$

Figure 10.2 shows the straight line model fit, $y = 0.275x - 2.054$, with the data plus error bars overlaid.

Mathematica provides a variety of simple ways to enter the **G, E, D** matrices and then we can evaluate the matrix of amplitudes, **A**, with the commands:

Ψ = **Transpose**[**G**].**Inverse**[**E**].**G**

A = **Inverse**[Ψ].**Transpose**[**G**].**Inverse**[**E**].**D**

10.3 Regression analysis

Least-squares fitting is often called *regression analysis* for historical reasons. For example, *Mathematica* provides the package called **Statistics 'Linear Regression'** for doing a linear least-squares fit. Regression analysis applies when we have, for example, two quantities like height and weight, or income and education and we want to predict one from information about the other. This is possible if there is a correlation between the two quantities, even if we lack a model to account for the correlation. Typically in these problems, there is little experimental error in the measurements compared to the intrinsic spread in their values. Thus, we can talk about the regression line for income

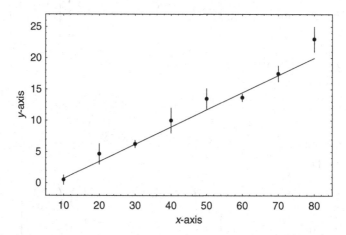

Figure 10.2 Straight line model fit with data plus errors overlaid. Clearly, the best fitting line is mostly determined by the data points with smallest error bars.

on education. The regression line is often called the *least-squares* line because it makes the sum of the squares of the residuals as small as possible.

In contrast, model fitting usually assumes that there is an underlying exact relationship between some measured quantities, and we are interested in the best choice of model parameters or in comparing competing models. In some areas, especially in the life sciences, the phenomena under study are sufficiently complex that good models are hard to come by and regression analysis is the name of the game.

The results from a regression analysis may make no physical sense but may point the way to a physical model. Consider the following simple example which assumes the experimenter is ignorant of an elementary geometrical relationship, that the area of a rectangle is equal to the product of the width and height. The experimenter fabricates many different shaped rectangles and determines the area of each rectangle by counting the number of very small squares that fit into the rectangle. Our experimenter then examines whether there is a correlation between rectangle area and perimeter. The resulting regression line would look like a line with considerable scatter, because the perimeter is not a good physical model for the area. In contrast, a plot of area versus the product of the width times the height would be an almost perfect straight line, limited only by the measurement accuracy of the width and height measurements.

10.4 The posterior is a Gaussian

We have succeeded in finding the most probable model parameters. Now we want to determine the shape of their joint probability distribution with an eye to specifying credible regions for each parameter. We will continue to work with the simple case where all the data errors are assumed to be IID so that $p(\{A_\alpha\}|D, I)$ is given by Equation (10.12),

$$p(\{A_\alpha\}|D, I) = Ce^{-Q(\{A_\alpha\})/2\sigma^2}. \tag{10.57}$$

Then maximizing $p(\{A_\alpha\}|D, I)$, corresponds to minimizing Q. Since we've already taken one derivative of Q (Equation (10.15)), let's see what happens when we take another. Define $\delta A_\alpha = A_\alpha - \hat{A}_\alpha$. Call the value of Q at the mode Q_{min}. Recall that the mode is the value that maximizes the probability density. Consider a Taylor series expansion of Q about the Q_{min}.

$$Q = Q_{min} + \sum_\alpha \frac{\partial Q}{\partial A_\alpha}\bigg|_{min} \delta A_\alpha + 1/2 \sum_{\alpha\beta} \frac{\partial^2 Q}{\partial A_\alpha \partial A_\beta}\bigg|_{min} \delta A_\alpha \delta A_\beta. \tag{10.58}$$

The first derivative is zero at the minimum and from Equation (10.8), it is clear there are no higher derivatives than the second. We are now in the position to write Q in a form that explicitly reveals the posterior distribution to be a multivariate Gaussian.

$$Q = Q_{min} + \Delta Q(A), \tag{10.59}$$

and

$$\Delta Q(A) = 1/2 \sum_{\alpha\beta} \frac{\partial^2 Q}{\partial A_\alpha \partial A_\beta}\bigg|_{\min} \delta A_\alpha \delta A_\beta. \tag{10.60}$$

Taking another derivative of Equation (10.15) and substituting from Equation (10.31), we get the equation[8]

$$\frac{\partial^2 Q}{\partial A_\alpha \partial A_\beta}\bigg|_{\min} = 2\vec{g}_\beta \cdot \vec{g}_\alpha = 2\psi_{\beta\alpha} = 2\psi_{\alpha\beta}. \tag{10.62}$$

Substituting this into Equation (10.60), we get the equation

$$\Delta Q(A) = \sum_{\alpha\beta} \delta A_\alpha \psi_{\alpha\beta} \delta A_\beta, \tag{10.63}$$

where ψ is a symmetric matrix. Note: the differential $d(\delta A_\alpha) = dA_\alpha$, so densities for the A_α are directly equal to densities for the δA_α. Thus,

$$p(\{A_\alpha\}|D, I) = C' \exp\left\{-\frac{\Delta Q}{2\sigma^2}\right\}, \tag{10.64}$$

where

$$C' = C \exp\left\{-\frac{Q_{\min}}{2\sigma^2}\right\}, \tag{10.65}$$

is an adjusted normalization constant. If we let $\delta \mathbf{A}$ be a column matrix of δA_α values, then Equation (10.64) can be written as

$$p(\{A_\alpha\}|D, I) = C' \exp\left\{-\frac{\delta \mathbf{A}^T \psi \delta \mathbf{A}}{2\sigma^2}\right\}. \tag{10.66}$$

---

[8] For $M = 2$, ψ is given by

$$\psi \equiv \begin{pmatrix} g_1^2 & \vec{g}_1 \cdot \vec{g}_2 \\ \vec{g}_2 \cdot \vec{g}_1 & g_2^2 \end{pmatrix}$$

$$= \frac{1}{2} \begin{pmatrix} \frac{\partial^2 Q}{\partial A_1^2} & \frac{\partial^2 Q}{\partial A_1 \partial A_2} \\ \frac{\partial^2 Q}{\partial A_2 \partial A_1} & \frac{\partial^2 Q}{\partial A_2^2} \end{pmatrix}. \tag{10.61}$$

Since ψ is symmetric there is a change of variable, $\delta\mathbf{A} = \mathbf{O}\delta\mathbf{X}$, that transforms[9] the quadratic form $\delta\mathbf{A}^T\psi\delta\mathbf{A}$ into the quadratic form $\delta\mathbf{X}^T\Lambda\delta\mathbf{X}$ (*Principal Axis Theorem*). Λ is a diagonal matrix whose diagonal elements are the eigenvalues of the ψ matrix and the columns of O are the orthonormal eigenvectors of ψ.

For the $M = 2$ case, we have

$$
\begin{aligned}
\Delta Q &= (\delta X_1 \; \delta X_2)\begin{pmatrix} \lambda_1 & 0 \\ 0 & \lambda_2 \end{pmatrix}\begin{pmatrix} \delta X_1 \\ \delta X_2 \end{pmatrix} \\
&= \lambda_1\delta X_1^2 + \lambda_2\delta X_2^2,
\end{aligned}
\tag{10.67}
$$

where λ_1 and λ_2 are the eigenvalues of ψ. They are all positive since ψ is positive definite.[10] Thus, $\Delta Q = k$ (a constant) defines the ellipse (see Figure 10.3),

$$
\frac{\delta X_1^2}{k/\lambda_1} + \frac{\delta X_2^2}{k/\lambda_2} = 1,
\tag{10.68}
$$

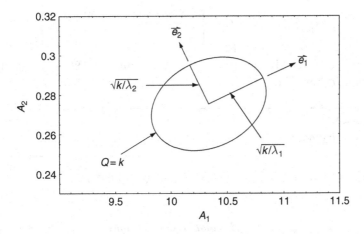

Figure 10.3 The contour in $A_1 - A_2$ parameter space for $\Delta Q = $ a constant k. It is an ellipse, centered at $(\hat{A}_1, \hat{A}_2)$, whose major and minor axes are determined by the eigenvalues λ and eigenvectors $\vec{e}$ of ψ. Note: dX_1 and dX_2 in Equation (10.67), are measured in the directions of the eigenvectors $\vec{e}_1$ and $\vec{e}_2$, respectively.

[9] In two dimensions the transformation $\mathbf{O}$ corresponds to a planar rotation followed by a reflection of the δX_2 axis. Since ψ is symmetric this is equivalent to a rotation of the axes.
[10] This would not be the case if the basis vectors were linearly dependent. In that case, ψ would be singular and ψ^{-1} would not exist.

with major and minor axes given by $\sqrt{k/\lambda_1}$ and $\sqrt{k/\lambda_2}$, respectively. From Equation (10.64), we see that $\Delta Q = k$ corresponds to a contour of constant posterior probability in the space of our two parameters.

Clearly, Figure 10.3 provides information about the joint credible region for the model parameters A_1 and A_2. We still need to compute what size of ellipse corresponds to, say, a 95% credible region. In the next section, we discuss how to find various summaries of a Gaussian posterior.

Problem: In Section 10.2.2, we fitted a straight line model to the data in Table 10.1. Find the eigenvalues of the corresponding Ψ matrix given by Equation (10.52) and which we repeat here.

$$\Psi = \frac{1}{\sigma^2} \begin{pmatrix} 76 & 3010 \\ 3010 & 152\,300 \end{pmatrix}, \tag{10.69}$$

where $\sigma^2 = 8.1$.

In that problem, the errors were not all the same, so we employed the data covariance matrix $\mathbf{E}^{-1}$. For that situation, Equation (10.66) becomes

$$p(\{A_\alpha\}|D,I) = C' \exp\left\{ -\frac{\sum_{\alpha\beta} \delta A_\alpha [\Psi]_{\alpha\beta} \delta A_\beta}{2} \right\}. \tag{10.70}$$

Since there are only two parameters (A_1, A_2), we can rewrite Equation (10.70) as

$$p(A_1, A_2|D,I) = C' \exp\left\{ -\frac{\sum_{\alpha=1}^{2} \sum_{\beta=1}^{2} \delta A_\alpha [\Psi]_{\alpha\beta} \delta A_\beta}{2} \right\}. \tag{10.71}$$

Solution: We can readily determine the eigenvalues of Ψ with the following *Mathematica* command:

Eigenvalues[Ψ]
{18 809.8, 2.03766}

So the two eigenvalues are $\lambda_1 = 18\,809.8$ and $\lambda_2 = 2.03766$.

10.4.1 Joint credible regions

A credible region is a locus of points of constant probability which surrounds a region containing a specified probability in the joint probability distribution. Figure 10.3 illustrated that in the two-parameter case, this locus is an ellipse defined by $\Delta Q = k$ where k is a constant. In this section, we will find out for what value of k the ellipse contains say 68.3% of the posterior joint probability. The results will be presented in a more general way so we can specify the credible region corresponding to $\Delta Q = k$ for an arbitrary number of model parameters, not just the $M = 2$ case.

We slightly simplify the notation for this subsection to connect with more familiar results, by writing the posterior density for the A_α as

$$p(\{A_\alpha\}|D, I) = Ce^{-\chi^2/2}, \tag{10.72}$$

where $\chi^2(\{A_\alpha\}) = Q/\sigma^2$. Let $\chi^2_{min} = Q_{min}/\sigma^2$, and $\Delta\chi^2 = \Delta Q/\sigma^2$. Then from Equation (10.64), we can write

$$p(\{A_\alpha\}|D, I) = C'e^{-\Delta\chi^2/2}. \tag{10.73}$$

By definition, the boundary of a joint credible region for all the amplitudes is defined by $\chi^2(\{A_\alpha\}) = \chi^2_{min} + \Delta\chi^2_{crit}$, where $\Delta\chi^2_{crit}$ is a constant chosen such that the region contains some specified probability, P. Our task is to find $\Delta\chi^2_{crit}$ such that

$$P = \int_{\Delta\chi^2 < \Delta\chi^2_{crit}} d^M A \, p(\{A_\alpha\}|D, I). \tag{10.74}$$

The result, which is given without proof, is

$$P = 1 - \frac{\gamma(M/2, \Delta\chi^2_{crit}/2)}{\Gamma(M/2)}. \tag{10.75}$$

This is the probability within the joint credible region for all the amplitudes corresponding to $\Delta\chi^2 < \Delta\chi^2_{crit}$. The quantity $\gamma(M/2, \Delta\chi^2_{crit}/2)$ is one form of the *incomplete gamma function*,[11] which is given by

$$\gamma(\nu/2, x) = \frac{1}{\Gamma(\nu/2)} \int_x^\infty e^{-t} t^{\frac{\nu}{2}-1} \, dt, \tag{10.76}$$

where $\nu = M$ is the number of degrees of freedom. Recall from Section 6.2 that the χ^2 distribution is a special case of a gamma distribution. In *Mathematica* $\gamma(\nu/2, \Delta\chi^2_{crit}/2)$ is given by the command

---

Gamma$[\nu/2, \Delta\chi^2_{crit}/2]$

Table 10.2 gives values for $\Delta\chi^2_{crit}$ as a function of P and ν obtained from Equation (10.75).

For example, the credible region containing probability $P = 68.3\%$, for $M = 2$ free model parameters, is bounded by a surface of constant $\chi^2 = \chi^2_{min} + \Delta\chi^2_{crit} = \chi^2_{min} + 2.3$. Note: in Table 10.2, the degrees of freedom $\nu = M$, the number of free model parameters.

---

[11] See *Numerical Recipes* by Press *et al.* (1992).

Table 10.2 *This table allows us to find the value of $\Delta\chi^2_{\text{crit}}$ in Equation (10.74) that defines the boundary of the joint posterior probability in ν model parameters that contains a specified probability P. Thus, the P = 68.3% joint probability boundary in two parameters corresponds to a $\Delta\chi^2_{\text{crit}} = 2.3$, where $\Delta\chi^2_{\text{crit}}$ is measured from χ^2_{min}, the value corresponding to the peak of the joint posterior probability.*

P	Degrees of Freedom ν					
	1	2	3	4	5	6
68.3%	1.00	2.30	3.53	4.72	5.89	7.04
90%	2.71	4.61	6.25	7.78	9.24	10.6
95.4%	4.00	6.17	8.02	9.70	11.3	12.8
99%	6.63	9.21	11.3	13.3	15.1	16.8
99.73%	9.00	11.8	14.2	16.3	18.2	20.1
99.99%	15.1	18.4	21.1	23.5	25.7	27.8

Question: Figure 10.3 shows an ellipse of constant probability for a two-parameter model defined by $\Delta Q = k$ (a constant). For what value of k does the ellipse contain a probability of 95.4%?

Answer: First we note that $\Delta Q = \sigma^2 \Delta\chi^2$. From Table 10.2, we obtain $\Delta\chi^2 = 6.17$ for $\nu = 2$ degrees of freedom (for a two-parameter model) and a probability of 95.4%. The desired value of $k = 6.17\sigma^2$.

Question: Suppose we fit a model with six linear parameters. We are really only interested in two of these parameters so we remove the other four by marginalization. The result is the posterior probability distribution for the two interesting parameters. Now suppose we want to plot the 95.4% credible region (ellipse) for these two parameters. How many degrees of freedom should be used when consulting Table 10.2?

Answer: 2.

Problem: In Section 10.2.2, we fitted a straight line model to the data in Table 10.1. Compute and plot the 95.4% joint credible region for the slope and intercept.

Solution: From Table 10.2, we obtain $\Delta\chi^2 = 6.17$ for $\nu = 2$ degrees of freedom (for a two-parameter model) and a probability of 95.4%. Now $\Delta\chi^2$ is related to the Ψ matrix by

$$\Delta\chi^2 = \sum_{\alpha=1}^{2}\sum_{\beta=1}^{2} \delta A_\alpha [\Psi]_{\alpha\beta} \delta A_\beta, \tag{10.77}$$

where

$$\Psi = \frac{1}{\sigma^2}\begin{pmatrix} 76 & 3010 \\ 3010 & 152\,300 \end{pmatrix} = \begin{pmatrix} 9.383 & 371.6 \\ 371.6 & 18\,802 \end{pmatrix}, \tag{10.78}$$

since $\sigma^2 = 8.1$. Combining Equations (10.77) and (10.78), we obtain

$$\Delta\chi^2 = (\delta A_1 \ \delta A_2) \begin{pmatrix} 9.383 & 371.6 \\ 371.6 & 18\,802 \end{pmatrix} \begin{pmatrix} \delta A_1 \\ \delta A_2 \end{pmatrix} = 6.17. \qquad (10.79)$$

It is convenient to change from rectangular coordinates $(\delta A_1, \delta A_2)$ to polar coordinates (r, θ). Equation (10.79) becomes

$$(r\cos\theta \ \ r\sin\theta) \begin{pmatrix} 9.383 & 371.6 \\ 371.6 & 18\,802 \end{pmatrix} \begin{pmatrix} r\cos\theta \\ r\sin\theta \end{pmatrix} = 6.17. \qquad (10.80)$$

Next, solve this equation for r for a set of values of θ, and by so doing map out the joint credible region. In *Mathematica* this can easily be accomplished by:

```
polarA [r_, θ_] := {{r * Cos[θ]}, {r * Sin[θ]}};
tpolarA[r_, θ_] := Transpose[polarA[r, θ]];
locus = Table[
    {NSolve[Flatten[tpolarA[r, θ].Ψ.polarA[r, θ]][[1]] = = 6.17, r]
    [[2]][[1, 2]], θ}, {θ, 0, 2π, Δθ}];
```

Finally, transform the r, θ values back to δA_1, δA_2 and convert them to A_1, A_2, where $A_1 = \delta A_1 + \hat{A}_1$ and $A_2 = \delta A_2 + \hat{A}_2$. Note: $\hat{A}_1$ and $\hat{A}_2$ are the most probable values of the intercept and slope, respectively. Figure 10.4(a) shows a plot of the resulting 95.4% joint credible region (dashed curve). The solid curve shows the 68.3% joint credible region derived in the same way.

What if we are interested in summarizing what the data and our prior information have to say about the slope, i.e., determining the marginal PDF, $p(A_2|D, I)$? In this

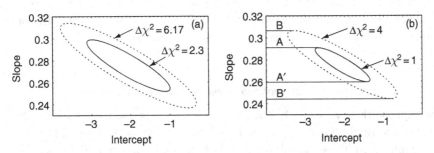

Figure 10.4 Panel (a) is a plot of the 95.4% (dashed) and 68.3% (solid) joint credible regions for the slope and intercept of the best-fit straight line to the data of Table 10.1. Panel (b) shows ellipses corresponding to $\Delta\chi^2 = 1.0$ and $\Delta\chi^2 = 4.0$. The two lines labeled A and A', which are tangent to the inner ellipse, define the 68.3% credible region for the marginal PDF, $p(A_2|D, I)$. The two lines labeled B and B', which are tangent to the outer ellipse, define the 95.4% credible region for $p(A_2|D, I)$.

case, we need to marginalize over all possible values of A_1. We will look at this question more fully in Section 10.5, but we can use the material of this section to obtain the 68.3% credible region for A_2 as follows. By good fortune it turns out that for Gaussian posteriors, in any number of dimensions, the marginal PDF is also equal to the projected distribution (projected PDF). What do we mean by the projected PDF for A_2? Imagine a light source located a great distance away along the A_1 axis, illuminating the 3-dimensional probability density function, thought of as an opaque mountain sitting on the A_1, A_2 plane. The height of the mountain at any particular A_1, A_2 is equal to $p(A_1, A_2|D, I)$. The shadow cast by this mountain on the plane defined by $A_1 = 0$ is called the *projected probability density function of A_2*.

To plot out the projected PDF, we can do the following. Each value of A_2 in our final plot corresponds to a line parallel to the A_1 axis. Vary A_1 along this line and find the maximum value of $p(A_1, A_2|D, I)$ along the line. If we raise the line to this height it will be tangent to the surface of the probability mountain for this A_2. This is the value of the projected PDF for that particular value of A_2.

Now consider the two lines shown in Figure 10.4(b), labeled A and A', which define the borders of the 68.3% credible region for A_2. You might naively expect these two lines to be tangent to the ellipse containing 68.3% of the joint probability, $p(A_1, A_2|D, I)$, as illustrated in Figure 10.4(a) for $\Delta\chi^2 = 2.3$. The correct answer is that they are tangent to the ellipse shown in Figure 10.4(b), corresponding to $\Delta\chi^2 = 1.0$, as indicated in Table 10.2 for one degree of freedom. The two lines, B and B', which define the 95.4% credible region, are tangent to the ellipse defined by $\Delta\chi^2 = 4.0$. In a like fashion we could locate the 68.3% and 95.4% credible region boundaries for A_1.

10.5 Model parameter errors

In Sections 10.2.1 and 10.2.2, we found the most probable values of linear model parameters. To complete the discussion, we need to specify the uncertainties of these parameters and introduce the *parameter covariance matrix*.

10.5.1 Marginalization and the covariance matrix

Now suppose that we are only interested in a subset of the model amplitudes (for example, one amplitude may describe an uninteresting mean background level, or, we may be interested in the probability density function of only one of the parameters). We can summarize the implications of the data for the interesting amplitudes by calculating the marginal distribution for those amplitudes, integrating the uninterest-ing *nuisance parameters* out of the full joint posterior. In this subsection we start by showing how to integrate out a single amplitude; the procedure can be repeated to remove more parameters. We then consider the special case of a model with only two parameters ($M = 2$) and see how this leads to an understanding of the parameter errors.

Suppose that the amplitude we want to marginalize out is A_1. Returning to the Q notation and IID Gaussian errors, the marginal distribution for the remaining amplitudes is then

$$
\begin{aligned}
p(A_2, \ldots, A_M | D, I) &= \int_{\Delta A_1} dA_1 \, p(\{A_\alpha\} | D, I) \\
&= C' \int_{\Delta A_1} dA_1 \, e^{-\Delta Q/2\sigma^2},
\end{aligned}
\tag{10.81}
$$

where

$$
\Delta Q(A) = \sum_{\alpha\beta} \delta A_\alpha \psi_{\alpha\beta} \delta A_\beta.
\tag{10.82}
$$

To perform the required integral, we first pull out the δA_1-dependent terms in ΔQ, writing

$$
\Delta Q = (\delta A_1)^2 \psi_{11} + 2\delta A_1 \sum_{\beta=2}^{M} \psi_{1\beta} \delta A_\beta + \sum_{\alpha,\beta=2}^{M} \delta A_\alpha \psi_{\alpha\beta} \delta A_\beta,
\tag{10.83}
$$

where δA_1 appears only in the first two terms.

Now we complete the square for δA_1 by adding and subtracting a term as follows:

$$
\begin{aligned}
\Delta Q ={}& (\delta A_1)^2 \psi_{11} + 2\delta A_1 \sum_{\beta=2}^{M} \psi_{1\beta} \delta A_\beta + \frac{1}{\psi_{11}} \left(\sum_{\beta=2}^{M} \psi_{1\beta} \delta A_\beta \right)^2 \\
&- \frac{1}{\psi_{11}} \left(\sum_{\beta=2}^{M} \psi_{1\beta} \delta A_\beta \right)^2 + \sum_{\alpha,\beta=2}^{M} \delta A_\alpha \psi_{\alpha\beta} \delta A_\beta \\
={}& \psi_{11} \left[\delta A_1 + \psi_{11}^{-1} \sum_{\beta=2}^{M} \psi_{1\beta} \delta A_\beta \right]^2 + \Delta Q_r.
\end{aligned}
\tag{10.84}
$$

By construction, δA_1 appears only in the squared term, and the terms depending on the remaining δA's make up the reduced quadratic form,

$$
\Delta Q_r = -\frac{1}{\psi_{11}} \left(\sum_{\beta=2}^{M} \psi_{1\beta} \delta A_\beta \right)^2 + \sum_{\alpha,\beta=2}^{M} \delta A_\alpha \psi_{\alpha\beta} \delta A_\beta.
\tag{10.85}
$$

Equation (10.81) can now be written

$$p(A_2, \ldots, A_M | D, I) = C' e^{-\Delta Q_r / 2\sigma^2} \int_{\Delta A_1} d(\delta A_1)$$

$$\times \exp \left[-\frac{\psi_{11}}{2\sigma^2} \left(\delta A_1 + \psi_{11}^{-1} \sum_{\beta=2}^{M} \psi_{1\beta} \delta A_\beta \right)^2 \right].$$
(10.86)

The integrand is a Gaussian in δA_1, with variance σ^2 / ψ_{11}. If the range of integration, ΔA_1, were infinite, the integral would merely be a constant (the normalization constant for the Gaussian, $\sigma \sqrt{2\pi / \psi_{11}}$). With a finite range, it can be written in terms of error functions with arguments that depend on A_2 through A_M. But as long as the prior range is large compared to $\sigma / \sqrt{\psi_{11}}$, this integral will be very nearly constant with respect to the remaining amplitudes. Thus, to a good approximation, the marginal distribution is

$$p(A_2, \ldots, A_M | D, I) = C'' e^{-\Delta Q_r / 2\sigma^2},$$
(10.87)

where $C'' = C' \sigma \sqrt{2\pi / \psi_{11}}$ is a new normalization constant. In the limit where $\Delta A_1 \to \infty$, this result is exact.

Again, it is useful to consider the special case of only two parameters, $M = 2$. In this case, after marginalizing out A_1 we are left with $p(A_2 | D, I)$. Let's evaluate this now. We can rewrite Equation (10.85),

$$\Delta Q_r = -\frac{1}{\psi_{11}} (\psi_{12} \delta A_2)^2 + \delta A_2 \psi_{22} \delta A_2$$

$$= \delta A_2^2 \left(\frac{\psi_{11} \psi_{22} - \psi_{12}^2}{\psi_{11}} \right).$$
(10.88)

Thus, we can write

$$p(A_2 | D, I) = C' \sigma \sqrt{2\pi / \psi_{11}} \, e^{-\delta A_2^2 / 2\sigma_2^2},$$
(10.89)

which is a Gaussian with variance σ_2^2 given by

$$\sigma_2^2 = \sigma^2 \left(\frac{\psi_{11}}{\psi_{11} \psi_{22} - \psi_{12}^2} \right).$$
(10.90)

Similarly, if we had marginalized out A_2 instead, we would have obtained a Gaussian PDF for $p(A_1 | D, I)$ with σ_1^2 given by

$$\sigma_1^2 = \sigma^2 \left(\frac{\psi_{22}}{\psi_{11} \psi_{22} - \psi_{12}^2} \right).$$
(10.91)

Notice that the variances for A_1 and A_2 can be written in terms of the elements of ψ^{-1}:

$$\psi^{-1} = \frac{1}{\psi_{11}\psi_{22} - \psi_{12}^2} \begin{pmatrix} \psi_{22} & -\psi_{12} \\ -\psi_{12} & \psi_{11} \end{pmatrix}. \tag{10.92}$$

Comparing with Equations (10.91) and (10.90), we can write

$$\sigma_1^2 = \sigma^2 [\psi^{-1}]_{11}, \tag{10.93}$$

$$\sigma_2^2 = \sigma^2 [\psi^{-1}]_{22}. \tag{10.94}$$

Thus, we have shown for the $M = 2$ case, that the matrix ψ^{-1} that we needed to solve for $\hat{A}_\alpha$ (see Equation (10.34)), when multiplied by the data variances (σ^2), also contains information about the errors of the parameters. In Section 10.5.3, we will generalize this result for the case of a linear model with an arbitrary number of parameters M, and show that

$$\sigma_\alpha^2 = \sigma^2 [\psi^{-1}]_{\alpha\alpha}. \tag{10.95}$$

The matrix $\mathbf{V} = \sigma^2 \psi^{-1}$ is given the name *parameter variance-covariance matrix* or simply the *parameter covariance matrix*.[12] We shall shortly define what we mean by covariance.

If we wish to summarize our posterior state of knowledge about the parameters with a few numbers, then we can write

$$A_1 = \hat{A}_1 \pm \sigma_1 \tag{10.96}$$

$$A_2 = \hat{A}_2 \pm \sigma_2. \tag{10.97}$$

Formally, the variance of A_1 is defined as the expectation value of the square-of-the-deviations from the mean μ_1;

$$\mathrm{Var}(A_1) = \langle (A_1 - \mu_1)^2 \rangle = \int_{\Delta A_1} dA_1 (A_1 - \mu_1)^2 p(A_1|D, I). \tag{10.98}$$

The idea of variance can be broadened to consider the simultaneous deviations of both A_α and A_β. The covariance is given by

$$\sigma_{\alpha\beta} = \int_{\Delta A_\alpha} \int_{\Delta A_\beta} dA_\alpha dA_\beta (A_\alpha - \mu_\alpha)(A_\beta - \mu_\beta) p(A_\alpha, A_\beta|D, I) \tag{10.99}$$

and is a measure of the correlation between the inferred parameters. If, for example, there is a high probability that overestimates of A_α are associated with overestimates

---

[12] If we are employing the covariance matrix, $\mathbf{E}$, for our knowledge of the measurement errors, then simply replace ψ^{-1} by $\mathbf{\Psi}^{-1}$, $\psi_{\alpha\beta}$ by $\Psi_{\alpha\beta}$ and drop all the factors of σ^2 in Equations (10.93), (10.94), (10.101), and (10.100).

of A_β, and underestimates of A_α associated with underestimates of A_β, then the covariance will be positive. Negative covariance (anti-correlation) implies that overestimates of A_α will be associated with underestimates of A_β. When the estimate of one parameter has little or no influence on the inferred value of the other, then the magnitude of the covariance will be negligible in comparison to the variance terms, $|\sigma_{\alpha\beta}| \ll \sqrt{\sigma_{\alpha\alpha}\sigma_{\beta\beta}} = \sqrt{\sigma_\alpha^2 \sigma_\beta^2}$.

By now you may have guessed that the covariance of the inferred parameters is given by σ^2 times the off-diagonal elements of the ψ^{-1} matrix.

$$\sigma_{\alpha\beta} = \sigma^2 [\psi^{-1}]_{\alpha\beta}. \tag{10.100}$$

Thus, for the $M = 2$ case, we have

$$\sigma_{12} = \sigma^2 [\psi^{-1}]_{12}$$
$$= \sigma^2 \left(\frac{-\psi_{12}}{\psi_{11}\psi_{22} - \psi_{12}^2} \right). \tag{10.101}$$

Referring to Figure 10.3, we see that the major axis of the elliptical credible region is inclined to the A_1 axis with a positive slope. This indicates a positive correlation between the parameters A_1 and A_2. A value of $\sigma_{12} = 0$ would correspond to a major axis which is aligned with the A_1 axis if $\sigma_1 > \sigma_2$.

10.5.2 Correlation coefficient

It is useful to summarize the correlation between estimates of any two parameters by a coefficient in the range from -1 to $+1$, where -1 indicates complete negative correlation, $+1$ indicates complete positive correlation, and 0 indicates no correlation. The correlation coefficient is defined by

$$\rho_{\alpha\beta} = \frac{\sigma_{\alpha\beta}}{\sqrt{\sigma_{\alpha\alpha}\sigma_{\beta\beta}}} = \frac{\sigma_{\alpha\beta}}{\sqrt{\sigma_\alpha^2 \sigma_\beta^2}}. \tag{10.102}$$

In the extreme case of $\rho = \pm 1$, the elliptical contours will be infinitely wide in one direction (with only information in the prior preventing this catastrophe) and oriented at an angle $\pm \tan^{-1} \left[\sqrt{[\Psi^{-1}]_{22}/[\Psi^{-1}]_{11}} \right]$. In this case, the parameter error bars σ_1 and σ_2 will be huge, saying that our individual estimates of A_1 and A_2 are completely unreliable, but we can still infer a linear combination of the parameters quite well. For ρ large and positive, the probability contours will all bunch up close to the line $A_2 = \text{intercept} + mA_1$, where $m = \sqrt{[\Psi^{-1}]_{22}/[\Psi^{-1}]_{11}}$. We can rewrite this as $A_2 - mA_1 = \text{intercept}$. Varying the intercept corresponds to motion perpendicular to this line. The concentration of probability contours implies that the probability density of the intercept is quite narrow. Since the intercept is equal to $A_2 - mA_1$, this indicates that the data contain a lot of information about the difference $A_2 - mA_1$. If ρ is large and negative, then we can infer the sum $A_2 + mA_1$.

Problem: In Section 10.2.2, we fitted a straight line model to the data in Table 10.1. Evaluate the errors for the joint and marginal posterior density functions for the intercept, A_1, and slope, A_2, from the diagonal elements of $\mathbf{V} = \mathbf{\Psi}^{-1} = (\mathbf{G}^T \mathbf{E}^{-1} \mathbf{G})^{-1}$ which is given by

$$\mathbf{\Psi}^{-1} = \sigma^2 \begin{pmatrix} 0.061 & -0.0012 \\ -0.0012 & 0.000030 \end{pmatrix}, \tag{10.103}$$

where $\sigma^2 = 8.1$.

Solution: Let σ_1 and σ_2 be the 1σ errors of A_1 and A_2. $\mathbf{\Psi}^{-1}$ is the variance-covariance matrix of the parameter errors. In this case, it includes the data covariance matrix $\mathbf{E}^{-1}$ so we need to use Equations (10.93) and (10.94) without the σ^2 term in front.

$$\sigma_1 = \sqrt{[\mathbf{\Psi}^{-1}]_{11}} = \sqrt{0.061\sigma^2} = 0.70, \tag{10.104}$$

$$\sigma_2 = \sqrt{[\mathbf{\Psi}^{-1}]_{22}} = \sqrt{0.000030\ \sigma^2} = 0.016, \tag{10.105}$$

and,

$$\begin{pmatrix} \hat{A}_1 \\ \hat{A}_2 \end{pmatrix} = \begin{pmatrix} -2.05 \pm 0.70 \\ 0.275 \pm 0.016 \end{pmatrix}. \tag{10.106}$$

The correlation coefficient is

$$\rho_{12} = \frac{\sigma_{12}}{\sqrt{\sigma_{11}\sigma_{22}}} = \frac{[\mathbf{\Psi}^{-1}]_{12}}{\sqrt{[\mathbf{\Psi}^{-1}]_{11}[\mathbf{\Psi}^{-1}]_{22}}} = \frac{-0.0012}{\sqrt{0.061 \times 0.00003}} = -0.885. \tag{10.107}$$

When there are only two parameters, it is more informative to give a contour plot of the joint posterior probability density function, which illustrates the correlation in the parameter error estimates. In Section 10.4.1, we showed how to compute the joint credible region for the slope and intercept and plotted two examples in Figure 10.4(a). This figure is repeated in the left panel of Figure 10.5. The two contours shown enclose 95.4% and 68.3% of the probability.

In the previous analysis, the intercept and slope are referenced to the origin of our data. It turns out that the size of the correlation coefficient depends on the origin we choose for the x-coordinate, as we demonstrate in Figure 10.6. In fact, if we shift the origin by just the right amount, call it x_w, we can eliminate the correlation altogether.

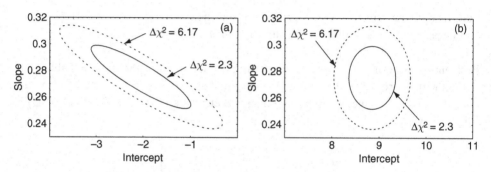

Figure 10.5 Panel (a) shows a contour plot of the joint posterior PDF $p(A_1, A_2|D, I)$. The dashed and solid contours enclose 95.4% and 68.3% of the probability, respectively. Panel (b) is the same but using the weighted average x-coordinate as the origin of the fit.

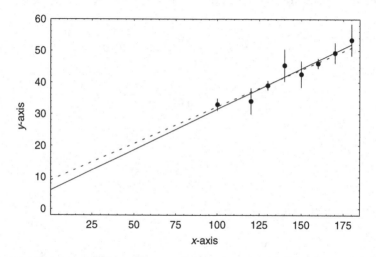

Figure 10.6 Two straight line fits to some data with an origin well outside the x range of the data. Clearly, any variation in the slope parameter will have a strongly correlated effect on the intercept and vice versa. If the x origin had been chosen closer to the middle of the data, then variations in slope would have a much smaller effect on the intercept.

From Equation (10.107) it is clear that $\rho_{12} = 0$ if $[\mathbf{\Psi}^{-1}]_{12} = 0$, and it is easy to show that the off-diagonal elements of $\mathbf{\Psi}^{-1}$ are zero if the off-diagonal elements of $\mathbf{\Psi}$ are zero. These modified basis functions are referred to as orthogonal basis functions. From Equations (10.47) and (10.46) we can write

$$[\mathbf{\Psi}]_{12} = \sum_{ij} g_{i1} [\mathbf{E}^{-1}]_{ij} \, g_{j2}. \qquad (10.108)$$

For the straight line model $g_{i1} = \{1, 1, \ldots, 1\}$ and $g_{j2} = \{x_1, x_2, \ldots, x_N\}$. Shifting the origin to x_w changes g_{j2} to $g'_{j2} = \{x_1 - x_w, x_2 - x_w, \ldots, x_N - x_w\}$. We can solve for x_w by setting

$$[\Psi]'_{12} = \sum_{ij} g_{i1} \, [E^{-1}]_{ij} \, g'_{j2} = 0. \tag{10.109}$$

From Equation (10.52), we can rewrite Equation (10.109) as

$$[\Psi]'_{12} = (1, 1, 1, \ldots, 1) \frac{1}{\sigma^2} \begin{pmatrix} 14 & 0 & 0 & \cdots & 0 \\ 0 & 3 & 0 & \cdots & 0 \\ 0 & 0 & 25 & \cdots & 0 \\ \cdot & \cdot & \cdot & \cdot & \cdot \\ 0 & 0 & \cdots & 0 & 2 \end{pmatrix} \begin{pmatrix} x_1 - x_w \\ x_2 - x_w \\ x_3 - x_w \\ \cdot \\ x_N - x_w \end{pmatrix}$$

$$= \frac{1}{\sigma^2} (14, 3, 25, \ldots, 2) \begin{pmatrix} x_1 - x_w \\ x_2 - x_w \\ x_3 - x_w \\ \cdot \\ x_N - x_w \end{pmatrix} \tag{10.110}$$

$$= (w_1, w_2, w_3, \ldots, w_N) \begin{pmatrix} x_1 - x_w \\ x_2 - x_w \\ x_3 - x_w \\ \cdot \\ x_N - x_w \end{pmatrix}$$

$$= w_1(x_1 - x_w) + w_2(x_2 - x_w) + \cdots + w_N(x_n - x_w) = 0,$$

where the data weights, w_i, are given by the diagonal elements of the inverse data covariance matrix E^{-1}. The solution of Equation (10.110) is given by

$$x_w = \frac{\sum w_i x_i}{\sum w_i}. \tag{10.111}$$

Panel (b) of Figure 10.5 shows the joint credible region for the parameters obtained using the weighted average, x_w, as the origin. The major and minor axes of the ellipse are now parallel to the parameter axes. The sensitivity of the analysis to the choice of origin arises from the model predictions' dependence on the origin.

Suppose that instead of a straight line model, we had wanted to fit a higher order polynomial to the data. The appropriate function to fit, to ensure the coefficients are uncorrelated, can be shown to be of the form

$$
\begin{aligned}
y(x_i) = A_1 + A_2(x_i - x_w) + A_3(x_i - \gamma_1)(x_i - \gamma_2) \\
+ A_4(x_i - \delta_1)(x_i - \delta_2)(x_i - \delta_3) + \cdots,
\end{aligned}
\tag{10.112}
$$

In the case of a polynomial with just the first 3 terms, we can compute γ_1 and γ_2 from the two equations

$$
[\mathbf{\Psi}]_{13}' = \sum_{ij} g_{i1} \, [\mathbf{E}^{-1}]_{ij} \, g_{j3}' = 0,
\tag{10.113}
$$

$$
[\mathbf{\Psi}]_{23}' = \sum_{jk} g_{j2}' \, [\mathbf{E}^{-1}]_{jk} \, g_{k3}' = 0,
\tag{10.114}
$$

where $g_{j2}' = \{x_1 - x_w, x_2 - x_w, \ldots, x_N - x_w\}$

and $g_{k3}' = \{(x_1 - \gamma_1)(x_1 - \gamma_2), (x_2 - \gamma_1)(x_2 - \gamma_2), \ldots, (x_N - \gamma_1)(x_N - \gamma_2)\}$. (10.115)

10.5.3 More on model parameter errors

In Sections 10.2.1 and 10.2.2, we found the most probable values of linear model parameters. To complete the discussion, we need to specify the uncertainties of these parameters. We made a start on this in Section 10.5.1 for the special case of a linear model with only two parameters ($M = 2$). We also introduced the concept of the covariance of two parameters and the correlation coefficient. In this section, we will generalize these results for an arbitrary value of M. In Section 10.4, we showed that the posterior probability distribution of the parameters $p(\{A_\alpha\}|D, I)$ is a multivariate Gaussian given by

$$
p(\{A_\alpha\}|D, I) \propto \exp\left[-\frac{1}{2\sigma^2} \sum_{\alpha\beta} \delta A_\alpha \psi_{\alpha\beta} \delta A_\beta\right].
\tag{10.116}
$$

If we use the more powerful matrix formulation which includes the data covariance matrix $\mathbf{E}$ (see Section 10.2.2), then we need to replace

$$
\frac{1}{2\sigma^2} \sum_{\alpha\beta} \delta A_\alpha \psi_{\alpha\beta} \delta A_\beta \quad \text{by} \quad \frac{1}{2} \sum_{\alpha\beta} \delta A_\alpha \Psi_{\alpha\beta} \delta A_\beta.
$$

Equation (10.116) becomes

$$
\begin{aligned}
p(\{A_\alpha\}|D, I) &\propto \exp\left[-\frac{1}{2} \sum_{\alpha\beta} \delta A_\alpha \Psi_{\alpha\beta} \delta A_\beta\right] \\
&= \exp\left[-\frac{1}{2} \sum_{\alpha\beta} (A_\alpha - \hat{A}_\alpha)[\mathbf{\Psi}]_{\alpha\beta}(A_\beta - \hat{A}_\beta)\right].
\end{aligned}
\tag{10.117}
$$

Now compare Equation (10.117) with Equation (10.41) for $p(D|\{A_\alpha\}, I)$ which is repeated here.

$$p(D|\{A_\alpha\}, I) \propto \exp\left[-\frac{1}{2}\sum_{ij}(d_i - f_i)[\mathbf{E}^{-1}]_{ij}(d_j - f_j)\right]. \qquad (10.118)$$

Both have the same form. In Equation (10.118), $\mathbf{E}$ is the data covariance matrix. By analogy the inverse of $\mathbf{\Psi}$, $\mathbf{\Psi}^{-1}$ is the model parameter covariance matrix. Thus, everything we need to know about the uncertainties with which the various parameters have been determined, and their correlations, is given by $\mathbf{\Psi}^{-1} = (\mathbf{G}^T\mathbf{E}^{-1}\mathbf{G})^{-1}$, a matrix which we previously computed (Section 10.2.2) in the determination of the most probable values of the parameters. The variance terms are given by the diagonal elements of $\mathbf{\Psi}^{-1}$ and the covariance terms by the off-diagonal elements. We see that the parameter errors depend on the data errors through $\mathbf{E}^{-1}$ but in a complicated way which depends on our choice of model basis functions.

In Section 10.2.2, we also saw that $\mathbf{E}^{-1}$ plays the role of a metric in the full N-dimensional vector space of the data. In a similar fashion, $\mathbf{\Psi}$ plays the role of a metric in the M-dimensional subspace spanned by the model functions.

10.6 Correlated data errors

In this section, we compute the mean of a data set for which the off-diagonal elements of the data covariance matrix, $\mathbf{E}$, are not all zero, i.e., the noise components are correlated. These correlations can be introduced by the experimental apparatus prior to the digitization of the data, or by subsequent software operations. Panel (a) of Figure 10.7 shows 100 simulated data samples of a mean, $\mu = 0.5$, with added IID Gaussian noise ($\sigma = 1$). Panel (b) shows the same data after a smoothing operation that replaces each original sample (d_i) by a weighted average (z_i) of the original sample and its nearest neighbors according to Equation (10.119).

$$z_i = \begin{cases} 0.75d_i + 0.25d_{i+1} & \text{for } i = 1 \\ 0.25d_{i-1} + 0.5d_i + 0.25d_{i+1} & \text{for } 1 < i < 100 \\ 0.25d_{i-1} + 0.75d_i & \text{for } i = 100. \end{cases} \qquad (10.119)$$

If the characteristic width of the signal component in the data is very broad[13] (in this example the signal is a DC offset), then the smoothing will have little effect on the signal component. However, it will introduce correlations into the independent noise components. These correlations need to be incorporated in the analysis. The dominant correlation in the smoothed data is with the nearest neighbor on either side. There is

---

[13] If the smoothing has a significant effect on the signal component then this can be accounted for in the signal model by a convolution operation, as discussed in Appendix B.4.3.

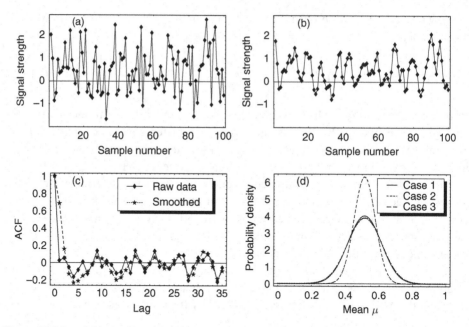

Figure 10.7 Panel (a) shows 100 independent samples of a mean value μ. In panel (b), the data have been smoothed using a running average that introduces correlations. Panel (c) compares the autocorrelation functions (ACF) for the raw data and smoothed data. Panel (d) compares the posterior density for μ for three cases. Case 1 is based on an analysis of the independent samples. The smoothed data results correspond to case 2 (assuming no correlations) and case 3 (including correlations).

also a weaker correlation with the next nearest neighbor[14]. A common tool for calculating correlations is the autocorrelation function (ACF), which was introduced in Section 5.13.2. To compute the ACF of the noise[15] we need a sample of data (without any signal present) which has been smoothed in the same way. Panel (c) of Figure 10.7 compares the autocorrelation functions for the raw data and smoothed data. For the smoothed data, the ACF yields a correlation coefficient $\rho_1 = 0.68$ for nearest neighbors (lag of 1), and $\rho_2 = 0.16$ for next nearest neighbors (lag of 2).

Panel (d) shows the Bayesian marginal posterior of the mean, $p(\mu|D, I)$, computed for three cases. In case 1, $p(\mu|D, I)$ was computed from the independent samples of panel (a) following the treatment of Section 9.2.1. In case 2, $p(\mu|D, I)$ was computed

---

[14] According to equation (5.63), we first subtract the mean of the noise data which can introduce a correlation between all the noise terms. If N is large this correlation is very weak and has been neglected in the current analysis.
[15] We can write

$$\sigma_{12} = \langle e_1 e_2 \rangle = \langle e_2 e_3 \rangle = \langle e_i e_{i+1} \rangle = \sigma^2 \rho(j = 1),$$

where $\rho(j = 1)$ is the value of the ACF for lag $j = 1$.

from the smoothed data and assuming no correlation. This second case results in a narrower posterior which is unwarranted because our state of information is unchanged from case 1, we have simply transformed the original data via a smoothing operation. In case 3, we incorporated information about the correlations introduced by the smoothing. This yielded essentially the same result as we obtained in case 1, as we would expect. In all three cases, we assumed the noise σ was an unknown nuisance parameter. We assumed a Jeffreys prior for σ and a uniform prior for μ.

For case 3, the likelihood was computed from Equation (10.41), which for the current problem can be written as

$$
p(D|\mu, \sigma, I) = \frac{1}{(2\pi)^{N/2}\sqrt{\det \mathbf{E}}} \exp\left[-\frac{1}{2}\sum_{ij}(d_i - f_i)[\mathbf{E}^{-1}]_{ij}(d_j - f_j)\right]
$$

$$
= \frac{1}{(2\pi)^{N/2}\sqrt{\det \mathbf{E}}} \exp\left[-\frac{1}{2}\delta\mathbf{Y}^T\mathbf{E}^{-1}\delta\mathbf{Y}\right],
$$

(10.120)

where $\delta\mathbf{Y}^T = \{(d_1 - f_1), (d_2 - f_2), \ldots, (d_N - f_N)\}$ is a vector of the differences between the measured and predicted data values. From the results of the ACF, the data covariance matrix, $\mathbf{E}$, is given by

$$
\mathbf{E} = \sigma^2
\begin{pmatrix}
1 & \rho_1 & \rho_2 & 0 & \cdots & 0 & 0 & 0 \\
\rho_1 & 1 & \rho_1 & \rho_2 & \cdots & 0 & 0 & 0 \\
\rho_2 & \rho_1 & 1 & \rho_1 & \cdots & 0 & 0 & 0 \\
\cdot & \cdot & \cdot & \cdot & \cdot & \cdot & \cdot & \cdot \\
0 & 0 & 0 & 0 & \cdots & \rho_2 & \rho_1 & 1
\end{pmatrix}
$$

$$
= \sigma^2
\begin{pmatrix}
1 & 0.68 & 0.16 & 0 & \cdots & 0 & 0 & 0 \\
0.68 & 1 & 0.68 & 0.16 & \cdots & 0 & 0 & 0 \\
0.16 & 0.68 & 1 & 0.68 & \cdots & 0 & 0 & 0 \\
\cdot & \cdot & \cdot & \cdot & \cdot & \cdot & \cdot & \cdot \\
0 & 0 & 0 & 0 & \cdots & 0.16 & 0.68 & 1
\end{pmatrix}.
$$

(10.121)

10.7 Model comparison with Gaussian posteriors

In this section, we are interested in comparing the probabilities of two linear models with different numbers of amplitude parameters. From our treatment of model comparison in Section 3.5, it is clear that the key quantity in model comparison is the evaluation of the global likelihood of a model. Calculation of the global likelihood requires integrating away all of the model parameters from the product of the prior and the likelihood. The integral required to calculate the global likelihood was given earlier as Equation (10.14), which we repeat here:

$$
\begin{aligned}
p(D|M_i, I) &= \frac{1}{\prod_\alpha \Delta A_\alpha} \frac{1}{\sigma^N (2\pi)^{N/2}} \int_{\Delta A} d^M A_\alpha e^{-(Q/2\sigma^2)} \\
&= \frac{1}{\prod_\alpha \Delta A_\alpha} \frac{1}{\sigma^N (2\pi)^{N/2}} e^{-(Q_{\min}/2\sigma^2)} \int_{\Delta A} d^M A_\alpha e^{-(\Delta Q/2\sigma^2)},
\end{aligned}
\tag{10.122}
$$

where we have used Equation (10.59) to expand Q.

We could do the remaining integral by repeating the process of the preceding Section 10.5.1 for each amplitude: complete the square and integrate, one amplitude at a time. This gets to be very tedious if there are a large number of parameters. A mathematically more elegant approach involves transforming to an orthonormal set of model basis functions. The result is given by

$$
\begin{aligned}
p(D|M_i, I) &= \left[\frac{(2\pi)^{M/2} \sqrt{\det \mathbf{V}}}{\prod_\alpha \Delta A_\alpha} \right] \frac{1}{\sigma^N (2\pi)^{N/2}} e^{-(\chi^2_{\min}/2)} \\
&= \Omega_M \, \mathcal{L}_{\max},
\end{aligned}
\tag{10.123}
$$

where $\mathbf{V}$ is the parameter covariance matrix. The quantity $\mathcal{L}_{\max}$ is the likelihood for the model at the mode, which is given by

$$
\mathcal{L}_{\max} = p(D|\hat{A}, M_i) = \frac{1}{\sigma^N (2\pi)^{N/2}} e^{-(\chi^2_{\min}/2)}
\tag{10.124}
$$

and the Occam factor for the model is

$$
\Omega_M = \frac{(2\pi)^{M/2} \sqrt{\det \mathbf{V}}}{\prod_\alpha \Delta A_\alpha}.
\tag{10.125}
$$

Assigning competing models equal prior probabilities, the posterior probability for a model will be proportional to Equation (10.123). The odds ratio in favor of one model over a competitor is simply given by the ratio of Equation (10.123) for the two models. Suppose model 1 has M_1 parameters, denoted A_α, and has a minimum χ^2 equal to $\chi^2_{1,\min}$. Suppose model 2 has M_2 parameters, denoted A'_α, and has a minimum χ^2 equal to $\chi^2_{2,\min}$. Then the odds ratio in favor of model 1 over model 2 is

$$
\begin{aligned}
O_{12} &= \frac{p(M_1|I)}{p(M_2|I)} \times \frac{p(D|M_1, I)}{p(D|M_2, I)} = 1 \times \frac{p(D|M_1, I)}{p(D|M_2, I)} \\
&= e^{\Delta \chi^2_{\min}/2} (2\pi)^{(M_1 - M_2)/2} \sqrt{\frac{\det \mathbf{V}_1}{\det \mathbf{V}_2}} \frac{\prod_{\alpha=1}^{M_2} \Delta A'_\alpha}{\prod_{\alpha=1}^{M_1} \Delta A_\alpha},
\end{aligned}
\tag{10.126}
$$

where $\mathbf{V}_1$ and $\mathbf{V}_2$ are the covariance matrices for the estimated parameters and $\Delta\chi^2_{\min} = \chi^2_{2,\min} - \chi^2_{1,\min}$. If the two models have some parameters in common, then the ratio of the prior ranges ΔA_α and $\Delta A'_\alpha$ for these parameters will cancel.

Problem: In Section 3.6, we considered two competing theories. One theory (M_1) predicted the existence of a spectral line with known Gaussian shape and location, and a line amplitude in the range 0.1 to 100 units. The other theory (M_2) predicted no spectral line (i.e., an amplitude $= 0$). We now re-analyze the 64 channel spectral line data given in Table 3.1 using the linear least-squares method discussed in this chapter. We will assume a uniform prior for the line amplitude predictions of M_1.[16]

Solution: First we calculate the most probable amplitude using Equation (10.34),

$$\hat{\mathbf{A}} = (\mathbf{G}^T\mathbf{E}^{-1}\mathbf{G})^{-1}\mathbf{G}^T\mathbf{E}^{-1}\mathbf{D} = \boldsymbol{\Psi}^{-1}\mathbf{G}^T\mathbf{E}^{-1}\mathbf{D}. \tag{10.127}$$

For M_1, the model prediction, f_i, is given by

$$f_i = A_1 g_i = A_1 \exp\left(-\frac{(\nu_i - \nu_o)^2}{2\sigma_L^2}\right), \tag{10.128}$$

and so there is only one basis function, g_i. Thus, $\mathbf{G}^T$ is given by

$$\mathbf{G}^T = (g_1, g_2, \dots, g_{64}). \tag{10.129}$$

Also, for this problem, the inverse data covariance matrix is a 64×64 matrix given by

$$\mathbf{E}^{-1} = \frac{1}{\sigma^2}\begin{pmatrix} 1 & 0 & 0 & \cdots & 0 \\ 0 & 1 & 0 & \cdots & 0 \\ 0 & 0 & 1 & \cdots & 0 \\ . & . & . & & . \\ 0 & 0 & \cdots & 0 & 1 \end{pmatrix} = \begin{pmatrix} 1 & 0 & 0 & \cdots & 0 \\ 0 & 1 & 0 & \cdots & 0 \\ 0 & 0 & 1 & \cdots & 0 \\ . & . & . & & . \\ 0 & 0 & \cdots & 0 & 1 \end{pmatrix}, \tag{10.130}$$

since the errors are independent and identically distributed with a $\sigma^2 = 1$.

The $\mathbf{D}$ matrix is a column matrix containing the 64 channel spectrometer measurements given in Table 3.1 of Section 3.6. Now that we have specified all the matrices, we can evaluate $\hat{\mathbf{A}}$ from Equation (10.127) using *Mathematica*.

Since M_1 has only one parameter, the parameter covariance matrix $\mathbf{V} = \boldsymbol{\Psi}^{-1} = (\mathbf{G}^T\mathbf{E}^{-1}\mathbf{G})^{-1}$ is a single number equal to the variance of $\hat{A}$. The final answer for $\hat{A}$ is

$$\hat{A} = 1.54 \pm 0.53. \tag{10.131}$$

Now we want to compute the odds ratio, O_{12}, in favor of M_1 over the competing model M_2. Since the two models were assigned equal prior probability, the odds are

---

[16] The diligent reader might object at this point that in Section 3.6, we gave a strong argument for using a Jeffreys prior for the line amplitude. Linear least-squares analysis is widely used in data analysis and we wanted to highlight its strengths and weaknesses which are discussed in the conclusions given at the end of the problem.

given by

$$O_{12} = \frac{p(M_1|I)}{p(M_2|I)} \times \frac{p(D|M_1, I)}{p(D|M_2, I)} = 1 \times \frac{p(D|M_1, I)}{p(D|M_2, I)}. \tag{10.132}$$

Model M_2 has no undetermined parameter and predicts the spectrum equals zero apart from noise. The quantity $p(D|M_2, I)$ is given by

$$p(D|M_2, I) = \frac{1}{\sigma^N (2\pi)^{N/2}} e^{-\chi^2_{2,\min}/2}, \tag{10.133}$$

where

$$\chi^2_{2,\min} = \sum_i^{N=64} \frac{d_i^2}{\sigma^2} = 57.13. \tag{10.134}$$

For model M_1 we can use Equation (10.123) with $M = 1$ parameter, yielding

$$p(D|M_1, I) = \left[\frac{(2\pi)^{1/2} \sqrt{\det \mathbf{V}}}{\Delta A_1} \right] \frac{1}{\sigma^N (2\pi)^{N/2}} e^{-\chi^2_{1,\min}/2} \tag{10.135}$$

$$= \Omega_1 \mathcal{L}_{\max},$$

where

$$\chi^2_{1,\min} = \sum_i^{N=64} \frac{(d_i - \hat{A}g_i)^2}{\sigma^2} = 48.49. \tag{10.136}$$

Equation (10.135) contains an Occam penalty, Ω_1, which penalizes M_1 for prior parameter space that gets ruled out by the data through the likelihood function. The penalty arises automatically from marginalizing over the prior range ΔA. In this case,

$$\Omega_1 = \left[\frac{(2\pi)^{1/2} \sqrt{\det \mathbf{V}}}{\Delta A_1} \right] = 0.0133. \tag{10.137}$$

Substituting Equations (10.133) and (10.135) into Equation (10.132), we get

$$O_{12} = \Omega_1 e^{(\chi^2_{2,\min} - \chi^2_{1,\min})/2} = 1.0. \tag{10.138}$$

Conclusions:

a) Not surprisingly, the results obtained here for $\hat{A}$ and O_{12} are the same as we got from the brute force analysis used in Section 3.6 for the uniform prior assumption for A. In the current problem, we were dealing with only one parameter (for M_1). Some problems involve a very large number of linear parameters and in these cases, the linear least-squares approach is very efficient because no integrals need to be performed.

b) In principle, linear least-squares analysis is only applicable for linear parameters with *uniform priors*. In Section 3.10, we learned that there are good reasons for distinguishing between location parameters, i.e., both positive and negative values are allowed, and scale parameters, which are always positive. We also learned that there are strong reasons for preferring a Jeffreys prior over a uniform prior when dealing with a scale parameter. Of course, in some problems, we are fortunate to have much more selective prior information about parameter values, e.g., based on the results of previous experiments.

   The reader may well ask why we chose the spectral line example where the amplitude is a scale parameter. Linear least-squares analysis is widely used in data analysis and we wanted to highlight its strengths and weaknesses. For many parameter estimation problems, the choice of prior is not too important because the posterior probability density is usually dominated by the likelihood function which is generally rather strongly peaked except when there are very little data.

c) In model selection problems, the choice of prior is much more critical. In Section 3.6, we addressed this question in considerable detail. We showed that using a more appropriate Jeffreys prior led to an odds ratio favoring M_1 which was a factor of ~ 11 larger than for the uniform prior assumption. The main message here is to only use the material on model comparison in Section 10.7 when dealing with parameters for which the choice of a uniform prior is appropriate.

10.8 Frequentist testing and errors

The results in this chapter have been developed from a Bayesian perspective. For comparison purposes, we now introduce a section on model testing and parameter errors from a frequentist perspective. My apologies to those of you who have your Bayesian hat on at this point and can't face the transition again. You can always skip over this section now and return to it if you want to answer question 3(e) in the problems at the end of this chapter. In Section 7.2.1, we discussed the use of the χ^2 statistic in hypothesis testing. Once we have determined the best set of model parameters, we can use the χ^2 statistic to test if the model is acceptable by attempting to reject the model at some confidence level. From Equation (10.41) we see that χ^2 for the fit[17] is given by

$$\chi^2 = \sum_{ij} (d_i - f_i)[\mathbf{E}^{-1}]_{ij}(d_j - f_j). \tag{10.139}$$

If the errors are independent, this reduces to the more familiar form:

$$\chi^2 = \sum_i \frac{(d_i - f_i)^2}{\sigma_i^2}. \tag{10.140}$$

---

[17] In the frequentist context, if the data errors are IID normal, then treated as a random variable, this quantity will have a χ^2 distribution with the number of degrees of freedom equal to $N - M$. If the data covariance matrix $\mathbf{E}^{-1}$ has non-zero off-diagonal elements, indicating correlations, then the number of degrees of freedom will be less than $N - M$.

If the model contains M parameters and there are N data points, then our confidence in rejecting the model is given by the *Mathematica* command:

---

1 - GammaRegularized $\left[\frac{N-M}{2}, \frac{\chi^2}{2}\right]$

Some words of caution are in order on the use of the above for rejecting a hypothesis. First **GammaRegularized** $[(N - M)/2, \chi^2/2]$ measures the significance of the test, which equals the area of the χ^2 distribution to the right of our measured value. Again, if the model is correct and the data errors are known to be IID normal, this area represents the fraction of hypothetical repeats of the experiment that are expected to fall in this tail area by chance. If this area is very small, then there are at least three possibilities: (1) we have underestimated the size of the data errors, (2) the model does a poor job of explaining the systematic structure in the data, or (3) the model is correct; the result is just a very unlikely statistical fluctuation. Because experimental errors are frequently underestimated, it is not uncommon to require a significance < 0.001 before rejecting a hypothesis.
Note: if the significance is very large, e.g., $\gg 0.5$, this is an indication that the data errors may have been overestimated.

Note: *Mathematica* provides a command called

ChiSquarePValue $[\chi^2, N - M]$

---

which has a maximum value of 0.5. This command can sometimes lead to confusion because it measures the area in either tail region. Thus, if the measured value of χ^2 is less than the number of degrees of freedom (the expectation value for the hypothetical reference distribution), we would not want to refer to our confidence in rejecting the model as **1-ChiSquarePValue**.

The above model test assumes we know the errors accurately. What if the scatter in the data from the best fitting model is considerably larger than the data errors used in the analysis but we are convinced that the model is correct? The other option is that we have underestimated the errors. Perhaps the model correctly describes some aspect of the data but in addition something else is going on. Frequently this is how we discover the presence of some new phenomenon: by looking for systematic effects in the residuals after subtracting off the best-fit model from the data.

In Section 10.5.1, we saw that information about the model parameter errors is contained in the covariance matrix, one form of which is given by

$$\mathbf{V} = \sigma^2 \psi^{-1} = \sigma^2 (\mathbf{G}^T \mathbf{G})^{-1}. \tag{10.141}$$

If we have underestimated the data errors (σ) then this will lead to our underestimating the parameter errors. We saw in Section 9.2.3 that in a Bayesian analysis, we can marginalize over any unknown data error and ensure that our parameter uncertainties properly reflect the size of the residuals between the best fitting model and data (see Figure 9.3).

A useful frequentist method for obtaining more robust parameter errors is based on assuming the model is correct and then adjusting all the assumed measurement errors by a factor k. The new value of χ^2 is then given by

$$\chi^2 = \sum_{i=1}^{N} \frac{(d_i - f_i)^2}{k^2 \sigma_i^2} = \frac{\chi_{\text{meas}}^2}{k^2}, \tag{10.142}$$

where χ_{meas}^2 is the value of χ^2 computed using the initial σ_i error estimates. The factor k is computed in the following way. When the model is valid, and the data errors are known, the expected value of χ^2 for the best choice of model parameters, χ_{expect}^2, is equal to the number of degrees of freedom $= N - M$, where $N = $ the number of data points, and $M = $ the number of fit parameters. The procedure then is to adjust the value of k so that χ^2 in Equation (10.142) is equal to $N - M$.

$$\chi^2 = \frac{\chi_{\text{meas}}^2}{k^2} = \chi_{\text{expect}}^2 = N - M. \tag{10.143}$$

The solution is

$$k = \sqrt{\frac{\chi_{\text{meas}}^2}{N - M}}. \tag{10.144}$$

Including a factor k is a good thing to do since often nature is more complicated than either the model or known measurement errors can account for. Increasing all the measurement errors from σ to $k\sigma$ corresponds to increasing the terms in the parameter covariance matrix (see Equation (10.141)) by k^2 or the parameter standard errors by k. The equivalent operation in Bayesian analysis would be to introduce k as a nuisance parameter and marginalize over our prior uncertainty in this parameter.

The method just described for obtaining more accurate error estimates assumed that the model was true. We cannot turn around and use these errors in a χ^2 hypothesis test to see if we can reject the model. In contrast, in a Bayesian analysis, we can allow for uncertainty in the value of k and also carry out a model selection operation after marginalizing over the unknown k. Recall that whenever we marginalize over a model parameter, this introduces an Occam penalty which penalizes the model for our prior uncertainty in the parameter in a quantitative fashion. If all the models that we are comparing in the model selection operation depend on this parameter in the same way, then these Occam penalties cancel out. If they depend on the parameter in different ways (e.g., models $p(C, S|D, I)$ and $p(C, \overline{S}|D, I)$ in Section 9.4), then the Occam penalties do not cancel.

10.8.1 Other model comparison methods

It is often useful to plot the variance of the residuals versus the number of model fitting parameters, M. For example, in the case of a polynomial model we can vary the number of parameters. Of course, for $M = 0$, the residual variance is just the data variance. We can characterize a model by how quickly the curve of residual variance versus M drops. Any variance curve which drops below another indicates a model which is better, in the sense that it achieves a better quality fit to the data with a given number of model functions. What one would expect to find is a very rapid drop as the systematic signal is taken up by the model, followed by a slow drop as additional model functions expand the noise. The total number of useful model parameters is determined by the break in this curve. In constructing the residual variance curves, we need to be aware that if we were to rearrange the order in which the *best* model terms are incorporated, we can always produce a curve that is above that of the same model but with a different order. We want to order the model parameters to produce the lowest residual variance curve before selecting the break point.

The F statistic can also be used to decide which basis functions are significant. Suppose our null hypothesis is that a model with M unknown parameters is the correct model. If the model's prediction for the ith data point is designated f_i, then

$$\chi_v^2 = \sum_{i=1}^{N} \frac{(d_i - f_i)^2}{\sigma^2} \tag{10.145}$$

has a χ_v^2 distribution with $v = N - M$ degrees of freedom. Consider the effect on χ^2 when an extra term is added to the model fitting function so the number of degrees of freedom is decreased by 1. If the simpler model is true then the effect of the extra term is to remove some of the noise variation. The expected decrease in χ^2 is the same as if we had not added the extra term but simply reduced N by one. Thus

$$\Delta\chi^2 = \chi_v^2 - \chi_{v-1}^2 \tag{10.146}$$

has a χ^2 distribution with 1 degree of freedom.

According to equation (6.35)

$$F = \frac{\Delta\chi^2}{\chi_{v-1}^2/(v-1)} \tag{10.147}$$

follows an F distribution[18] with 1 and $v - 1$ degrees of freedom. If the simpler model is correct you expect to get an f ratio near 1.0. If the ratio is much greater than 1.0, there are two possibilities:

1. The more complicated model is correct.
2. The simpler model is correct, but random scatter led the more complicated model to fit better. The P-value tells you how rare this coincidence would be.

---

[18] Since σ^2 is a common factor in the calculation of both $\Delta\chi^2$ and χ_{v-1}^2 we can rewrite equation (10.147) as

$$F = (v - 1)\frac{\sum_{i=1}^{N}(d_i - f_i)^2 - \sum_{i=1}^{N}(d_i - f_{+i})^2}{\sum_{i=1}^{N}(d_i - f_{+i})^2} \tag{10.148}$$

where f_{+i} = the predicted value for the model with the extra term. Thus it is not necessary to know σ^2 to carry out this F-test.

If the P-value is small enough (e.g., $\leq 5\%$), reject the simpler model. Otherwise, conclude that there is no compelling evidence to reject the simpler model.

As an example, we use the F-test to compare models M_1 (line exists) and M_2 (no line exists) in the spectral line problem of Section 3.6. The best fit for M_1 (63 degrees of freedom) yielded a $\chi^2_{\nu=63} = 48.49$. For M_2, with 64 degrees of freedom, $\chi^2_{\nu=64} = 57.13$. Substituting these values into equation (10.147) yields $f = 11.23$. This corresponds to a P-value $= 0.14\%$. On the basis of this F-test, we can reject the simpler model M_2 at a 99.86% confidence level.

10.9 Summary

Here we briefly summarize the main results of this chapter:

1. We saw how the Bayesian treatment leads to the familiar method of least-squares when we are interested in the question of the most probable set of model parameters (see Equation (10.34)), assuming an IID normal distribution for our knowledge of the measurement errors and a flat prior for each parameter.

2. We then relaxed the IID requirement for our knowledge of the measurement errors by introducing $\mathbf{E}$, the covariance matrix for the errors. Equation (10.48) gives the revised solution for the most probable set of parameters. Weighted linear least-squares can be seen as a special case of this equation.

3. A full description of our knowledge of the model parameters is given by the joint posterior distribution for the parameters. For a linear model, and a flat prior for each parameter, this distribution is particularly simple, namely a multivariate Gaussian. Equation (10.75) or Table 10.2 defines the boundary, $\Delta\chi^2_{\text{crit}}$, of a (joint) credible region for one or more of the parameters that contains a specified probability. Also, it turns out that for Gaussian posteriors, in any number of dimensions, the marginal PDF is also equal to the projected distribution (projected PDF).

4. A useful summary of the parameter errors is given by the model parameter covariance matrix, $\mathbf{V} = \sigma^2 \psi^{-1}$. If we are employing the covariance matrix, $\mathbf{E}$, for our knowledge of the measurement errors, then simply replace ψ^{-1} by $\Psi^{-1} = (\mathbf{G}^T \mathbf{E}^{-1} \mathbf{G})^{-1}$, $\psi_{\alpha\beta}$ by $\Psi_{\alpha\beta}$ and drop all the factors of σ^2 in Equations (10.93), (10.94), (10.100), and (10.101).

5. The parameter covariance matrix also provides information about the correlation between the estimates of any pair of model parameters, which is conveniently expressed by the correlation coefficient (see Equation (10.102)), ρ, ranging between ± 1. If ρ is close to ± 1 then it will not be possible to estimate reliably the two parameters separately, but we can still infer a linear combination of the parameters quite well. This comes about because the model basis functions are not orthogonal. At the end of Section 10.5.2, we show how to construct an orthogonal polynomial model.

6. The key quantity in Bayesian model comparison is the global likelihood of a model. Calculation of the global likelihood requires integrating away all of the model parameters. The final result regarding Bayesian model selection is usually expressed as an odds ratio, which is given by Equation (10.126). It is important to remember that this equation assumes uniform parameter priors and prior boundaries well removed from the peak of the posterior. Where these assumptions do not hold, the necessary marginalizations must in general be

carried out numerically and the resulting odds ratio can be very different. See Section 3.6 for a detailed example of this latter point.

7. Common frequentist methods for model testing and estimating robust parameter errors are discussed in Section 10.8.

10.10 Problems

1. Fit a straight line model to the data given in Table 10.3, where $\overline{d_i}$ is the average of n_i data values measured at x_i. The probability of the individual d_i measurements is normal with $\sigma = 4.0$, regardless of the x_i value.

   a) Give the slope and intercept of the best-fit line together with their errors.
   b) Plot the best-fit straight line together with the data values and their errors.
   c) Give the parameter covariance matrix.
   d) Repeat (a) and (c) but this time use the average x-coordinate as the origin. Comment on the differences between the covariance matrices in (c) and (d).

2. Compute and plot the ellipse that defines the 68.3% and 95.4% joint credible region for the slope and intercept, for the data given in Table 10.3 (see question 1). The shape of this ellipse depends on the x-coordinate origin used in the fit (see Figure 10.5). Use the average x-coordinate as the origin. See the section of the *Mathematica* tutorial entitled "Joint Credible Region Contouring."

3. Table 10.4 gives measurements of ozone partial pressure, y, in millibars in each of 15 atmospheric layers where each layer, x, is approximately 2 km in height. The layers have been scaled for convenience from -7 to $+7$.
   Use the least-squares method to fit the data with

   (i) a quadratic model: $y(x) = A_1 + A_2 x + A_3 x^2$
   (ii) a cubic model: $y(x) = A_1 + A_2 x + A_3 x^2 + A_4 x^3$

Table 10.3 *Data table*

x_i	$\overline{d_i}$	n_i
10	0.387	14
20	5.045	3
30	7.299	25
40	6.870	2
50	16.659	3
60	13.951	22
70	16.781	5
80	20.323	2

Table 10.4 *Measurements of ozone partial pressure, y, in millibars in each of 15 atmospheric layers where each layer, x, is approximately 2 km in height. The layers have been scaled for convenience from −7 to +7.*

Layer	Pressure	Layer	Pressure	Layer	Pressure	Layer	Pressure
−7	53.8	−5	73.2	−2	97.4	3	93.6
−7	53.3	−5	75.6	−2	98.3	3	86.2
−7	54.8	−5	76.2	−1	102.8	3	87.9
−7	54.6	−5	72.7	−1	96.9	3	89.5
−7	53.7	−4	79.4	−1	98.2	4	74.8
−7	55.2	−4	81.1	0	98.9	4	82.3
−7	55.7	−4	85.2	0	96.1	4	76.9
−7	54.1	−4	83	0	99.6	4	81.2
−6	63.8	−4	84.1	0	91.4	5	73.6
−6	64.2	−4	82.8	1	101.1	5	65.4
−6	66.9	−3	90.3	1	94.6	5	67.1
−6	67.2	−3	84.2	1	95.9	6	60.2
−6	65.4	−3	88.3	2	92.3	6	54.9
−6	67.3	−3	86	2	96.6	6	50.8
−5	71.8	−2	93.2	2	98.5	7	44.7
						7	38.5

Please include the following items as part of your solution:

a) In this problem, you don't know that the raw data errors are normally distributed, or even if the variance is the same from one layer to the next. Explain how you can take advantage of the Central Limit Theorem (CLT) in this problem.

Note: real data are seldom as nice as we would like. For some ozone layers there are fewer than five data values (the approximate number recommended for applying the CLT), so you may want to combine data for some of the layers where this is a problem. Of course, combining layers results in lower structural resolution.

Note: you must provide a table of the ozone values and computed errors you actually used in your model fitting. Explain how you computed the errors.

b) Determine the parameters for each model and the variance-covariance matrix. Quote an error for each parameter and explain what your errors mean.

c) Compare the models with the data by plotting the model fits on the same graph as your data. Include error bars (as you have determined them to be) on the data points used for fitting.

d) Compute the Bayesian odds ratio of the cubic model to the quadratic model. For the purpose of this calculation, assume the prior information warrants a

flat prior probability for each parameter with ranges ΔA_α given by:
$\Delta A_1 = 100$, $\Delta A_2 = \Delta A_3 = 10$, and $\Delta A_4 = 1$.
Explain in words what you conclude from this.

e) Calculate the frequentist χ^2 goodness-of-fit statistic and the P-value (significance) for each model. The confidence in rejecting the model $= 1 - $P-value. Explain what you conclude from these goodness-of-fit results.

4. Repeat the analysis of the ozone data as described in the previous problem, but this time adopt the following different strategy: instead of rebinning the data to take advantage of the CLT, use the original binning as given in Table 10.4. According to the MaxEnt principle (see Section 8.7.4), unless we have prior information that justifies the use of some other sampling distribution, then use a Gaussian sampling distribution. It makes the least assumptions about the information we don't have and will lead to the most conservative estimates. Use Equation (9.51) to estimate σ of each layer. In contrast to the approach proposed in the previous problem, we do not have to sacrifice the resolution of the original data through rebinning.

11

Nonlinear model fitting

11.1 Introduction

In the last chapter, we learned that the posterior distribution for the parameters in a linear model with Gaussian errors and flat priors is itself a multivariate Gaussian. The topology for this distribution in the multi-dimensional parameter space is very simple. In contrast, even for flat priors, the topology of the posterior for a nonlinear model can be very complex with many hills and valleys.

Examples of nonlinear models:

1. $f_i = A_1 \cos \omega t_i + A_2 \sin \omega t_i$
   where A_1, A_2 are linear parameters,
   and ω is a nonlinear parameter.

2. $f_i = A_1 + A_2 \exp \left\{ -\dfrac{(x_i - C_1)^2}{2\sigma_1^2} \right\} + A_3 \exp \left\{ -\dfrac{(x_i - C_2)^2}{2\sigma_2^2} \right\}$
   where A_1, A_2, A_3 are linear parameters,
   and $C_1, C_2, \sigma_1^2, \sigma_2^2$ are nonlinear parameters.

In this chapter, we will let θ represent the set of all parameters both linear and nonlinear and $\hat{\theta}$ the most probable set of the parameters. Again, the problem is to find the most probable set of parameters together with an estimate of their errors. (Of course, if the posterior has several maxima of comparable magnitude then it doesn't make sense to talk about a single best set of parameters.) The Bayesian solution to the problem is very simple in principle but can be very difficult in practice. The calculations require integrals over the parameter space which can be difficult to evaluate.

The brute force approach is as follows: for a one-parameter model, the most robust way is to plot the posterior or χ^2. This entails division of the parameter range into a finite number of grid points. As long as there are enough grid points to cover the prior range (a few hundred is usually adequate), this will usually work. It doesn't matter whether the posterior PDF is asymmetric, multi-modal or differentiable.

This approach can easily be extended to two parameters. It is also easy to compute marginal distributions. One need only add up the probabilities in the θ_1 or θ_2 direction, as appropriate.

After the two-parameter case, however, this approach rapidly becomes impractical. In fact, the number of calculations is $\sim (100)^M$, where M is the number of parameters. For example:

2 parameters might take 100 milliseconds to compute
5 parameters might take one day to compute
11 parameters might take the age of the universe to compute.

Fortunately, the last fifteen years have seen remarkable developments in practical algorithms for performing Bayesian calculations (Loredo, 1999). They can be grouped into three families: asymptotic approximations; methods for moderate dimensional models; and methods for high dimensional models. In this chapter, we will mainly be concerned with solutions that assume the posterior distribution for the parameters can be approximated by a multivariate Gaussian. We will first illustrate this in a simulation and then focus on methods for efficiently finding the most probable set of parameters and their covariance matrix.

In the following chapter, we will give an introduction to Markov chain Monte Carlo algorithms which facilitate full Bayesian calulations for nonlinear models involving very large numbers of parameters.

11.2 Asymptotic normal approximation

Expressed in frequentist language, asymptotic theory tells us that the maximium likelihood estimator becomes more unbiased, more normally distributed and of smaller variance as the sample size becomes larger (see Lindley, 1965). In other words, as the sample size increases, the nonlinear problem asymptotically approaches a linear problem. From a Bayesian perspective, the posterior distribution for the parameters asymptotically approaches a multivariate normal (Gaussian) distribution. We will illustrate this with a simulation.

We simulated data sets with different numbers of data points by randomly sampling a nonlinear model, represented by $f(x|\alpha)$, which has one nonlinear parameter α. We also added independent Gaussian noise to each data point with a mean of zero and a $\sigma = 2$. The data values are described by the equation

$$y_i = f(x_i|\alpha = 2/3) + e_i. \tag{11.1}$$

Figure 11.1 illustrates a set of $N = 12$ simulated data points together with a plot of the known model prediction. Of course, the data points differ from this model because of the added noise. We then carry out a Bayesian analysis of the simulated data, assuming we know the mathematical form of the model but not the value of the model parameter, α. Our goal is to infer the posterior PDF for α assuming a flat prior. The

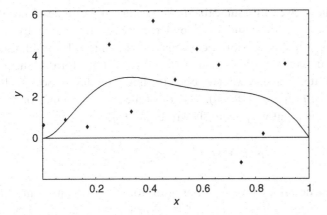

Figure 11.1 A simulated set of 12 data points for a nonlinear model with the one parameter $\alpha = 2/3$ (solid line) plus added Gaussian noise.

steps involved in calulating the posterior should now be fairly familiar to the reader (e.g., Section 10.2). The resulting PDFs are graphed in Figure 11.2 for four data sets of different size, N. It is apparent from this simulation that for small data sets, the posterior exhibits multiple peaks, but as N increases, the posterior approaches a Gaussian shape with a decreasing variance. The conclusion is not affected by the choice of prior; in large samples, the data totally dominate the priors and the result converges on a value of $\alpha = 2/3$, the value used to simulate the data. For a nonlinear model with M parameters, the joint posterior for the parameters asymptotically approaches an M-dimensional multivariate Gaussian as the number of data points becomes much greater than the number of unknown parameters.

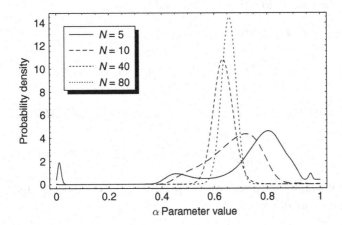

Figure 11.2 The Bayesian posterior density function for the nonlinear model parameter for four simulated data sets of different size ranging from $N = 5$ to $N = 80$. The $N = 5$ case has the broadest distribution and exhibits four maxima.

In what follows, we will assume that in the vicinity of the mode of the joint posterior, the product of the prior and likelihood can be approximated by a multivariate Gaussian. We want to develop a convenient mathematical formulation to describe an approximate multivariate Gaussian. We start with one form of the posterior for a true multivariate Gaussian we developed for linear models in Section 10.4 which we repeat here (see Equation (10.66)), only this time we let A stand for the set of linear model parameters that we previously wrote as $\{A_\alpha\}$.

$$p(A|D, M, I) = C' \exp\left\{-\frac{\delta\mathbf{A}^T\psi\delta\mathbf{A}}{2\sigma^2}\right\}. \tag{11.2}$$

This equation describes the joint posterior for a set of linear model parameters assuming flat priors for the parameters. When we use the more powerful matrix formulation which includes the data covariance matrix $\mathbf{E}$ (see Section 10.5.3), then we replace ψ by $\mathbf{\Psi}$ and rewrite Equation (11.2) as

$$p(A|D, M, I) = C'e^{-\frac{1}{2}(\delta\mathbf{A}^T\mathbf{\Psi}\delta\mathbf{A})}. \tag{11.3}$$

The term C' is the value of the posterior at the mode, which can be written as the product of the prior times the maximum value of the likelihood. The exponential term in Equation (11.3) describes the variation of the likelihood about the mode which has the form of a multivariate Gaussian. Thus, we can rewrite Equation (11.2) as

$$p(A|D, M, I) \propto p(A|M, I)p(D|A, M, I) = p(A|M, I)\mathcal{L}(A)$$
$$= p(\hat{A}|M, I)\mathcal{L}(\hat{A})\exp\left[-\frac{1}{2}\sum_{\alpha\beta}(A_\alpha - \hat{A}_\alpha)[\mathbf{\Psi}]_{\alpha\beta}(A_\beta - \hat{A}_\beta)\right]. \tag{11.4}$$

Now take the natural logarithm of both sides.

$$\ln[p(A|M, I)\mathcal{L}(A)] = \ln[p(\hat{A}|M, I)\mathcal{L}(\hat{A})]$$
$$+ \left[-\frac{1}{2}\sum_{\alpha\beta}(A_\alpha - \hat{A}_\alpha)[\mathbf{\Psi}]_{\alpha\beta}(A_\beta - \hat{A}_\beta)\right]. \tag{11.5}$$

We can show that $\mathbf{\Psi}$ is a matrix of second derivatives of $\ln[p(A|M, I)\mathcal{L}(A)]$ at $A = \hat{A}$.

$$\Psi_{\alpha\beta} = -\frac{\partial^2}{\partial A_\alpha \partial A_\beta}\ln[p(A|M, I)\mathcal{L}(A)] \quad (\text{at } A = \hat{A}). \tag{11.6}$$

For the nonlinear model case, we will represent the set of model parameters by θ and write an equation analogous to (11.4).

$$p(\theta|D, M, I) \approx p(\hat{\theta}|M, I)\mathcal{L}(\hat{\theta})\exp\left[-\frac{1}{2}\sum_{\alpha\beta}(\theta_\alpha - \hat{\theta}_\alpha)[\mathbf{I}]_{\alpha\beta}(\theta_\beta - \hat{\theta}_\beta)\right], \tag{11.7}$$

where $\mathbf{I}$ is called the *Fisher information matrix* and is the nonlinear problem analog of Ψ in the linear case. The approximate sign in the above equation is there because the posterior is only approximately a multivariate Gaussian at the mode. We can rewrite Equation (11.7) as

$$\ln[p(\theta|M,I)\mathcal{L}(\theta)] = \ln\left[p(\hat{\theta}|M,I)\mathcal{L}(\hat{\theta})\right] + \left[-\frac{1}{2}\sum_{\alpha\beta}(\theta_\alpha - \hat{\theta}_\alpha)[\mathbf{I}]_{\alpha\beta}(\theta_\beta - \hat{\theta}_\beta)\right]. \quad (11.8)$$

$\mathbf{I}$ is a matrix of second derivatives of $\ln[p(\theta|M,I)\mathcal{L}(\theta)]$ at $\theta = \hat{\theta}$.

$$\mathbf{I}_{\alpha\beta} = -\frac{\partial^2}{\partial\theta_\alpha\partial\theta_\beta}\ln[p(\theta|M,I)\mathcal{L}(\theta)] \quad (\text{at } \theta = \hat{\theta}). \quad (11.9)$$

Recall that Ψ^{-1} is the covariance matrix of the parameters in the linear problem. Ψ^{-1} provides a measure of how wide or spread out the Gaussian is. If the posterior in the nonlinear problem is not Gaussian, but is unimodal (single peak), then $\mathbf{I}^{-1}$ does not give the variances and covariances of the posterior distribution. However, it may give a good estimate of them, and is probably easier to calculate than the integrals required to get the variances and covariances.

A difficulty arising in these computations is that it has not been possible to present guidelines for how large the sample size must be for asymptotic properties to be closely approximated. In Section 11.4, we will assume the approximation is good enough, and focus on useful schemes for finding the most probable parameters, $\hat{\theta}$. But first we will investigate another useful type of approximation that allows us to obtain a better estimate of the desired Bayesian quantities without having to perform complicated integrals. These kinds of approximation originated with Laplace, so they are called *Laplacian approximations*.

11.3 Laplacian approximations

11.3.1 Bayes factor

Suppose we want to compute the Bayes factor for model comparison (Section 3.5). In this case, we need to compute the global likelihood, $p(D|M,I)$, by integrating over all the model parameters (also required for the normalization constant in parameter estimation). We can evaluate this from Equation (11.7).

$$p(D|M,I) = \int d\theta\, p(\theta|M,I)\mathcal{L}(\theta)$$

$$\approx p(\hat{\theta}|M,I)\mathcal{L}(\hat{\theta})\int d\theta\, \exp\left[-\frac{1}{2}(\delta\theta^T\mathbf{I}\delta\theta)\right], \quad (11.10)$$

where $[\delta\theta]_\alpha = (\theta_\alpha - \hat{\theta}_\alpha)$. We can use the principal axis theorem to make a change of variables according to $\delta\theta = \mathbf{O}\delta\mathbf{X}$, that transforms $\delta\theta^T\mathbf{I}\delta\theta$ to $\delta\mathbf{X}^T\Lambda\delta\mathbf{X}$, where Λ is a diagonal matrix of eigenvalues of the I matrix. The columns of O are the orthonormal eigenvectors of I. Let $\lambda_1, \lambda_2, \ldots, \lambda_M$ be the eigenvalues of I. Then we can write

$$\mathcal{I} = \int d\theta \; \exp\left[-\frac{1}{2}(\delta\theta^T I \delta\theta)\right]$$

$$= J \int dX \; \exp\left[-\frac{1}{2}(\delta X^T \Lambda \delta X)\right], \tag{11.11}$$

where $J = \det O$, is the Jacobian of the transformation, $\int d\theta = J \int dX$. Since the columns of O are orthonormal $J = 1$.

For the $M = 2$ case,

$$\mathcal{I} = \int dX_\alpha \exp\left[-\frac{\lambda_\alpha \delta X_\alpha^2}{2}\right] \int dX_\beta \exp\left[-\frac{\lambda_\beta \delta X_\beta^2}{2}\right]$$

$$= (\sqrt{2\pi})^2 \frac{1}{\sqrt{\lambda_\alpha}} \frac{1}{\sqrt{\lambda_\beta}} \int dX_\alpha \frac{1}{\sqrt{2\pi} \, 1/\sqrt{\lambda_\alpha}} \exp\left[-\frac{\delta X_\alpha^2}{2/\lambda_\alpha}\right]$$

$$\times \int dX_\beta \frac{1}{\sqrt{2\pi} \, 1/\sqrt{\lambda_\beta}} \exp\left[-\frac{\delta X_\beta^2}{2/\lambda_\beta}\right] \tag{11.12}$$

$$= (\sqrt{2\pi})^2 \frac{1}{\sqrt{\lambda_\alpha}} \frac{1}{\sqrt{\lambda_\beta}}.$$

For the general case of arbitrary M, we have,

$$\mathcal{I} = (2\pi)^{M/2} \frac{1}{\sqrt{\prod_\mu \lambda_\mu}}. \tag{11.13}$$

We can express our result for $\mathcal{I}$ in terms of the $\det I$ by writing $I = O\Lambda O^T$. Then

$$\det I = \det O^T \times \det \Lambda \times \det O$$

$$= 1 \times \prod_\mu \lambda_\mu \times 1 = \prod_\mu \lambda_\mu. \tag{11.14}$$

Substituting Equation (11.14) into Equation (11.13), we obtain

$$\mathcal{I} = (2\pi)^{M/2}(\det I)^{-1/2}. \tag{11.15}$$

Even if the multivariate Gaussian approximation is not exact, but the posterior distribution has a single dominant peak located away from the prior boundary of the parameter space, then the use of Equation (11.15) provides a useful Laplacian approximation. Thus, the global likelihood can be written as

$$p(D|M, I) \approx p(\hat{\theta}|M, I)\mathcal{L}(\hat{\theta})(2\pi)^{M/2}(\det I)^{-1/2}. \tag{11.16}$$

In the case of a perfect Gaussian approximation and a uniform parameter prior, Equation (11.16) reduces to Equation (10.123). In Section 11.4, we will discuss how to locate the best set of parameters, $\hat{\theta}$.

11.3.2 Marginal parameter posteriors

We can also use the Laplacian approximation in Equation (11.15) to do the integral needed to eliminate nuisance parameters. Suppose we want to obtain the marginal probability distribution for one of the parameters which we will label θ.[1] We need to remove the remaining parameters which we label collectively as ϕ. Instead of integrating over the ϕ parameters, we construct a "profile" function for θ, found by maximizing[2] the prior $\times$ the likelihood over ϕ for each choice of θ: $f(\theta) = \max_\phi p(\theta, \phi | M, I) \mathcal{L}(\theta, \phi)$. The profile function is a projection of the posterior onto the θ axis. Finding a maximum is generally much faster than computing the integrals. An efficient method of finding the maximum, starting from a good guess, is discussed in Section 11.5. We can construct an approximate marginal distribution for θ by multiplying $f(\theta)$ by a factor that accounts for the volume of ϕ space:

$$p(\theta | D, M, I) \propto f(\theta) [\det \mathbf{I}(\theta)]^{-1/2}, \tag{11.17}$$

where $\mathbf{I}(\theta)$ is the information matrix of the nuisance parameters, with θ held fixed.

To illustrate how different the marginal and projected distributions can be, consider a hypothetical joint probability distribution for the parameters θ and ϕ as shown in panel (a) of Figure 11.3. The projected and marginal distributions for θ are shown in

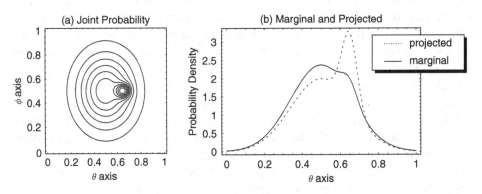

Figure 11.3 Comparison of the projected and marginal probability distribution for the θ parameter.

---

[1] More generally, θ can represent a subset of one or more parameters of interest, with the remainder considered as nuisance parameters.

[2] This process is easy to visualize if there are only two parameters A_1 and A_2. The joint probability distribution, $p(\bar{A}_1, A_2 | D, I)$, is a three-dimensional space with A_1 and A_2 as the x, y-axes and probability as the z-axis. Each choice of A_2 (i.e., $A_2 = $ constant) corresponds to a vertical slice through the probability mountain. We then vary A_1 until we find the maximum value of the probability in that slice. Repeat this process for all possible choices of parameter A_2 and record the probability $p(\bar{A}_1, A_2 | D, I)$. The resulting PDF is a function of A_2 and can be seen to be the projection of the joint probability mountain onto the A_2 axis.

panel (b). Although the peak probability occurs near $\theta = 0.65$, more probability resides in the broad plateau to the left of the peak and this is indicated by the marginal distribution. The value of the marginal, $p(\theta|D, I)$, for any particular choice of θ, is proportional to the integral over ϕ, i.e., the area under a slice of the joint probability distribution for θ fixed. Clearly, this area can be approximated by the peak height of the slice times a characteristic width of the probability distribution in the slice. In this two-parameter problem, the projected distribution is converted to an approximation of the true marginal by multiplying by the factor $[\det \mathbf{I}(\theta)]^{-1/2}$ in Equation (11.17), which gives the scale of the width of the distribution in the ϕ direction for the particular value of θ. Recall from Equation (11.14) the $\det \mathbf{I}(\theta)$ is equal to the product of the eigenvalues of $\mathbf{I}(\theta)$. At this point, it might be useful to refer back to Figure 10.3, which shows how the eigenvalues of the corresponding ψ matrix in the linear model problem give information on the scale of the width of the posterior.

We explore the Laplacian marginal distribution further in the following example: consider a nonlinear model of the form $f(x|\alpha, \beta) = x^{\alpha-1}(1 - x)^{\beta-1}$ for $0 < x$ and $\alpha, \beta > 1$. We constructed a simulated data set for 12 values of the independent variable, x, using this nonlinear model with $\alpha = 6, \beta = 3$, and added independent Gaussian noise with a mean of zero and a $\sigma = 0.005$. The data values are described by the equation

$$y_i = f(x_i|\alpha = 6, \beta = 3) + e_i. \tag{11.18}$$

The results of this simulation are shown in the four panels of Figure 11.4. Panel (a) shows the simulated data (diamonds) and model (solid curve). Panel (b) shows a contour plot of the Bayesian joint posterior probability of α and β, which differs significantly from a multivariate Gaussian. Panel (c) compares the projected or profile function (dots) and Bayesian marginal probability density for β (dashed). Panel (d) is the same as (c) but with the Laplacian approximation of the marginal overlaid, illustrating the close agreement with the true marginal density. You will have to look very closely to see any difference. The difference between the derived most probable values of $\beta = 3.2$ and the true value of $\beta = 3$ is simply a consequence of the added noise.

The Laplacian marginal distribution can perform remarkably well even for modest amounts of data, despite the fact that one might expect the underlying Gaussian approximation to be good only to order $1/\sqrt{N}$, the usual rate of asymptotic convergence to a Gaussian. The Laplacian approximations are good to order $1/N$ or higher. For more details on this point, see Tierney and Kadane (1986).

11.4 Finding the most probable parameters

In this section, we will assume flat priors for the model parameters and focus on methods for finding the peak of the likelihood. Again, we assume the data are given by

$$d_i = f_i + e_i,$$

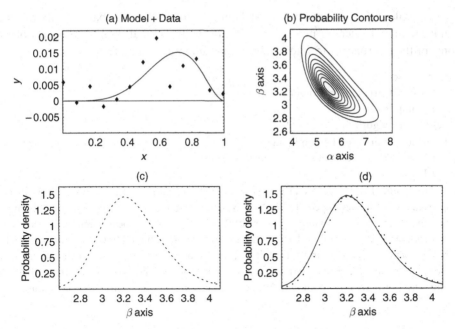

Figure 11.4 The figure provides a demonstration of the Laplacian approximation to the marginal posterior for a model parameter. Panel (a) shows the simulated data (diamonds) and model (solid curve), which has two nonlinear parameters α and β. (b) shows a contour plot of the Bayesian joint posterior probability of α and β, which differs significantly from a multivariate Gaussian. (c) compares the projected or profile function (dots) and Bayesian marginal probability density for β (dashed). (d) is the same as (c) but with the Laplacian approximation of the marginal overlaid, illustrating the close agreement with the true marginal density.

where f_i represents the model function and assume our knowledge of noise e_i leads us to assume Gaussian errors.

Then the likelihood is given by

$$L(\theta) = p(D|\theta, M, I) = C \exp\left[-\frac{1}{2} \sum_{i,j=1}^{N} (d_i - f_i)[\mathbf{E}^{-1}]_{ij}(d_j - f_j) \right], \qquad (11.19)$$

where $\mathbf{E} =$ covariance matrix of measurement errors. If the errors are independent, $\mathbf{E}$ is diagonal with entries equal to σ_i^2. In this case

$$p(D|\theta, M, I) = (2\pi)^{-N/2} \left\{ \prod_{i=1}^{N} \sigma_i^{-1} \right\} \exp\left[-\frac{1}{2} \sum_{i=1}^{N} \frac{(d_i - f_i)^2}{\sigma_i^2} \right]$$

$$= C \exp\left[-\frac{\chi^2(\theta)}{2} \right]. \qquad (11.20)$$

In general, $\chi^2(\theta)$ may have many local minima but only one global minimum. For a nonlinear model, there is no general solution to the global minimization problem. Some of the approaches to finding the global minimum are as follows:

1. Random search techniques
   a) Monte Carlo exploration of parameter space
   b) Simulated annealing
   c) Genetic algorithm
2. Home in on minimum from initial guess
   a) Levenberg Marquardt (iterative linearization)
   b) Downhill simplex
3. Combination of above. MINUIT is a very powerful Fortran-based function minimization and error analysis tool developed at CERN. It is designed to find the minimum value of a multi-parameter function and analyze the shape of the function around the minimum. The principal application is for statistical analysis, working on χ^2 or log-likelihood functions, to compute the best-fit parameter values and uncertainties, including correlations between the parameters. MINUIT contains code for carrying out a combination of the above items 1(a), 1(b), 2(a) and 2(b). For more information on MINUIT see: http://wwwinfo.cern.ch/asdoc/minuit.

11.4.1 Simulated annealing

The idea of using a temperature parameter in optimization problems started to become popular with the introduction of the simulated annealing (SA) method by Kirkpatrick et al. (1983). It is based on a thermodynamic analogy to growing a crystal starting with the material in a liquid state called a melt. When a melt is slowly cooled, the atoms will achieve the lowest energy crystal state (i.e., global minimum), whereas if it is rapidly cooled, it will reach a higher energy amorphous state.

Kirkpatrick et al. (1983) proposed a computer imitation of thermal annealing for use in optimization problems. In one version of simulated annealing, we construct a modified posterior probability distribution $p_T(\theta|D, I)$ which is given by

$$p_T(\theta|D, I) = \exp\left\{\frac{\ln[p(\theta|D, I)]}{T}\right\}, \qquad (11.21)$$

which contains a temperature parameter T. For $T = 1$, $p_T(\theta|D, I)$ is equal to the true posterior distribution for θ. For higher temperatures, $p_T(\theta|D, I)$ is a flatter version of $p(\theta|D, I)$. The basic scheme involves an exploration of the parameter space by a series of random changes in the current θ_c estimate of the solution

$$\theta_{\text{next}} = \theta_c + \Delta\theta, \qquad (11.22)$$

where $\Delta\theta$ is chosen by a random number generator. The proposed update is always considered advantageous if it yields a higher $p_T(\theta|D, I)$, but bad moves are sometimes accepted. This occasional allowance of retrograde steps provides a mechanism for escaping entrapment in local maxima. The process starts off with T large so the acceptance rate for unrewarding changes is high. The value of T is gradually decreased towards $T = 1$ as the number of iterations gets larger and the acceptance rate of unrewarding changes drops. This general scheme, of always accepting an uphill step while sometimes accepting a downhill step, has become known as the Metropolis algorithm (Metropolis *et al.*, 1953). At each value of T the Metropolis algorithm is used to explore the parameter space. The Metropolis algorithm and the related Metropolis–Hasting algorithms are described in more detail in Section 12.2.

Assuming a flat prior for θ, it is frequently the case that $p(\theta|D, I) \propto \exp\{-\chi^2/2\}$. Simulated annealing works well for a χ^2 topology like that shown in Figure 11.5, where there is an underlying trend towards a global minimum.

11.4.2 Genetic algorithm

Genetic algorithms are a class of search techniques inspired from the biological process of evolution by means of natural selection (Holland, 1992). They can be used to construct numerical optimization techniques that perform robustly in parameter search spaces with complex topology.

Consider the following generic modeling task: a model that depends on a set of adjustable parameters is used to fit a given dataset; the task consists in finding the single parameter set that minimizes the difference between the model's predictions and the data. The genetic algorithm consists of the following steps.

1. Start by generating a set ("population") of trial solutions, usually by choosing random values for all model parameters.
2. Evaluate the goodness-of-fit ("fitness") of each member of the current population (through a χ^2 measure with the data, for example).
3. Select pairs of solutions ("parents") from the current population, with the probability of a given solution being selected made proportional to that solution's fitness. Breed the two solutions selected in (2) and produce two new solutions ("offspring").

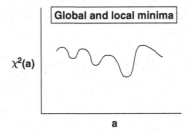

Figure 11.5 Sample topology of χ^2 for a nonlinear model with one parameter labeled **a**.

4. Repeat steps (2)–(3) until the number of offspring produced equals the number of individuals in the current population.
5. Use the new population of offspring to replace the old population.
6. Repeat steps (1) through (5) until some termination criterion is satisfied (e.g., the best solution of the current population reaches a goodness-of-fit exceeding some preset value).

Superficially, this may look like some peculiar variation of a Monte Carlo theme. There are two crucial differences: first, the probability of a given solution being selected to participate in a breeding event is made proportional to that solution's fitness (step 2); better trial solutions breed more often, the computational equivalent of natural selection. Second, the production of new trial solutions from existing ones occurs through breeding. This involves encoding the parameters defining each solution as a string-like structure ("chromosome"), and performing genetically inspired operations of crossover and mutation to the pair of chromosomes encoding the two parents, the end result of these operations being two new chromosomes defining the two offspring. Applying the reverse process of decoding those strings into solution parameters completes the breeding process and yields two new offspring solutions that incorporate information from both parents.

If you want to try out the genetic algorithm and watch a demonstration, check out the following web site: http://www.hao.ucar.edu/public/research/si/pikaia/pikaia.html#sec2. PIKAIA (pronounced "pee-kah-yah") is a general purpose function optimization Fortran-77 subroutine based on a genetic algorithm. PIKAIA was written by Paul Charbonneau and Barry Knapp (Charbonneau, 1995; Charbonneau and Knapp, 1995) both at the High Altitude Observatory, a scientific division of the National Center for Atmospheric Research in Boulder, Colorado. The above web site lists other useful references.

11.5 Iterative linearization

In this section, we will develop the equations needed for understanding the *Levenberg–Marquardt method* which is discussed in Section 11.5.1. This is a widely used and efficient scheme for homing in on the best set of nonlinear model parameters, $\hat{\theta}$, starting from an initial guess of their values. Start with a Taylor series expansion of χ^2 about some point in θ parameter space represented by θ_c (standing for θ_{current}) and keep only the first three terms:

$$\chi^2(\theta) \approx \chi^2(\theta_c) + \sum_k \frac{\partial \chi^2(\theta_c)}{\partial \theta_k} \delta\theta_k + \frac{1}{2} \sum_{kl} \frac{\partial^2 \chi^2(\theta_c)}{\partial \theta_k \, \partial \theta_l} \delta\theta_k \delta\theta_l, \qquad (11.23)$$

where

$$\delta\theta = \theta - \theta_c. \qquad (11.24)$$

For a linear model, χ^2 is quadratic so there are no higher derivatives. Let

$$\kappa_{kl} = \frac{1}{2} \frac{\partial^2 \chi^2(\theta_c)}{\partial \theta_k \, \partial \theta_l}$$

be called the *curvature matrix*. On the topic of nomenclature, in nonlinear analysis literature, the Hessian (**H**) matrix is frequently mentioned and is related to our curvature matrix by $\mathbf{H} = 2\boldsymbol{\kappa}$. In matrix form, Equation (11.23) becomes

$$\chi^2(\theta) \approx \chi^2(\theta_c) + \nabla\chi^2(\theta_c)\delta\theta + \delta\theta^T \kappa \delta\theta. \tag{11.25}$$

Take the gradient of both sides of Equation (11.25)

$$\nabla\chi^2(\theta) \approx \nabla\chi^2(\theta_c) + \boldsymbol{\kappa}\,\delta\theta. \tag{11.26}$$

The left hand side is the gradient at location $\delta\theta$ away from θ_c.

Now consider the special case where $\delta\theta$ takes us from θ_c to $\hat{\theta}$, the best set of parameter values. At $\delta\theta = \hat{\theta} - \theta_c, \chi^2 = \chi^2_{\min}$. In this case,

$$\nabla\chi^2(\hat{\theta}) = \nabla\chi^2_{\min} = 0 \tag{11.27}$$

$$\boldsymbol{\kappa}\,\delta\theta = -\nabla\chi^2(\theta_c) \tag{11.28}$$

or

$$\hat{\theta} = \theta_c - \boldsymbol{\kappa}^{-1}\nabla\chi^2(\theta_c) \tag{11.29}$$

where $\boldsymbol{\kappa}^{-1}$ = inverse of the curvature matrix. For a linear model, χ^2 is exactly a quadratic and thus $\boldsymbol{\kappa}$ is constant independent of θ_c.

For a nonlinear model, we expect that sufficiently close to $\chi^2_{\min}, \chi^2$ will be approximately quadratic so we should be able to ignore higher order terms in the Taylor expansion. Equation (11.29) should provide a reasonable approximation if θ_c is close to $\hat{\theta}$.

This suggests an iterative algorithm:

1. Start with a good guess θ_1 of $\hat{\theta}$.
2. Evaluate gradient $\nabla\chi^2(\theta_1)$ and curvature matrix $\boldsymbol{\kappa}(\theta_1)$.
3. Calculate improved estimate using Equation (11.29).
4. Repeat process until gradient $= 0$.

When $\nabla\chi^2(\theta_c) = 0$ then $\boldsymbol{\kappa}^{-1}$ = information matrix. Thus, the covariances of the parameters are to a good approximation given by

$$\sigma_{kl} = [\boldsymbol{\kappa}^{-1}]_{kl}. \tag{11.30}$$

If Equation (11.29) provides a poor approximation to the shape of the model function at θ_c, then all we can do is to step down the gradient.

$$\theta_{\text{next}} = \theta_c - \text{constant}\nabla\chi^2(\theta_c), \tag{11.31}$$

where the constant is small enough not to exhaust the downhill direction (more on the constant later). Note: if you are planning on writing your own program for iterative linearization, see the useful tips on computing the gradient and curvature (Hessian) matrices given in Press (1992).

11.5.1 Levenberg–Marquardt method

We can rewrite Equation (11.28) as a set of M simultaneous equations for $k = 1, \ldots, M$

$$\sum_{l=1}^{M} \kappa_{kl} \delta\theta_l = \Omega_k, \tag{11.32}$$

where $\Omega_k = -\partial \chi^2(\theta_c)/\partial\theta_k$; and for $M = 2$,

$$\kappa_{11}\delta\theta_1 + \kappa_{12}\delta\theta_2 = \Omega_1,$$
$$\kappa_{21}\delta\theta_1 + \kappa_{22}\delta\theta_2 = \Omega_2.$$

We can also rewrite Equation (11.31) as

$$\delta\theta_l = \text{constant} \times \Omega_l. \tag{11.33}$$

Equations (11.32) and (11.33) are central to the discussion of the Levenberg–Marquardt method which follows.

Far from $\chi^2_{\min}$, use Equation (11.33) which corresponds to stepping down the direction of steepest descent on a scale set by the constant. Close to $\chi^2_{\min}$, use Equation (11.32) which allows us to jump directly to the minimum.

What sets the scale of the constant in Equation (11.33)? Note: $\Omega_l = -\partial\chi^2/\partial\theta_l$ has dimensions of $1/\theta_l$ which may have dimensions (e.g., m). Each component Ω_l may have different dimensions. The constant of proportionality between Ω_l and $\delta\theta_l$ must therefore have dimensions of θ_l^2. Looking at κ, there is only one obvious quantity with the above dimension and that is $1/\kappa_{ll}$, the reciprocal of the diagonal element. But the scale might be too big, so divide it by an adjustable non-dimensional fudge factor γ.

$$\delta\theta_l = \frac{1}{\gamma\kappa_{ll}}\Omega_l \tag{11.34}$$

or

$$\gamma\kappa_{ll}\delta\theta_l = \Omega_l. \tag{11.35}$$

The next step is to combine Equations (11.32) and (11.35) by defining a new curvature matrix κ'

$$\kappa'_{kk} = \kappa_{kk}(1 + \gamma) \tag{11.36}$$

and

$$\kappa'_{kl} = \kappa_{kl} \quad (k \neq l).$$ (11.37)

The new equation is

$$\sum_{l=1}^{M} \kappa'_{kl} \delta\theta_l = \Omega_k.$$ (11.38)

If γ is large, κ' is forced into being dominated by the diagonal elements and becomes Equation (11.33). If $\gamma \to 0$, Equation (11.38) $\to$ Equation (11.32). The basis of the method is that when θ_c is far from $\hat{\theta}$, then Equation (11.33) representing the steepest descent is best. When θ_c is close to $\hat{\theta}$, then Equation (11.32) is best.

The Levenberg–Marquardt method employs Equation (11.38) which can switch between these two desirable states (Equations (11.32) and (11.33)) by varying γ. Recall Equation (11.32) can jump to the $\chi^2_{\min}$ in one step if the approximation is valid.

11.5.2 Marquardt's recipe

1. Compute $\chi^2(\theta_1)$ for guess of $\hat{\theta}$.
2. Pick a small value of $\gamma \approx 0.001$.
3. Solve Equation (11.38) for $\delta\theta$ and evaluate $\chi^2(\theta_1 + \delta\theta)$.
4. If $\chi^2(\theta_1 + \delta\theta) \geq \chi^2(\theta_1)$ increase γ by factor of 10 and go to (3).
5. If $\chi^2(\theta_1 + \delta\theta) < \chi^2(\theta_1)$, decrease γ by a factor of 10, update trial solution. $\theta_2 \leftarrow \theta_1 + \delta\theta$.
6. Repeat steps (3) to (5) until the solution converges.

Since κ plays the role of a metric on the M-dimensional subspace spanned by the model functions, the Levenberg–Marquardt method is referred to as a *variable metric approach*. The matrix κ is the same as the ψ matrix in the linear model case.

All that is necessary is a condition for stopping the iteration. Iterating to convergence or machine accuracy is generally wasteful and unnecessary since the minimum at best is only a statistical estimate of the parameter θ. Recall from our earlier discussion of joint credible regions in Section 10.4.1, that changes in χ^2 by an amount $\ll 1$ are never statistically meaningful. For $M = 2$ parameters, the probability ellipse defined by $\Delta\chi^2 = 2.3$ away from $\chi^2_{\min}$ encompasses 68.3% of the joint PDF. For $M = 1$ the corresponding $\Delta\chi^2 = 1$. These considerations suggest that, in practice, stop iterating on the 1st or 2nd iteration that χ^2 decreases by an amount $\ll 1$.

Once the minimum is found, set $\gamma = 0$ and compute the variance-covariance matrix

$$\mathbf{V} = \boldsymbol{\kappa}^{-1}$$

to obtain the estimated standard errors of the fitted parameters. *Mathematica* uses the Levenberg–Marquardt method in **NonlinearRegress** analysis. Subroutines are also available in Press (1992). If the posterior has several maxima of comparable magnitude, then in this case it doesn't make sense to talk about a single best $\hat{\theta}$.

11.6 *Mathematica* **example**

In this example, we illustrate the solution of a simple nonlinear model fitting problem using *Mathematica*'s **NonlinearRegress** which implements the Levenberg–Marquardt method. The data consist of one or possibly two spectral lines sitting on an unknown constant background. The measurement errors are assumed to be IID normal with a $\sigma = 0.3$. Model 1 assumes the spectrum contains a single spectral line while model 2 assumes two spectral lines. The raw data and measurement errors are shown in panel (a) of Figure 11.6, together with the best fitting model 1 shown by the solid curve. The parameter values for the best fitting model 1 were obtained with the **NonlinearRegress** command as illustrated in Figure 11.7. The arguments to the command are as follows:

1. **data** is a list of (x, y) pairs of data values where the x value is a frequency and the y value a signal strength.
2. **model**[f] is the mathematical form of the model for the spectrum signal strength as a function of frequency, f. This is given by

$$\mathbf{model}[f\_] := a0 + a1 \ \mathbf{line}[f, f1]$$

where,

$$\mathbf{line}[f\_, f1\_] := \frac{\mathrm{Sin}[2\pi(f - f1)/\Delta f]}{2\pi(f - f1)/\Delta f},$$

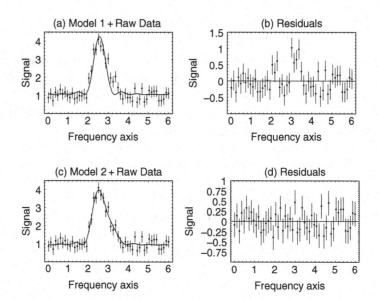

Figure 11.6 Two nonlinear models fitted to simulated spectral line data. Panel (a) shows the raw data and the best fit (solid curve) for model 1 which assumes a single line. Panel (b) illustrates the model 1 fit residuals. Panel (c) shows the best-fit model 2 compared to the data. Panel (d) shows the model 2 residuals.

```
result = Nonlinear [Regressdata,
 model[f], {f}, {{a0, 1.2}, {a1,4}, {f1, 2.6}}, Weights->wt,
  RegressionReport->
{BestFitParameters, ParameterCITable,
   AsymptoticCovarianceMatrix, FitResiduals, BestFit},
ShowProgress-> True]
```

Iteration:1 ChiSquared:128.05290737029276` Parameters:{1.2, 4., 2.6}

Iteration:2 ChiSquared:67.07120991521835` Parameters:{1.04667, 3.1995, 2.58491}

Iteration:3 ChiSquared:66.77321228639481` Parameters:{1.04602, 3.20506, 2.57661}

Iteration:4 ChiSquared:66.74836021101748` Parameters:{1.04588, 3.20626, 2.5742}

Iteration:5 ChiSquared:66.74632730538346` Parameters:{1.04587, 3.20636, 2.57351}

Iteration:6 ChiSquared:66.7461616482541` Parameters:{1.04587, 3.20636, 2.57332}

{BestFitParameters→ {a0→1.04587, a1→3.20637, f1→2.57326},

$$
\text{ParameterCITable}\rightarrow
\begin{array}{cccc}
 & \text{Estimate} & \text{Asymptotic SE} & \text{CI} \\
a0 & 1.04587 & 0.0545299 & \{0.936233,\ 1.15551\} \\
a1 & 3.20637 & 0.190783 & \{2.82277,\ 3.58996\}' \\
f1 & 2.57326 & 0.0204051 & \{2.53224,\ 2.61429\}
\end{array}
$$

$$
\text{AsymptoticCovarianceMatrix}\rightarrow
\begin{pmatrix}
0.00297351 & -0.00434888 & -1.20476\times10^{-6} \\
-0.00434888 & 0.036398 & 1.68676\times10^{-6} \\
1.20476\times10^{-6} & 1.68676\times10^{-6} & 0.000416367
\end{pmatrix},
$$

FitResiduals→ {-0.209178, 0.0139153, -0.351702, 0.108264, -0.081711, 0.232977,
 -0.0502895, 0.0859089, -0.0988917, -0.181999, 0.0500372, -0.33713, -0.329429,
 -0.218716, -0.151456, 0.0154221, -0.264555, 0.496109, 0.256117, 0.616037, -0.194287,
 -0.0907203, -0.448441, 0.12935, -0.0518219, 1.02491, 0.559758, 0.660035, 0.980723,
 0.291933, -0.133344, -0.389732, -0.0151427, -0.186632, -0.166834, -0.388142,
 0.297471, -0.477721, -0.287358, -0.331853, -0.401507, -0.263538, 0.20304, 0.102339,
 0.166333, 0.178261, -0.337149, -0.339727, -0.258125, 0.100323, 0.062864},

$$
\text{BestFit}\rightarrow 1.04587 + \frac{0.182741\ \text{Sin}[4.18879\ (-2.57326 + f)]^2}{(-2.57326 + f)^2}\ \}
$$

Figure 11.7 Example of the use of the *Mathematica* command **NonlinearRegress** to fit model 1 to the spectral data.

where $\Delta f = 1.5$. Note: **line** $[f, f1]$ becomes indeterminate for $(f - f1) = 0$. To avoid the likelihood of this condition occurring in **NonlinearRegress**, set the initial estimate of $f1$ to a non-integer number.

3. f is the independent variable frequency.

4. The third item is a list of the unknown model parameters and initial estimates. Since **NonlinearRegress** uses the Levenberg–Marquardt method, it is important that the initial estimates land you somewhere in the neighborhood of the global minimum of χ^2, where

$$
\chi^2 = \sum_{i=1}^{N} \frac{(d_i - \mathbf{model}[f_i])^2}{\sigma_i^2} = \sum_{i=1}^{N} \mathbf{wt}_i (d_i - \mathbf{model}[f_i])^2.
$$

5. **wt** is an optional list of weights to be assigned to the data points, where $\mathbf{wt}_i = 1/\sigma_i^2$.

6. **RegressionReport** is a list of options for the output of **NonlinearRegress**.

7. **ShowProgress** $\rightarrow$ **True** shows the value of χ^2 achieved after each iteration of the Levenberg–Marquardt method and the parameter values at that step.

The full **NonlinearRegress** command together with its arguments is shown in bold face type in Figure 11.7. The output, shown in normal type face, indicates that the minimum χ^2 achieved for model 1 was 66.7.[3] Below that is a list of the various **RegressionReport** items. The second item lists the parameter values, the asymptotic standard error for each parameter and the frequentist confidence interval (95% by default) for each parameter. The asymptotic error for each parameter is equal to the square root of the corresponding diagonal element in the **AsymptoticCovarianceMatrix**. The use of these errors is based on the assumption that in the vicinity of the mode, the joint posterior probability density function for the parameters is a good approximation to a multivariate Gaussian (see Section 11.2). The **AsymptoticCovarianceMatrix** $= \mathbf{I}^{-1}$, the inverse of the observed information matrix. Note: the **AsymptoticCovarianceMatrix** elements, as given by **NonlinearRegress**, have been scaled by a factor $k^2 = 1.39$ where k is given by Equation (10.144). This leads to more robust parameter errors but we must correct for this later on when we compute the Bayesian odds ratio for comparing models 1 and 2. The values of χ^2 quoted in the output of Figure 11.7 have not been modified by the k factor and thus $\chi^2_{min} = 66.7$ is the minimum value calculated on the basis of the input measurement error $\sigma = 0.3$.

Panel (b) of Figure 11.6 shows the residuals after subtracting model 1 from the data. There is clear evidence for another spectral line at about 3.6 on the frequency axis. On the basis of these residuals, a second model was constructed, consisting of two spectral lines sitting on a constant background. Model 2 has the mathematical form:

$$\text{model}[\, f, \ f\_]:=a0+a1 \ \text{line}[f, \ f1]+a2 \ \text{line}[\, f, \ f2].$$

Panel (c) shows the best fitting model 2. The residuals shown in Panel (d) appear to be consistent with the measurement errors and show no evidence for any further systematic signal component. The output from *Mathematica*'s **NonlinearRegress** command for model 2 is shown in Figure 11.8.

11.6.1 Model comparison

Here, we compute the Bayesian odds ratio given by

$$O_{21} = \frac{p(M_2|D,I)}{p(M_1|D,I)} = \frac{p(M_2|I)}{p(M_1|I)} \times \frac{p(D|M_2,I)}{p(D|M_1,I)}$$

$$= \frac{p(M_2|I)}{p(M_1|I)} \times \text{Bayes factor.} \tag{11.39}$$

---

[3] Here, we evaluate the frequentist theory confidence in rejecting model 1. Model 1 has three fit parameters so the number of degrees of freedom $= N - M = 51$ data points $-3 = 48$; thus the confidence is $= 1 - \textbf{GammaRegularized}\left[\frac{N-M}{2}, \frac{\chi^2}{2}\right] = 0.96$.

```
result = NonlinearRegress[data,
model2[f], {f}, {{a0, 1.2}, {a1, 3}, {f1, 2.6}, {a2, 1}, {f2, 3.5}},Weights-> wt,
  RegressionReport ->
{BestFitParameters, ParameterCITable, BestFit},
ShowProgress -> True]

Iteration:1 ChiSquared:113.44567855780089` Parameters:{1.2, 3., 2.6, 1., 3.5}

Iteration:2 ChiSquared:44.87058091255775`
  Parameters:{0.996707, 3.16903, 2.55205, 0.447086, 3.22451}

Iteration:3 ChiSquared:39.478968989091` Parameters:{0.958289, 3.08133, 2.51609, 0.866957, 2.98474}

Iteration:4 ChiSquared:31.202371941916066`
  Parameters:{0.934233, 2.99456, 2.48976, 1.14837, 3.12047}

Iteration:5 ChiSquared:30.26136004999436`
  Parameters:{0.933864, 2.94356, 2.48483, 1.20244, 3.06137}

Iteration:6 ChiSquared:30.236264668012137`
  Parameters:{0.932706, 2.93206, 2.48383, 1.22363, 3.06401}

{BestFitParameters → {a0→0.932717, a1→2.93172, f1→2.48379, a2→1.22387, f2→3.0638},
```

		Estimate	Asymptotic SE	CI
	a0	0.932717	0.0406876	{0.850817,1.01462}
ParameterCITable →	a1	2.93172	0.182306	{2.56476,3.29869}
	f1	2.48379	0.0253238	{2.43281,2.53476}
	a2	1.22387	0.182139	{0.857242,1.5905}
	f2	3.0638	0.0606851	{2.94164,3.18595}

BestFit →

$$0.932717 + \frac{0.0697522\ \text{Sin}[4.18879(-3.0638 + f)]^2}{(-3.0638+f)^2} + \frac{0.167088\ \text{Sin}[4.18879(-2.48379 + f)]^2}{(-2.48379 + f)^2}$$

Figure 11.8 The output from *Mathematica*'s **NonlinearRegress** command for model 2.

We will use the Laplace approximation for the Bayes factor described in Section 11.3.1 which expresses the global likelihood, given by Equation (11.16), in terms of the determinant of the information matrix, **I**.

$$p(D|M_i, I) \approx p(\hat{\theta}|M_i, I)\mathcal{L}(\hat{\theta})(2\pi)^{M/2}(\det \mathbf{I})^{-1/2}$$
$$= \frac{1}{\prod_\alpha \Delta\theta_\alpha} \frac{1}{\sigma^N (2\pi)^{N/2}} e^{-\chi^2_{\min}/2}(2\pi)^{M/2}(\det \mathbf{I})^{-1/2}. \quad (11.40)$$

Let **V** stand for the parameter asymptotic covariance matrix in the nonlinear problem, so

$$(\det \mathbf{I})^{-1/2} = \sqrt{\det \mathbf{V}}. \quad (11.41)$$

Equation (11.40) assumes uniform priors for the model parameters, where $\Delta\theta_\alpha$ is the prior range for parameter θ_α. In the current problem, we assume the prior ranges for the parameters are known to within a factor of three of the initial estimates used in **NonlinearRegress**, i.e., $3\theta_\alpha - \theta_\alpha/3 = 2.667\theta_\alpha$.

Let $\mathbf{V}^*$ be the asymptotic covariance matrix elements returned by *Mathematica*'s **NonlinearRegress** command. Recall that *Mathematica* scales the asymptotic covariance matrix elements by a factor k^2, to allow for more robust parameter errors, where k is given by Equation (10.144). We need to remove this factor for use in Equation (11.40), by multiplying the asymptotic covariance matrix provided by *Mathematica* by $1/k^2$, before computing its determinant, i.e.,

$$\mathbf{V} = \frac{1}{k^2} \, \mathbf{V}^*. \tag{11.42}$$

We can extract $\mathbf{V}^*$ from **result**, the name given to the result of the **NonlinearRegress** command. For model 1 (see Figure 11.7) the covariance matrix was the third item in the **RegressionReport**. Thus, $\mathbf{V}^* = \textbf{result}[[3, 2]][[1]]$ is the desired matrix.[4] The value of O_{21} derived from Equations (11.39), (11.40), (11.41), and (11.42), assuming equal prior probabilities for the two models, is 1.4×10^5.

11.6.2 Marginal and projected distributions

Finally, we will compute the Laplacian approximation to the Bayesian marginal probability density function $p(\theta|D, M, I)$ and compare it to the frequentist projected probability, which we refer to as the profile function, $f(\theta)$, according to Equation (11.17). We illustrate this calculation for the $a2$ parameter. The Laplacian marginal is the profile function $f(a2)$ times the factor

$$[\det \ \mathbf{I}(a2)]^{-1/2} = \sqrt{\det \ \mathbf{V}(a2)} = \sqrt{\det \ \frac{1}{k^2} \mathbf{V}^*(a2)}. \tag{11.43}$$

The quantity $\mathbf{V}^*(a2)$ is the asymptotic covariance matrix evaluated by **NonlinearRegress** obtained by fixing $a2$ and minimizing χ^2, with all the other parameters free to vary. This can be done using a simple Do loop to repeatedly run **NonlinearRegress** for different values of $a2$. For an example of this, see the nonlinear fitting section of the *Mathematica* tutorial. The profile function is given by

$$f(a2) \propto \exp\left(\frac{-\chi^2_{\min}(a2)}{2}\right). \tag{11.44}$$

Let $\hat{k}$ be the value of k in Equation (11.43) for the fit corresponding to the most probable set of parameters. If $\hat{k} > 1$, this indicates that the data errors may have been underestimated. An approximate way[5] to take account of this, when computing the marginal parameter PDF, is to modify Equations (11.43) and (11.44) as follows:

$$[\det \ \mathbf{I}(a2)]^{-1/2} = \sqrt{\det \ \mathbf{V}(a2)} = \sqrt{\det \ \frac{\hat{k}^2}{k^2} \mathbf{V}^*(a2)} \tag{11.45}$$

$$f(a2) \propto \exp\left(\frac{-\chi^2_{\min}(a2)}{2 \, \hat{k}^2}\right). \tag{11.46}$$

---

[4] The quantity **result[[3, 2]]** is $\mathbf{V}^*$ expressed in *Mathematica*'s **MatrixForm**. To compute the determinant of this matrix we need to extract the argument of **MatrixForm** which is given by **result[[3, 2]][[1]]**.
[5] A fully Bayesian way of handling this would be to treat k as a parameter and marginalize over a prior range for k.

The resulting projected and marginal PDF for $a2$ are shown in Figure 11.9 and are clearly quite different. It is a common frequentist practice to improve on the asymptotic standard errors of a parameter by finding the two values of the parameter for which the projected $\chi^2 = \chi^2_{\min} + 1$, in analogy to the linear model case (see Table 10.2). For example, see the command MINOS in the MINUIT software (James, 1998). As we have discussed earlier, the Bayesian marginal distribution should be strongly preferred over the projected, and the Laplacian approximation provides a quick way of estimating the marginal.

Finally we can readily compute the Bayesian 95% credible region for $a2$ from the marginal distribution and compare with the 95% confidence interval returned by **NonlinearRegress**. They are:

$$\text{Bayesian 95\% credible region} = (0.74, 1.68)$$
$$\text{frequentist 95\% confidence interval} = (0.86, 1.60).$$

11.7 Errors in both coordinates

In Section 4.8.2, we derived the likelihood function applicable to the general problem of fitting an arbitrary model when there are independent errors in both coordinates. For the special case of a straight line model (see also Gull, 1989b) the likelihood function, $p(D|M, A, B, I)$, is given by Equation (4.60), which we repeat here after replacing y_i by d_i.

$$
p(D|M, A, B, I) = (2\pi)^{-N/2} \left(\prod_{i=1}^{N} \left(\sigma_i^2 + B^2 \sigma_{xi}^2 \right)^{-1/2} \right)
$$

$$
\times \exp \left\{ \sum_{i=1}^{N} \frac{-(d_i - m(x_{i0}|A, B))^2}{2 \left(\sigma_i^2 + B^2 \sigma_{xi}^2 \right)} \right\}.
$$

(11.47)

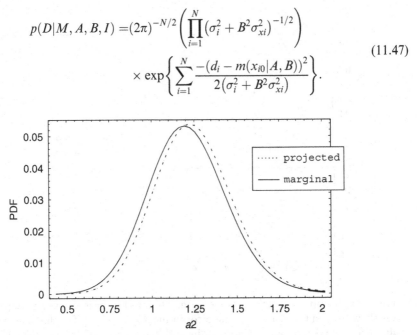

Figure 11.9 Comparison of the projected and Laplacian marginal PDF for the $a2$ parameter.

Here, A and B are model parameters representing the intercept and slope. It is apparent that when there are errors in both coordinates, the problem has become nonlinear in the parameters.

Problem: In Section 10.2.2, we fitted a straight line model to the data given in Table 10.1 using the method of least-squares. This time assume that the x_i coordinates are uncertain with an uncertainty descibed by a Gaussian PDF with a $\sigma_{xi} = 3$. Using the likelihood given in Equation (11.47), compute the marginal PDF for the intercept (A) and the marginal for the slope (B), and compare the results to the case where $\sigma_{xi} = 0$. Assume uniform priors with boundaries well outside the region with significant likelihood.

Solution: Since we are assuming flat priors, the joint posterior $p(A, B|D, M, I)$ is directly proportional to the likelihood. The marginal PDF for the intercept is given by

$$p(A|D, M, I) = \int dB \, p(A, B|D, M, I)$$

$$\propto p(A|M, I) \int dB \, p(B|M, I) \, p(D|M, A, B, I). \tag{11.48}$$

We can write a similar equation for the marginal PDF for the slope. The upper two panels of Figure 11.10 show plots of the two marginals for two cases. The solid curves correspond to $\sigma_{xi} = 0$ (no uncertainty in x_i values), and the dashed curves to $\sigma_{xi} = 3$. The uncertainty in the x_i values results in broader and shifted marginals.

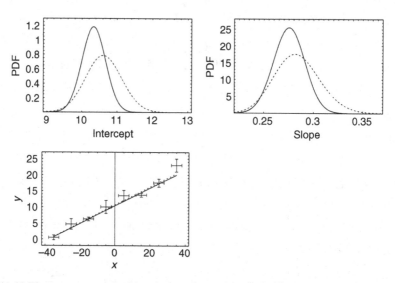

Figure 11.10 The top two panels show the marginal PDFs for the intercept and slope. The solid curves show the result when there is no error in the x-coordinate. The dashed curves are the result when there are errors in both coordinates. The lower panel shows the corresponding best-fit straight lines.

The lower panel of Figure 11.10 shows the most probable straight line fits for the two cases. The likelihood function given by Equation (11.47) contains σ_{xi} in two terms. In both terms, it is multiplied by the slope parameter B. The effect of the first term is to favor smaller values of B. The effect of the second term is to decrease the relative weight given to measurements with a smaller σ_i, i.e., this causes the points to be given more equal weight. In this particular case, the best fitting line has a slope and intercept which are slightly larger when $\sigma_{xi} = 3.0$.

11.8 Summary

The problem of finding the best set of parameters for a nonlinear model can be very challenging, because the posterior distribution for the parameters can be complex with many hills and valleys. As the sample size increases, the posterior asymptotically approaches a multivariate normal distribution (Section 11.2). Unfortunately, there are no clear guidelines for how large the sample size must be. The goal is to find the global maximum in the posterior, or equivalently, the minimum in χ^2. A variety of methods are discussed, including random search techniques like simulated annealing and the genetic algorithm. The other main approach is to home in on the minimum in χ^2 from a good initial guess using an iterative linearization technique like Levenberg–Marquardt (Sections 11.5 and 11.5.1), the method used in *Mathematica*'s **Non-linearRegress** command.

Once the minimum is located, the parameter errors can be approximately estimated from $\mathbf{I}^{-1}$, the inverse of the information matrix (Section 11.2). $\mathbf{I}^{-1}$ is analogous to the parameter covariance matrix in linear model fitting. Improved error estimates can be obtained from the Laplacian approximation to the marginal posterior distribution for any particular parameter (see Section 11.3.2). For model comparison problems, Section 11.3.1 describes a useful Laplacian approximation for the global likelihood, $p(D|M, I)$, that is needed in calculating the Bayes factor. Section 11.6 and the section entitled, "Nonlinear Least-Squares Fitting" in the accompanying *Mathematica* tutorial, provide useful templates for the analysis of typical nonlinear model fitting problems.

The data from some experiments have errors in both coordinates which can turn a linear model fitting problem into a nonlinear problem. This issue was discussed earlier in Section 4.8.2, and a particular example of fitting a straight line model was treated in Section 11.7.

11.9 Problems

Nonlinear Model Fitting

(See "Nonlinear Least-Squares Fitting" in the *Mathematica* tutorial.)

Table 11.1 gives a frequency spectrum consisting of 100 pairs of frequency and voltage (x, y). From measurements when the signal was absent, the noise is known to be IID normal with a standard deviation $\sigma = 0.3$ voltage units. The spectrum is

Table 11.1 *The table contains a frequency spectrum consisting of 100 pairs of frequency and voltage.*

f (Hz)	V	f (Hz)	V	f (Hz)	V	f (Hz)	V
1.00	1.391	5.25	5.537	9.50	3.113	13.75	2.038
1.17	1.000	5.42	6.091	9.67	3.293	13.92	2.585
1.34	0.552	5.59	6.163	9.84	3.139	14.09	2.492
1.51	1.249	5.76	5.365	10.01	2.840	14.26	2.193
1.68	0.534	5.93	5.916	10.18	3.119	14.43	1.866
1.85	1.386	6.10	5.530	10.35	3.311	14.60	1.571
2.02	0.971	6.27	4.552	10.52	4.347	14.77	1.779
2.19	0.901	6.44	3.833	10.69	4.819	14.94	1.542
2.36	0.851	6.61	3.756	10.86	4.378	15.11	1.562
2.53	1.334	6.78	3.055	11.03	4.544	15.28	1.666
2.70	0.549	6.95	3.009	11.20	4.562	15.45	0.904
2.87	1.373	7.12	2.855	11.37	5.662	15.62	1.074
3.04	0.997	7.29	2.357	11.54	4.479	15.79	1.530
3.21	1.231	7.46	2.732	11.71	5.373	15.96	0.747
3.38	1.586	7.63	1.836	11.88	4.883	16.13	0.945
3.55	2.244	7.80	1.918	12.05	4.678	16.30	1.301
3.72	1.914	7.97	1.534	12.22	5.100	16.47	1.323
3.89	2.467	8.14	2.238	12.39	3.868	16.64	0.919
4.06	2.609	8.31	2.623	12.56	4.132	16.81	1.320
4.23	3.036	8.48	2.275	12.73	3.702	16.98	0.915
4.40	3.581	8.65	2.408	12.90	3.267	17.15	0.814
4.57	4.073	8.82	2.701	13.07	3.323	17.32	0.983
4.74	5.010	8.99	2.659	13.24	3.413	17.49	1.158
4.91	4.989	9.16	3.224	13.41	2.762	17.66	0.917
5.08	4.940	9.33	2.237	13.58	2.418	17.83	1.355

thought to consist of two or more narrow lines which are broadened by the instrumental response of the detector which is well described by a Gaussian with a $\sigma_L = 1.0$ frequency unit. In addition, there is an unknown constant offset. Use a model for the signal consisting of a sum of Gaussians plus a constant offset of the form

$$y(x_i) = A_0 + A_1 \exp\left(-\frac{(x_i - C_1)^2}{2\sigma_L^2}\right) + A_2 \exp\left(-\frac{(x_i - C_2)^2}{2\sigma_L^2}\right) + \cdots$$

In this problem, refer to the model with two lines as model 2, that with three lines as model 3, etc.

The objective of this assignment is to determine the most probable model and the best estimates of the model parameters and their errors. Find the most likely number of lines by fitting progressively more Gaussians, examining the residuals after each trial. The following items are required as part of your solution:

1. Plot the raw data together with error bars. **NonlinearRegress** in *Mathematica* uses the Levenberg–Marquardt method which requires good initial guesses of the parameter values. For each model, provide a table of your initial guess of each parameter value.

2. For each choice of model, give a table of the best-fit parameters and their errors as derived from the asymptotic covariance matrix. Also list the covariance matrix. Note: If you are using *Mathematica*'s **NonlinearRegress**, remember that it computes an asymptotic covariance matrix that is scaled by a factor k^2. This is an attempt to obtain more robust parameter errors based on assuming the model is correct and then adjusting all the assumed measurement errors by a factor k (explained in Section 11.6; see also Equation (10.142) and discussion). For each choice of model, compute the factor k.

3. Plot each model on top of the data with error bars.

4. For each model, plot the residuals and decide whether there is evidence for another line to be fitted. Estimate the parameters of the line from the residuals and then generate a new model to fit to the data that includes the new line together with the earlier lines. Note: the residuals may suggest the presence of multiple lines. It is best to add only the strongest one to your next model. Some of the minor features in the residuals will disappear as the earlier model lines re-adjust their best locations in response to the addition of the one new line.

5. For each model, calculate the χ^2 goodness-of-fit and the frequentist P-value (significance), which represents the fraction of hypothetical repeats of the experiment that are expected to fall in the tail area by chance if the model is correct. The confidence in rejecting the model $= 1 - $ P-value. Explain what you conclude from these goodness-of-fit results.

6. For each model, compute the Laplacian estimate of the global likelihood for use in the model selection problem. Compute the odds ratio for model (i) compared to model ($i - 1$). Assume a uniform prior for each model amplitude parameter, with a range of $\pm$ a factor of 3 of your initial guess, A_g, for the parameter, i.e., $3A_g - A_g/3 = 2.667A_g$. Assume a uniform prior for each model line center frequency parameter within the range 1 to 17 frequency units.

7. For your best model, compute and plot (on the same graph) the projected probability and the Laplacian approximation to the marginal probability for $A3$, the amplitude of the third strongest line. Again, see the *Mathematica* tutorial for an example.

12

Markov chain Monte Carlo

12.1 Overview

In the last chapter, we discussed a variety of approaches to estimate the most probable set of parameters for nonlinear models. The primary rationale for these approaches is that they circumvent the need to carry out the multi-dimensional integrals required in a full Bayesian computation of the desired marginal posteriors. This chapter provides an introduction to a very efficient mathematical tool to estimate the desired posterior distributions for high-dimensional models that has been receiving a lot of attention recently. The method is known as *Markov Chain Monte Carlo* (MCMC). MCMC was first introduced in the early 1950s by statistical physicists (N. Metropolis, A. Rosenbluth, M. Rosenbluth, A. Teller, and E. Teller) as a method for the simulation of simple fluids. Monte Carlo methods are now widely employed in all areas of science and economics to simulate complex systems and to evaluate integrals in many dimensions. Among all Monte Carlo methods, MCMC provides an enormous scope for dealing with very complicated systems. In this chapter we will focus on its use in evaluating the multi-dimensional integrals required in a Bayesian analysis of models with many parameters.

The chapter starts with an introduction to Monte Carlo integration and examines how a Markov chain, implemented by the Metropolis–Hastings algorithm, can be employed to concentrate samples to regions with significant probability. Next, *tempering* improvements are investigated that prevent the MCMC from getting stuck in the region of a local peak in the probability distribution. One such method called *parallel tempering* is used to re-analyze the spectral line problem of Section 3.6. We also demonstrate how to use the results of parallel tempering MCMC for model comparison. Although MCMC methods are relatively simple to implement, in practice, a great deal of time is expended in optimizing some of the MCMC parameters. Section 12.8 describes one attempt at automating the selection of these parameters. The capabilities of this automated MCMC algorithm are demonstrated in a re-analysis of an astronomical data set used to discover an extrasolar planet.

312

12.2 Metropolis–Hastings algorithm

Suppose we can write down the joint posterior density,[1] $p(X|D, I)$, of a set of model parameters represented by X. We now want to calculate the expectation value of some function $f(X)$ of the parameters. The expectation value is obtained by integrating the function weighted by $p(X|D, I)$.

$$\langle f(X) \rangle = \int f(X)p(X|D, I)dX = \int g(X)dX. \tag{12.1}$$

For example, if there is only one parameter and we want to compute its mean value, then $f(X) = X$. Also, we frequently want to compute the marginal probability of a subset X_A of the parameters and need to integrate over the remaining parameters designated X_B. Unfortunately, in many cases, we are unable to perform the integrals required in a reasonable length of time. In this section, we develop an efficient method to approximate the desired integrals, starting with a discussion of Monte Carlo integration. Given a value of X, the discussion below assumes we can compute the value of $g(X)$.

In straight Monte Carlo integration, the procedure is to pick n points, uniformly randomly distributed in a multi-dimensional volume (V) of our parameter space X. The volume must be large enough to contain all regions where $g(X)$ contributes significantly to the integral. Then the basic theorem of Monte Carlo integration estimates the integral of $g(X)$ over the volume V by

$$\langle f(X) \rangle = \int_V g(X)dX \approx V \times \langle g(X) \rangle \pm V \times \sqrt{\frac{\langle g^2(X) \rangle - \langle g(X) \rangle^2}{n}}, \tag{12.2}$$

where

$$\langle g(X) \rangle = \frac{1}{n}\sum_{i=1}^{n} g(X_i); \quad \langle g^2(X) \rangle = \frac{1}{n}\sum_{i=1}^{n} g^2(X_i). \tag{12.3}$$

There is no guarantee that the error is distributed as a Gaussian, so the error term is only a rough indicator of the probable error. When the random samples X_i are independent, the law of large numbers ensures that the approximation can be made as accurate as desired by increasing n. Note: n is the number of random samples of $g(X)$, not the size of the fixed data sample. The problem with Monte Carlo integration is that too much time is wasted sampling regions where $p(X|D, I)$ is very small. Suppose in a one-parameter problem the fraction of the time spent sampling regions of high probability is 10^{-1}. Then in an M-parameter problem, this fraction could easily fall to 10^{-M}. A variation of the simple Monte Carlo described above, which involves reweighting the integrand and adjusting the sample rules (known as "importance sampling"), helps considerably but it is difficult to design the reweighting for large numbers of parameters.

---

[1] In the literature dealing with MCMC, it is common practice to write $\pi(X)$ instead of $p(X|D, I)$.

In general, drawing samples independently from $p(X|D, I)$ is not currently computationally feasible for problems where there are large numbers of parameters. However, the samples need not necessarily be independent. They can be generated by any process that generates samples from the *target distribution, $p(X|D, I)$*, in the correct proportions. All MCMC algorithms generate the desired samples by constructing a kind of random walk in the model parameter space such that the probability for being in a region of this space is proportional to the posterior density for that region. The random walk is accomplished using a Markov chain, whereby the new sample, X_{t+1}, depends on the previous sample X_t according to an entity called the *transition probability* or *transition kernel, $p(X_{t+1}|X_t)$*. The transition kernel is assumed to be time independent. The remarkable property of $p(X_{t+1}|X_t)$ is that after an initial burn-in period (which is discarded) it generates samples of X with a probability density equal to the desired posterior $p(X|D, I)$.

How does it work? There are two steps. In the first step, we pick a proposed value for X_{t+1} which we call Y, from a *proposal distribution, $q(Y|X_t)$*, which is easy to evaluate. As we show below, $q(Y|X_t)$ can have almost any form. To help in developing your intuition, it is perhaps convenient to contemplate a multivariate normal (Gaussian) distribution for $q(Y|X_t)$, with a mean equal to the current sample X_t. With such a proposal distribution, the probability density decreases with distance away from the current sample.

The second step is to decide on whether to accept the candidate Y for X_{t+1} on the basis of the value of a ratio r given by

$$r = \frac{p(Y|D, I)}{p(X_t|D, I)} \frac{q(X_t|Y)}{q(Y|X_t)}, \tag{12.4}$$

where r is called the *Metropolis ratio*. If the proposal distribution is symmetric, then the second factor is $= 1$. If $r \geq 1$, then we set $X_{t+1} = Y$. If $r < 1$, then we accept it with a probability $= r$. This is done by sampling a random variable U from Uniform(0, 1), a uniform distribution in the interval 0 to 1. If $U \leq r$ we set $X_{t+1} = Y$, otherwise we set $X_{t+1} = X_t$. This second step can be summarized by a term called the *acceptance probability* $\alpha(X_t, Y)$ given by

$$\alpha(X_t, Y) = \min(1, r) = \min\left(1, \frac{p(Y|D, I)}{p(X_t|D, I)} \frac{q(X_t|Y)}{q(Y|X_t)}\right). \tag{12.5}$$

The MCMC method as initially proposed by Metropolis *et al.* in 1953, considered only symmetric proposal distributions, having the form $q(Y|X_t) = q(X_t|Y)$. Hastings (1970) generalized the algorithm to include asymmetric proposal distributions and the generalization is commonly referred to as the Metropolis–Hastings algorithm. There are now many different versions of the algorithm.

The Metropolis–Hastings algorithm is extremely simple:

1. Initialize X_0; set $t = 0$.
2. Repeat {Obtain a new sample Y from $q(Y|X_t)$
   Sample a Uniform(0,1) random variable U
      If $U \leq r$ set $X_{t+1} = Y$ otherwise set $X_{t+1} = X_t$ Increment t}

Example 1:

Suppose the posterior is a Poisson distribution, $p(X|D, I) = \lambda^X e^{-\lambda}/X!$. For our proposal distribution $q(Y|X_t)$, we will use a simple random walk such that:

1. Given X_t, pick a random number $U_1 \sim U(0, 1)$
2. If $U_1 > 0.5$, propose $Y = X_t + 1$ otherwise $Y = X_t - 1$
3. Compute the Metropolis ratio $r = p(Y|D, I)/p(X_t|D, I) = \lambda^{Y-X_t} X_t!/Y!$
4. Acceptance/rejection: $U_2 \sim U(0, 1)$
    Accept $X_{t+1} = Y$ if $U_2 \leq r$ otherwise set $X_{t+1} = X_t$

Figure 12.1 illustrates the results for the above simple MCMC simulation using a value of $\lambda = 3$ and starting from an initial $X_0 = 10$ which is far out in the tail of the posterior. Panel (a) shows a sequence of 1000 samples from the MCMC. It is clear that the samples quickly move from our starting point far out in the tail to the vicinity of the posterior mean. Panel (b) compares a histogram of the last 900 samples from the MCMC with the true Poisson posterior which is indicated by the solid line. The

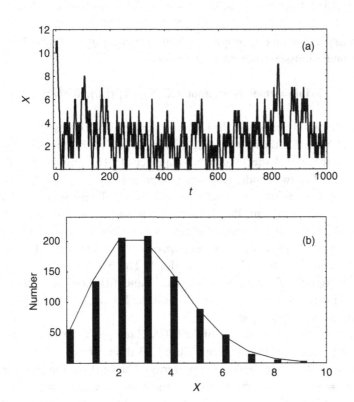

Figure 12.1 The results from a simple one-dimensional Markov chain Monte Carlo simulation for a Poisson posterior for X. Panel (a) shows a sequence of 1000 samples from the MCMC. Panel (b) shows a comparison of the last 900 MCMC samples with the true posterior indicated by the solid curve.

agreement is very good. We treated the first 100 samples as an estimate of the burn-in period and did not use them.

Example 2:

Now consider a MCMC simulation of samples from a joint posterior $p(X_1, X_2|D, I)$ in two parameters X_1 and X_2, which has a double peak structure. Note: if we want to refer to the tth time sample of the ith parameter from a Markov chain, we will do so with the designation $X_{t,i}$. We define the posterior in *Mathematica* with the following commands.

Needs["Statistics 'MultinormalDistribution' "]

dist1 = MultinormalDistribution [{0, 0}, {{1, 0}, {0, 1}}]
The first argument {0, 0} indicates the multinormal distribution is centered at 0,0.
The second argument {{1, 0}, {0, 1}} gives the covariance of the distribution.
dist2 = MultinormalDistribution[{4, 0}, {{2, 0.8}, {0.8, 2}}]
Posterior = 0.5 (PDF[dist1, {X₁, X₂}]+ PDF[dist2, {X₁, X₂}])
The factor of 0.5 ensures the posterior is normalized to an area of one.

In this example, we used a proposal density function $q(Y_1, Y_2|X_1, X_2)$ which is a two-dimensional Gaussian (normal) distribution.

$$\textbf{[MultinormalDistribution[\{X_1, X_2\}, \{\{\sigma_1^2, 0\}, \{0, \sigma_2^2\}\}]]}$$

The results for 8000 samples of the posterior generated with this MCMC are shown in Figure 12.2. Note that the first 50 samples were treated as the burn-in period and are not included in this plot. Panel (a) shows a sequence of 7950 samples from the MCMC with $\sigma_1 = \sigma_2 = 1$. The two model parameters represented by X_1 and X_2 could be very different physical quantities each characterized by a different scale. In that case, σ_1 and σ_2 could be very different. Panel (b) shows the same points with contours of the posterior overlaid. The distribution of sample points matches the contours of the true posterior very well. Panel (c) shows a comparison of the true marginal posterior (solid curve) for X_1 and the MCMC marginal (dots). The MCMC marginal is simply a normalized histogram of the X_1 sample values. Panel (d) shows a comparison of the true marginal posterior (solid curve) for X_2 and the MCMC marginal (dots). In both cases, the agreement is very good.

We also investigated the evolution of the MCMC samples for proposal distributions with different values of σ. Panel (a) in Figure 12.3 shows the case for a $\sigma \sim 1/10$ the scale of the smallest features in the true posterior. The starting point for each simulation was at $X_1 = -4.5, X_2 = 4.5$. In this case, the burn-in period is considerably longer and it appears that a larger number of samples would be needed to do justice to the posterior which is indicated by the contours. Panel (b) illustrates the case for $\sigma = 1$, the value used for Figure 12.2. Panel (c) uses a $\sigma \sim 10$ times the scale of the smallest features in the posterior. From the density of the points it appears that we have used a

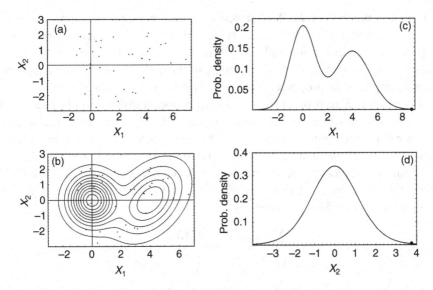

Figure 12.2 The results from a two-dimensional Markov chain Monte Carlo simulation of a double peaked posterior. Panel (a) shows a sequence of 7950 samples from the MCMC. Panel (b) shows the same points with contours of the posterior overlaid. Panel (c) shows a comparison of the marginal posterior (solid curve) for X_1 and the MCMC marginal (dots). Panel (d) shows a comparison of the marginal posterior (solid curve) for X_2 and the MCMC marginal (dots).

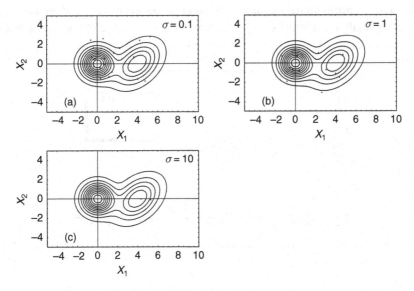

Figure 12.3 A comparison of the samples from three Markov chain Monte Carlo runs using Gaussian proposal distributions with differing values of the standard deviation: (a) $\sigma = 0.1$, (b) $\sigma = 1$, (c) $\sigma = 10$. The starting point for each run was at $X_1 = -4.5$ and $X_2 = 4.5$.

much smaller number of MCMC samples. In fact we used the same number of samples. Recall that in MCMC we carry out a test to decide whether to accept the new proposal (see discussion following Equation (12.4)). If we fail to accept the proposal, then we set $X_{t+1} = X_t$. Thus, many of the points in panel (c) are repeats of the same sample as the proposed sample was rejected on many occasions.

It is commonly agreed that finding an ideal proposal distribution is an art. If we restrict the conversation to Gaussian proposal distributions then the question becomes what is the optimum choice of σ? As mentioned earlier, the samples from a MCMC are not independent, but exhibit correlations. In Figure 12.4, we illustrate the correlations of samples corresponding to the three choices of σ used in Figure 12.3 by plotting the *autocorrelation functions* (ACFs) for X_2. The ACF, $\rho(h)$, which was introduced in Section 5.13.2, is given by

$$\rho(h) = \frac{\sum_{\text{overlap}} [(X_t - \overline{X})(X_{t+h} - \overline{X})]}{\sqrt{\sum_{\text{overlap}} (X_t - \overline{X})^2} \times \sqrt{\sum_{\text{overlap}} (X_{t+h} - \overline{X})^2}}, \qquad (12.6)$$

where X_{t+h} is a shifted version of X_t and the summation is carried out over the subset of samples that overlap. The shift h is referred to as the *lag*. It is often observed that $\rho(h)$ is roughly exponential in shape so we can model the ACF

$$\rho(h) \sim \exp\{-\frac{h}{\tau_{\text{exp}}}\}. \qquad (12.7)$$

The autocorrelation time constant, τ_{exp}, reflects the convergence speed of the MCMC sampler and is approximately equal to the interval between independent samples. In general, the smaller the value of τ_{exp} the better, i.e., the more efficient, the MCMC

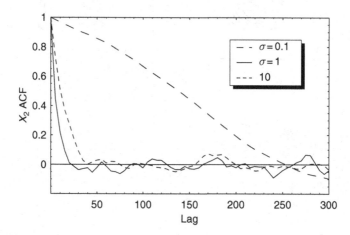

Figure 12.4 A comparison of the autocorrelation functions for three Markov chain Monte Carlo runs using Gaussian proposal distributions with differing values of the standard deviation: $\sigma = 0.1, \sigma = 1, \sigma = 10$.

sampler is. Examination of Figure 12.4 indicates that of the three choices of σ chosen above, $\sigma = 1.0$ leads to the smallest values of τ_{exp} for X_2. Of course, in this example just considered, we have set $\sigma_{X_1} = \sigma_{X_2}$. In general, they will not be equal. Related to the optimum choice of σ is the average rate at which proposed state changes are accepted, called the *acceptance rate*. Based on empirical studies, Roberts, Gelman, and Gilks (1997) recommend calibrating the acceptance rate to about 25% for a high-dimensional model and to about 50% for models of one or two dimensions. The acceptance rates corresponding to our three choices of σ in Figure 12.3 are 95%, 63%, and 5%, respectively.

A number of issues arise from a consideration of these two simple examples. How do we decide: (a) the length of the burn-in period, (b) when to stop the Markov chain, and (c) what is a suitable proposal distribution? For a discussion of these points, the reader is referred to a collection of review and application papers (Gilks, Richardson, and Spiegelhalter 1996). For an unpublished 1996 roundtable discussion of informal advice for novice practitioners, moderated by R. E. Kass, see *www.amstat.org/ publications/tas/kass.pdf*. The treatment of MCMC given in this text is intended only as an introduction to this topic.

Loredo (1999) gives an interesting perspective on the relationship between the development of MCMC in statistics and certain computational physics techniques. Define a function $\Lambda(X) = -\ln[p(X|I)\,p(D|X,I)]$. Then the posterior distribution can be written as $p(X|D,I) = e^{-\Lambda(X)}/Z$, where $Z = \int dX\, e^{-\Lambda(X)}$. Evaluation of the posterior resembles two classes of problems familiar to physicists: evaluating Boltzmann factors and partition functions in statistical mechanics, and evaluating Feynman path weights and path integrals in Euclidean quantum field theory. For a discussion of some useful modern extensions of the Metropolis algorithm that are particularly accessible to physical scientists, see Liu (2001) and the first section of Toussaint (1989). A readable tutorial for statistics students is available in Chib and Greenberg (1995).

12.3 Why does Metropolis–Hastings work?

Remarkably, for a wide range of proposal distributions $q(Y|X)$, the Metropolis–Hastings algorithm generates samples of X with a probability density which converges on the desired target posterior $p(X|D,I)$, called the *stationary distribution* of the Markov chain. For the distribution of X_t to converge to a stationary distribution, the Markov chain must have three properties (Roberts, 1996). First, it must be *irreducible*. That is, from all starting points, the Markov chain must be able to (eventually) jump to all states in the target distribution with positive probability. Second it must be *aperiodic*. This stops the chain from oscillating between different states in a regular periodic movement. Finally the chain must be *positive recurrent*. This can be expressed in terms of the existence of a stationary distribution $\pi(X)$, say, such that if an initial value X_0 is sampled from $\pi(X)$, then all subsequent iterates will also be distributed according to $\pi(X)$.

To see that the target distribution is the stationary distribution of the Markov chain generated by the Metropolis–Hastings algorithm, consider the following: suppose we start with a sample X_t from the target distribution. The probability of drawing X_t from the posterior is $p(X_t|D, I)$. The probability that we will draw and accept a sample X_{t+1} is given by the transition kernel, $p(X_{t+1}|X_t) = q(X_{t+1}|X_t)\, \alpha(X_t, X_{t+1})$, where $\alpha(X_t, X_{t+1})$ is given by Equation (12.5). The joint probability of X_t and X_{t+1} is then given by

$$
\begin{aligned}
\text{Joint probability}(X_t, X_{t+1}) &= p(X_t|D, I)\, p(X_{t+1}|X_t) \\
&= p(X_t|D, I)\, q(X_{t+1}|X_t)\alpha(X_t, X_{t+1}) \\
&= p(X_t|D, I)\, q(X_{t+1}|X_t) \min\left(1, \frac{p(X_{t+1}|D, I)\, q(X_t|X_{t+1})}{p(X_t|D, I)\, q(X_{t+1}|X_t)}\right) \\
&= \min(p(X_t|D, I)\, q(X_{t+1}|X_t), p(X_{t+1}|D, I)q(X_t|X_{t+1})) \\
&= p(X_{t+1}|D, I)\, q(X_t|X_{t+1})\alpha(X_{t+1}, X_t) \\
&= p(X_{t+1}|D, I)\, p(X_t|X_{t+1}).
\end{aligned}
\tag{12.8}
$$

Thus, we have shown

$$
p(X_t|D, I)\, p(X_{t+1}|X_t) = p(X_{t+1}|D, I)\, p(X_t|X_{t+1}),
\tag{12.9}
$$

which is called the *detailed balance equation*.

In statistical mechanics, detailed balance occurs for systems in thermodynamic equilibrium.[2] In the present case, the condition of detailed balance means that the Markov chain generated by the Metropolis–Hastings algorithm converges to a stationary distribution.

Recall from Equation (12.8) that $p(X_t|D, I)p(X_{t+1}|X_t)$ is the joint probability of X_t and X_{t+1}. We will now integrate this joint probability with respect to X_t, making use of Equation (12.9), and demonstrate that the result is simply the marginal probability distribution of X_{t+1}.

$$
\begin{aligned}
\int p(X_t|D, I)\, p(X_{t+1}|X_t)dX_t &= \int p(X_{t+1}|D, I)p(X_t|X_{t+1})\, dX_t \\
&= p(X_{t+1}|D, I) \int p(X_t|X_{t+1})\, dX_t \\
&= p(X_{t+1}|D, I).
\end{aligned}
\tag{12.10}
$$

Thus, we have shown that once a sample from the stationary target distribution has been obtained, all subsequent samples will be from that distribution.

---

[2] It may help to consider the following analogy: suppose we have a collection of hydrogen atoms. The number of atoms making a transition from excited state t to state $t + 1$ in 1 s is given by $N \times p(t) \times p(t + 1|t)$, where N equals the total number of atoms, $p(t)$ is the probability of an atom being in state t, and $p(t + 1|t)$ is the probability that an atom in state t will make a transition to state $t + 1$ in 1 s. Similarly the number making transitions from $t + 1$ to t in 1 s is given by $N \times p(t + 1) \times p(t|t + 1)$. In thermodynamic equilibrium, the rate of transition from t to $t + 1$ is equal to the rate from $t + 1$ to t, so

$$
p(t) \times p(t + 1|t) = p(t + 1) \times p(t|t + 1).
$$

12.4 Simulated tempering

The simple Metropolis–Hastings algorithm outlined in Section 12.2 can run into difficulties if the target probability distribution is multi-modal. The MCMC can become stuck in a local mode and fail to fully explore other modes which contain significant probability. This problem is very similar to the one encountered in finding a global minimum in a nonlinear model fitting problem. One solution to that problem was to use simulated annealing (see Section 11.4.1) by introducing a temperature parameter T. The analogous process applied to drawing samples from a target probability distribution (e.g., Geyer and Thompson, 1995) is often referred to as *simulated tempering* (ST). In annealing, the temperature parameter is gradually decreased. In ST, we create a discrete set of progressively flatter versions of the target distribution using a temperature parameter. For $T = 1$, the distribution is the desired target distribution which is referred to as the cold sampler. For $T \gg 1$, the distribution is much flatter. The basic idea is that by repeatedly heating up the distribution (making it flatter), the new sampler can escape from local modes and increase its chance of reaching all regions of the target distribution that contain significant probability. Typical inference is based on samples drawn from the cold sampler and the remaining observations discarded. Actually, in Section 12.7 we will see how to use the samples from the hotter distributions to evaluate Bayes factors in model selection problems.

Again, let $p(X|D, I)$ be the target posterior distribution we want to sample. Applying Bayes' theorem, we can write this as

$$p(X|D, I) = C\, p(X|I) \times p(D|X, I),$$

where $C = 1/p(D|I)$ is the usual normalization constant which is not important at this stage and will be dropped. We can construct other flatter distributions as follows:

$$
\begin{aligned}
\pi(X|D, \beta, I) &= p(X|I)p(D|X, I)^{\beta} \\
&= p(X|I)\, \exp(\beta\, \ln[p(D|X, I)]), \quad \text{for } 0 < \beta < 1.
\end{aligned}
\tag{12.11}
$$

Rather than use a temperature which varies from 1 to infinity, we prefer to use its reciprocal which we label β and refer to as the tempering parameter. Thus β varies from 1 to zero. We will use a discrete set of β values labeled $\{1, \beta_2, \cdots, \beta_m\}$, where $\beta = 1$ corresponds to the cold sampler (target distribution) and β_m corresponds to our hottest sampler which is generally much flatter. This particular formulation is also convenient for our later discussion on determining the Bayes factor in model selection problems. Rather than describe ST in detail, we will describe a more efficient related algorithm called *parallel tempering* in the next section.

12.5 Parallel tempering

Parallel tempering (PT) is an attractive alternative to simulated tempering (Liu, 2001). Again, multiple copies of the simulation are run in parallel, each at a different temperature (i.e., a different $\beta = 1/\mathcal{T}$). One of the simulations, corresponding to $\beta = 1/\mathcal{T} = 1$, is the desired target probability distribution. The other simulations correspond to a ladder of higher temperature distributions indexed by i. Let $n\beta$ equal the number of parallel MCMC simulations. At intervals, a pair of adjacent simulations on this ladder are chosen at random and a proposal made to swap their parameter states. Suppose simulations β_i and β_{i+1} are chosen. At time t, simulation β_i is in state $X_{t,i}$ and simulation β_{i+1} is in state $X_{t,i+1}$. If the swap is accepted by the test given below then these states are interchanged. In the example discussed in Section 12.6, we specify that on average, a swap is proposed after every n_s iterations ($n_s = 30$ was used) of the parallel simulations in the ladder. This is done by choosing a random number, $U_1 \sim$ Uniform[0,1], at each time iteration and proposing a swap only if $U_1 \leq 1/n_s$. If a swap is to be proposed, we use a second random number to pick one of the ladder simulations i in the range $i = 1$ to $(n\beta - 1)$, and propose swapping the parameter states of i and $i + 1$. A Monte Carlo acceptance rule determines the probability for the proposed swap to occur. Accept the swap with probability

$$r = \min\left\{1, \frac{\pi(X_{t,i+1}|D,\beta_i,I)\ \pi(X_{t,i}|D,\beta_{i+1},I)}{\pi(X_{t,i}|D,\beta_i,I)\ \pi(X_{t,i+1}|D,\beta_{i+1},I)}\right\}, \tag{12.12}$$

where $\pi(X|D,\beta,I)$ is given by Equation (12.11). We accept the swap if $U_2 \sim$ Uniform[0,1] $\leq r$.

This swap allows for an exchange of information across the population of parallel simulations. In the higher temperature simulations, radically different configurations can arise, whereas in lower temperature states, a configuration is given the chance to refine itself. By making exchanges, we can capture and improve the higher probability configurations generated by the population by putting them into lower temperature simulations. Some experimentation is needed to refine suitable choices of β_i values. Adjacent simulations need to have some overlap to achieve a sufficient acceptance probability for an exchange operation.

12.6 Example

Although MCMC really comes into its own when the number of model parameters is very large, we will apply it to the toy spectral line problem we analyzed in Section 3.6, because we can compare with our earlier results. The objective of that problem was to test two competing models, represented by M_1 and M_2, on the basis of some spectral line data. Only M_1 predicts the existence of a particular spectral line. In the simplest version of the problem, the line frequency and shape is exactly predicted by M_1; the only quantity which is uncertain is the line strength T expressed in temperature units. The odds ratio in favor of M_1 was found to be 11:1 assuming a Jeffreys prior for the line strength. We also computed the most probable line strength. In Section 3.9, we investigated how our conclusions would be altered if the line frequency were uncertain, i.e., it could occur anywhere between channels 1 to 44. In that case, the odds ratio favoring M_1 dropped from 11:1 to $\approx$ 1:1, assuming a uniform prior for the line center frequency. Below, we apply both the Metropolis–Hastings and parallel tempering versions of MCMC to the problem of estimating the marginal posteriors of the line strength and center frequency to compare with our previous results. In Section 12.7, we will employ parallel tempering to compute the Bayes factor needed for model comparison.

Metropolis–Hastings results

In this section, we will draw samples from $p(X|D, M_1, I)$, where X is a vector representing the two parameters of model M_1, namely the line strength T and the line center frequency ν expressed as channel number. We use a Jeffreys prior for T in the range $T_{\min} = 0.1$ mK to $T_{\max} = 100$ mK. We assume a uniform prior for ν in the range channel 1 to 44. The steps in the calculation are as follows:

1. Initialize X_0; set $t = 0$.
   In this example we set $X_0 = \{T_0 = 5, \nu_0 = 30\}$
2. Repeat {

   a) Obtain a new sample Y from $q(Y|X_t)$
      $Y = \{T', \nu'\}$
      we set $q(T'|T_t) = $ **Random[NormalDistribution[$T_t, \sigma_T = 1.0$]]**
      and $q(\nu'|\nu_t) = $ **Random[NormalDistribution[$\nu_t, \sigma_f = 1.0$]]**
   b) Compute the Metropolis ratio

   $$r = \frac{p(Y|D, M_1, I)}{p(X_t|D, M_1, I)} = \frac{p(T', \nu'|M_1, I)\ p(D|M_1, T', \nu', I)}{p(T_t, \nu_t|M_1, I)\ p(D|M_1, T_t, \nu_t, I)}$$

   where $p(D|M_1, T, \nu, I)$ is given by Equations (3.44) and (3.41).
   The priors $p(T, \nu|M_1, I) = p(T|M_1, I)\ p(\nu|M_1, I)$ are given by Equations (3.38) and (3.33).
      Note: if T' or ν' lie outside the prior boundaries set $r = 0$.
   c) Acceptance/rejection: $U \sim$ **U(0, 1)**
   d) Accept $X_{t+1} = Y$ if $U \leq r$, otherwise set $X_{t+1} = X_t$
   e) Increment t}

Figure 12.5 shows results for 10^5 iterations of a Metropolis–Hastings Markov chain Monte Carlo. Panel (a) shows every 50th value of parameter ν, expressed as a channel number, and panel (c) the same for parameter T. It is clear that the ν values move quickly to a region centered on channel 37 with occasional jumps to a region centered on channel 24 and only one jump to small channel numbers. The T parameter can be seen to fluctuate between 0.1 and ~3.5 mK. Panels (b) and (d) show a blow-up of the first 500 iterations. It is apparent from these panels that the burn-in period is very short, < 50 iterations for a starting state of $T = 5$ and $\nu = 30$.

Figure 12.6 shows distributions of the two parameters. In panel (a), the joint distribution of T and ν is apparent from the scatter plot of every 20th iteration obtained

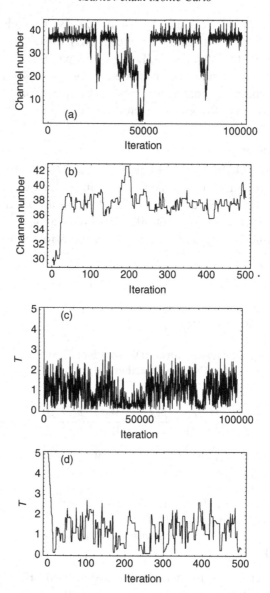

Figure 12.5 Results for 10^5 iterations of a Metropolis–Hastings Markov chain Monte Carlo. Panel (a) shows every 50th value of parameter ν and panel (c) the same for parameter T. Panels (b) and (d) show a blow-up of the first 500 iterations.

after dropping the burn-in period consisting of the first 50 iterations. To obtain the marginal posterior density for the ν parameter, we simply plot a histogram of all the ν values (post burn-in) normalized by dividing by the sum of the ν values multiplied by the width of each bin. This is shown plotted in panel (b) together with our earlier marginal distribution (solid curve) computed by numerical integration. It is clear that

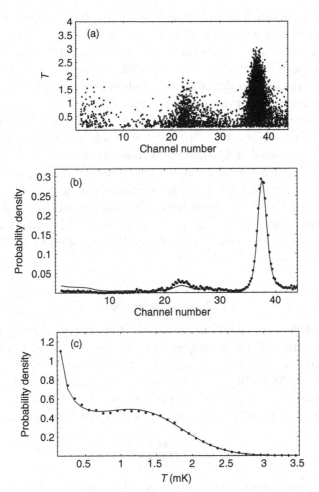

Figure 12.6 Results for the spectral line problem using a Metropolis–Hastings Markov chain Monte Carlo analysis. Panel (a) is a scatter plot of the result for every 20th iteration in the two model parameters, channel number and line strength T. Panel (b) shows the marginal probability density for channel number (points) compared to our earlier numerical integration result indicated by the solid curve. Panel (c) shows the marginal probability density for line strength T (points) compared to our earlier numerical integration result indicated by the solid curve.

10^5 iterations of Metropolis–Hastings does a good job of defining the dominant peak of the probability distribution for ν but does a poor job of capturing two other widely separated islands containing significant probability. On the other hand, it is clear from panel (c) that it has done a great job of defining the distribution of T.

Parallel tempering results

We also analyzed the spectral line data with a parallel tempering (PT) version of MCMC described in Section 12.5. We used five values for the tempering parameter, β, uniformly spaced between 0.01 and 1.0, and ran all five chains in parallel. At intervals (on average every 50 iterations) a pair of adjacent simulations on this ladder are chosen at random and a proposal made to swap their parameter states. We used the same starting state of $T = 5, \nu = 30$ and executed 10^5 iterations. The final results for the $\beta = 1$, corresponding to the target distribution, are shown in Figures 12.7 and 12.8. The acceptance rate for this simulation was 37%.

Comparing panel (a) of Figures 12.7 and 12.5, we see that the PT version visits the two low-lying regions of ν probability much more frequently than the Metropolis–Hastings version. Comparing the marginal densities of Figures 12.8 and 12.6 we see that the PT marginal density for ν is in better agreement with the expected results indicated by the solid curves. For both versions, the marginal densities for T are in excellent agreement with the expected result. In more complicated problems, we often cannot conveniently compute the marginal densities by another method. In this case, it is useful to compare the results from a number of PT simulations with different starting parameter states.

12.7 Model comparison

So far we have demonstrated how to use MCMC to compute the marginal posteriors for model parameters. In this section, we will show how to use the results of parallel tempering to compute the Bayes factor used in model comparison (Skilling, 1998; Goggans and Chi, 2004). In the toy spectral line problem of Section 3.6, we were interested in computing the odds ratio of two models M_1 and M_2 which from Equation (3.30) is equal to the prior odds times the Bayes factor given by

$$B_{12} = \frac{p(D|M_1, I)}{p(D|M_2, I)}, \tag{12.13}$$

where $p(D|M_1, I)$ and $p(D|M_2, I)$ are the global likelihoods for the two models. In the version of this problem analyzed in Section 12.6, M_1 has two parameters ν and T. For independent priors,

$$p(D|M_1, I) = \int d\nu \, p(\nu|M_1, I) \int dT \, p(T|M_1, I) p(D|M_1, \nu, T, I). \tag{12.14}$$

In what follows, we will generalize the model parameter set to an arbitrary number of parameters which we represent by the vector X.

To evaluate $p(D|M_1, I)$, using parallel tempering MCMC, we first define a partition function

$$Z(\beta) = \int dX \, p(X|M_1, I) \, p(D|M_1, X, I)^\beta$$

$$= \int dX \exp\{\ln[p(X|M_1, I)] + \beta \ln[p(D|M_1, X, I)]\}, \tag{12.15}$$

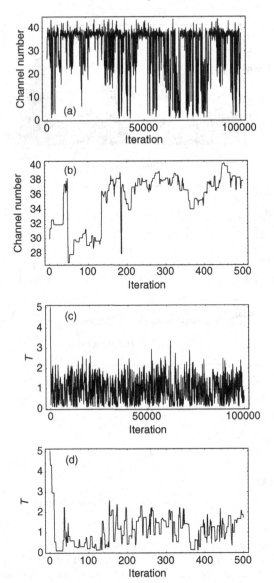

Figure 12.7 Results for 10^5 iterations of a parallel tempering Markov chain Monte Carlo. Panel (a) shows every 50th value of parameter ν and panel (c) the same for parameter T. Panels (b) and (d) show a blow-up of the first 500 iterations.

where β is the tempering parameter introduced in Section 12.4. Now take the derivative of $\ln[Z(\beta)]$.

$$\frac{d}{d\beta}\ln[Z(\beta)] = \frac{1}{Z(\beta)}\frac{d}{d\beta}Z(\beta) \qquad (12.16)$$

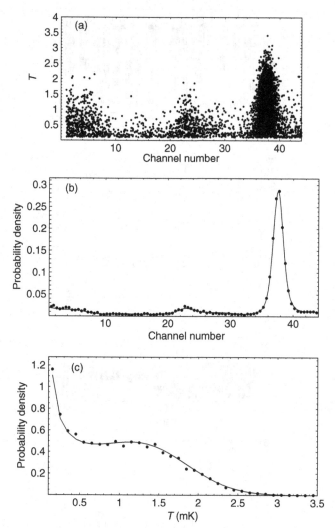

Figure 12.8 Results for the spectral line problem using a Markov chain Monte Carlo analysis with parallel tempering. Panel (a) is a scatter plot of the result for every 20th iteration in the two model parameters, channel number and line strength T. Panel (b) shows the marginal probability density for channel number (points) compared to our earlier numerical integration result indicated by the solid line. Panel (c) shows the marginal probability density for line strength T (points) compared to our earlier numerical integration result indicated by the solid line.

$$
\begin{aligned}
\frac{d}{d\beta} Z(\beta) &= \int dX \, \ln[p(D|M_1, X, I)] \\
&\quad \times \exp\{\ln[p(X|M_1, I)] + \beta \ln[p(D|M_1, X, I)]\} \\
&= \int dX \, \ln[p(D|M_1, X, I)] \, p(X|M_1, I) \, p(D|M_1, X, I)^{\beta}.
\end{aligned}
\tag{12.17}
$$

Substituting Equation (12.17) into Equation (12.16), we obtain

$$\frac{d}{d\beta}\ln[Z(\beta)] = \frac{\int dX \ln[p(D|M_1,X,I)] \, p(X|M_1,I) \, p(D|M_1,X,I)^\beta}{\int dX \, p(X|M_1,I) \, p(D|M_1,X,I)^\beta}$$

$$= \langle \ln[p(D|M_1,X,I)] \rangle_\beta, \tag{12.18}$$

where $\langle \ln[p(D|M_1,X,I)] \rangle_\beta$ is the expectation value of the $\ln[p(D|M_1,X,I)]$. This quantity is easily evaluated from the MCMC results which consist of sets of X_t samples, one set for each value of the tempering parameter β. Let $\{X_{t,\beta}\}$ represent the samples for tempering parameter β.

$$\langle \ln[p(D|M_1,X,I)] \rangle_\beta = \frac{1}{n}\sum_t \ln[p(D|M_1,X_{t,\beta},I)], \tag{12.19}$$

where n is the number of samples in each set after the burn-in period. From Equation (12.18) we can write

$$\int_0^1 d\ln[Z(\beta)] = \ln[Z(1)] - \ln[Z(0)]$$

$$= \int d\beta \, \langle \ln[p(D|M_1,X,I)] \rangle_\beta. \tag{12.20}$$

Now from Equation (12.15)

$$Z(1) = \int dX \, p(X|M_1,I) \, p(D|M_1,X,I) = p(D|M_1,I), \tag{12.21}$$

and

$$Z(0) = \int dX \, p(X|M_1,I). \tag{12.22}$$

From Equations (12.20) and (12.21) we can write

$$\ln[p(D|M_1,I)] = \ln[Z(0)] + \int d\beta \langle \ln[p(D|M_1,X,I)] \rangle_\beta. \tag{12.23}$$

For a normalized prior, $Z(0) = 1$ and Equation (12.23) becomes

$$\ln[p(D|M_1,I)] = \int d\beta \langle \ln[p(D|M_1,X,I)] \rangle_\beta. \tag{12.24}$$

Armed with Equation (12.24) we are now in a position to evaluate the Bayes factor given by Equation (12.13), which is at the heart of model comparison.

Returning to the spectral line problem,

$$\langle \ln[p(D|M_1,\nu,T,I)] \rangle_\beta = \frac{1}{n}\sum_t \ln[p(D|M_1,\nu_{t,\beta},T_{t,\beta},I)]. \tag{12.25}$$

We evaluated Equation (12.25) for the five values of $\beta = 0.01, 0.2575, 0.505, 0.7525,$ 1.0 used in the PT MCMC analysis of Section 12.6. The results were $-97.51, -87.1937, -86.4973, -85.9128, -85.1565$, respectively. We then evaluated the integral in Equation (12.24) by generating an interpolating function and integrating the interpolating function in the interval 0 to 1. This yielded $\ln[p(D|M_1, I)] = -87.4462$. A more sophisticated interpolation of the results yielded $\ln[p(D|M_1, I)] = -87.3369$. Model M_2 had no free parameters and $p(D|M_2, I) = 1.133 \times 10^{-38}$ from Equation (3.49). The resulting Bayes factors for the two interpolations were $B_{1,2} = 0.93$ and 1.04, respectively. This should be compared to $B_{1,2} = 1.06$ obtained from our earlier solution to this problem.

12.8 Towards an automated MCMC

As the number of model parameters increases, so does the time required to choose a suitable σ value for each of the parameter proposal distributions. Suitable means that MCMC solutions, starting from different locations in the prior parameter space, yield equilibrium distributions of model parameter values that are not significantly different, in an acceptable number of iterations. Generally this involves running a series of chains, each time varying σ for one or more of the parameter proposal distributions, until the chain appears to converge on an equilibrium distribution with a proposal acceptance rate, λ, that is reasonable for the number of parameters involved, e.g., approximately 25% for a large number of parameters (Roberts, Gelman, and Gilks, 1997). This is especially time consuming if each parameter corresponds to a different physical quantity, so that the σ values can be very different. In this section, we describe one attempt at automating this process, which we apply to the detection of an extrasolar planet using some real astronomical data.

Suppose we are dealing with M parameters that are represented collectively by $\{X_\alpha\}$. Let σ_α represent the characteristic width of a symmetric proposal distribution for X_α. We will assume Gaussian proposal distributions but the general approach should also be applicable to other forms of proposal distributions. To automate the MCMC, we need to incorporate a control system that makes use of some form of error signal to steer the selection of the $\{\sigma_\alpha\}$.

For a manually controlled MCMC, a useful approach is to start with a large value of σ_α, approximately one tenth of the prior uncertainty of that parameter. In a PT MCMC, this will normally be sufficient to provide access to all areas with significant probability within the prior range, but may result in a very small acceptance rate for the $\beta = 1$ member of the PT MCMC chain. By running a number of smaller iteration chains, each time perturbing one or more of the $\{\sigma_\alpha\}$, it soon becomes clear which parameters are restraining the acceptance rate from a more desirable level. Larger $\{\sigma_\alpha\}$ values yield larger jumps in parameter proposal values. The general approach of refining the $\{\sigma_\alpha\}$ towards smaller values is analogous to an annealing operation. The refinement is terminated when the target proposal acceptance rate is reached.

In the automated version of this process described below, the error signal used for the control system is the difference between the current acceptance rate and a target acceptance rate. The control system steers the proposal σ's to desirable values during the burn-in stage of a single parallel tempering MCMC run. Although inclusion of the control system may result in a somewhat longer burn-in period, there is a huge saving in time because it eliminates many trial runs to manually establish a suitable set of $\{\sigma_\alpha\}$. In addition the control system error monitor provides another indication of the length of the burn-in period. In practice, it is important to repeat the operation for a few different choices of initial parameter values, to ensure that the MCMC results converge.

The automatic parallel tempering MCMC (APT MCMC) algorithm contains major and minor cycles. During the major cycles the current set of $\{\sigma_\alpha\}$ are used for n_1 iterations. The acceptance rate achieved during this major cycle is compared to the target acceptance rate. If the difference (control system error signal), ϵ, is greater than a chosen threshold, tol_1, then a set of minor cycles, one cycle of n_2 iterations for each σ_α, are employed to explore the sensitivity of the acceptance rate to each σ_α. The $\{\sigma_\alpha\}$ are updated and another major cycle run. If tol_1 is set $= 0$, then the minor cycles are always performed after each major cycle. At this point, the reader might find it useful to examine the evolution of the error signal, and the $\{\sigma_\alpha\}$, for the examples shown in Figures 12.12 and 12.13. One can clearly see the expected Poisson fluctuations in the error signal after the $\{\sigma_\alpha\}$ stabilize. For these examples we set $\text{tol}_1 = 1.5\sqrt{\lambda n_1}$ to reduce the number of minor cycles. Normally the control system is turned off after ϵ is less than some threshold, tol_2. Typically $\text{tol}_2 = \sqrt{\lambda n_1}$.

Full details of the control system are not included here as it is considered experimental and in a process of evolution. The latest version is included in the *Mathematica* tutorial in the section entitled "Automatic parallel tempering MCMC," along with useful default values for the algorithm parameters and input data format. Figure 12.9 provides a summary of the inputs and outputs for the APT MCMC algorithm. In the following section we demonstrate the behavior of the algorithm with a set of astronomical data used to detect an extrasolar planet.

12.9 Extrasolar planet example

In this section, we will apply the automated parallel tempering MCMC described in Section 12.8 to some real astronomical data, which were used to discover (Tinney *et al.*, 2003) an extrasolar planet orbiting a star with a catalog number HD 2039. Although light from the planet is too faint to be detected, the gravitational tug of the planet on the star is sufficient to produce a measurable Doppler shift in the velocity of absorption lines in the star's spectrum. By fitting a Keplerian orbit to the measured radial velocity data, v_i, it is possible to obtain information about the orbit and a lower limit on the mass of the unseen planet. The predicted model radial velocity, f_i, for a particular orbit is given below, and involves six unknowns. The geometry of a stellar orbit with respect to the observer is shown in Figure 12.10. The points labeled F, P,

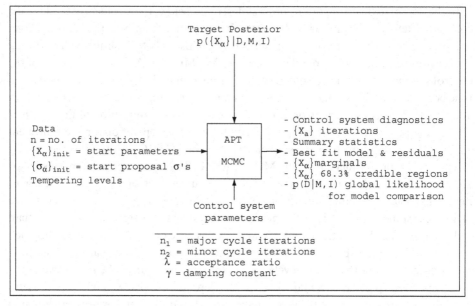

Figure 12.9 An overview schematic of the inputs and outputs for the automated parallel tempering MCMC.

and S, are the location of the focus of the elliptical orbit, periastron, and the star's position at time t_i, respectively.

$$f_i = V + K[\cos\{\theta(t_i + t_0) + \omega\} + e\cos\omega],\qquad(12.26)$$

where

$V =$ the systematic velocity of the system.
$K =$ velocity amplitude $= 2\pi P^{-1}(1 - e^2)^{-1/2} a \sin i.$
$P =$ the orbital period.
$a =$ the semi-major axis of the orbit.
$e =$ the eccentricity of the elliptical orbit.
$i =$ the inclination of the orbit as defined in Figure 12.10.
$\omega =$ the longitude of periastron, angle LFA in Figure 12.10.
$\chi =$ the fraction of an orbit prior to the start of data-taking that periastron occurred at. Thus, $t_0 = \chi P =$ the number of days prior to $t_i = 0$ that the star was at periastron, for an orbital period of P days. At $t_i = 0$, the star is at an angle AFB from periastron. $\theta(t_i + t_0) =$ the angle (AFS) of the star in its orbit relative to periastron at time t_i.

The dependence of θ on t_i, which follows from the conservation of angular momentum, is given by the solution of

$$\frac{d\theta}{dt} - \frac{2\pi[1 + e\cos\theta(t_i + t_0)]^2}{P(1 - e^2)^{3/2}} = 0.\qquad(12.27)$$

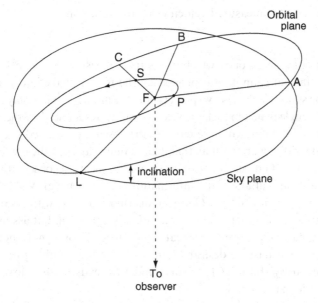

Figure 12.10 The geometry of a stellar orbit with respect to the observer. The sky plane is perpendicular to the dashed line connecting the star and the observer.

To fit Equation (12.26) to the data, we need to specify the six model parameters, P, K, V, e, ω, χ.

The measured radial velocities and their errors are shown Figure 12.11. As we have discussed before, it is good idea not to assume that the quoted measurement errors are the only error component in the data.

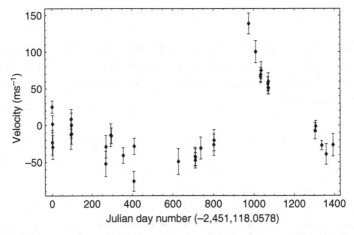

Figure 12.11 HD 2039 radial velocity measurements plotted from the data given in Tinney *et al.*, (2003).

We can represent the measured velocities by the equation

$$v_i = f_i + e_i, \tag{12.28}$$

where e_i is the component of v_i which arises from measurement errors plus any real signal in the data that cannot be explained by the model prediction f_i. For example, suppose that the star actually has two planets, and the model assumes only one is present. In regard to the single planet model, the velocity variations induced by the second planet act like an additional unknown noise term. In the absence of detailed knowledge of the effective noise distribution, other than that it has a finite variance, the maximum entropy principle tells us that a Gaussian distribution would be the most conservative choice (i.e., maximally non-committal about the information we don't have). We will assume the noise variance is finite and adopt a Gaussian distribution for e_i with a variance σ_i^2.

In a Bayesian analysis where the variance of e_i is unknown, but assumed to be the same for all data points, we can treat σ as an unknown nuisance parameter. Marginalizing over σ has the desirable effect of treating anything in the data that can't be explained by the model as noise and this leads to the most conservative estimates of model parameters.

In the current problem, the quoted measurement errors are not all the same. We let $s_i =$ the experimenter's estimate of σ_i, prior to fitting the model and examining the model residuals. The σ_i values are not known, but the s_i values are our best initial estimates. They also contain information on the relative weight we want to associate with each point. Since we do not know the absolute values of the σ_i, we introduce a parameter called the noise scale parameter, b, to allow for this.[3] It could also be called a noise weight parameter. Several different definitions of b are possible including $\sigma_i^2 = b s_i^2$ and $\sigma_i = b s_i$. The definition we use here is given by

$$\frac{1}{\sigma_i^2} = \frac{b}{s_i^2}. \tag{12.29}$$

Again marginalizing over b has the desirable effect of treating anything in the data that can't be explained by the model as noise, leading to the most conservative estimates of orbital parameters. Since b is a scale parameter, we assume a Jeffreys prior (see Section 3.10).

---

[3] Note added in proof:
A better choice for parameterizing any additional unknown noise term is to rewrite equation (12.28) as

$$v_i = f_i + e_i + e_0$$

where e_i is the noise component arising from known but unequal measurement errors, and e_0 is the additional unknown noise term. From the arguments given above, we can characterize the combination of $e_i + e_0$ by a Gaussian distribution with variance $= \sigma_i^2 + \sigma_0^2$. With this form of parameterization we would marginalize over σ_0 instead of b. The one advantage of using b is that it can allow for the possibility that the measurement errors have been overestimated.

$$p(b|I) = \frac{1}{b \ln \frac{b_{max}}{b_{min}}}, \qquad (12.30)$$

with $b_{max} = 2$ and $b_{min} = 0.1$. We also compute $p(b|D, \text{Model}, I)$. If the most probable estimate of $b \approx 1$, then the one-planet model is doing a good job accounting for everything that is not noise based on the s_i estimates. If $b < 1$, then either the model is not accounting for significant real features in the data or the initial noise estimates, s_i, were low.

12.9.1 Model probabilities

In this section, we set up the equations needed to (a) specify the joint posterior probability of the model parameters (parameter estimation problem) for use in the MCMC analysis, and (b) decide if a planet has been detected (model selection problem). To decide if a planet has been detected, we will compare the probability of $M_1 \equiv$ "the star's radial velocity variations are caused by one planet" to the probability of $M_0 \equiv$ "the radial velocity variations are consistent with noise." From Bayes' theorem, we can write

$$p(M_1|D, I) = \frac{p(M_1|I)\, p(D|M_1, I)}{p(D|I)} = C\, p(M_1|I)\, p(D|M_1, I), \qquad (12.31)$$

where

$$p(D|M_1, I) = \int dP \int dK \int dV \int de \int d\chi \int d\omega \int db\, p(P, K, V, e, \chi, \omega, b|M_1, I) \times p(D|M_1, P, K, V, e, \chi, \omega, b, I). \qquad (12.32)$$

The joint prior for the model parameters, assuming independence, is given by

$$p(P, K, V, e, \chi, \omega, b|M_1, I) = \frac{1}{P \ln\left(\frac{P_{max}}{P_{min}}\right)} \frac{1}{K \ln\left(\frac{K_{max}}{K_{min}}\right)} \frac{1}{(V_{max} - V_{min})}$$
$$\times \frac{1}{(e_{max} - e_{min})} \frac{1}{2\pi} \frac{1}{b \ln\left(\frac{b_{max}}{b_{min}}\right)}. \qquad (12.33)$$

Note: we have assumed a uniform prior for χ in the range 0 to 1, so $p(\chi|M_1, I) = 1$.

$$p(D|M_1, P, K, V, e, \chi, \omega, b, I) = Ab^{N/2} \times \exp\left[-\frac{b}{2} \sum_{i=1}^{N} \frac{(v_i - f_i)^2}{s_i^2}\right], \qquad (12.34)$$

where

$$A = (2\pi)^{-N/2} \left[\prod_{i=1}^{N} s_i^{-1}\right]. \qquad (12.35)$$

For the purposes of estimating the model parameters, we will assume a prior uncertainty in b in the range $b_{min} = 0.1$ and $b_{max} = 2$.

When it comes to comparing the probability of M_1 to M_0, or to a model which assumes there are two planets present, we will set $b = 1$ and perform the model comparison based on the errors quoted in Tinney *et al.*, (2003). The probability of M_0 is given by

$$p(M_0|D,I) = C\ p(M_0|I)p(D|M_0,I), \tag{12.36}$$

where

$$
\begin{aligned}
p(D|M_0,I) &= \int db \int dV\ p(V,b|D,M_0,I) \\
&= \int db \int dV\ p(V,b|M_0,I)\ p(D|M_0,V,b,I),
\end{aligned}
\tag{12.37}
$$

$$p(V|M_0,I) = \frac{1}{(V_{max} - V_{min})}\frac{1}{b\ln\left(\frac{b_{max}}{b_{min}}\right)}, \tag{12.38}$$

and

$$p(D|M_0,V,b,I) = (2\pi)^{-N/2}\left[\prod_{i=1}^{N} s_i^{-1}\right]b^{\frac{N}{2}}\exp\left[-\frac{b}{2}\sum_{i=1}^{N}\frac{(v_i - V)^2}{s_i^2}\right]. \tag{12.39}$$

The integral over V in Equation (12.37) can be performed analytically yielding

$$
\begin{aligned}
p(D|M_0,I) = A\sqrt{\frac{\pi}{2}}W^{-1/2}\int db\ b^{\frac{N-3}{2}}\exp\left[-\frac{bW}{2}\sum_{i=1}^{N}(\overline{v_w^2} - (\overline{v_w})^2)^2\right] \\
\times\ [\mathrm{erf}(u_{max}) - \mathrm{erf}(u_{min})],
\end{aligned}
\tag{12.40}
$$

where

$$\overline{v_w} = \sum_{i=1}^{N} w_i\ v_i, \tag{12.41}$$

$$\overline{v_w^2} = \sum_{i=1}^{N} w_i\ v_i^2, \tag{12.42}$$

$$w_i = 1/s_i^2, \tag{12.43}$$

$$W = \sum_{i=1}^{N} w_i, \tag{12.44}$$

$$u_{max} = \left(\frac{bW}{2}\right)^{-1/2}(V_{max} - \overline{v_w}), \qquad (12.45)$$

$$u_{min} = \left(\frac{bW}{2}\right)^{-1/2}(V_{min} - \overline{v_w}). \qquad (12.46)$$

In conclusion, Equations (12.31) and (12.34) are required for the parameter estimation part of the problem, and Equations (12.32) and (12.40) answer the model selection part of the problem. Equation (12.32) is evaluated from the results of the parallel tempering chains according to the method discussed in Section 12.7.

12.9.2 Results

The APT MCMC algorithm described in Section 12.8 was used to re-analyze the measurements of Tinney *et al.* (2003). Figures 12.12 and 12.13 show the diagnostic information output by the MCMC control system for two runs of the APT MCMC algorithm that use different starting values for the parameters and different starting values for the proposal σ's. The top left panel shows the evolution of the control system error for 100 000 iterations. Even for the best set of $\{\sigma_\alpha\}$, the control system error will exhibit statistical fluctuations of order $\sqrt{\lambda n_1}$ which will result in fluctuations of $\{\sigma_\alpha\}$ throughout the run. Recall, $\lambda =$ the target acceptance fraction and $n_1 =$ the number of iterations in major cycles (see Section 12.8). These fluctuations are of no consequence since the equilibrium distribution of parameter values is insensitive to small fluctuations in $\{\sigma_\alpha\}$. To reduce the time spent in perturbing $\{\sigma_\alpha\}$ values, we set a threshold on the control system error of $1.5\sqrt{\lambda n_1}$. When the error is less than this value no minor cycles are executed. Normally, the control system is disabled the first time the error is $<1.5\sqrt{\lambda n_1}$. This was not done in the two examples shown in order to illustrate the behavior of the control system and evolution of the $\{\sigma_\alpha\}$. For the two runs, the error drops to a level consistent with the minimum threshold set for initiating a change in $\{\sigma_\alpha\}$ in 8000 and 9600 iterations, respectively. The other six panels exhibit the evolution of the $\{\sigma_\alpha\}$ to relatively stable values. Table 12.1 compares the starting and final values for two APT MCMC runs with a set of $\{\sigma_\alpha\}$ values arrived at manually. The starting parameter values for the two APT MCMC runs are shown in Table 12.2. Control system parameters were: $sc_{max} = 0.1$, $n_1 = 1000$, $n_2 = 100$, $\lambda = 0.25$, and a damping factor, $\gamma = 1.6$. sc_{max} specifies the maximum scaling of $\{\sigma_\alpha\}$ to be used in a minor cycle. Tempering β values used were $\{0.01, 0.2575, 0.505, 0.7525, 1\}$. β values are chosen to give $\simeq 50\%$ swap acceptance between adjacent levels.

Figure 12.14 shows the iterations of the six model parameters, P, K, V, e, χ, ω, for the 100 000 iterations of APT MCMC 1. Only every 100th value is plotted. The plot for K shows clear evidence that parallel tempering is doing its job, enabling regions of significant probability to be explored apart from the biggest peak region. A conservative burn-in period of 8000 samples was arrived at from an examination of the

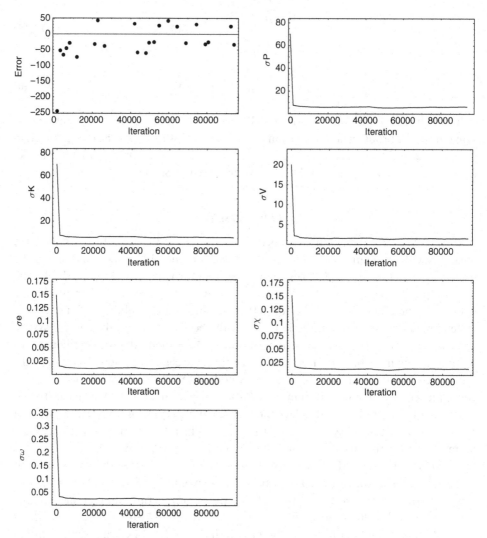

Figure 12.12 The upper left panel shows the evolution of the APT MCMC control system error versus iteration number for the first run. The other six panels exhibit the evolution of the Gaussian parameter proposal distribution σ's.

control system error, shown in the upper left panel of Figure 12.12, and the parameter iterations of Figure 12.14.

The level of agreement between two different MCMC runs can be judged from a comparison of the marginal distributions of the parameters. Figures 12.15 and 12.16 show the posterior marginals for the six model parameters, P, K, V, e, χ, ω, and the noise scale parameter b for APT MCMC 1 and APT MCMC 2, respectively. The final

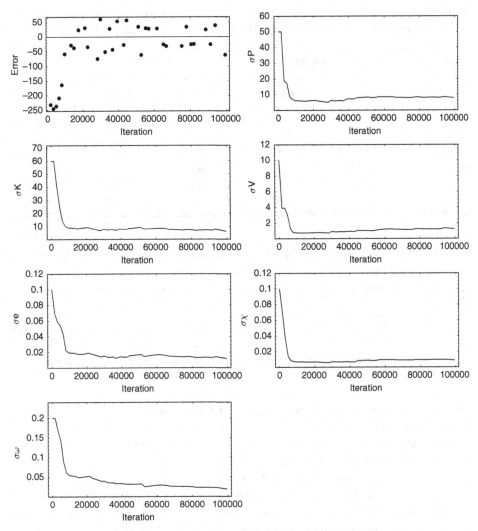

Figure 12.13 The upper left panel shows the evolution of the APT MCMC control system error versus iteration number for the second run. The other six panels exhibit the evolution of the Gaussian parameter proposal distribution σ's.

model parameter values are given in Table 12.3, along with values of $a \sin i$, $M \sin i$, and the Julian date of periastron passage, that were derived from the parameter values.

$$a \sin i(\text{km}) = 1.38 \times 10^5 KP\sqrt{1 - e^2}, \tag{12.47}$$

where K is in units of m s^{-1} and P is in days.

$$M \sin i = 4.91 \times 10^{-3}(M_\ast)^{2/3} KP^{1/3}\sqrt{1 - e^2}, \tag{12.48}$$

Table 12.1 *Comparison of the starting and final values of proposal distribution σ's for two automatic parallel tempering MCMC runs, to manually derived values.*

Proposal	APT MCMC 1		APT MCMC 2		Manual
σ	Start	Final	Start	Final	Final
σP (days)	70	6.2	50	7.8	10
σK(m s^{-1})	70	5.7	60	6.0	5
σV(m s^{-1})	20	1.5	10	1.2	2
σe	0.15	0.012	0.1	0.012	0.005
$\sigma \chi$	0.15	0.013	0.1	0.009	0.007
$\sigma \omega$	0.3	0.023	0.2	0.019	0.05

Table 12.2 *Starting parameter values for the two automatic parallel tempering MCMC runs.*

Trial	P	K	V	e	χ	ω	b
1	950	80	−2	0.4	0.0	0.0	1.0
2	1300	250	5	0.2	0.0	0.0	1.0

where M is the mass of the planet measured in Jupiter masses, and M_* is the mass of the star in units of solar masses.

One important issue concerns what summary statistic to use to represent the best estimate of the parameter values. We explore the question of a suitable robust summary statistic further in Section 12.10. In Table 12.3, the final quoted parameter values correspond to the MAP values. The median values are shown in brackets below. The error bars correspond to the boundaries of the 68.3% credible region of the marginal distribution. The MAP parameter values for APT MCMC 1 were used to construct the model plotted in panel (a) of Figure 12.17. The residuals are shown in panel (b).

Figure 12.18 shows the posterior probability distributions for $a \sin i$, $M \sin i$, and the Julian date of periastron passage, that are derived from the MCMC samples of the orbital parameters.

The Bayes factors, $p(D|M_1, I)/p(D|M_0, I)$, determined from the two APT MCMC runs were 1.4×10^{14} and 1.6×10^{14}. Clearly, both trials overwhelmingly favor M_1 over M_0.

The upper panel of Figure 12.19 shows a comparison of the marginal and projected probability density functions for the velocity amplitude, K, derived from the APT MCMC parameter samples. To understand the difference, it is useful to examine the strong correlation that is evident between K and orbital eccentricity in the lower panel.

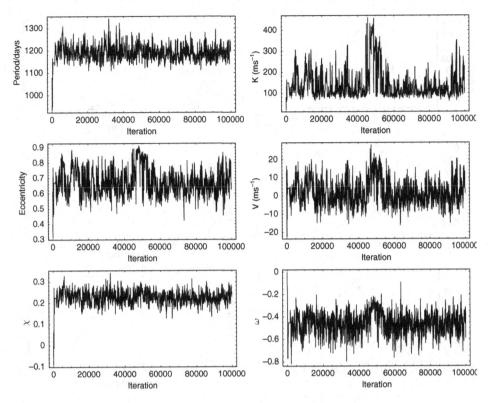

Figure 12.14 The figure shows every 100th APT MCMC iteration of the six model parameters, P, K, V, e, χ, ω.

Not only is the density of samples much higher at low K values, but the characteristic width of the K sample distribution is also much broader, giving rise to an enhancement in the marginal beyond that seen in the projected.

Finally, even though the 68.3% credible region contains $b = 1$, we decided to analyze the best-fit residuals, shown in the lower panel of Figure 12.17, to see what probability theory had to say about the evidence for another planet.[4] The APT MCMC program was re-run on the residuals to look for evidence of a second planet in the period range 2 to 500 days, $K = 1$ to 40 m s^{-1}, $V = -10$ to 10 m s^{-1}, $e = 0$ to 0.95, $\chi = 0$ to 1, and $\omega = -\pi$ to π. The most probable orbital solution had a period of 11.90 ± 0.02 days, $K = 18^{+9}_{-15}$ m s^{-1}, $V = -2.7^{+2.4}_{-1.6}$ m s^{-1}, eccentricity $= 0.626^{+0.16}_{-0.18}$, $\omega = 156^{+2}_{-4}$ deg, periastron passage $= 1121 \pm 1$ days (JD $-2,450,000$), and an $M \sin i = 0.14^{+0.07}_{-0.04}$. Figure 12.20 shows this orbital solution overlaid on the residuals for two cycles of phase. Note: the second cycle is just a repeat of the first. The computed Bayes factor $p(D|M_2, I)/p(D|M_1, I) = 0.7$. Assuming *a priori* that

---

[4] Note: a better approach would be to fit a two-planet model to the original radial velocity data.

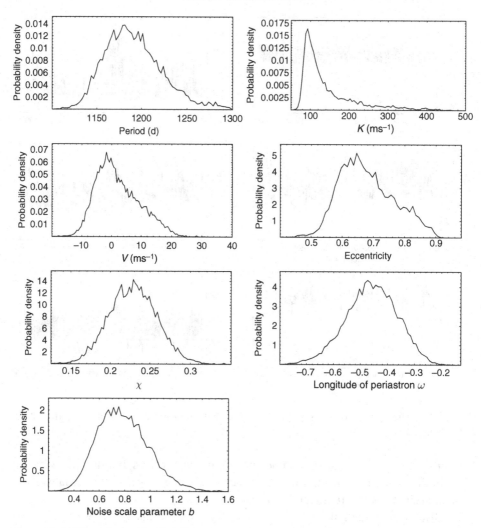

Figure 12.15 The marginal probabilities for the six model parameters, P, K, V, e, χ, ω, and the noise scale parameter b for the run APT MCMC 1.

$p(M_2|I) = p(M_1|I)$, this result indicates that it is more probable that the orbital solution for the residuals arises from fitting a noise feature than from the existence of a second planet. Thus, there is insufficient evidence at this time to claim the presence of a second planet.

12.10 MCMC robust summary statistic

In the previous section, the best estimate of each model parameter is based on the maximum *a posteriori* (MAP) value. It has been argued, e.g., Fox and Nicholls (2001),

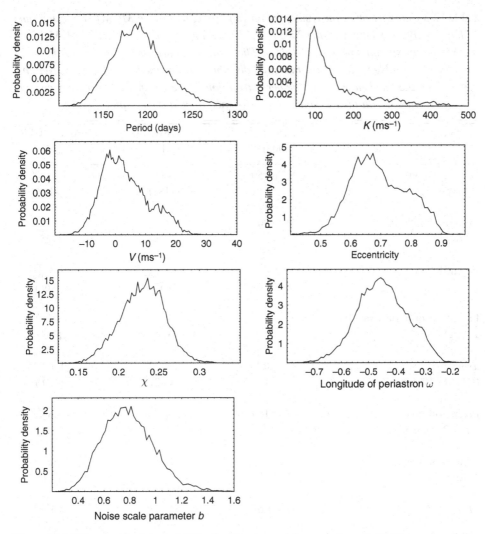

Figure 12.16 The marginal probabilities for the six model parameters, P, K, V, e, χ, ω, and the noise scale parameter b for the run APT MCMC 2.

that MAP values are sometimes unrepresentative of the bulk of the posterior probability. Fox and Nicholls were considering the reconstruction of degraded binary images. The current problem is very different but the issue remains the same: what choice of summary statistic to use? Two desirable properties are: a) that it be representative of the marginal probability distribution, and b) the set of summary parameter values provides a good fit to the data. Here, we consider three other possible choices of summary statistic. They are the mean, the median, and the marginal posterior mode (MPM), all of which satisfy point (a). In repeated APT MCMC runs, it was found that

Table 12.3 *Comparison of the results from two parallel tempering MCMC Bayesian runs with the analysis of Tinney et al. (2003). The values quoted for the two APT MCMC runs are MAP (maximum a posterior) values. The error bars correspond to the boundaries of the 68.3% credible region of the marginal distribution. The median values are given in brackets on the line below. Note: the periastron time and error quoted by Tinney et al. is identical with their P value and is assumed to be a typographical error.*

Parameter	Tinney *et al.* (2003)	APT MCMC 1	APT MCMC 2
Orbital period P (days)	1183 ± 150	1188^{+28}_{-35} (1188)	1177^{+36}_{-21} (1188)
Velocity amplitude K (m s^{-1})	130 ± 20	106^{+46}_{-29} (115)	116^{+56}_{-39} (125)
Eccentricity e	0.67 ± 0.1	$0.63^{+0.12}_{-0.06}$ (0.67)	$0.65^{+0.15}_{-0.06}$ (0.68)
Longitude of periastron ω (deg)	333 ± 15	333^{+6}_{-5} (334)	332^{+8}_{-3} (334)
$a \sin i$ (units of 10^6 km)	1.56 ± 0.3	$1.35^{+0.4}_{-0.3}$ (1.42)	$1.4^{+0.4}_{-0.4}$ (1.66)
Periastron time $(JD - 2,450,000)$	1183 ± 150	864^{+18}_{-58} (845)	856^{+52}_{-28} (844)
Systematic velocity V (m s^{-1})		-0.7^{+8}_{-6} (0.8)	1.4^{+7}_{-7} (2.1)
$M \sin i$ (M_J)	4.9 ± 1.0	$4.2^{+1.2}_{-1.0}$ (4.5)	$4.5^{+1.3}_{-1.3}$ (4.7)
RMS about fit	15	13.8 (14.1)	14.0 (14.0)

the MPM solution provided a relatively poor fit to the data, while the mean was somewhat better, and in all cases, the median provided a good fit – almost as good as the MAP fits. One example of the fits is shown in Figure 12.21. The residuals were as follows: (a) 14.0 m s^{-1} (MAP), (b) 16.1 (mean), (c) 18.7 (MPM), and (d) 14.0 (median).

In the previous example the Bayes factor favored the one-planet model, M_1, compared to the no-planet model, M_0, by a factor of approximately 10^{14}. It is also

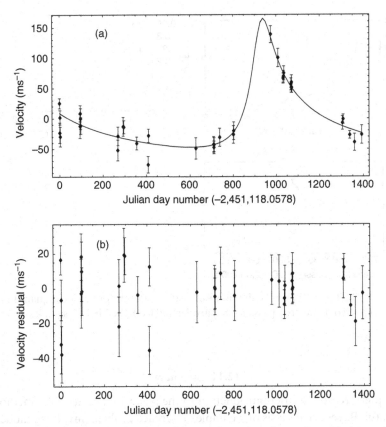

Figure 12.17 Panel (a) shows the raw data with error bars plotted together with the model radial velocity curve using the MAP (maximum *a posteriori*) summary statistic. Panel (b) shows the radial velocity residuals.

interesting to compare the four different summary statistics in the case where the Bayes factor is close to 1, as we found for the toy spectral line problem in Section 12.7, i.e., neither model is preferred. Figure 12.22 shows a comparison of the fits obtained using (a) the MAP, (b) the mean, (c) the MPM, and (d) the median. Both the MAP and median summary statistic placed the model line at the actual location of the simulated spectral line (channel 37). The MAP achieved a slightly lower RMS residual (RMS = 0.87) compared to the median (RMS = 0.89). The mean statistic performed rather poorly and the MPM not much better.

The conclusion, based on the current studies, is that the median statistic provides a robust alternative to the common MAP statistic for summarizing the posterior distribution. Unfortunately, the median was not one of the statistics considered in Fox and Nicholls (2001).

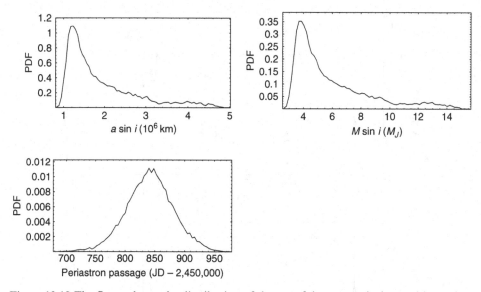

Figure 12.18 The figure shows the distribution of three useful astronomical quantities; $a \sin i$, $M \sin i$ and epoch of periastron passage, that are derived from the MCMC samples of the orbital parameters.

12.11 Summary

This chapter provides a brief introduction to the powerful role MCMC methods can play in a full Bayesian analysis of a complex inference problem involving models with large numbers of parameters. We have only demonstrated their use for models with a small to a moderate number of parameters, where they can easily be compared with results from other methods. These comparisons will provide a useful introduction and calibration of these methods for readers wishing to handle more complex problems. For the examples considered, the median statistic proved to be a robust alternative to the common MAP statistic for summarizing the MCMC posterior distribution.

The most ambitious topic treated in this chapter dealt with an experimental new algorithm for automatically annealing the σ values for the parameter proposal distributions in a parallel tempering Markov chain Monte Carlo (APT MCMC) calculation. This was applied to the analysis of a set of astronomical data used in the detection of an extrasolar planet. Existing analyses are based on the use of nonlinear least-squares methods which typically require a good initial guess of the parameter values (see Section 11.5). Frequently, the first indication of a periodic signal comes from a periodogram analysis of the data. As we show in the next chapter, a Bayesian analysis based on prior information of the shape of the periodic signal can frequently do a better job of detection than the ordinary Fourier power spectrum, otherwise known as the Schuster periodogram. In the extrasolar planet Kepler problem, the mathematical form of the signal is well known and is built into the Bayesian analysis. The APT

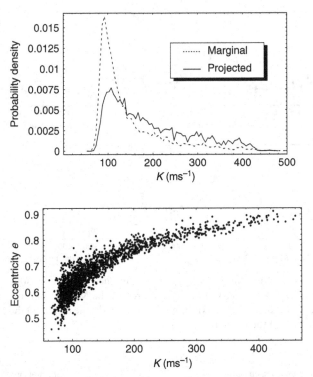

Figure 12.19 The upper panel shows a comparison of the marginal and projected probability density functions for the velocity amplitude, K. The lower panel illustrates the strong correlation between K and orbital eccentricity.

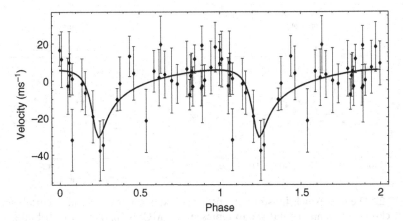

Figure 12.20 The figure shows the most probable orbital solution to the data residuals (for two cycles of phase), after removing the best fitting model of the first planet.

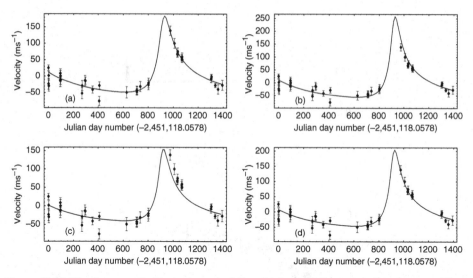

Figure 12.21 The four panels illustrate typical fits obtained in the extrasolar planet problem using different choices of summary statistic to represent the MCMC parameter distributions. They correspond to: (a) the MAP (maximum *a posteriori*), (b) the mean, (c) the MPM (marginal posterior mode), and (d) the median.

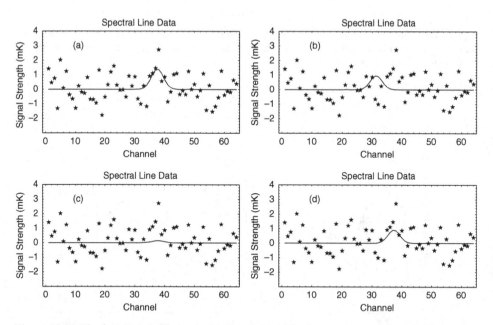

Figure 12.22 The four panels illustrate the fits obtained in the toy spectral line problem using different choices of summary statistic to represent the MCMC parameter distributions. They correspond to: (a) the MAP (maximum *a posteriori*), (b) the mean, (c) the MPM (marginal posterior mode), and (d) the median.

MCMC algorithm implemented in Section 12.9 is thus effective for both detecting and characterizing the orbits of extrasolar planets. Another advantage is that a good initial guess of the orbital parameter values is not required, which allows for the earliest possible detection of a new planet. Moreover, the built-in Occam's razor in the Bayesian analysis can save a great deal of time in deciding whether a detection is believable.

Finally, it is important to remember that the MCMC techniques described in this chapter are basically tools to allow us to evaluate the integrals needed for a full Bayesian analysis of some problem of interest. The APT MCMC algorithm discussed in the context of the extrasolar planet problem can readily be modified to tackle other very different problems.

12.12 Problems

1. In Section 12.6, we used both the Metropolis–Hastings and parallel tempering (PT) versions of MCMC to re-analyze the toy spectral line problem of Section 3.6. A program to perform the PT calculations is given in the Markov chain Monte Carlo section of the *Mathematica* tutorial. Use this program to analyze the spectrum given in Table 12.4, for $n = 20\,000$ to $50\,000$ iterations, depending on the speed of your computer (try it out first with only one tempering level, i.e., $\beta = 1$). As part of your solution, include figures like 12.7 and 12.8, and compute the quasi–Monte Carlo estimate of the Bayes factor, used to compare the two competing models. Explain how you arrived at your choice for the number of burn-in samples.

   The prior information is the same as that assumed in Section 12.6. Theory predicts the spectral line has a Gaussian shape with a line width $\sigma_L = 2$ frequency channels, and line center between channels 1 and 44. The noise in each channel is known to be Gaussian with a $\sigma = 1.0\,\text{mK}$ and the spectrometer output is in units of mK.

2. Repeat the analysis of problem 1 only this time assume the line width is also uncertain. Adopt a uniform prior for the line width (σ_L), with upper and lower bounds of 0.5 and 4 frequency channels, respectively. You will need to modify the parallel tempering MCMC program to allow for the addition of the line width parameter. You will also need to specify prior bounds on σ_L and a starting σ for the line width in the Gaussian proposal distribution (typically 10% of the prior range). Your solution should include a plot of the marginal probability distribution for each of the three parameters and a calculation of the Bayes factor for comparing the two models. Justify your choice for the number of burn-in samples.

3. (Difficult problem) In Section 11.6, we illustrated the solution of a simple nonlinear model fitting problem using *Mathematica*'s **NonlinearRegress**, which implements the Levenberg–Marquardt method. In this problem we want to analyze the same spectral line data (Table 12.5) using the experimental APT MCMC software given in the *Mathematica* tutorial and discussed in Section 12.8. It will yield a fully Bayesian solution to the problem without the need to assume the asymptotic normal approximation, or, assume the Laplacian approximations for computing the Bayes factor and marginals. In general, MCMC solutions come into their own for

Table 12.4 *Spectral line data consisting of 64 frequency channels obtained with a radio astronomy spectrometer. The output voltage from each channel has been calibrated in units of effective black body temperature expressed in mK. The existence of negative values arises from receiver channel noise which gives rise to both positive and negative fluctuations.*

ch. #	mK	ch. #	mK	ch. #	mK	ch. #	mK
1	0.82	17	−0.90	33	−0.03	49	−0.72
2	−2.07	18	0.33	34	1.47	50	0.38
3	0.38	19	0.80	35	1.70	51	0.02
4	0.99	20	−1.42	36	1.89	52	−1.26
5	−0.12	21	0.28	37	4.55	53	1.35
6	−1.35	22	−0.42	38	3.59	54	−0.04
7	−0.20	23	0.12	39	2.02	55	−1.45
8	0.36	24	0.14	40	0.21	56	1.48
9	0.78	25	−0.63	41	0.05	57	−1.16
10	1.01	26	−1.77	42	0.54	58	−0.40
11	0.44	27	−0.67	43	−0.09	59	0.01
12	0.34	28	0.55	44	−0.61	60	0.29
13	1.58	29	1.98	45	2.49	61	−1.35
14	0.08	30	−0.08	46	0.07	62	−0.21
15	0.38	31	1.16	47	−1.45	63	−1.67
16	−0.71	32	0.48	48	0.56	64	0.70

Table 12.5 *Spectral line data consisting of 51 pairs of frequency and signal strength (mK) measurements.*

f	mK	f	mK	f	mK	f	mK
0.00	0.86	1.56	0.97	3.12	1.95	4.68	1.39
0.12	1.08	1.68	0.97	3.24	1.75	4.80	0.64
0.24	0.70	1.80	1.06	3.36	2.03	4.92	0.79
0.36	1.16	1.92	0.85	3.48	1.42	5.04	1.27
0.48	0.98	2.04	1.94	3.60	1.06	5.16	1.17
0.60	1.32	2.16	2.34	3.72	0.79	5.28	1.23
0.72	1.05	2.28	3.55	3.84	1.11	5.40	1.23
0.84	1.17	2.40	3.53	3.96	0.88	5.52	0.71
0.96	0.96	2.52	4.11	4.08	0.88	5.64	0.71
1.08	0.86	2.64	3.72	4.20	0.68	5.76	0.80
1.20	1.12	2.76	3.52	4.32	1.39	5.88	1.16
1.32	0.79	2.88	2.78	4.44	0.62	6.00	1.12
1.44	0.86	3.00	3.03	4.56	0.80		

higher dimensional problems but it is desirable to gain experience working with simpler problems.

Modify the APT MCMC software to analyze this data for the two models described in Section 11.6.

In *Mathematica*, model 1 has the form:

model$[a0\_, a1\_, f1\_] := a0 + a1$ **line**$[f1]$

where

line$[f\,1\_] := \dfrac{\mathrm{Sin}[2\pi(f-f1)/\Delta f]}{2\pi(f-f1)/\Delta f}$ and $\Delta f = 1.5$.

Model 2 has the form:

model$[a0\_, a1\_, a2\_, f1\_, f2\_] := a0 + a1$ **line**$[f1] + a2$ **line**$[f2]$,

where $f\,2$ is assumed to be the higher frequency line.

Adopt uniform priors for all parameters and assume a prior range from 0 to 10 for $a0$, $a1$ and $a2$. For the frequency parameters use a prior range from 1.0 to 5.0. Note: for a multi-spectral line model, each spectral peak is free to move through the full frequency range which can result in the occurrence of degenerate peaks in the joint posterior. It is therefore necessary to redefine the parameters after the MCMC iterations are terminated, in such a way that $a1$ and $f1$ correspond to the lowest frequency spectral line parameters, etc. Please refer to section 5 of a recent paper that discusses a suitable multi-frequency prior (P. C. Gregory, *Monthly Notices of the Royal Astronomical Society* **374**, p. 1321, 2007).

13

Bayesian revolution in spectral analysis[1]

13.1 Overview

Science is all about identifying and understanding organized structures or patterns in nature. In this regard, periodic patterns have proven especially important. Nowhere is this more evident than in the field of astronomy. Periodic phenomena allow us to determine fundamental properties like mass and distance, enable us to probe the interior of stars through the new techniques of stellar seismology, detect new planets, and discover exotic states of matter like neutron stars and black holes. Clearly, any fundamental advance in our ability to detect periodic phenomena will have profound consequences in our ability to unlock nature's secrets. The purpose of this chapter is to describe advances that have come about through the application of Bayesian probability theory,[2] and provide illustrations of its power through several examples in physics and astronomy. We also examine how non-uniform sampling can greatly reduce some signal aliasing problems.

13.2 New insights on the periodogram

Arthur Schuster introduced the *periodogram* in 1905, as a means for detecting a periodicity and estimating its frequency. If the data are evenly spaced, the periodogram is determined by the Discrete Fourier Transform (DFT), thus justifying the use of the DFT for such detection and measurement problems. In 1965, Cooley and Tukey introduced the Fast Discrete Fourier Transform (FFT), a very efficient method of implementing the DFT that removes certain redundancies in the computation and greatly speeds up the calculation of the DFT. A detailed treatment of the DFT and FFT is given in Appendix B.

The Schuster periodogram was introduced largely for intuitive reasons, but in 1987, Jaynes provided a formal justification by applying the principles of Bayesian inference

---

[1] The term "spectral analysis" has been used in the past to denote a wider class of problems than will be considered in this chapter. For a brief introduction to stochastic spectrum estimation, see Appendix B.13.4.

[2] The first three sections of this chapter are a revised version of an earlier paper by the author (Gregory, 2001), which is reproduced here with the permission of the American Institute of Physics.

as follows: suppose we are analyzing data consisting of samples of a continuous function contaminated with additive independent Gaussian noise with a variance of σ^2. Jaynes showed that, presuming the possible periodic signal is sinusoidal (but with unknown amplitude, frequency, and phase), the Schuster periodogram exhausts all the information in the data relevant to assessing the possibility that a signal is present, and to estimating the frequency and amplitude of such a signal. The periodogram is essentially the squared magnitude of the FFT and can be defined as

$$\text{periodogram} = C(f_n) = \frac{1}{N}\left|\sum_{k=0}^{N-1} d_k\, e^{i2\pi n \Delta f k T}\right|^2 \tag{13.1}$$

$$= \frac{1}{N}|FFT|^2.$$

In an FFT, the frequency interval, $\Delta f = 1/T$, where T is the duration of the data set. The quantity $C(f_n)$ is indeed fundamental to spectral analysis but not because it is itself a satisfactory spectrum estimator. Jaynes showed that the probability for the frequency of a periodic sinusoidal signal is given approximately by[3]

$$p(f_n|D, I) \propto \exp\left\{\frac{C(f_n)}{\sigma^2}\right\}. \tag{13.2}$$

Thus, the proper algorithm to convert $C(f_n)$ to $p(f_n|D, I)$ involves first dividing $C(f_n)$ by the noise variance and then exponentiating. This naturally suppresses spurious ripples at the base of the periodogram, usually accomplished with linear smoothing; but does it by attenuation rather than smearing, and therefore does not sacrifice any precision. The Bayesian nonlinear processing of $C(f_n)$ also yields, when the data give evidence for them, arbitrarily sharp spectral peaks. Since the peak in $p(f_n|D, I)$ can be much sharper than the peak in $C(f_n)$, it is necessary to zero pad the FFT to obtain a sufficient density of points in $C(f_n)$ for use in Equation (13.2) to accurately define a peak in $p(f_n|D, I)$.

Figure 13.1 provides a demonstration of these properties for a simulated data set consisting of a single sine wave plus additive Gaussian noise given by Equation (13.3).

$$y = A\cos 2\pi ft + \text{Gaussian noise (mean} = 0, \sigma = 1), \tag{13.3}$$

where $A = 1, f = 0.1$ Hz. The upper panel shows 64 simulated data points computed from Equation (13.3), with one-σ error bars. The middle panel is the Fourier power spectral density or periodogram, computed for this data according to Equation (13.1).[4] The sinusoidal signal is clearly indicated by the prominent peak.

---

[3] Bretthorst (2000) derives the exact result for non-uniformly sampled data which involves an analogous nonlinear transformation of the Lomb–Scargle periodogram (Lomb, 1976; Scargle, 1982, 1989). Bretthorst (2001) also shows how to generalize the Lomb–Scargle periodogram for the case of a non-stationary sinusoid. This is discussed further in Section 13.5.

[4] The 64 points were zero padded to provide a total of 512 points for the FFT. See Appendix B.11 for more details on zero padding.

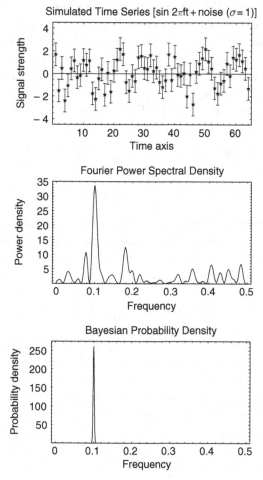

Figure 13.1 Comparison of conventional (middle panel) and Bayesian analysis (lower panel) of a simulated time series (upper panel).

The *signal-to-noise ratio* (S/N), defined as the ratio of the RMS signal amplitude to the noise σ, was 0.7 in the above simulation. If we repeated the simulation with a larger S/N ratio, the main peak would increase in relation to the noise peaks and we would start to notice side lobes emerging associated with the finite duration of the data set (rectangular window function, see Appendix B.7.1). However, a well-known property of the periodogram is that the width of any spectral peak depends only on the duration of the data set and not on the signal-to-noise level. Various methods have been used to determine the accuracy to which the peak frequency can be determined, but, as we see below, the Bayesian posterior probability for the signal frequency provides this information directly.

The lower panel of Figure 13.1 shows the Bayesian probability density for the period of the signal, derived from Equation (13.2). As the figure demonstrates, the spurious noise features are suppressed and the width of the spectral peak is much narrower than the peak in the periodogram. In a Bayesian analysis, the width of spectral peak, which reflects the accuracy of the frequency estimate, is determined by the duration of the data, the S/N, and the number of data points. More precisely, the standard deviation of the spectral peak, δf, for a S/N > 1, is given by

$$\delta f \approx \left(1.6\frac{S}{N}T\sqrt{N}\right)^{-1} \text{Hz},\qquad(13.4)$$

where T = the data duration in s, and N = the number of data points in T. To improve the accuracy of the estimate, the two most important factors are how long we sample (the T dependence) and the signal-to-noise ratio.

Equation (13.2) assumes that the noise variance is a known quantity. In some situations, the noise is not well understood, i.e., our state of knowledge is less certain. Even if the measurement apparatus noise is well understood, the data may contain a greater complexity of phenomena than the current signal model incorporates. In such cases, Equation (13.2) is no longer relevant, but again, Bayesian inference can readily handle this situation by treating the noise variance as a nuisance parameter with a prior distribution reflecting our uncertainty in this parameter. We saw how to do that when estimating the mean of a data set in Section 9.2.3. The resulting posterior can be expressed in the form of a Student's t distribution. The corresponding result for estimating the frequency of a single sinusoidal signal (Bretthorst, 1988) is given approximately[5] by

$$p(f_n|D, I) \propto \left[1 - \frac{2C(f_n)}{N\overline{d^2}}\right]^{\frac{2-N}{2}},\qquad(13.5)$$

where N is the number of data values and $\overline{d^2} = (1/N)\sum_i d_i^2$ is the mean square average of the data values. The analysis assumes any DC component in the data has been removed. If σ is not well known, then it is much safer to use Equation (13.5) than Equation (13.2) because Equation (13.5) will treat anything that cannot be fitted by the model as noise. This leads to more conservative estimates.

A corollary of Jaynes' analysis is that for any other problem (e.g., non-sinusoidal light curve, non-Gaussian noise, or non-uniform sampling) use of the FFT is not optimal; more information can be extracted from the data if we use more sophisticated statistics. Jaynes made this point himself, and it has been amply demonstrated in the work of Bretthorst (1988), who has applied similar methods to signal detection and estimation problems with non-sinusoidal models with Gaussian noise probabilities.

---

[5] Note: Equations (13.2) and (13.5) do not require the data to be uniformily sampled provided that: 1) the number of data values N is large, 2) there is no constant (DC) component in the data, and 3) there is no evidence of a low frequency.

In the following sections, we will consider two general classes of spectral problems: (a) those for which we have strong prior information of the signal model, and (b) those for which we have no specific prior information about the signal.

13.2.1 How to compute $p(f|D,I)$

In Section 13.2, we saw that the periodogram, $C(f_n)$, follows naturally from Bayesian probability theory[6] when our prior information indicates there is a single sine wave present in the data and we want to compute the $p(f_n|D, I)$. Equation (13.2) gives the relationship between $p(f_n|D, I)$ and $C(f_n)$ if the noise σ is known, and Equation (13.5) applies when the noise σ is unknown. The value of the periodogram at a set of discrete frequencies, indexed by n, is given by

$$
\begin{aligned}
C(f_n) &= \frac{1}{N} \left| \sum_{j=1}^{N} d_j e^{i2\pi f_n t_j} \right|^2 \\
&= \frac{1}{N} |\text{FFT}|^2 \\
\text{or } C(n) &= \frac{1}{N} \left| \sum_{j=1}^{N} d_j e^{i2\pi \frac{(n-1)(j-1)}{N}} \right|^2 \\
&= \frac{|H_n|^2}{N},
\end{aligned}
\tag{13.6}
$$

where H_n is the FFT or DFT transform defined by Equations (B.49) and (B.55) in Appendix B.

We illustrate the calculations in more detail by comparing the Bayesian $p(n|D, I)$ to the one-sided PSD (given in Appendix B.13.2) for two simulated time series shown in Figure 13.2. The time series consist of 64 samples at one-second intervals of

$$
d_j = A \cos 2\pi f t_j + \text{Gaussian noise } (\sigma = 1).
\tag{13.7}
$$

The simulated data for two different choices of signal amplitude ($A = 0.8$ and $A = 10$), corresponding to low and high signal-to-noise ratios, are shown in the two top panels of Figure 13.2. In the computation of the FFT, we take N time samples at intervals of T seconds and compute N transform points H_n. The value, $n = 0$, corresponds to the FFT at zero frequency, and $n = N/2$ to the value at the Nyquist frequency $= 1/(2T)$. Values of n between $N/2 + 1$ to $N - 1$ correspond to values of the FFT for negative frequencies. In Appendix B.13.1 we show that $|H_n|^2 T/N$ is the two-sided PSD (two-sided periodogram) with units of power Hz^{-1}.

---

[6] In general, for a different signal model or noise model, Bayesian inference will lead to an equation involving a different function or statistic of the data for computing the probability of the signal frequency.

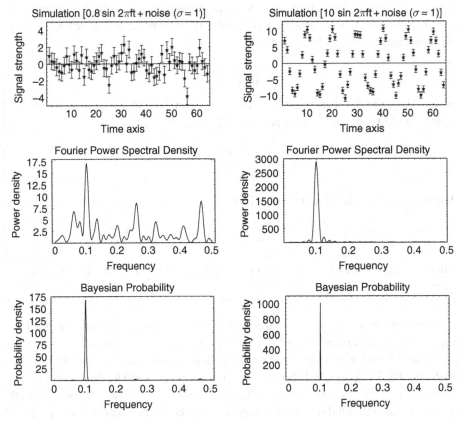

Figure 13.2 Comparison of conventional (middle panels) and Bayesian analysis (bottom panels) of two simulated time series (top panels).

In the computation of the Bayesian $p(n|D, I)$, $C(n)$ is just the positive frequency part of $|H_n|^2/N$.

$$C(n) = \frac{|H_n|^2}{N} \quad \text{for } n = 0, 1, \ldots, \frac{N}{2}. \tag{13.8}$$

Both $C(n)$ and σ^2 have units of power and thus their ratio is dimensionless.

In general, $p(n|D, I)$ will be very narrow when $C(n)/\sigma^2 > 1$ because of the exponentiation occurring in Equations (13.2) or (13.5). Thus, to accurately define $p(n|D, I)$ we need to zero pad the FFT to obtain a sufficient density of H_n points to accurately define the $p(n|D, I)$ peak. Zero padding is used to obtain higher frequency resolution in the transform and is discussed in detail in Appendix B.11. In the zero padding case, Equation (13.8) becomes

$$C(n) = \frac{|H_n|^2}{N_{\text{orig}}} \quad \text{for } n = 0, 1, \ldots, \frac{N_{\text{zp}}}{2}, \tag{13.9}$$

where N_{orig} is the number of original time series samples and N_{zp} is the total number of points including the added zeros. For analysis of the time series in Figure 13.2, we zero padded to produce $N_{zp} = 512$ points.

Box 13.1

Note: *Mathematica* uses a slightly different definition of H_n to that given in Equation (13.6), which we designate by $[H_n]_{Math}$.

$$[H_n]_{Math} = \frac{1}{\sqrt{N}} \sum_{j=1}^{N} d_j e^{i2\pi \frac{(n-1)(j-1)}{N}}$$

The modified version of Equation (13.9) is

$$C(n) = \frac{N_{zp}}{N_{orig}} |[H_n]_{Math}|^2 \quad \text{for } n = 1, 2, \ldots, \frac{N_{zp}}{2} + 1,$$

where $[H_n]_{Math} = $ **Fourier[data]**, **data** is a list of d_j values, and the zero frequency corresponds to the $n = 1$ term.

The bottom two panels of Figure 13.2 show the Bayesian posterior $p(n|D, I)$ computed from Equation (13.2) for the two simulations. The middle panels show the one-sided PSD for comparison. For the weak signal-to-noise simulation shown on the left, the PSD display exhibits many spurious noise peaks. The Bayesian $p(n|D, I)$ shows a single strong narrow peak while the spurious noise features have been strongly suppressed. Keep in mind that both quantities were computed from the same FFT of the time series. The comparison serves to show how much of an improvement can be obtained by a Bayesian estimation of the period over the intuitive PSD spectrum estimator even for a RMS signal-to-noise ratio of ≈ 0.6.

The three panels on the right hand side of Figure 13.2 illustrate the corresponding situation for a RMS signal-to-noise ratio of ≈ 7. In this case, we can clearly see the side lobes adjacent to the main peak which arise from using a rectangular data windowing function. In a conventional analysis, these side lobes are reduced by employing a data windowing function which reduces the relative importance of data at either end and results in a broadening of the spectral peak. The Bayesian analysis suppresses both the side lobes and spurious ripples by attenuation, and results in a very much narrower spectral peak. Also, because we have computed $p(n|D, I)$ directly, we can readily compute the accuracy of our spectral peak frequency estimate.

13.3 Strong prior signal model

Larry Bretthorst (1988, 1990a, b, c, 1991) extended Jaynes' work to more complex signal models with additive Gaussian noise and revolutionized the analysis of Nuclear Magnetic Resonance (NMR) signals. In NMR free-induction decay, the signal consists of a sum of exponentially decaying sinusoids of different frequency and decay rate. The top two panels of Figure 13.3 illustrate the quadrature channel measurements in an NMR free-induction decay experiment. In this example, the S/N is very high. The middle panel illustrates the conventional absorption spectrum based on an

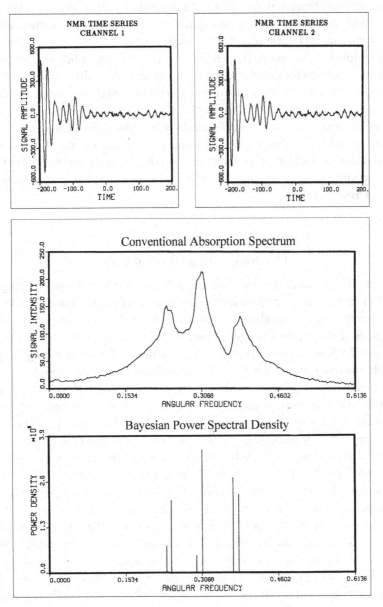

Figure 13.3 Comparison of conventional analysis (middle panel) and Bayesian analysis (bottom panel) of the two-channel NMR time series (top two panels). (Figure credit G. L. Bretthorst, reproduced by permission from the American Institute of Physics.)

FFT of the data, which shows three obvious spectral peaks with an indication of further structure in the peaks. The bottom panel illustrates Bretthorst's Bayesian analysis of this NMR data, which clearly isolates six separate peaks. The resolution is so good that the six peaks appear as delta functions in this figure. A similar improvement was obtained in the estimation of the decay rates. The Bayesian analysis provides much more reliable and informative results when prior knowledge of the shape of the signal and noise statistics are incorporated.

The question of how many frequencies are present, and what are the marginal PDFs for the frequencies and decay rates, can readily be addressed in the Bayesian framework using a Markov chain Monte Carlo computation. We saw how to do this in Sections 12.5 to 12.9. Frequently, the physics of the problem provides additional information about the relationships between pairs of frequencies, which can be incorporated as useful prior information. Varian Corporation now offers an expert analysis package with their new NMR machines based on Bretthorst's Bayesian algorithm. The manual for this software is available online at http://bayesiananalysis.wustl.edu/.

13.4 No specific prior signal model

In this case, we are addressing the detection and measurement of a periodic signal in a time series when we have no specific prior knowledge of the existence of such a signal or of its characteristics, including its shape. For example, an extraterrestrial civilization might be transmitting a repeating pattern of information either intentionally or unintentionally. What scheme could we use to optimally detect such a signal after we have made our best guess at a suitable wavelength of observation? Bayesian inference provides a well-defined procedure for solving any inference problem including questions of this kind. However, to proceed with the calculation, it is necessary to assume a model or family of models which is capable of approximating a periodic signal of arbitrary shape. A very useful Bayesian solution to the problem of detecting a signal of unknown shape was worked out by the author in collaboration with Tom Loredo (Gregory and Loredo, 1992, 1993, 1996), for the case of event arrival time data.

The Gregory–Loredo (GL) algorithm was initially motivated by the problem of detecting periodic signals (pulsars) in X-ray astronomy data. In this case, the time series consisted of individual photon arrival times where the appropriate sampling distribution is the Poisson distribution. To address the periodic signal detection problem, we compute the ratio of the probabilities (odds) of two models M_{Per} and M_1. Model M_{Per} is a family of periodic models capable of describing a background plus a periodic signal of arbitrary shape. Each member of the family is a histogram with m bins, with m ranging from 2 to some upper limit, typically 12. Three examples are shown in Figure 13.4. The prior probability that M_{Per} is true is divided equally among the members of this family. Model M_1 assumes the data are consistent with a

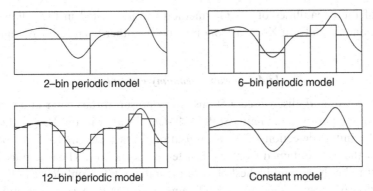

2–bin periodic model   6–bin periodic model

12–bin periodic model   Constant model

Figure 13.4 Three of the four panels show members of M_{Per}, a family of histogram (piecewise constant) periodic signal models, with $m = 2, 6$, and 12 bins, respectively. The constant rate model, M_1, is a special case of M_{Per}, with $m = 1$ bin, and is illustrated in the bottom right panel.

constant event rate. M_1 is a special case of M_{Per}, with $m = 1$ bin. Model M_1 is illustrated in the bottom right panel of Figure 13.4.

The Bayesian calculation automatically incorporates a quantified Occam's penalty, penalizing models with a larger number of bins for their greater complexity.[7] The calculation thus balances model simplicity with goodness-of-fit, allowing us to determine both whether there is evidence for a periodic signal, and the optimum number of bins for describing the structure in the data. The parameter space for the m-bin periodic model consists of the unknown period, an unknown phase (position of the first bin relative to the start of the data), and m histogram amplitudes describing the signal shape. A remarkable feature of this particular signal model is that the search in the m shape parameters can be carried out analytically, permitting the method to be computationally tractable. Further research is underway to investigate computationally tractable ways of incorporating additional desirable features into the signal model, such as variable bin widths to allow for a reduction in the number of bins needed to describe certain types of signal.

The solution in the Poisson case yields a result that is intuitively very satisfying. The probability for the family of periodic models can be shown to be approximately inversely proportional to the entropy (Gregory and Loredo, 1992) of any significant organized periodic structure found in the search parameter space. What structure is significant is determined through built-in quantified Occam's penalties in the calculation. Of course, structure with a high degree of organization corresponds to a state of low entropy. In the absence of knowledge about the shape of the signal, the method identifies the most organized significant periodic structure in the model parameter space.

---

[7] The Occam penalty becomes so large for $m \geq 12$, that the data are generally not good enough to make it worthwhile including periodic models with larger values of m.

Some of the capabilities of the GL method are illustrated in the following two examples, one taken from X-ray astronomy and the other from radio astronomy.

13.4.1 X-ray astronomy example

In 1984, Seward *et al.* discovered a 50 ms X-ray pulsar at the center of a previously known radio supernova remnant, SNR 0540-693, located in the Large Magellanic Cloud. The initial detection of X-ray pulsations was from an FFT periodogram analysis of the data obtained from the Einstein Observatory. The true pulsar signal turned out to be the second highest peak in the initial FFT. Confidence in the reality of the signal was established from FFT runs on other data sets. The pulsar was re-observed with the ROSAT Observatory by Seward and colleagues, but this time, an FFT search failed to detect the pulsar. In Gregory and Loredo (1996), we used the GL method on a sample ROSAT data set of 3305 photons provided by F. Seward. The data spanned an interval of 116 341 s and contained many gaps.

In the first instance, we incorporated the prior information on the period, period derivative and their uncertainties, obtained from the earlier detection with the Einstein Observatory. The Gregory–Loredo method provides a calculation of the global odds ratio defined as the ratio of the probability for the family of periodic models to the probability of a constant rate model, regardless of the exact shape, period and phase of the signal. The resulting odds ratio of 2.6×10^{11} indicates near certainty in the presence of a periodic signal.

It is interesting to consider whether we would still claim the detection of a periodic signal if we did not have the prior information derived from the earlier detection. Thus, in the second instance, we assume a prior period search range extending from the rotational breakup period of a neutron star (≈ 1.5 ms), to half the duration of the data. This gives an odds ratio of 4.5×10^5. This is greatly reduced due to the much larger Occam penalty associated with not knowing the period. But this still provides overwhelming evidence for the presence of a periodic signal, despite the fact that it was undetected by FFT techniques.

In their paper, Seward *et al.* (1984) used another method commonly employed in X-ray astronomy, called *period folding* (also known as *epoch folding*), to obtain the pulsar light curve and a best period. Period folding involves dividing the trial period into m bins (typically five) and binning the data modulo the trial period for a given trial phase. The χ^2 statistic is used to decide at some significance level, whether a constant model can be rejected, and thus indirectly infer the presence of a periodic signal. In Seward *et al.* (1984), their period uncertainty was estimated from the half-width of the χ^2 peak, which is sometimes used as a naive estimate of the accuracy of the frequency estimate. Figure 13.5 shows a comparison of the largest frequency peak comparing the GL marginal probability density for f to the period folding $\langle \chi^2 \rangle_\phi$ statistic. The width of the GL marginal probability density for f is more than an order of magnitude smaller.

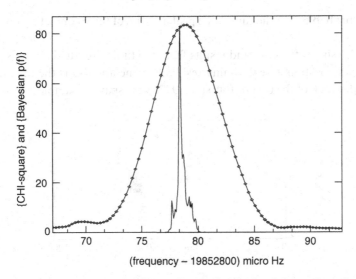

Figure 13.5 Close-up of largest frequency peak comparing the Gregory–Loredo probability density for f to the period folding $\langle \chi^2 \rangle_\phi$ statistic (diamonds). The $\langle \chi^2 \rangle_\phi$ statistic versus trial frequency results from epoch folding analysis using $m = 5$ bins (Gregory and Loredo, 1996).

13.4.2 Radio astronomy example

In 1999, the author generalized the GL algorithm to the Gaussian noise case. Application of the method to a radio astronomy data set has resulted in the discovery of a new periodic phenomenon (Gregory, 1999, 2002; Gregory et al., 1999; Gregory and Neish, 2002) in the X-ray and radio emitting binary, LS I +61°303. LS I +61°303 is a remarkable tenth magnitude binary star (Gregory and Taylor, 1978; Hutchings and Crampton, 1981) that exhibits periodic radio outbursts every 26.5 days (Taylor and Gregory, 1982), which is the binary orbital period. The radio, infrared, optical, X-ray and γ-ray data indicate that the binary consists of a rapidly rotating massive young star, called a *Be star*, together with a neutron star in an eccentric orbit.

The Be star exhibits a dense equatorial wind and the periodic radio outbursts are thought to arise from variations in wind accretion by the neutron star in its eccentric orbit. Some of the energy associated with the accretion process is liberated in the form of outbursts of radio emission. One puzzling feature of the outbursts has been the variablity of the orbital phase of the outburst maxima, which can range over 180 degrees of phase. In addition, the strength of the outburst peaks was known to vary on time scales of approximately 4 years (Gregory et al., 1989; Paredes et al., 1990).

Armed with over twenty years of data, we (Gregory, 1999; Gregory et al., 1999) applied Bayesian inference to assess a variety of hypotheses to explain the outburst timing residuals and peak flux density variations. The results for both the outburst peak flux density and timing residuals demonstrated a clear 1667-day periodic modulation in both quantities. The periodic modulation model was found to be $\sim 3 \times 10^3$

times more probable than the sum of the probabilities of three competing non-periodic models.

Figure 13.6 shows the data and results from the timing residual analysis. Panel (a) shows the radio outburst peak timing residuals.[8] The abscissa is the time interval in days from the peak of the first outburst in 1977. Very sparsely sampled measurements

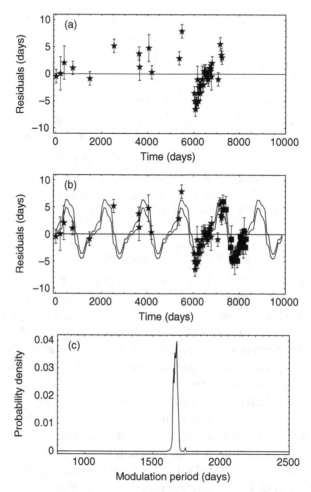

Figure 13.6 Panel (a) shows the outburst timing residuals. A comparison of the predicted outburst timing residuals with the data versus time is shown in panel (b). The solid curves show the estimated mean light curve, ± 1 standard deviation. The new data are indicated by a filled box symbol. Panel (c) shows the probability for the modulation period of LS I $+61°303$.

---

[8] The timing residuals depend on the assumed orbital period which is not accurately known independent of the radio data. The GL algorithm was modified to compute the joint probability distribution of the orbital and modulation periods. Only the marginal distribution for the modulation period is shown.

were obtained from the initial discovery in 1977 until 1992. However, beginning in January 1994, Ray *et al.* (1997) performed detailed monitoring (several times a day) with the National Radio Astronomy Observatory Green Bank Interferometer. With such sparsely sampled data, the eye is unable to pick out any obvious periodicity. Panel (c) shows the Bayesian marginal probability density for the modulation period. The single well-defined peak provides clear evidence for a periodicity of approximately 1667 days.

Subsequent monitoring of the binary star system has confirmed and refined the orbital and modulation period. Panel (b) shows a comparison of the predicted outburst timing residuals with the data versus time. The solid curves show the estimated mean light curve, ± 1 standard deviation. The new data, indicated by a shaded box, nicely confirm the periodic modulation model. This discovery has contributed significantly to our understanding of Be star winds.

The *Mathematica* tutorial includes a Markov chain Monte Carlo version of the GL algorithm, for the Gaussian noise case, in the section entitled, "MCMC version of the Gregory–Loredo algorithm."

13.5 Generalized Lomb–Scargle periodogram

In Section 13.2, we introduced Jaynes' insights on the periodogram from probability theory and discussed how to compute $p(f|D, I)$ in more detail in Section 13.2.1. Bretthorst (2000, 2001) generalized Jaynes' insights to a broader range of single-frequency estimation problems and sampling conditions and removed the need for the approximations made in the derivation of Equations (13.2) and (13.5). In the course of this development, Bretthorst established a connection between the Bayesian results and an existing frequentist statistic known as the Lomb–Scargle periodigram (Lomb, 1976; Scargle, 1982, 1989), which is a widely used replacement for the Schuster periodogram in the case of non-uniform sampling. We will summarize Bretthorst's Bayesian results in this section. In particular, his analysis allows for the following complications:

1. Either real or quadrature data sampling. Quadrature data involve measurements of the real and imaginary components of a complex signal. The top two panels of Figure 13.3 show an example of quadrature signals occurring in NMR.
   Let $d_R(t_i)$ denote the real data at time t_i and $d_I(t'_i)$ denote the imaginary data at time t'_i. There are N_R real samples and N_I imaginary samples for a total of $N = N_R + N_I$ samples.
2. Uniform or non-uniform sampling and for quadrature data with non-simultaneous sampling.
   The analysis does not require the t_i and t'_i to be simultaneous and successive samples can be unequally spaced in time.
3. Allows for a non-stationary single sinusoid model of the form

$$d_R(t_i) = A\cos(2\pi f t_i - \theta)Z(t_i) + B\sin(2\pi f t_i - \theta)Z(t_i) + n_R(t_i), \qquad (13.10)$$

where A and B are the cosine and sine amplitudes, and $n_R(t_i)$ denotes the noise at t_i. The function $Z(t_i)$ describes an arbitrary modulation of the amplitude, e.g., exponential decay as exhibited in NMR signals. If $Z(t)$ is a function of any parameters, those parameters are assumed known, e.g., the exponential decay rate. $Z(t)$ is sometimes called a *weighting function* or *apodizing* function.

The corresponding signal model for the imaginary channel is given by

$$d_I(t'_j) = A\cos(2\pi f t'_j - \theta)Z(t'_j) + B\sin(2\pi f t'_j - \theta)Z(t'_j) + n_I(t'_j). \qquad (13.11)$$

The angle θ is defined in such a way as to make the cosine and sine functions orthogonal on the discretely sampled times. This corresponds to the condition

$$
\begin{aligned}
0 = &\sum_{i=1}^{N_R} \cos(2\pi f t_i - \theta)\sin(2\pi f t_i - \theta)Z(t_i)^2 \\
&- \sum_{j=1}^{N_I} \sin(2\pi f t'_j - \theta)\cos(2\pi f t'_j - \theta)Z(t'_j)^2.
\end{aligned} \qquad (13.12)
$$

The solution of Equation (13.12) is given by

$$\theta = \frac{1}{2}\tan^{-1}\left[\frac{\sum_{i=1}^{N_R}\sin(4\pi f t_i)Z(t_i)^2 - \sum_{j=1}^{N_I}\sin(4\pi f t'_j)Z(t'_j)^2}{\sum_{i=1}^{N_R}\cos(4\pi f t_i)Z(t_i)^2 - \sum_{j=1}^{N_I}\cos(4\pi f t'_j)Z(t'_j)^2}\right]. \qquad (13.13)$$

Note: if the data are simultaneously sampled, $t_i = t'_j$, then the orthogonal condition is automatically satisfied so $\theta = 0$.

4. The noise terms $n_R(t_i)$ and $n_I(t_i)$ are assumed to be IID Gaussian with an unknown σ. Thus, σ is a nuisance parameter, which is assumed to have a Jeffreys prior. By marginalizing over σ, any variability in the data that is not described by the model is assumed to be noise.

The final Bayesian expression for $p(f|D,I)$, after marginalizing over the amplitudes A and B (assuming independent uniform priors), is given by

$$p(f|D,I) \propto \frac{1}{\sqrt{C(f)S(f)}}[N\overline{d^2} - \overline{h^2}]^{\frac{2-N}{2}}, \qquad (13.14)$$

where the mean-square data value, $\overline{d^2}$, is defined as

$$\overline{d^2} = \frac{1}{N}\left[\sum_{i=1}^{N_R}d_R(t_i)^2 + \sum_{j=1}^{N_I}d_I(t'_j)^2\right]. \qquad (13.15)$$

The term $\overline{h^2}$ is given by

$$\overline{h^2} = \frac{R(f)^2}{C(f)} + \frac{I(f)^2}{S(f)}, \qquad (13.16)$$

where

$$R(f) \equiv \sum_{i=1}^{N_R} d_R(t_i) \cos(2\pi f t_i - \theta) Z(t_i) - \sum_{j=1}^{N_I} d_I(t_j') \sin(2\pi f t_j' - \theta) Z(t_j'), \qquad (13.17)$$

$$I(f) \equiv \sum_{i=1}^{N_R} d_R(t_i) \sin(2\pi f t_i - \theta) Z(t_i) + \sum_{j=1}^{N_I} d_I(t_j') \cos(2\pi f t_j' - \theta) Z(t_j'), \qquad (13.18)$$

$$C(f) \equiv \sum_{i=1}^{N_R} \cos^2(2\pi f t_i - \theta) Z(t_i)^2 + \sum_{j=1}^{N_I} \sin^2(2\pi f t_j' - \theta) Z(t_j')^2 \qquad (13.19)$$

and

$$S(f) \equiv \sum_{i=1}^{N_R} \sin^2(2\pi f t_i - \theta) Z(t_i)^2 + \sum_{j=1}^{N_I} \cos^2(2\pi f t_j' - \theta) Z(t_j')^2. \qquad (13.20)$$

13.5.1 Relationship to Lomb–Scargle periodogram

If the sinusoidal signal is known to be stationary ($Z(t_i)$ is a constant) and the data are entirely real, then Equations (13.17) to (13.20) greatly simplify. In this case, the quantity $\overline{h^2}$ given by Equation (13.16) corresponds to the Lomb–Scargle periodogram; however, we now see this statistic in a new light. The Bayesian expression for $p(f|D, I)$ (Equation (13.14)) involves a nonlinear processing of the Lomb–Scargle periodogram, analogous to the nonlinear processing of the Schuster periodogram in Equation (13.5). In fact, Bretthorst showed that for uniformly sampled quadrature data and a stationary sinusoid, the Lomb–Scargle periodogram reduces to a Schuster period-ogram, the power spectrum of the data. For real data, Equations (13.2) and (13.5) are only approximately true. As we will demonstrate in Figure 13.7, Equation (13.5) can provide an excellent approximation to $p(f|D, I)$ for uniformly sampled real data and a stationary sinusoid, and the Schuster periodogram is much faster to compute than the Lomb–Scargle periodogram.

Equations (13.14) to (13.20) provide the exact answer for $p(f|D, I)$ for a much wider range of problems and involve a generalized version of the Lomb–Scargle statistic.

13.5.2 Example

In this example, we compare the Schuster periodogram to the Lomb–Scargle period-ogram, for the time series simulation involving a stationary sinusoid model and

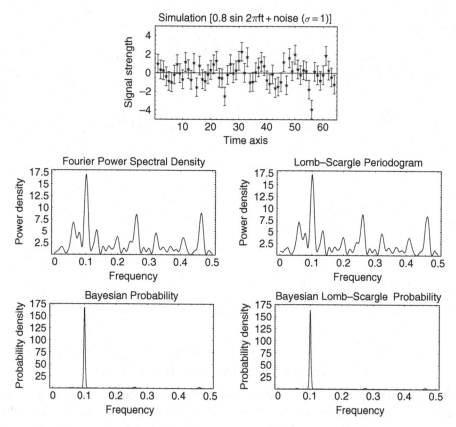

Figure 13.7 The middle two panels compare the Fourier power spectral density (Schuster periodogram) and Lomb–Scargle periodogram for the uniformly sampled time series simulation shown in the top panel. The bottom two panels compare the Bayesian counterparts for the same time series.

uniformily sampled real data that we used in Section 13.2.1. In the first, which is illustrated in Figure 13.7, the data are uniformly sampled. The top panel shows the time series and the two middle panels show the Fourier power spectral density (Schuster periodogram) and Lomb–Scargle periodogram of this time series. The corresponding Bayesian $p(f|D, I)$ probability densities are shown in the bottom two panels. Clearly, for this example, the Schuster periodogram provides an excellent approximation to the Lomb–Scargle periodogram, and is much faster to compute. Recall that the width of the spectral peak in the Bayesian $p(f|D, I)$ depends on the signal-to-noise ratio (SNR). Even for a moderate SNR, the spectral peak can become very narrow, requiring a large number of evaluations of the Lomb–Scargle statistic at very closely spaced frequencies.

The second example, which is illustrated in Figure 13.8, makes use of the same time series, but has 14 samples removed creating gaps in the otherwise uniform sampling. These gaps can clearly be seen in the top right panel. To compute the Schuster periodogram, some assumption must be made regarding the data in the gaps, to achieve the uniform sampling required for the calculation of the FFT. In the top left panel, the missing data have been filled in with values equal to the time series average. In the calculation of the Lomb–Scargle periodogram, only the actual data are used. The two bottom panels again illustrate the corresponding Bayesian $p(f \,|\, D, I)$ probability densities. In this case, it is clear that the Bayesian generalization of the Lomb–Scargle periodogram does a better job.

In the latter example, the data are uniformly sampled apart from the gaps. In the next section we will explore the issue of non-uniform sampling in greater detail.

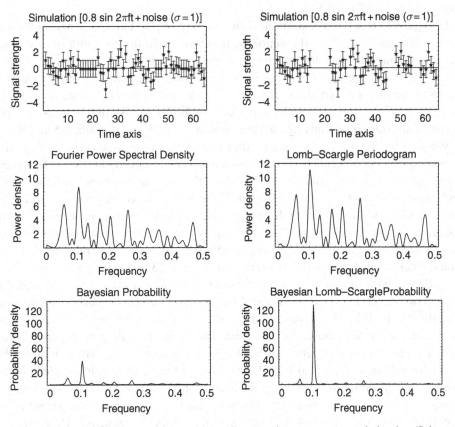

Figure 13.8 The middle two panels compare the Fourier power spectral density (Schuster periodogram) and Lomb–Scargle periodogram for a time series with significant data gaps, as shown in the top right panel. The bottom two panels compare the Bayesian counterparts for the same time series.

13.6 Non-uniform sampling

In many cases, the data available are not uniformly sampled in the coordinate of interest, e.g., time. In some cases this introduces complications, but on the flip side, there is a distinct advantage. Non-uniform sampling can eliminate the common problem of aliasing (Bretthorst 1988, 2000a). In this section, we explore this effect with a demonstration.

We start with a uniform time series of 32 points containing a sinusoidal signal plus additive independent Gaussian noise. The data are described by the following equation:

$$d_k = 2\cos(2\pi fkT) + \text{noise } (\sigma = 1) \quad \text{with } f = 1.23\text{Hz}, \tag{13.21}$$

where T is the sample interval and k is an index running for 32 points. In this demonstration, $T = 1$ s. At 1.23 Hz, the signal frequency is well above the Nyquist frequency $= 1/(2T) = 0.5$ Hz.

In Figure 13.9 we demonstrate how the aliasing arises. The top panel shows the Fourier Transform (FT) of the sampling. It is convenient to show both positive and negative frequencies which arise in the mathematics of the FT. The middle panel shows the FT of the signal together with the Nyquist frequency. The bottom panel shows the resulting convolution. There are three aliased signals at $f = 0.23, f = 0.77$, and $f = 1.77$, only one of which, at $f = 0.23$, is below the Nyquist frequency. For deeper understanding of this figure, the reader is referred to Appendix B.5 and B.6.

We start by computing the Fourier transform and Bayesian posterior probability density for the signal frequency of the initial uniformly sampled data. We will replace some of the samples by samples taken at times that are not an integer multiple of T, and explore how the spectrum is altered. Since we will be considering non-uniform samples, we make use of the Lomb–Scargle periodogram, discussed in Section 13.5, to compute the power spectrum. We also display the Bayesian posterior probability density for the signal frequency using Bretthorst's Bayesian generalization of the Lomb–Scargle algorithm, also discussed in Section 13.5. Figure 13.10 shows the evolution of both quantities as the number of non-uniform samples is increased. In the top two panels (a), the original signal frequency at 1.23 Hz is clearly seen together with three aliased signals. In the second row (b), one uniformly sampled data point has been replaced by one non-uniform sample. The Lomb–Scargle periodogram shows only a slight change, but remarkably, the Bayesian probability density has clearly distinguished the real signal at 1.23 Hz. As more and more uniformly sampled points are replaced, the amplitudes of the aliases in the Lomb–Scargle periodogram decrease.

Notice that for the non-uniform sampling used in this demonstration, no alias occurs up to frequencies ≤ 4 times the effective Nyquist frequency. Of course, it must be true that the aliasing phenomenon returns at sufficiently high frequencies. If the sampling times t_k, although non-uniform, are all integer multiples of some small interval Δt, then signals at frequencies $> 1/(2\Delta t)$ will still be aliased.

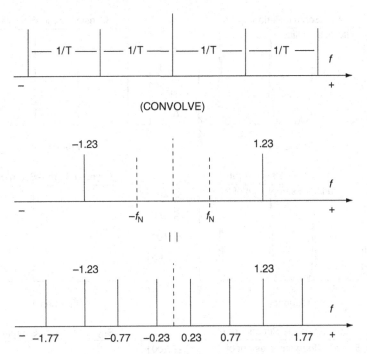

Figure 13.9 How aliasing arises. Uniform sampling at an interval T in the time domain corresponds to convolution in the frequency domain. The top panel shows the Fourier Transform (FT) of the sampling. The middle panel shows the FT of the signal together with the Nyquist frequency. The bottom panel shows the resulting convolution. There are 3 aliased signals at $f = 0.23$, $f = 0.77$, and $f = 1.77$, only one of which, at $f = 0.23$, is below the Nyquist frequency.

Consideration of Figure 13.11 shows why aliasing does not occur for the non-uniform sampling. In this example, we first generated four data points (filled boxes) by sampling a 1.23 Hz sinusoidal signal, with no noise, at one-second intervals. The figure shows four sinusoids corresponding to the 1.23 Hz signal and the 3 aliases at 0.23, 0.77 and 1.77 Hz, which all pass through these uniformly sampled data points. Next, we replaced the first uniform sample by one which is non-uniformly sampled in time (star). In this case, only the 1.23 Hz sinusoid passes through all four points.

There is clearly an advantage to employing non-uniform sampling which needs to be considered as part of the experimental design. As Figure 13.11 clearly demonstrates, even the addition of a small number of non-uniform samples (only one required in this 32-point time series) to an otherwise uniformly sampled data set is sufficient to strongly suppress aliasing in the Bayesian posterior probability density for signal frequency.

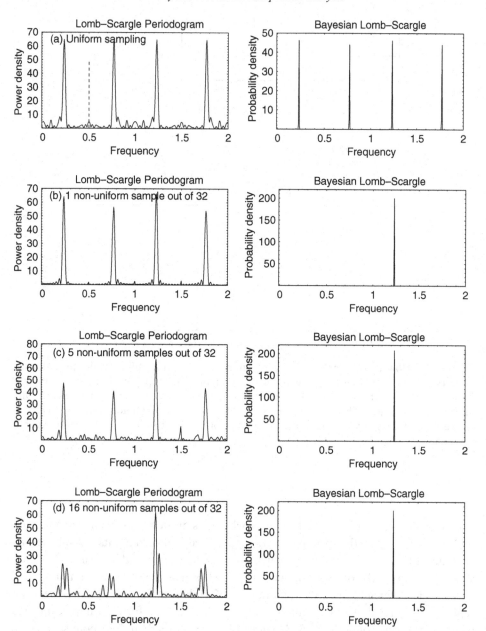

Figure 13.10 Evolution of the Lomb–Scargle periodogram (left) and Bretthorst's Bayesian generalization of the Lomb–Scargle periodogram (right), with increasing number of non-uniform samples in the time series. Notice how sensitive the Bayesian result is to a change of only one sample from a uniform interval to a non-uniform interval.

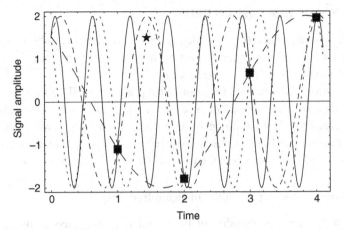

Figure 13.11 An illustration of how four different frequencies can all pass through the same set of four uniformly sampled data points (boxes) but only one passes through all the points when one sample is relaced by a non-uniform sample (star).

13.7 Problems

Table 13.1 is a simulated times series consisting of a single sinusoidal signal with additive IID Gaussian noise. In this problem, you will compare the usual one-sided power spectral density discussed in Appendix B to the Bayesian posterior probability density for the frequency of the model sinusoid.

Table 13.1 *The table contains 64 samples of a simulated times series consisting of a single sinusoidal signal with additive IID Gaussian noise.*

t (s)	mK	t (s)	mK	t (s)	mK	t (s)	mK
1	0.474	17	−0.865	33	−0.225	49	0.369
2	0.281	18	0.206	34	−1.017	50	0.695
3	1.227	19	−0.926	35	0.817	51	1.291
4	−1.523	20	2.294	36	−2.064	52	0.978
5	−0.831	21	0.786	37	−0.103	53	−0.592
6	−0.978	22	0.522	38	1.878	54	−0.986
7	0.169	23	−1.04	39	0.625	55	−1.005
8	0.04	24	−0.181	40	1.418	56	−1.268
9	0.76	25	−1.47	41	0.464	57	−0.571
10	0.847	26	−1.837	42	−1.182	58	1.128
11	0.106	27	0.523	43	−1.319	59	0.64
12	−1.814	28	0.605	44	1.354	60	0.144
13	−1.16	29	−1.595	45	−1.784	61	−1.468
14	0.249	30	−0.413	46	−0.989	62	−0.71
15	−1.054	31	1.275	47	−1.52	63	−1.486
16	−0.359	32	−1.644	48	1.239	64	−0.129

Part 1: Fast Fourier Transform and PSD

a) Use an FFT to determine the one-sided power spectral density (PSD), as defined by Equation (B.102) in Appendix B. Plot both the raw data and your spectrum and determine the period of the strongest peak.

b) To obtain a more accurate determination of the peak in the PSD, add zeros to the end of the input data so that the total data set (data + appended zeros) is 1024 points and recompute the PSD. Note: although the number of spectral points will increase, the $1/N$ normalization term in Equation (B.102) still refers to the original number of data points. Plot the new PSD. Do you expect the width of the peak to be affected by zero padding?

Part 2: Bayesian posterior probability of signal frequency

The Bayesian posterior probability of the signal frequency, assuming a model of a single harmonic signal plus independent Gaussian noise, is given by Equation (13.2). If the noise is not well understood, then it is safer to use the Student's t form of Equation (13.5) which treats anything that cannot be fitted by the model as noise and leads to more conservative parameter estimates. Since we are evaluating $p(n|D,I)$ at n discrete frequencies, we rewrite Equation (13.5) as

$$p(n|D,I) = \frac{\left[1 - \frac{2C(n)}{N_{\text{orig}}\overline{d^2}}\right]^{\frac{2-N_{\text{orig}}}{2}}}{\sum_0^{N_{zp}/2}\left[1 - \frac{2C(n)}{N_{\text{orig}}\overline{d^2}}\right]^{\frac{2-N_{\text{orig}}}{2}}}, \tag{13.22}$$

where $\overline{d^2} = \frac{1}{N_{\text{orig}}}\sum_i d_i^2$.

In Equation (13.22), the frequency associated with any particular value of n is given by $f_n = n/NT$. T equals the sample interval in time and N is the total number of samples. The value $n = 0$ corresponds to zero frequency. The quantity $C(n)$ is the positive frequency part of the two-sided periodogram (two-sided PSD) given by Equation (13.6), which we rewrite as

$$C(n) = \frac{|H_n|^2}{N}, \quad \text{for} \quad n = 0, 1, \ldots, \frac{N}{2}. \tag{13.23}$$

In general, $p(n|D,I)$ will be very narrow when $C(n)/\sigma^2 > 1$ because of the exponentiation occurring in Equation (13.2). Thus, to accurately define $p(n|D,I)$, we need to zero pad the FFT to obtain a sufficient density of H_n points to accurately define the $p(n|D,I)$ peak. Zero padding is discussed in detail in Appendix B. In the zero padding case, Equation (13.23) becomes

$$C(n) = \frac{|H_n|^2}{N_{\text{orig}}}, \quad \text{for} \quad n = 0, 1, \ldots, \frac{N_{zp}}{2}, \tag{13.24}$$

where N_{orig} is the number of original time series samples and N_{zp} is the total number of points including the added zeros.

(a) Compute and plot $p(n|D, \sigma, I)$ from a zero padded FFT of the time series given above.

(b) Measure the width of the peak at half height of $p(n|D, \sigma, I)$ and compare the width of the PSD peak at half height.

(c) Plot the natural logarithm of the Bayesian $p(n|D, I)$ and compare its shape to the PSD.

Note: *Mathematica* uses a slightly different definition of H_n to that given in Equation (B.51), which we designate by $[H_n]_{\text{Math}}$. The modified version of Equation (13.24) is given by:

$$C(n) = \frac{N_{zp}}{N_{orig}} |[H_n]_{\text{Math}}|^2, \quad \text{for} \quad n = 0, 1, \ldots, \frac{N_{zp}}{2}. \tag{13.25}$$

14

Bayesian inference with Poisson sampling

14.1 Overview

In many experiments, the basic data consist of a set of discrete events distributed in space, time, energy, angle or some other coordinate. They include macroscopic events like a traffic accident or the location of a star. They also include microscopic events such as the detection of individual particles or photons in time or position. In experiments of this kind, our prior information often leads us to model the probability of the data (likelihood function) with a Poisson distribution. See Section 4.7 for a derivation of the Poisson distribution, and Section 5.7.2 for the relationship between the binomial and Poisson distributions.

For temporally distributed events, the Poisson distribution is given by

$$p(n|r, I) = \frac{(rT)^n e^{-rT}}{n!}. \tag{14.1}$$

It relates the probability that n discrete events will occur in some time interval T to a positive real-valued Poisson process event rate r. When n and rT are large, the Poisson distribution can be accurately approximated by a Gaussian distribution. Here, we will be concerned with situations where the Gaussian approximation is not good enough and we must work directly with the Poisson distribution.

In this chapter, we employ Bayes' theorem to solve the following inverse problem: compute the posterior PDF for r given the data D and prior information I. We divide this into three common problems:

1. How to infer a Poisson rate r.
2. How to infer a signal in a known background.
3. Analysis of ON/OFF data, where ON is the signal + background and OFF is a just the background. The background is only known imprecisely from the OFF measurement.

The treatment is similar to that given by Loredo (1992), but also includes a treatment of the source detection question in the ON/OFF measurement problem.

In the above three problems, the Poisson rate is assumed to be constant in the ON or OFF source position. In Section 14.5, we consider a simple radioactive decay problem in which r varies significantly over the duration of the data.

14.2 Infer a Poisson rate

The simplest problem is to infer the rate r from a single measurement of n events. From Bayes' theorem:

$$p(r|n, I) = \frac{p(r|I)p(n|r, I)}{p(n|I)}.$$ (14.2)

The prior information I must specify both $p(r|I)$ and the likelihood function $p(n|r, I)$. In this case, the latter is just the Poisson distribution.

$$I \begin{cases} \text{likelihood:} & p(n|r, I) = [(rT)^n e^{-rT}]/n! \\ \text{prior:} & p(r|I) = ? \end{cases}$$

Our first guess at $p(r|I)$ is a Jeffreys prior since r is a scale parameter. However, the scale invariance argument is not valid if r may vanish. Instead, we adopt a uniform prior for r based on the following argument: intuition suggests ignorance of r corresponds to not having any prior preference for seeing any particular number of counts, n. In situations where it is desirable to use the Poisson distribution, the prior range for n is frequently small, so it is reasonable to use a uniform prior:

$$p(n|I) = \text{constant}.$$

But $p(n|I)$ is also the denominator in Equation (14.2), so we can write

$$p(n|I) = \int_0^\infty dr\, p(r|I)\, p(n|r, I)$$

$$= \frac{1}{n!T} \int_0^\infty d(rT)\, p(r|I)(rT)^n e^{-rT}.$$ (14.3)

For $p(n|I)$ to be constant, it is necessary that

$$\int_0^\infty d(rT)\, p(r|I)(rT)^n e^{-rT} \propto n!$$

but

$$\int_0^\infty dx\, x^n\, e^{-x} = \Gamma(n+1) = n!$$ (14.4)

which implies that $p(r|I) = \text{constant}$. Use

$$p(r|I) = \frac{1}{r_{\max}}, \quad 0 \le r \le r_{\max}.$$ (14.5)

Then

$$p(n|I) = \frac{1}{Tr_{\max}} \int_0^{r_{\max}} d(rT)(rT)^n e^{-rT}$$

$$= \frac{1}{Tr_{\max}} \frac{\gamma(n+1, r_{\max}T)}{n!},$$ (14.6)

where $\gamma(n, x) = \int_0^x dy\ y^{n-1} e^{-y}$ is one form of the incomplete gamma function.[1] Now substitute Equations (14.1), (14.5), and (14.6) into Equation (14.2) to obtain the posterior $p(r|n, I)$.

$$p(r|n, I) = \frac{T(rT)^n e^{-rT}}{n!} \times \frac{n!}{\gamma(n+1, r_{\max} T)}, \quad 0 \le r \le r_{\max}. \qquad (14.7)$$

For $r_{\max} T \gg n$, then $\gamma(n+1, r_{\max} T) \simeq \Gamma(n+1) = n!$ and Equation (14.7) simplifies to

$$p(r|n, I) = \frac{T(rT)^n e^{-rT}}{n!}, \quad r \ge 0. \qquad (14.8)$$

14.2.1 Summary of posterior

$$p(r|n, I) = \frac{T(rT)^n e^{-rT}}{n!}, \quad r \ge 0$$

$$\text{mode:} \quad r = n/T,$$
$$\text{mean:} \quad <r> = (n+1)/T,$$
$$\text{sigma:} \quad \sigma_r = \sqrt{n+1}/T.$$

Figure 14.1 illustrates the shape of the posterior $p(r|n, I)$, divided by the time interval T, for four different choices of n ranging from $n = 0$ to 100. In each case, the count interval $T = 1$ s. As n increases, the $p(r|n, I)$ becomes more symmetrical and gradually approaches a Gaussian (shown by the dotted line) with the same mode and standard deviation. For $n = 10$, the Gaussian approximation is still a poor fit. By $n = 100$, the Gaussian approximation provides a good fit near the mode but still departs noticeably in the wings.

The 95% credible region can be found by solving for the two values r_{high} and r_{low} which satisfy the two conditions:

$$p(r_{\text{high}}|n, I) = p(r_{\text{low}}|n, I),$$

and

$$\int_{r_{\text{low}}}^{r_{\text{high}}} p(r|n, I)\, dr = 0.95.$$

For $n = 1$ the credible region is

$$p\left(\frac{0.042}{T} \le r \le \frac{4.78}{T}\right) = 0.95. \qquad (14.9)$$

---

[1] See Press *et al.* (1992).

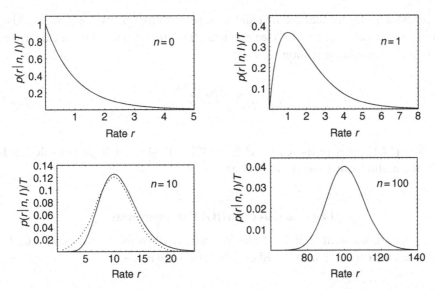

Figure 14.1 The posterior PDF, $p(r|n, I)$, divided by the time interval T, plotted for four different values of n. For comparison, a Gaussian with the same mode and standard deviation is shown by the dotted curve for the $n = 10$ and $n = 100$ cases.

14.3 Signal + known background

In this case, the measured rate consists of two components, one due to a signal of interest, s, and the other a known background rate, b.

$$r = s + b \quad \begin{cases} s = \text{signal rate} \\ b = \text{known background rate.} \end{cases}$$

Since we are assuming the background rate is known,

$$p(s|n, b, I) = p(r|n, b, I).$$

We can now use Equation (14.8) of the previous section for $p(r|n, b, I)$, and replace r by $s + b$. The result is

$$p(s|n, b, I) = C \frac{T[(s + b)T]^n e^{-(s+b)T}}{n!} \tag{14.10}$$

$$C^{-1} = \frac{e^{-bT}}{n!} \int_0^\infty d(sT)(s + b)^n \, T^n e^{-sT}. \tag{14.11}$$

The constant C ensures that the area under the probability density function $= 1$. Using a binomial expansion of $(s + b)^n$ (see Equation (D.7) in Appendix D), we can arrive at the following simple expression for C^{-1}:

$$C^{-1} = \sum_{i=0}^{n} \frac{(bT)^i e^{-bT}}{i!}. \qquad (14.12)$$

Equation (14.10) was proposed by Helene (1983, 1984) as a Bayesian solution for analyzing multichannel spectra in nuclear physics.

14.4 Analysis of ON/OFF measurements

In this section, we want to infer the source rate, s, when the background rate, b, is imprecisely measured. This is called an ON/OFF measurement.

$$\text{OFF} \rightarrow \text{detector pointed off source to measure } b$$
$$\text{ON} \rightarrow \text{detector pointed on source to measure } s + b.$$

The usual approach is to assume

$$\text{OFF} \rightarrow \hat{b} \pm \sigma_b$$
$$\text{ON} \rightarrow \hat{r} \pm \sigma_r,$$

where $\hat{b} = N_{\text{off}}/T$ and $\sigma_b = \sqrt{N_{\text{off}}}/T$ and $\hat{r} = N_{\text{on}}/T$ and $\sigma_r = \sqrt{N_{\text{on}}}/T$. Then $\hat{s} = \hat{r} - \hat{b}$ and the variance $\sigma_s^2 = \sigma_r^2 + \sigma_b^2$. This procedure works well for the Poisson case provided both s and b are large enough that the Poisson is well approximated by a Gaussian. But when *either* or *both* of the rates are small, the procedure fails. This can lead to negative estimates of s and/or error bars extending into non-physical negative values. This is a big problem in γ-ray and ultra-high energy astrophysics, where data are very sparse. First consider the OFF measurement:

$$p(b|N_{\text{off}}, I_b) = \frac{T_{\text{off}}(bT_{\text{off}})^{N_{\text{off}}} e^{-bT_{\text{off}}}}{N_{\text{off}}!}. \qquad (14.13)$$

For the ON measurement, we can write the joint probability of the source and background rate:

$$p(s, b|N_{\text{on}}, I) = \frac{p(s, b|I)p(N_{\text{on}}|s, b, I)}{p(N_{\text{on}}|I)}$$
$$= \frac{p(s|b, I)p(b|I)p(N_{\text{on}}|s, b, I)}{p(N_{\text{on}}|I)}. \qquad (14.14)$$

Note: the prior information, I, includes information about the background OFF measurement in addition to the model M_{s+b}, which asserts that the Poisson rate in the ON measurement is equal to $s + b$. We can express this symbolically by $I = N_{\text{off}}$, I_b, M_{s+b}. In the parameter estimation part of the problem, we will estimate the value of the source rate s in the model M_{s+b}. Following this, we will evaluate a model selection problem to compare the probability of model M_{s+b}, which assumes a source is present, to the simpler model M_b, which asserts that the Poisson rate in the ON source measurement is equal to b, i.e., no source is present.

14.4.1 Estimating the source rate

The likelihood for the ON measurement is the Poisson distribution for a source with strength $s + b$:

$$p(N_{\text{on}}|s, b, I) = \frac{[(s + b)T_{\text{on}}]^{N_{\text{on}}} e^{-(s+b)T_{\text{on}}}}{N_{\text{on}}!}. \tag{14.15}$$

We will again assume a constant prior for s, so we write $p(s|b, I) = 1/s_{\text{max}}$. The prior for b is simply the posterior from the background measurement, given by Equation (14.13). Combining Equations (14.13), (14.14) and (14.15), we can compute the joint posterior for s and b. To find the posterior for s alone, independent of the background, we just marginalize with respect to b.

$$p(s|N_{\text{on}}, I) = \int_0^{b_{\text{max}}} db\ p(s, b|N_{\text{on}}, I). \tag{14.16}$$

The exact integral can be calculated after expanding the binomial, $(s + b)^{N_{\text{on}}}$ and making use of the incomplete gamma function to evaluate the integrals, as we did in Section 14.2. The details of this calculation are given in Appendix D. The result is

$$p(s|N_{\text{on}}, I) = \sum_{i=0}^{N_{\text{on}}} C_i \frac{T_{\text{on}}(sT_{\text{on}})^i e^{-sT_{\text{on}}}}{i!}, \tag{14.17}$$

where

$$C_i \approx \frac{\left(1 + \frac{T_{\text{off}}}{T_{\text{on}}}\right)^i \frac{(N_{\text{on}}+N_{\text{off}}-i)!}{(N_{\text{on}}-i)!}}{\sum_{j=0}^{N_{\text{on}}} \left(1 + \frac{T_{\text{off}}}{T_{\text{on}}}\right)^j \frac{(N_{\text{on}}+N_{\text{off}}-j)!}{(N_{\text{on}}-j)!}}. \tag{14.18}$$

Note:

$$\sum_{i=0}^{N_{\text{on}}} C_i = 1.$$

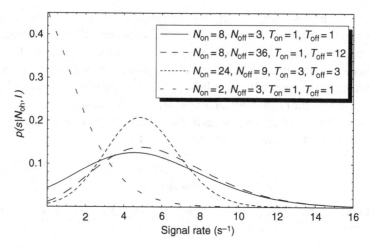

Figure 14.2 The posterior probability density of the source rate, $p(s|N_{on}, I)$, plotted for four different combinations of N_{on}, N_{off}, T_{on}, and T_{off}.

Figure 14.2 shows plots of the posterior probability density of the source rate $p(s|N_{on}, I)$ for four different combinations of N_{on}, N_{off}, T_{on}, and T_{off}. Notice that even in the case that $N_{on} < N_{off}$, the value of $p(s|N_{on}, I)$ is always zero for non-physical negative values of s. It is also clear that increasing the ON measurement time and/or the OFF measurement time sharpens the definition of $p(s|N_{on}, I)$.

We can gain a better understanding for the meaning of the complicated C_i term in Equation (14.17) by evaluating $p(s|N_{on}, I)$ for a restatement of the problem. The background information I includes N_{off}, the number of background events measured in the off-source measurement as well as T_{on} and T_{off}. For the state of information corresponding to N_{on}, I, we can use Bayes' theorem to compute $p(i|N_{on}, I)$, the probability that i of the on-source events are due to the source and $N_{on} - i$ are due to the background. Clearly, i is an integer that can take on values from 0 to N_{on}. We can then obtain the posterior probability for s, from the joint probability $p(s, i|N_{on}, I)$, by marginalizing over i as follows:

$$p(s|N_{on}, I) = \sum_{i=0}^{N_{on}} p(s, i|N_{on}, I) = \sum_{i=0}^{N_{on}} p(i|N_{on}, I)p(s|i, N_{on}, I)$$

$$= \sum_{i=0}^{N_{on}} p(i|N_{on}, I) \frac{T_{on}(sT_{on})^i e^{-sT_{on}}}{i!},$$

(14.19)

where we have used Equation (14.8) to evaluate $p(s|i, N_{on}, I)$. Comparing Equation (14.17) with (14.19), we can write $C_i = p(i|N_{on}, I)$, the probability that i of the ON measurement events are due to the source.

We are now able to interpret Equation (14.17) in the following useful way: Bayes' theorem estimates s by taking a weighted average of the posteriors one would obtain attributing $i = 0, 1, 2, \ldots, N_{on}$ events to the source. The weight C_i is equal to the probability of attributing i of the on-source events to the source, or equivalently, attributing $N_{on} - i$ events to the background, assuming M_{s+b} is true.

Now suppose our question changes from estimating the source rate, s, to the question of how confidently we can claim to have detected the source in the ON measurement. We might be tempted to reason as follows: for the source to have been detected in the ON measurement, then at least one of the N_{on} photons must have been from the source. The probability of attributing at least one of the N_{on} photons to the source is just the sum of C_i terms for $i = 1$ to N_{on}, which is given by

$$p(i \geq 1 | N_{on}, I) \approx \frac{\sum_{i=1}^{N_{on}} \left(1 + \frac{T_{off}}{T_{on}}\right)^i \frac{(N_{on}+N_{off}-i)!}{(N_{on}-i)!}}{\sum_{j=0}^{N_{on}} \left(1 + \frac{T_{off}}{T_{on}}\right)^j \frac{(N_{on}+N_{off}-j)!}{(N_{on}-j)!}}. \tag{14.20}$$

In Figure 14.3, we have plotted $p(i \geq 1 | N_{on}, I)$ versus N_{on} for two different background OFF measurements. In the first case, $N_{off} = 3$ counts in a $T_{off} = 1$ s. In the second case, $N_{off} = 36$ counts in a $T_{off} = 12$ s. In many experiments, the cost and/or effort required to obtain ON measurements (e.g., particle accelerator beam ON) is much greater than for OFF measurements. As our knowledge of the background rate improves, we see that $p(i \geq 1 | N_{on}, I)$ decreases for $N_{on} \leq 3$, the expected number of background photons, and increases above this. For $N_{on} = 8$ and $N_{off} = 3$ counts in $T_{off} = 1$ s, the probability is 95.6%. This rises to 98.9% for $N_{off} = 36$ counts in $T_{off} = 12$ s.

What is wrong with setting the probability of source detection equal to $p(i \geq 1 | N_{on}, I)$? The answer is that the C_i probabilities are based on assuming that

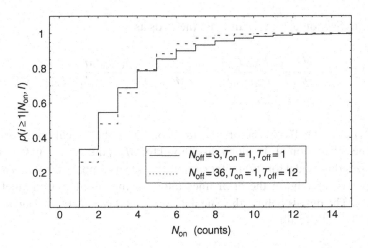

Figure 14.3 The probability of attributing at least one of the N_{on} photons to the source, $p(i \geq 1 | N_{on}, I)$, assuming model M_{s+b} is true, versus the number of counts in the ON measurement, for two different durations of the background OFF measurement.

the model M_{s+b} is true. They do not account for the extra complexity of M_{s+b} when compared to an intrinsically simpler model, M_b, which asserts that the Poisson rate in the ON source measurement is equal to b, i.e., no source is present. To answer the source detection question, we need to compare the probabilities of these two models, which we do next. This approach automatically introduces an Occam penalty which penalizes M_{s+b} for its greater complexity.

14.4.2 Source detection question

In the parameter estimation problem above, we assumed the truth of model M_{s+b} and estimated the value of s. Here, we will address the source detection (model selection) problem: "Have we detected a source in the ON measurement?" To answer this, we will compute the odds ratio, $O_{\{s+b,b\}}$, of two models M_{s+b} and M_b, which have the following meaning:

$M_b \equiv$ "the ON measurement is solely due to the Poisson background rate, b." The prior probability for b is derived from the OFF measurement.
$M_{s+b} \equiv$ "the Poisson rate in the ON source measurement is equal to $s + b$." The prior for s is a constant in the range 0 to $s_{\max}$. Again, the prior probability for b is derived from the OFF measurement.

In our earlier work, our background information was represented by $I = N_{\text{off}}, I_b, M_{s+b}$. In the current model selection problem, we will use the abbreviation

$$I_{\text{off}} = N_{\text{off}}, I_b. \tag{14.21}$$

According to Section 3.14, we can write the odds as

$$O_{\{s+b,b\}} = \frac{p(M_{s+b}|N_{\text{on}}, I_{\text{off}})}{p(M_b|N_{\text{on}}, I_{\text{off}})} = \frac{p(M_{s+b}|I_{\text{off}})}{p(M_b|I_{\text{off}})} \frac{p(N_{\text{on}}|M_{s+b}, I_{\text{off}})}{p(N_{\text{on}}|M_b, I_{\text{off}})} \tag{14.22}$$
$$= \text{prior odds} \times B_{\{s+b,b\}},$$

where $B_{\{s+b,b\}}$ is the Bayes factor, the ratio of the global likelihoods for the two models. The calculation of the global likelihood for M_{s+b} introduces an Occam factor that penalizes this model for its greater complexity when compared to M_b. The Occam factor depends directly on the prior uncertainty in the additional parameter s (see Section 3.5). The details behind the calculation of the Bayes factor are again given in Appendix D.2. The result is

$$B_{\{s+b,b\}} \approx \frac{N_{\text{on}}!}{s_{\max} T_{\text{on}} (N_{\text{on}} + N_{\text{off}})!} \sum_{i=0}^{N_{\text{on}}} \frac{(N_{\text{on}} + N_{\text{off}} - i)!}{(N_{\text{on}} - i)!} \left(1 + \frac{T_{\text{off}}}{T_{\text{on}}}\right)^i. \tag{14.23}$$

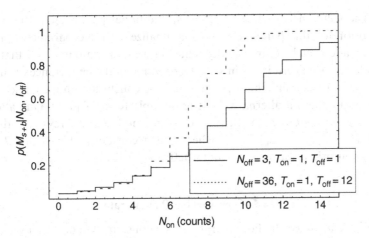

Figure 14.4 The probability that model M_{s+b} is true, $p(M_{s+b}|N_{on}, I_{off})$, versus the number of counts in the ON measurement, for two different durations of the background OFF measurement.

In what follows, we will assume a prior odds ratio of 1, so $O_{\{s+b,b\}} = B_{\{s+b, b\}}$. Since $p(M_{s+b}|N_{on}, I_{off}) + p(M_b|N_{on}, I_{off}) = 1$, we can express $p(M_{s+b}|N_{on}, I_{off})$ in terms of the odds ratio:

$$p(M_{s+b}|N_{on}, I_{off}) = \frac{1}{(1 + 1/O_{\{s+b, b\}})}. \tag{14.24}$$

In Figure 14.4, we have plotted $P(M_{s+b}|N_{on}, I_{off})$ versus N_{on} for two different background OFF measurements, assuming an $s_{max} = 30$. In the first case, $N_{off} = 3$ counts in a $T_{off} = 1$ s. In the second case, $N_{off} = 36$ counts in a $T_{off} = 12$ s. For a given value of N_{on}, the probability that a source is detected decreases for $N_{on} \le 3$, the expected number of background photons, and increases above this. For $N_{on} = 8$ and $N_{off} = 3$ counts in $T_{off} = 1$ s, the probability is 61.0%. This rises to 74.6% for $N_{off} = 36$ counts in $T_{off} = 12$ s. These are significantly lower than the corresponding probabilities for $p(i \ge 1|N_{on}, I)$ as shown in Figure 14.3.

Sensitivity to prior information

We close by reminding the reader that the conclusions of any Bayesian analysis are always conditional on the truth of our prior information. It is clear that in the source detection (model selection) problem, the Bayes factor (Equation (14.23)) is very sensitive to the choice of the prior upper boundary,[2] s_{max}. Halving the value of s_{max} causes the Bayes factor to increase by a factor of two. It is useful to consider the uncertainty in s_{max} as introducing a systematic error into our conclusion. As discussed in Section 3.6, we can readily allow for the effect of systematic error in a Bayesian

---

[2] We met this issue before in Section 3.8.1, for a completely different spectral line problem.

analysis. The solution is to treat s_{max} as an additional parameter in the problem, choose a prior for this parameter and marginalize over the parameter. This will introduce an additional Occam penalty reducing the odds ratio in a way that quantitatively reflects our uncertainties in s_{max}. Depending on the importance of the result, it may be useful to examine the dependence of the conclusion on the choice of prior for s, by considering an alternative but reasonable form of prior. One alternative choice worth considering in this case is a modified Jeffreys of the form $p(s|I) = 1/\{(s+a)\ln[(a+s_{max})/a]\}$, where a is a constant. This *modified Jeffreys* looks like a uniform prior for $s < a$ and a Jeffreys for $s > a$.

14.5 Time-varying Poisson rate

So far, we have assumed the Poisson rate, r, is a constant. We now analyze a simple problem in which r is a function of time.

$I \equiv$ "We want to estimate the half-life, τ, of a radioactive sample. The sample count rate is given by

$$r(t|r_0, \tau) = r_0 2^{-t/\tau}, \qquad (14.25)$$

where r_0 is the count rate at $t = 0$. Assume a uniform prior for r_0, and an independent Jeffreys prior for τ."

The data, D, are a simulated list of times,[3] $\{t_i\}$, for $N = 60$ measured Geiger counter clicks, for a radioactive sample with a half-life of 30 s. To make use of the full resolution of the data, we will work with the individual event times.

$$D = \{1.44, 1.64, 2.55, 2.88, 2.9, 3.27, 4.39, 5.01, 5.08, 5.11, 5.33, 5.4, 5.45,$$
$$5.58, 5.79, 6.17, 7.84, 7.86, 8.8, 8.9, 11.71, 11.73, 11.78, 14.88, 14.96,$$
$$15.61, 18.95, 19.42, 20.11, 20.28, 21.46, 21.52, 23.62, 24.21, 24.38,$$
$$24.39, 25.76, 27.92, 28.92, 29.28, 29.74, 30.04, 31.34, 32.08, 34.62,$$
$$35.04, 35.38, 36.43, 36.94, 38.97, 40.66, 41.62, 42.69, 43.02, 43.36,$$
$$45.11, 47.38, 49.65, 50.52, 51.22\}$$

From Bayes' theorem we can write

$$p(r_0, \tau|D, I) \propto p(r_0|I) \, p(\tau|I)p(D|r_0, \tau, I). \qquad (14.26)$$

The prior ranges' upper and lower boundaries for r_0 and τ are assumed to lie well outside the region of interest defined by the likelihood function. Thus, for our current parameter estimation problem, we can write

$$p(\tau|D, I) \propto \int_{r_0} dr_0 \frac{1}{\tau} \, p(D|r_0, \tau, I). \qquad (14.27)$$

---

[3] See Problem 6 for hints on how to simulate your own data set.

The likelihood can be calculated as follows: divide the observation period, T, into small time intervals, Δt, each containing either one event (counter click) or no event. We assume that Δt is sufficiently small that the average rate in the interval Δt is approximately equal to the rate at any time within the interval.

From the Poisson distribution, $p_0(t)$, the probability of no event in Δt is given by

$$p_0(t) = e^{-r(t)\Delta t}, \tag{14.28}$$

and the probability of one event is given by

$$p_1(t) = r(t)\Delta t e^{-r(t)\Delta t}. \tag{14.29}$$

If N and M are the number of time intervals in which one event and no events are detected, respectively, then the likelihood function is given by

$$
\begin{aligned}
p(D|r(t), I) &= \prod_{i=1}^{N} p_1(t_i) \prod_{j=1}^{M} p_0(t_j) \\
&= \Delta t^N \left[\prod_{i=1}^{N} r(t_i) \right] \exp\left[-\sum_{j=1}^{N+M} r(t_j)\Delta t \right] \\
&= \Delta t^N \left[\prod_{i=1}^{N} r(t_i) \right] \exp\left[-\int_T dt\, r(t) \right].
\end{aligned}
\tag{14.30}
$$

Note: we have replaced the sum of $r(t_j)\Delta t$ over all the observed intervals by the integral of the rate over the intervals, with the range of integration $T = (N + M)\Delta t$. The Δt^N factor in the likelihood cancels with the same factor which appears in the denominator of Bayes' theorem, so the result does not depend on the size of Δt, and is well-behaved even in the limit where Δt becomes infinitesimal.

Now use Equation (14.25) to substitute for $r(t)$ in Equation (14.30):

$$
\begin{aligned}
p(D|r_0, \tau, I) &= \Delta t^N \left[\prod_{i=1}^{N} r_0 2^{-t_i/\tau} \right] \exp\left[-\int_T dt\, r_0 2^{-t/\tau} \right] \\
&= \Delta t^N r_0^N 2^{\left(-\frac{1}{\tau}\sum_{i=1}^{N} t_i\right)} \exp\left[-\frac{r_0\tau}{\ln 2}\left(1 - 2^{-T/\tau}\right) \right],
\end{aligned}
\tag{14.31}
$$

where T is the duration of the time series data.

The marginal posterior probability density for the half-life can be obtained by substituting Equation (14.31) into Equation (14.27), and then evaluated numerically.[4] The result is shown in Figure 14.5.

---

[4] Since there are only two parameters, the joint probability distribution can be quickly evaluated at a finite number of two-dimensional grid points. Each marginal distribution can be obtained by summing the results for grid points along the other parameter.

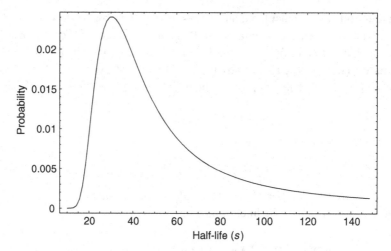

Figure 14.5 The marginal posterior for the half-life of a radioactive sample.

What if the rate were to vary with time in some unknown fashion, perhaps periodically? For an interesting treatment of this class of problem, see Gregory and Loredo (1992, 1993, and 1996) and Loredo (1992).

14.6 Problems

1. The results of an ON/OFF measurement are $N_{on} = 5$ counts, $N_{off} = 90$ counts, $T_{on} = 1$ s, $T_{off} = 100$ s. Plot $p(s)$, the posterior probability of the source rate.
2. Using the data given in Problem 1, compute the probability of attributing i of the on-source events to the source. Plot a graph of this probability for the range $i = 0$ to 5.
3. Repeat Problem 2, only this time, compute the probability of attributing j of the on-source events to the background. Plot a graph of this probability for the range $j = 0$ to 5.
4. For the radioactive counter times given in Section 14.5, compute and plot the marginal posterior PDF for the initial count rate r_0. Assume a uniform prior for r_0 in the range 0 to 5 counts s^{-1} and a Jeffreys prior for τ in the range 2 to 300 s.
5. Compute and plot the marginal posterior PDF for the radioactive sample half-life, based on the first 30 counter times given in Section 14.5. Assume a uniform prior for r_0 in the range 0 to 5 counts s^{-1} and a Jeffreys prior for τ in the range 2 to 300 s.
6. Simulate your own radioactive decay time series ($N = 100$ count times) for an initial decay rate of one count per second and a half-life of 40 s. Divide the decay into 20 bins. For each bin, use Equation (5.62) to generate a list of Poisson time intervals for a Poisson rate corresponding to the time of the middle of the bin. Convert the time intervals to a sequence of count times and add the start time of the corresponding bin. You can use *Mathematica*'s **Select[***list***,#1** < *bin boundary time* **&]** command to select out the times from any particular list that are less than the end time of the corresponding bin. Use **ListPlot[]** to plot a graph of your time series.

Appendix A
Singular value decomposition

Frequently, the solution of a linear least-squares problem using the normal equations of Section 10.2.2,

$$\hat{\mathbf{A}} = (\mathbf{G}^T \mathbf{E}^{-1} \mathbf{G})^{-1} \mathbf{G}^T \mathbf{E}^{-1} \mathbf{D} = \mathbf{\Psi}^{-1} \mathbf{G}^T \mathbf{E}^{-1} \mathbf{D}, \tag{A.1}$$

fails because a zero pivot occurs in the matrix calculation because $\mathbf{\Psi}$ is singular. If the matrix is sufficiently close to singular, the answer becomes extremely sensitive to round-off errors, in which case you typically get fitted A_α's with very large amplitudes that are delicately balanced to almost precisely cancel out. Here is an example of a nearly singular matrix:

$$\begin{pmatrix} 1.0 & 1.0 \\ 1.0 & 1.0001 \end{pmatrix}. \tag{A.2}$$

This arises when the data do not clearly distinguish between two or more of the basis functions provided.

The solution is to use singular value decomposition (SVD). When some combination of basis functions is irrelevant to the fit, SVD will drive the amplitudes of these basis functions down to small values rather than pushing them up to delicately canceling infinities.

How does SVD work?
First, we need to restate the least-squares problem slightly differently. In least-squares, we want to minimize

$$\chi^2 = \sum_i \frac{(d_i - f_i)^2}{\sigma_i^2} = \sum_i \left\{ \frac{d_i}{\sigma_i} - \frac{f_i}{\sigma_i} \right\}^2, \tag{A.3}$$

where

$$f_i(A) = \sum_{\alpha=1}^{M} A_\alpha g_{i\alpha}. \tag{A.4}$$

In matrix form, we can write

$$\chi^2 = |\mathbf{X}\,\mathbf{A} - \mathbf{b}|^2,$$ (A.5)

where X is the *design matrix* given by

$$\mathbf{X} = \begin{pmatrix} \frac{g_1(x_1)}{\sigma_1} & \cdot & \frac{g_M(x_1)}{\sigma_1} \\ \cdot & \cdot & \cdot \\ \frac{g_1(x_N)}{\sigma_N} & \cdot & \frac{g_M(x_N)}{\sigma_N} \end{pmatrix},$$ (A.6)

and

$$\mathbf{b} = \begin{pmatrix} \frac{d_1}{\sigma_1} \\ \cdot \\ \frac{d_N}{\sigma_N} \end{pmatrix}.$$ (A.7)

The problem is to find **A** which minimizes Equation (A.5).

Any rectangular matrix can be written in reduced SVD form as follows[1] (see any good linear algebra text for a proof):

$$\begin{array}{ccccc} \mathbf{X} & = & \mathbf{U}^T & \mathbf{D} & \mathbf{V}, \\ (N \times M) & & (N \times M)(M \times M)(M \times M) & & \end{array}$$ (A.9)

where the columns of **U** are orthonormal and are the eigenvectors of $\mathbf{X}\,\mathbf{X}^T$. The columns of **V** are orthonormal and are the eigenvectors of $\mathbf{X}^T\mathbf{X}$ and $\mathbf{X}^T\mathbf{X} = \mathbf{G}^T\mathbf{E}^{-1}\mathbf{G}$. The elements of the diagonal matrix, **D**, are called the *singular values* of **X**. These singular values, $\omega_1, \omega_1, \ldots, \omega_M$, are the square roots of the non-zero eigenvalues of both $\mathbf{X}^T\mathbf{X}$ and $\mathbf{X}\,\mathbf{X}^T$.

$$\mathbf{D} = \begin{pmatrix} \omega_1 & 0 & \cdot & 0 \\ 0 & \omega_2 & \cdot & 0 \\ \cdot & \cdot & \cdot & \cdot \\ \cdot & \cdot & \cdot & \cdot \\ 0 & 0 & \cdot & \omega_M \end{pmatrix}$$ (A.10)

The number of singular values is equal to the rank of **X**.

The three matrices U, D, V can be obtained in *Mathematica* with the command

$$\boxed{\{\mathbf{U}, \mathbf{D}, \mathbf{V}\} = \mathbf{SingularValues}[\mathbf{X}]}$$

---

[1] In some texts, Equation (A.9) is written in the form

$$\mathbf{X} = \mathbf{U}\,\mathbf{D}\,\mathbf{V}^T.$$ (A.8)

In a least-squares problem, the design matrix, $\mathbf{X}$, does not have an inverse because it is not a square matrix, but we can use SVD to construct a pseudo-inverse, $\mathbf{X}^+$, which provides the best solution in a least-squares sense to Equation (A.5) in terms of the basis functions that the data can distinguish between. We will designate that solution $\mathbf{A}^+$, which is given by

$$\mathbf{A}^+ = \mathbf{X}^+\mathbf{b}. \tag{A.11}$$

The pseudo-inverse, in the *Mathematica* convention, is given by

$$\mathbf{X}^+ = \mathbf{V}^T\mathbf{D}^{-1}\mathbf{U} = \mathbf{V}^T\left[\text{diag}\left(\frac{1}{\omega_\alpha}\right)\right]\mathbf{U}. \tag{A.12}$$

Before using Equation (A.11), it is desirable to investigate the singular values of the design matrix $\mathbf{X}$. If any singular values ω_α are close to zero, set $1/\omega_\alpha = 0$ for that α. This corresponds to throwing away one or more basis functions that the data can not decide on. The *condition number* of a matrix is defined by

$$
\begin{aligned}
\text{condition number} &= \frac{\text{maximum eigenvalue}}{\text{minimum eigenvalue}} \\
&= \left(\frac{\text{maximum singular value}}{\text{minimum singular value}}\right)^2.
\end{aligned}
\tag{A.13}
$$

The matrix becomes ill-conditioned if the reciprocal of its condition number approaches the floating point accuracy.

The **PseudoInverse** command in *Mathematica* allows one to specify a **Tolerance** option for throwing away basis functions whose singular values are $<$ **Tolerance** multiplied by the maximum singular value. Equation (A.11) becomes

$$\mathbf{A}^+ = \textbf{PseudoInverse}[\mathbf{X}, \textbf{Tolerance} -> t].\mathbf{b}$$

Note: use of the **Tolerance** option can lead to strange results when fitting polynomials. This is because the range (scale of changes) of the different basis functions can be very different, e.g., the x^3 will have a much larger range than say the x term. The different scales of the basis functions make a simple comparison of singular values difficult. In this case, it is better to rescale x so it lies in the range 0 to 1 or -1 to $+1$ before computing the singular values.

Appendix B
Discrete Fourier Transforms

B.1 Overview

The operations of convolution, correlation and Fourier analysis play an important role in data analysis and experiment simulations. These operations on digitally sampled data are efficiently carried out with the *Discrete Fourier Transform* (DFT), and in particular with a fast version of the DFT called the *Fast Fourier Transform* (FFT). In this section, we introduce the DFT and FFT and explore their relationship to the analytic Fourier transform and Fourier series. We investigate Fourier deconvolution of a noisy signal with an optimal Weiner filter. We also learn how to minimally sample data without losing any information (Nyquist theorem) and about the aliasing that occurs when a waveform is sampled at an insufficient rate. Since the DFT is an approximation to the analytic Fourier transform, we learn how to zero pad a time series to obtain accurate Fourier amplitudes, and to remove spurious end effects in discrete convolution. Finally, we explore two commonly used approaches to spectral analysis and how to reduce the variance of spectral density estimators.

B.2 Orthogonal and orthonormal functions

Before we consider the problem of representing a function in terms of a sum of orthogonal basis functions, we review the more familiar problem of representing a vector in terms of a set of orthogonal unit vectors. The vector **F** can be represented as the vector sum

$$\mathbf{F} = F_x \hat{i}_x + F_y \hat{i}_y + F_z \hat{i}_z, \tag{B.1}$$

where the $\hat{i}$'s are unit vectors along 3 mutually perpendicular axes. Because the unit vectors satisfy the relation $\hat{i}_x \cdot \hat{i}_y = \hat{i}_y \cdot \hat{i}_z = \hat{i}_z \cdot \hat{i}_x = 0$, they are said to be an *orthogonal set*. In addition,

$$\hat{i}_x \cdot \hat{i}_x = \hat{i}_y \cdot \hat{i}_y = \hat{i}_z \cdot \hat{i}_z = 1, \tag{B.2}$$

so they are called an *orthonormal set*. They are not the only orthonormal set which can be used to represent **F**. Any orthonormal coordinate system can be used. For example, in spherical coordinates,

$$\mathbf{F} = F_r \hat{i}_r + F_\theta \hat{i}_\theta + F_\phi \hat{i}_\phi. \tag{B.3}$$

In summary, we can represent the vector $\mathbf{F}$ by

$$\mathbf{F} = \sum_{n=1}^{N} F_n \hat{i}_n, \tag{B.4}$$

where the orthonormal set of basis vectors satisfies the condition

$$\begin{aligned} \hat{i}_m \cdot \hat{i}_n = \delta_{m,n} = 1 \quad & m = n \\ = 0 \quad & m \neq n. \end{aligned} \tag{B.5}$$

To find the scalar component along $\hat{i}_m$, take the scalar product of $\hat{i}_m$ with $\mathbf{F}$.

$$F_m = \mathbf{F} \cdot \hat{i}_m. \tag{B.6}$$

In an analogous fashion, we would like to represent a function $y(t)$ in terms of an orthonormal set of basis functions,

$$y(t) = \sum_{n=1}^{N} Y_n \phi_n(t). \tag{B.7}$$

We need to define the equivalent of the scalar product for use with functions, which is called the *inner product* for two functions. It is easy to show that

$$\frac{1}{\pi} \int_{-\pi}^{+\pi} \sin mt \sin nt \, dt = \delta_{m,n}. \tag{B.8}$$

If this relationship is to be satisfied, then the inner product between two functions should be defined as $\int_{-\pi}^{\pi} x(t) y(t) \, dt$. Thus, if

$$y(t) = \sum_{n=1}^{N} Y_n \, \phi_n(t) = \sum_{n=1}^{N} Y_n \frac{\sin nt}{\sqrt{\pi}}, \tag{B.9}$$

then the inner product of $y(t)$ and $\phi_m(t)$ is

$$\int y(t) \phi_m(t) dt = \sum_{n=1}^{N} Y_n \int_{-\pi}^{+\pi} \frac{\sin nt}{\sqrt{\pi}} \frac{\sin mt}{\sqrt{\pi}} dt$$
$$= \sum_{n=1}^{N} Y_n \delta_{m,n} = Y_m. \tag{B.10}$$

The next question is whether any function can be represented by an orthonormal series like Equation (B.9). Since all terms on the right side of Equation (B.9) are periodic in t, with period 2π, their sum will also be periodic. If the original function $y(t)$ is periodic as well, over the same period, then this series representation will be valid for all values of t. Otherwise, the series will only represent the function $y(t)$ in the range $-\pi < t < \pi$.

What about the question of completeness? Returning to the vector analogy: in general, a set of unit vectors is not complete if it is possible to find a vector belonging to the space which is orthogonal to every vector in the set, i.e., 3 basis vectors required for 3-dimensional space. How many dimensions does our function space have? It is clear from Equation (B.9) that for values $n > N$, it is possible to find a function $\sin nt$ which is orthogonal to all members. Furthermore, even if N is infinite, $\cos nt$ is orthogonal to every member of the set for any value of n. A complete set must contain at least an infinite number of functions of the form $\sin nt$ and $\cos nt$. We will say more about completeness later when we discuss the Nyquist sampling theorem.

Many well-known sets of functions exhibit relationships similar to Equation (B.9). Such function sets, not sinusoidal and usually not periodic, can be used to form orthogonal series. In general, the inner product can be defined in the interval $a < t < b$ as follows:

$$\int_a^b x(t)\, y^*(t)\, \omega(t)dt, \tag{B.11}$$

where $y^*(t)$ is the complex conjugate of $y(t)$ and $\omega(t)$ is a weighting function. A set of functions $\phi_n(t)$ is an orthogonal set over the range $a < t < b$ if

$$\Phi_n \cdot \Phi_m = \int_a^b \phi_n(t)\phi_m^*(t)\omega(t)dt = k_n\, \delta_{m,n}. \tag{B.12}$$

The set is orthonormal if $k_n = 1$ for all n.

Examples of useful orthogonal functions are:

1. $1, \cos x, \sin x, \cos 2x, \sin 2x, \ldots$ used in a Fourier series
2. Legendre polynomials
3. Spherical harmonics
4. Bessel functions
5. Chebyshev or Tschebyscheff polynomials
6. Laguerre polynomials
7. Hermite polynomials

B.3 Fourier series and integral transform

In the case of the Fourier series, the limits $-\pi$ to $+\pi$ correspond to the period of the function. The limits can be made arbitrary by setting $t = 2\pi t'/T$. Then

$$\frac{1}{\pi}\int_{-\pi}^{+\pi} \sin nt \sin mt\; dt = \frac{2}{T}\int_{-T/2}^{+T/2} \sin(2\pi n\, f_0 t') \sin(2\pi m\, f_0 t')dt', \tag{B.13}$$

where $f_0 = 1/T$.

B.3.1 Fourier series

The Fourier series representation of $y(t)$ is given by

$$y(t) = \sum_{n=0}^{\infty} [a_n \cos 2\pi n \, f_0 t + b_n \sin 2\pi n \, f_0 t]. \tag{B.14}$$

To find the coefficients a_n and b_n of $y(t)$, compute the inner product of $y(t)$ with the cosine and sine basis functions. This is analogous to finding the component of a vector in Equation (B.6).

$$a_n = \frac{2}{T} \int_{-T/2}^{T/2} y(t) \cos 2\pi n \, f_0 t \, dt, \tag{B.15}$$

and,

$$b_n = \frac{2}{T} \int_{-T/2}^{T/2} y(t) \sin 2\pi n \, f_0 t \, dt, \tag{B.16}$$

for $n = 0, 1, \ldots$

Exponential notation

We will rewrite Equation (B.14) using the common exponential notation, where

$$\cos 2\pi n \, f_0 t = \frac{1}{2} (e^{i2\pi n f_0 t} + e^{-i2\pi n f_0 t})$$

$$\sin 2\pi n \, f_0 t = \frac{1}{2i} (e^{i2\pi n f_0 t} - e^{-i2\pi n f_0 t}).$$

Equation (B.14) becomes

$$y(t) = a_0 + \frac{1}{2} \left[\sum_{n=1}^{\infty} (a_n - ib_n) e^{i2\pi n f_0 t} + \sum_{n=1}^{\infty} (a_n + ib_n) e^{-i2\pi n f_0 t} \right]. \tag{B.17}$$

To simplify the expression, negative values of n are introduced. Thus, we can rewrite Equation (B.17) as

$$y(t) = a_0 + \frac{1}{2} \left[\sum_{n=1}^{\infty} (a_n - ib_n) e^{-i2\pi(-n)f_0 t} + \sum_{n=1}^{\infty} (a_n + ib_n) e^{-i2\pi n f_0 t} \right]$$

$$= a_0 + \frac{1}{2} \left[\sum_{n=-1}^{-\infty} (a_{|n|} - ib_{|n|}) e^{-i2\pi n f_0 t} + \sum_{n=1}^{\infty} (a_n + ib_n) e^{-i2\pi n f_0 t} \right] \tag{B.18}$$

$$= \sum_{n=-\infty}^{\infty} Y_n e^{-i2\pi n f_0 t},$$

where

$$Y_n = \begin{cases} \frac{1}{2}(a_n + ib_n), & n > 0 \\ a_0, & n = 0 \\ \frac{1}{2}(a_{|n|} - ib_{|n|}), & n < 0. \end{cases} \tag{B.19}$$

In Equation (B.18), we expanded $y(t)$ in terms of the $e^{-i2\pi n f_0 t}$ basis set. Alternatively we could have expanded it in terms of the $e^{i2\pi n f_0 t}$ basis set, in which case we would write

$$y(t) = \sum_{n=-\infty}^{\infty} Y'_n e^{i2\pi n f_0 t}, \tag{B.20}$$

where

$$Y'_n = \begin{cases} \frac{1}{2}(a_n - ib_n), & n > 0 \\ a_0, & n = 0 \\ \frac{1}{2}(a_{|n|} + ib_{|n|}), & n < 0. \end{cases} \tag{B.21}$$

Both conventions exist in the literature, but we will use the convention specified by Equations (B.18) and (B.19) to be consistent with the default definitions for the Discrete Fourier Transform (discussed in Section B.7) used in *Mathematica*.

B.3.2 Fourier transform

In the Fourier series, the Fourier frequency components are separated by $f_0 = 1/T$. In the limit as $T \to \infty$, the Fourier components, Y_n, become a continuous function $Y(f)$ where $Y(f)$ is called the *Fourier transform* of $y(t)$.

$$y(t) = \int_0^\infty [g(f)\cos 2\pi f t + k(f)\sin 2\pi f t] df. \tag{B.22}$$

If we define $Y(f)$ by

$$Y(f) = \begin{cases} \frac{1}{2}\{g(f) + ik(f)\}, & f > 0 \\ g(0), & f = 0 \\ \frac{1}{2}\{g(|f|) - ik(|f|)\}, & f < 0, \end{cases} \tag{B.23}$$

then Equation (B.22) can be rewritten as

$$y(t) = \int_{-\infty}^{+\infty} Y(f)e^{-i2\pi f t} df \quad \text{where} \quad Y(f) = \int_{-\infty}^{+\infty} y(t)\, e^{i2\pi f t} dt. \tag{B.24}$$

Designate Fourier transform pairs by

$$y(t) \Longleftrightarrow Y(f). \tag{B.25}$$

Units: If t is measured in seconds, then f is in units of cycles s^{-1} = hertz. If t is measured in minutes, then f is in cycles per minute. Some common Fourier transform pairs are illustrated in Figure B.1.

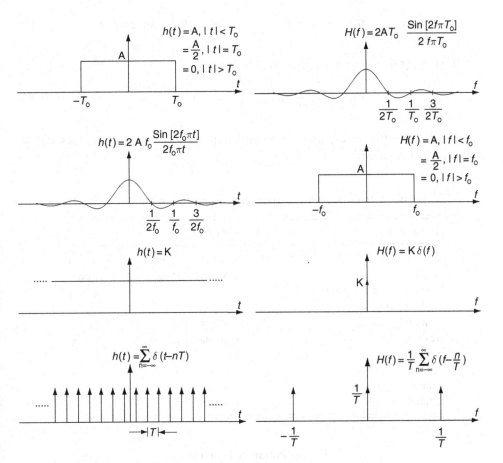

Figure B.1 Some common Fourier transform pairs.

Note: we normally associate the analysis of periodic functions such as a square wave with Fourier series rather than Fourier transforms. We can show that a Fourier transform reduces to a Fourier series whenever the function being transformed is periodic.

Example:
Consider the FT of a pulse time waveform

$$h(t) = \begin{cases} A, & |t| < T_0 \\ A/2, & t = \pm T_0 \\ 0, & |t| > T_0. \end{cases} \qquad (B.26)$$

The value of the function at a discontinuity must be defined to be the mid-value if the inverse Fourier transform is to hold (Brigham, 1988).

$$H(f) = \int_{-T_0}^{T_0} A e^{i2\pi ft} dt = A \int_{-T_0}^{T_0} \cos 2\pi\, ft dt + iA \int_{-T_0}^{T_0} \sin 2\pi\, ft dt. \tag{B.27}$$

The final integral $= 0$ since the integral is odd:

$$\Rightarrow H(f) = \left[\frac{A}{2\pi f} \sin 2\pi\, ft\right]_{-T_0}^{T_0} = 2AT_0 \frac{\sin 2\pi T_0 f}{2\pi T_0 f}. \tag{B.28}$$

Table B.1 gives the correspondence between important symmetry properties in time and frequency domains.

Table B.1 *Correspondence of symmetry properties in the two domains.*

If $h(t)$ is . . .	then $H(f)$ is . . .
real	real part even imaginary part odd
imaginary	real part odd imaginary part even
real even and imaginary odd	real
real odd and imaginary even	imaginary
real and even	real and even
real and odd	imaginary and odd
imaginary and even	imaginary and even
imaginary and odd	real and odd
complex and even	complex and even
complex and odd	complex and odd

B.4 Convolution and correlation

We previously considered some fundamental properties of the FT. However, there exists a class of FT relationships whose importance outranks those previously considered. These properties are the *convolution* and *correlation* theorems. The importance of the convolution operation in science is discussed in Section B.4.3.

Convolution integral

$$y(t) = \int_{-\infty}^{+\infty} s(\tau)h(t - \tau)d\tau = s(t) * h(t) \tag{B.29}$$

or alternatively,

$$y(t) = \int_{-\infty}^{+\infty} h(\tau)\, s(t - \tau)d\tau, \tag{B.30}$$

where the symbol $*$ in Equation (B.29) stands for convolution. The convolution procedure is illustrated graphically in Figure B.2.

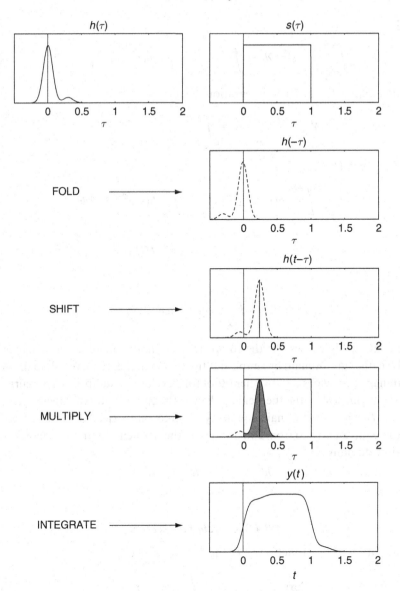

Figure B.2 Graphical illustration of convolution.

B.4.1 Convolution theorem

The convolution theorem is one of the most powerful tools in modern scientific analysis. According to the convolution theorem, the FT of the convolution of two functions is equal to the product of the FT of each function separately.

$$h(t) * s(t) \Longleftrightarrow H(f)\, S(f). \qquad (B.31)$$

Proof :

$$\int_{-\infty}^{+\infty} y(t)e^{i2\pi ft}dt = \int_{-\infty}^{+\infty} \left[\int_{-\infty}^{+\infty} s(\tau)h(t-\tau)d\tau \right] e^{i2\pi ft}dt. \qquad (B.32)$$

Now interchange the order of integration:

$$Y(f) = \int_{-\infty}^{+\infty} s(\tau) \left[\int_{-\infty}^{+\infty} h(t-\tau)e^{i2\pi ft}dt \right] d\tau. \qquad (B.33)$$

Let $r = (t - \tau)$. Then,

$$\left[\int_{-\infty}^{+\infty} h(t-\tau)e^{i2\pi ft}dt \right] = \int_{-\infty}^{\infty} h(r)e^{i2\pi f(r+\tau)}dr \qquad (B.34)$$

$$= e^{i2\pi f\tau}H(f). \qquad (B.35)$$

Therefore,

$$Y(f) = H(f) \int_{-\infty}^{\infty} s(\tau)e^{i2\pi f\tau}d\tau = H(f)\,S(f). \qquad (B.36)$$

This relationship allows one the complete freedom to convolve mathematically (or visually) in the time domain by simple multiplication in the frequency domain. Among other things, it provides a convenient tool for developing additional FT pairs.

Figure B.3 illustrates the theorem applied to the convolution of a rectangular pulse of width $2T_0$ with a bed of nails (an array of uniformly-spaced delta functions).

We can equivalently go from convolution in the frequency domain to multiplication in the time domain.

$$h(t)s(t) \longleftrightarrow H(f) * S(f). \qquad (B.37)$$

B.4.2 Correlation theorem

The correlation of two functions $s(t)$ and $h(t)$ is defined by

$$\text{Corr}\,(s,h) = z(t) = \int_{-\infty}^{\infty} s(\tau)h(\tau+t)d\tau. \qquad (B.38)$$

It is useful to compare Equation (B.38) with Equation (B.29) for convolution. Convolution involves a folding of $h(\tau)$ before shifting, while correlation does not. According to the correlation theorem,

$$z(t) \Longleftrightarrow H(f)S^*(f) = Z(f). \qquad (B.39)$$

Thus, Corr $(s,h) \Longleftrightarrow H(f)S^*(f)$ are an FT pair.
Compare with the convolution: $s(t) * h(t) \Longleftrightarrow S(f)H(f)$.

CONVOLVE IN TIME DOMAIN   MULTIPLY IN FREQUENCY DOMAIN

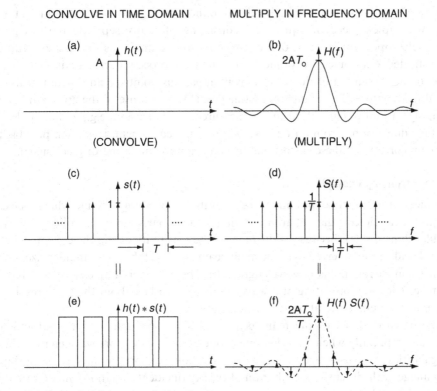

Figure B.3 Example of the convolution theorem.

Note: if $s(t)$ is a real and even function, $S(f)$ is real and $S^*(f) = S(f)$. Thus, in this case, $\text{Corr}\,(s, h) = s(t) * h(t) = $ convolution.

B.4.3 Importance of convolution in science

The goal of science is to infer how nature works based on measurements or observations.

$$\text{Nature} \Rightarrow \overset{\text{Measurements}}{\text{Apparatus}} \Rightarrow \text{Observation.}$$

Unfortunately, all measurement apparatus introduces distortions which need to be understood. Often, the most exciting questions of the day require pushing the measurement equipment to its very limits where the distortions are most extreme. Of course, some of these distortions can be approximately calculated from theory, like the diffraction effects of a telescope or microscope, but others need to be measured. Are there any general principles that help us to understand these distortions that we can apply to any measurement process? The answer is yes for any linear measurement process where the output is linearly related to the input signal, even if the apparatus is a very complex piece of equipment consisting of many separate parts, e.g., a radio telescope consisting of one or more parabolic antennas and a room full of sophisticated electronics. Any linear measurement process corresponds mathematically to a convolution of the measurement apparatus point spread function with the signal from nature. The point spread function is the response of the apparatus to an input signal that is unresolved in the measurement dimension, e.g., a short pulse in the time dimension. From an understanding of the equipment, it is often possible to partially correct for these distortions to better approximate the original signal.

Radio astronomy example

The simplest radio telescope consists of a parabolic collecting antenna which focuses, amplifies, and detects the radiation arriving within a narrow cone of solid angle (two angular dimensions). Because of diffraction, the sensitivity within the cone may have several peaks often referred to as the main beam and side lobes. The angular size of the main beam in radians is $\approx$ wavelength/telescope diameter. For example, a 100 m diameter telescope operating at a wavelength of 3 cm has about the same resolving power as the human eye at optical wavelengths.

The detailed shape of the main beam and side lobes may be very difficult to calculate, especially when the telescope is operated at very short wavelengths where irregularities in the telescope surface and gravitational distortions are most important. Any image of the intensity distribution of the sky (incident radiation), made with this instrument, will be blurred by this diffraction pattern or point spread function. Fortunately, the point spread function can be measured provided it is stable. This can be achieved by observing a strong "point" source of very small angular extent, much smaller than the main beam of the telescope.

The use of an unresolved "point" source to measure the telescope point spread function, and the blurring effect the point spread function has on a model extended source, are illustrated in Figure B.4 for one angular coordinate, θ. The dashed curve in the upper left panel shows the response of the telescope as a function of θ that we wish to measure. In this example, it consists of a main beam and a strong secondary side lobe. The solid curve represents the radio intensity distribution of an unresolved point source. The source is fixed in position but the telescope response, defined by the location of the center of the main beam and represented by θ_0, can be steered. By scanning the telescope (varying θ_0) across the point source, we can map out the telescope point spread function as shown in the lower left panel. One can see that the response of the telescope to the point source (point spread function) in θ_0 is the mirror image of the telescope response in θ.

Thus, to simulate an observation of an extended source, the model galaxy, we need to obtain the telescope response in θ by folding the measured point spread function in θ_0 about the main beam axis. Then for each pointing position of the telescope, we multiply the telescope response in θ times the galaxy intensity distribution and

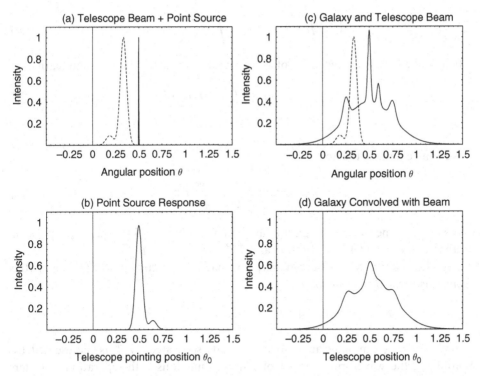

Figure B.4 A simulation of the response of a radio telescope to an unresolved "point" source and an extended source in one angular sky coordinate, θ. The dashed curve in the upper panels is the telescope sensitivity as a function of θ. The solid curve in the upper left represents the intensity distribution of a point source, and in the right panel, a model galaxy intensity distribution. The lower panels are the measured telescope output versus θ_0, the telescope pointing position.

integrate. The lower right panel shows the results of such a convolution with our model galaxy intensity distribution.

The convolution theorem provides what is often a simpler way of computing the measured galaxy intensity distribution. Just Fourier transform the telescope sensitivity in θ and the galaxy intensity distribution, multiply the two transforms, and then take the inverse Fourier transform. The inverse of convolution, called deconvolution, is demonstrated in Section B.10.1.

B.5 Waveform sampling

In many practical situations, we only obtain samples of some continuous function. How could we go about sampling the continuous voltage function, $v(t)$ to obtain a sample at $t = \tau$? One way would be to convert $v(t)$ to a frequency $f(t) = kv(t)$ and count the number of cycles (N) in some short time interval τ to $\tau + \triangle T$.

$$N = \int_{\tau}^{\tau + \Delta T} f(t)dt = \int_{\tau}^{\tau + \Delta T} k \, v(t)dt. \tag{B.40}$$

If $s(\tau)$, the time averaged value of $v(t)$ around $t = \tau$, is the desired sample, then

$$s(\tau) = \frac{N}{k \, \Delta T} = \frac{\int_{\tau}^{\tau + \Delta T} k \, v(t)dt}{\int_{\tau}^{\tau + \Delta T} k \, dt}. \tag{B.41}$$

We can generalize this to

$$s(\tau) = \frac{\int_{-\infty}^{+\infty} k(t - \tau) \, v(t)dt}{\int_{-\infty}^{+\infty} k(t - \tau)dt}, \tag{B.42}$$

where $k(t)$ is some suitable weighting function. One choice of $k(t)$ is a square pulse of width ΔT and height $1/\Delta T$. In this case, $\int_{-\infty}^{+\infty} k(t - \tau)dt = 1$.

The ideal choice which we can only approach in practice is $\rightarrow k(t) = \delta(t)$ (the *impulse* or *Dirac delta function*)

$$\delta(t - \tau) = 0 \quad \text{for} \quad t \neq \tau \quad \text{and} \quad \int_{-\infty}^{+\infty} \delta(t - \tau)dt = 1. \tag{B.43}$$

In most texts, sampling at uniform intervals separated by T is represented by multiplying the waveform by a set of impulse functions with separation T (often referred as a bed of nails).

Note: the FT of a bed of nails is another bed of nails such that

$$\triangle(t) = \sum_{n=-\infty}^{\infty} \delta(t - nT) \Longleftrightarrow \triangle(f) = \frac{1}{T} \sum_{n=-\infty}^{\infty} \delta\left(f - \frac{n}{T}\right), \tag{B.44}$$

where the area integral of one nail in the frequency domain $= 1/T$.

We can use the convolution theorem to illustrate (see Figure B.5) how to determine the FT of a sampled waveform. The FT of the sampled waveform is then a periodic function where one period is equal to, within the constant $(1/T)$, the FT of the continuous function $h(t)$. Notice that in this situation, we have not lost any information about the original continuous $h(t)$. By picking out one period of the transform, we can reconstruct identically the continuous waveform by the inverse FT.

B.6 Nyquist sampling theorem

Consider what would happen in Figure B.5 if the sampling interval were made larger. In the frequency domain, the separation $(= 1/T)$ between impulse functions of $S(f)$ would decrease. Because of this decreased spacing of the frequency impulses, their convolution with the frequency function $H(f)$ results in overlapping waveforms as illustrated in panel (f) of Figure B.6.

MULTIPLY IN TIME DOMAIN   CONVOLVE IN FREQUENCY DOMAIN

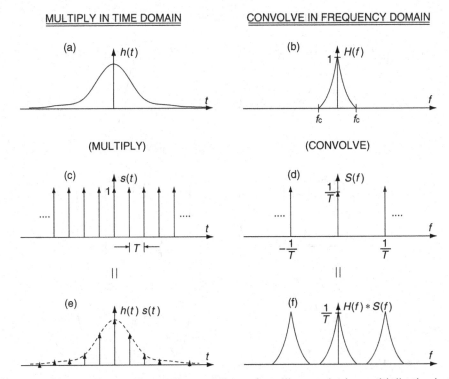

Figure B.5 The Fourier transform of a sampled waveform illustrated using multiplication in the time domain and convolution in the frequency domain.

In this case, we can no longer recover an undistorted simple period which is identical with the FT of the continuous function $h(t)$. This distortion of a sampled waveform is known as *aliasing*. It arises because the original waveform was not sampled at a sufficiently high rate. For a given sampling interval, T, the *Nyquist frequency* is defined as $1/(2T)$. If the waveform that is being sampled contains frequency components above the Nyquist frequency, they will give rise to aliasing.

Examination of Figure B.6(b) and (d) indicates that convolution overlap will occur until the separation of the impulses of $S(f)$ is increased to $1/T = 2f_c$, where f_c is the highest frequency. Therefore, the sampling interval T must be $\leq 1/(2f_c)$.

The *Nyquist sampling theorem* states that if the Fourier transform of a function $h(t)$ is zero for all $|f| \geq f_c$, then the continuous function $h(t)$ can be uniquely determined from a knowledge of its sampled values at intervals of $T \leq 1/(2f_c)$. If $H(f) = 0$ for $|f| \geq f_c$ then we say that $H(f)$ is band-limited. In practice, it is a good idea to use a smaller sample interval $T \approx 1/(4f_c)$.

Conversely, if $h(t)$ is time-limited, that is $h(t) = 0$ for $|t| \geq T_c$ then $h(t)$ can be uniquely reconstructed from samples of $H(f)$ at intervals $\triangle f = 1/(2T_c)$.

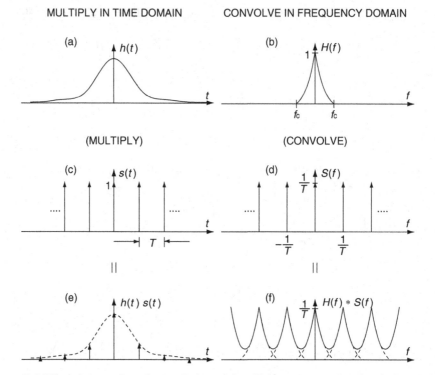

MULTIPLY IN TIME DOMAIN   CONVOLVE IN FREQUENCY DOMAIN

(a) $h(t)$

(b) $H(f)$

(MULTIPLY)   (CONVOLVE)

(c) $s(t)$

(d) $S(f)$

(e) $h(t)\,s(t)$

(f) $H(f)*S(f)$

Figure B.6 When the waveform is sampled at an insufficient rate, overlapping (referred to as aliasing) occurs in the transform domain.

B.6.1 *Astronomy example*

In this example, taken from radio astronomy, we are interested in determining the intensity distribution, $b(\theta, \phi)$ of a galaxy with the Very Large Array (VLA). The position of any point in the sky is specified by the two spherical coordinates θ, ϕ.

The VLA is an aperture synthesis radio telescope consisting of twenty-seven 25 m diameter dish antennas which can be moved along railway tracks to achieve a variety of relative spacings up to a maximum of 21 km. By this means, it can make images with an angular resolution equivalent to that of a telescope with a 21 km diameter aperture (0.08 arcseconds at $\lambda = 1$ cm). The signal from each antenna is cross-correlated separately with the signals from all other antennas while all antennas track the same source. It can be shown that each cross-correlation is directly proportional to a two-dimensional Fourier component of $b(\theta, \phi)$. If there are N antennas, there are $N(N-1)/2$ correlation pairs. The VLA records $27(27-1)/2 = 351$ Fourier components simultaneously.

The FT of $b(\theta, \phi)$ is equal to $B(u, v)$. The quantities u and v are called *spatial frequencies* and have units $= 1/\theta$ where θ is in radians (dimensionless). Let $u = x/\lambda$ and $v = y/\lambda$ be the components of the projected separation of any pair of antennas on a plane perpendicular to the line of sight to the distant radio source in units of the observing wavelength.

In practice, one wants to measure the minimum number of Fourier components necessary to reconstruct $b(\theta, \phi)$, i.e., move the dish antennas along the railway tracks as little as possible. This is where the sampling theorem comes in handy.

If the galaxy is known to have a finite angular width $\Delta\theta = \Delta\phi = \Delta\psi$, then from the sampling theorem, this means we only need to sample in u and v at intervals of

$$\Delta u = \Delta v \le \frac{1}{\Delta\psi}. \tag{B.45}$$

Note: in this problem, $\Delta\psi$ is the equivalent to $2f_c$ in Figure B.6 in the time frequency problem.

Thus, if $\Delta\psi \equiv 10\,\text{arcseconds} = 4.8 \times 10^{-5}$ radians,

$$\Delta u = \frac{\Delta x}{\lambda} = 20\,833 = \Delta v = \frac{\Delta y}{\lambda}. \tag{B.46}$$

If the wavelength of observation, $\lambda = 6\,\text{cm}$, then $\Delta x = 1.25\,\text{km}$, which means that the increment in antenna spacing required for complete reconstruction of $b(\theta, \phi)$ is 1.25 km.

Since the antennas are 25 m in diameter, they could in principle be spaced at intervals of $\sim 25\,\text{m}$ and at that increment in spacing, we could obtain all the Fourier components (coefficients) necessary to reconstruct (synthesize) the image of a source of angular extent ≤ 8.3 arcmin. Because each antenna will shadow its neighbor at very close spacings, the limiting angular size that can be completely reconstructed is smaller than this.

B.7 Discrete Fourier Transform

B.7.1 Graphical development

The approach here is to develop the *Discrete FT* (abbreviated DFT) from a graphical derivation based on the waveform sampling and the convolution theorem, following the treatment given by Brigham (1988). The steps in this derivation are illustrated in Figure B.7.

Panel (a) of Figure B.7 shows the continuous FT pair $h(t)$ and $H(f)$. To obtain discrete samples, we multiply $h(t)$ by the bed of nails shown in the time domain which corresponds to convolving in the f domain, with the result shown in (c). Note: the effect of sampling is to create a periodic version of $H(f)$. To represent the fact that we only want a finite number of samples, we multiply in the time domain by the rectangular window function of width T_0 of panel (d), which corresponds to convolving with its frequency transform. This has side lobes, which produce an undesirable ripple in our transform as seen in panel (e). One way to reduce the ripple is to use a tapered window function instead of a rectangle.

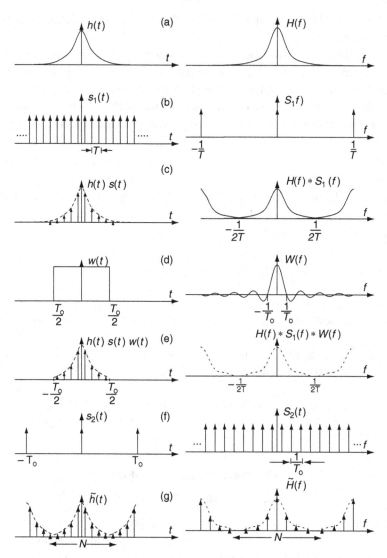

Figure B.7 Graphical development of the Discrete Fourier Transform. See discussion in Section B.7.1.

Finally, to manipulate a finite number of samples in the frequency domain, we multiply in the frequency domain by a bed of nails at a frequency interval $\Delta f = 1/T_0$ as shown in panels (f) and (g). After taking the DFT of our N samples of $h(t)$, we obtain an N-sample approximation of the $H(f)$ as shown in panel (g).

Note 1: sampling in the frequency domain results in a periodic version of $h(t)$ in the time domain. It is very important to be aware of this periodic property when executing convolution operations using the DFT.

Note 2: the location of the rectangular window function $w(t)$ is very important. Its width T_0 equals NT, where T is the sample interval. If $w(t)$ had been located so that a sample value coincided with each end-point, the rectangular function would be $N + 1$ sample values, and the convolution of $h(t)s(t)w(t)$ with the impulses spaced at intervals of T_0 as shown in panels (f) and (g) would result in time domain aliasing.

B.7.2 Mathematical development of the DFT

We now estimate the FT of a function from a finite number of samples. Suppose we have N equally spaced samples $h_k \equiv h(kT)$ at an interval of T seconds, where $k = 0, 1, 2, \ldots, N - 1$. From the discussion of the Nyquist sampling theorem, for a sample interval T, we can only obtain useful frequency information for $|f| < f_c$. We seek estimates at the discrete values

$$f_n \equiv \frac{n}{NT}, \quad n = -\frac{N}{2}, \ldots, -1, 0, 1, \ldots, \frac{N}{2}, \tag{B.47}$$

where the upper and lower limits are $\pm f_c$. Counting $n = 0$, this range corresponds to $N + 1$ values of frequency, but only N values will be unique.

$$
\begin{aligned}
H(f_n) &= \int_{-\infty}^{+\infty} h(t) \, e^{i2\pi f_n t} dt \\
&\approx \sum_{k=0}^{N-1} h(kT) \, e^{i2\pi f_n kT} \, T \\
&= T \sum_{k=0}^{N-1} h_k \, e^{i2\pi n \Delta f kT} \\
&= T \sum_{k=0}^{N-1} h_k \, e^{i2\pi nk/N} \\
&= TH(n\Delta f) = TH\left(\frac{n}{NT}\right) = TH_n,
\end{aligned}
\tag{B.48}
$$

where $T =$ the sample interval and h_k for $k = 0, \ldots, N - 1$, are the sample values of the truncated $h(t)$ waveform.

H_n is defined as the DFT of h_k and given by

$$H_n = \sum_{k=0}^{N-1} h_k \, e^{i2\pi nk/N}. \tag{B.49}$$

Defined in this way, H_n does not depend on the sample interval T.

The relationship between the DFT of a set of samples of a continuous function $h(t)$ at interval T, and the continuous FT of $h(t)$ can be written as

$$H(f_n) = TH_n. \tag{B.50}$$

We can show that since Equation (B.49) for H_n is periodic, there are only N distinct complex values computable. To show this, let $n = r + N$, where r is an arbitrary integer from 0 to $N - 1$.

$$H_n = H\left(\frac{n}{NT}\right) = \sum_{k=0}^{N-1} h_k \, e^{i2\pi k(r+N)/N}$$

$$= \sum_{k=0}^{N-1} h_k \, e^{i2\pi kr/N} \, e^{i2\pi k}$$

$$= \sum_{k=0}^{N-1} h_k \, e^{i2\pi kr/N} \quad\quad\quad (B.51)$$

$$= H\left(\frac{r}{NT}\right) = H_r,$$

since $e^{i2\pi k} = \cos(2\pi k) + i\sin(2\pi k) = 1$ for k an integer.

Until now, we have assumed that the index n varies from $-N/2$ to $N/2$. Since H_n is periodic, it follows that $H_{-n} = H_{N-n}$ so $H_{-N/2} = H_{N/2}$ and thus we only need N values of n. It is customary to let n vary from 0 to $N - 1$. Then $n = 0$ corresponds to the DFT at zero frequency and $n = N/2$ to the value at $\pm f_c$. Values of n between $N/2 + 1$ and $N - 1$ correspond to values of the DFT for negative frequencies from $-(N/2 - 1), -(N/2 - 2), \ldots, -1$. Thus, to display the DFT in the same way as an analytic transform is displayed ($-f$ on the left and $+f$ on the right), it is necessary to reorganize the DFT frequency values.

B.7.3 Inverse DFT

Again, our starting point is the integral FT:

$$h(kT) = h_k = \int_{-\infty}^{+\infty} H(f) \, e^{-i2\pi fkT} \, df$$

$$h(kT) \approx \sum_{n=0}^{N-1} H(f_n) \, e^{-i2\pi f_n kT} \, \Delta f. \quad\quad (B.52)$$

Now substitute $H(f_n) = TH_n$ and $\Delta f = (1/NT)$:

$$h_k = \sum_{n=0}^{N-1} TH_n \, e^{-i2\pi f_n kT} \, \Delta f$$

$$h_k = \frac{1}{N} \sum_{n=0}^{N-1} H_n \, e^{-i2\pi f_n kT}. \quad\quad (B.53)$$

Note: the definition of DFT pair given in *Mathematica* is the more symmetrical form

$$H_n = \frac{1}{\sqrt{N}} \sum_{k=1}^{N} h_k \ e^{i2\pi(n-1)(k-1)/N} \tag{B.54}$$

$$h_k = \frac{1}{\sqrt{N}} \sum_{n=1}^{N} H_n \ e^{-i2\pi(n-1)(k-1)/N}, \tag{B.55}$$

and the zero frequency corresponds to the $n = 1$ term.

Box B.1

The Fourier transform of a list of real or complex numbers, represented by u_i, is given by the *Mathematica* command
Fourier [{$u_1, u_2, \cdots, u_n$}].
The inverse Fourier transform is given by
InverseFourier [{$v_1, v_2, \cdots, v_n$}].
Mathematica can find Fourier transforms for data in any number of dimensions. In n dimensions, the data is specified by a list nested n levels deep. For two dimensions, often used in image processing, the command is
Fourier [{{$u_{11}, u_{12}, \cdots, u_{1n}$}, {$u_{21}, u_{22}, \cdots, u_{2n}$}, $\cdots$ }].
An example of the use of the FFT in the convolution and deconvolution of an image is given in the accompanying *Mathematica* tutorial.

B.8 Applying the DFT

We have already developed the relationship between the discrete and continuous Fourier transforms. Here, we explore the mechanics of applying the DFT to the computation of Fourier transforms and Fourier series. The primary concern is one of correctly interpreting these results.

B.8.1 DFT as an approximate Fourier transform

To illustrate the application of the DFT to the computation of Fourier transforms, consider Figure B.8. Figure B.8(a) shows the real function $f(t)$, given by

$$f(t) = \begin{cases} 0, & t < 0 \\ e^{-t}, & t \geq 0. \end{cases} \tag{B.56}$$

We wish to compute by means of the DFT an approximation to the Fourier transform of this function.

The first step in applying the discrete transform is to choose the number of samples N and the sample interval T. For $T = 0.25$, we show the samples of $f(t)$ within the dashed rectangular window function in Figure B.8(b). Note: the start of the window

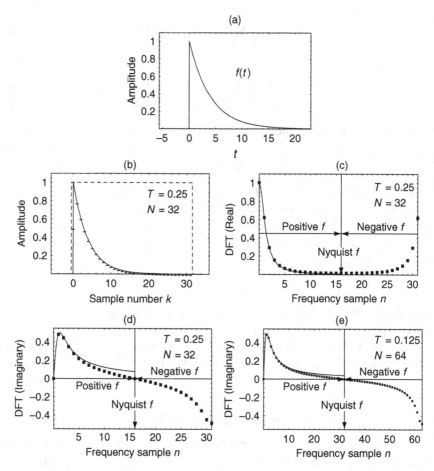

Figure B.8 A 32-point DFT of the function $f(t)$. The function itself is plotted in panel (a). Panel (b) illustrates the location of the 32 time samples within the rectangular window function (dashed box). Panel (c) compares the real part of the DFT to the continuous Fourier transform shown by the solid curve. The imaginary part of the DFT is compared to the continuous case (solid curve) in panel (d). Panel (e) illustrates the improved agreement obtained by halving the sample interval in time and doubling the number of samples.

function occurs $T/2$ ahead of the first sample so that there are only N samples within the window, as discussed in Section B.7.1. Also, the value of the function at a discontinuity must be defined to be the mid-value if the inverse Fourier transform is to hold. Since the DFT assumes the function is periodic, we set this value equal to the average of the function value at both ends to avoid the discontinuity at $t = 0$.

We next compute the Fourier transform using the DFT approximation

$$H(f_n) \approx T H_n = T H \left(\frac{n}{NT} \right), \tag{B.57}$$

where

$$H\left(\frac{n}{NT}\right) = \sum_{k=0}^{N-1} \left[e^{-kT}\right] e^{i2\pi nk/N}; \quad n = 0, 1, \ldots, N-1. \tag{B.58}$$

Note: the scale factor T in Equation (B.57), is required to produce equivalence between the continuous and discrete transforms. These results are shown in panels (c) and (d) of Figure B.8. In Figure B.8(c), we show the real part of the Fourier transform as computed by Equation (B.58). The index $n = 0$ corresponds to zero frequency or the DC term, which is proportional to the data average. Note: the real part of the discrete transform is symmetrical about $n = N/2$, the Nyquist frequency sample. The real part of a Fourier transform of a real function is even and the imaginary part of the transform is odd. In Figure B.8(b), the results for the real part for $n > N/2$ are simply negative frequency results. For $T = 0.25\,\mathrm{s}$, the physical frequency associated with frequency sample $n = N/2$ is $1/(2T) = 2\,\mathrm{Hz}$. Sample $(N/2) + 1 = 17$ corresponds to a negative frequency $= -(N/2 - 1)/(NT) = -1.875\,\mathrm{Hz}$, and sample $n = 31$ corresponds to the frequency $= -1/(NT) = -0.125\,\mathrm{Hz}$.

The conventional method of displaying results of the discrete Fourier transform is to graph the results of Equation (B.58) as a function of the parameter n. As long as we remember that those results for $n > N/2$ actually relate to negative frequency results, then we should encounter no interpretation problems.

In panel (d) of Figure B.8, we illustrate the imaginary part of the Fourier transform and the discrete transform. As shown, the discrete transform approximates rather poorly the continuous transform for the higher frequencies. To reduce this error, it is necessary to decrease the sample interval T and increase N. Panel (e) shows the improved agreement obtained by halving T and doubling N to 64 samples. We note that the imaginary function is odd with respect to $n = N/2$. Again, those results for $n > N/2$ are to be interpreted as negative frequency results.

In summary, applying the discrete Fourier transform to the computation of the Fourier transform only requires that we exercise care in the choice of T and N and interpret the results correctly. For a worked example of the DFT using *Mathematica*, see the section entitled "Exercise on DFT, Zero Padding and Nyquist Sampling," in the accompanying *Mathematica* tutorial.

B.8.2 Inverse discrete Fourier transform

Assume that we are given the continuous real and imaginary frequency functions considered in the previous discussion and that we wish to determine the corresponding time function by means of the inverse discrete Fourier transform

$$h(kT) = \triangle f \sum_{n=0}^{N-1} [R(n\triangle f) + iI(n\triangle f)] e^{-i2\pi nk/N} \quad \text{for } k = 0, 1, \ldots, N-1, \tag{B.59}$$

where $\triangle f$ is the sample interval in frequency. Assume $N = 32$ and $\triangle f = 1/8$.

Since we know that $R(f)$, the real part of the complex frequency function, must be an even function, then we fold $R(f)$ about the frequency $f = 2.0\,\mathrm{Hz}$, which corresponds to the sample point $n = N/2$. As shown in Figure B.9(a), we simply sample the frequency function up to the point $n = N/2$ and then fold these values about $n = N/2$ to obtain the remaining samples.

In Figure B.9(b), we illustrate the method of determining the N samples of the imaginary part of the frequency function. Because the imaginary frequency function is odd, we must not only fold about the sample value $N/2$ but also flip the results. To preserve symmetry, we set the sample at $n = N/2$ to zero.

Computation of Equation (B.59) with the sampled function illustrated in Figures B.9(a) and (b) yields the inverse discrete Fourier transform. The result is a complex function whose imaginary part is approximately zero and whose real part is as shown in panel (c). We note that at $k = 0$, the result is approximately equal to the correct mid-value and reasonable agreement is obtained for all but the results for k large. Improvement can be obtained by reducing $\triangle f$ and increasing N.

The key to using the discrete inverse Fourier transform for obtaining an approximation to continuous results is to specify the sampled frequency functions correctly.

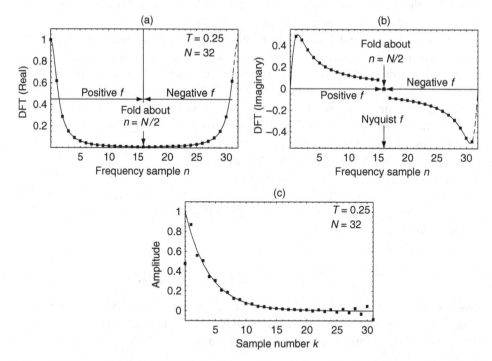

Figure B.9 Panels (a) and (b) illustrate the sampling of the real and imaginary parts of the continuous Fourier transform in readiness for computing the inverse DFT which is shown in panel (c).

Figures B.9(a) and (b) illustrate this correct method. One should observe the scale factor Δf which was required to give a correct approximation to continuous inverse Fourier transform results.

B.9 The Fast Fourier Transform

The FFT (Cooley and Tukey, 1965) is a very efficient method of implementing the DFT that removes certain redundancies in the computation and greatly speeds up the calculation of the DFT. Consider the DFT

$$A(n) = \sum_{k=0}^{N-1} x(k) e^{i2\pi nk/N} \quad (n = 0, 1, \ldots, N-1), \tag{B.60}$$

where we have replaced kT by k and n/NT by n for convenience of notation.

Let $w = e^{i2\pi/N}$. Equation (B.60) can be written in matrix form:

$$\begin{bmatrix} A(0) \\ A(1) \\ A(2) \\ A(3) \end{bmatrix} = \begin{bmatrix} w^0 & w^0 & w^0 & w^0 \\ w^0 & w^1 & w^2 & w^3 \\ w^0 & w^2 & w^4 & w^6 \\ w^0 & w^3 & w^6 & w^9 \end{bmatrix} \begin{bmatrix} x(1) \\ x(2) \\ x(3) \\ x(4) \end{bmatrix} \tag{B.61}$$

$$\text{or } \mathbf{A}(n) = \mathbf{w}^{nk}\,\mathbf{x}(k). \tag{B.62}$$

It is clear from the matrix representation that since $\mathbf{w}$ and possibly $\mathbf{x}(k)$ are complex, then N^2 complex multiplications and $(N)(N-1)$ complex additions are necessary to perform the required matrix computation. The FFT owes its success to the fact that the algorithm reduces the number of multiplications from N^2 to $N \log_2 N$.

For example, if $N = 1024 = 2^{10}$

$$N^2 = 2^{20} \text{ operations in DFT}$$

$$N \log_2 N = 2^{10} \times 10 \text{ operations in FFT}.$$

This amounts to a factor of 100 reduction in computer time and round-off errors are also reduced.

How does it work?

The FFT takes an N-point transform and splits it into two $N/2$-point transforms. This is already a saving, since $2(N/2)^2 < N^2$. The $N/2$-point transforms are not computed, but each split into two $N/4$-point transforms. It takes $\log_2 N$ of these splittings, so that generating the N-point transform takes a total of approximately $N \log_2 N$ operations rather than N^2.

The mathematics involves a splitting of the data set $x(k)$ into odd and even labeled points, $y(k)$ and $z(k)$.

$$\text{Let } y(k) = x(2k) \quad \text{for} \quad k = 0, 1, \dots, N/2 - 1$$
$$z(k) = x(2k + 1). \tag{B.63}$$

Equation (B.60) can be rewritten as:

$$A(n) = \sum_{k=0}^{N/2-1} \left\{ y(k)e^{i4\pi nk/N} + z(k)e^{i2\pi n(2k+1)/N} \right\}$$

$$= \sum_{k=0}^{N/2-1} y(k)e^{i4\pi nk/N} + e^{i2\pi n/N} \sum_{k=0}^{N/2-1} z(k)e^{i4\pi nk/N}. \tag{B.64}$$

This will still generate the whole set $A(n)$ if n is allowed to vary over the full range $(0 \leq n \leq N - 1)$. First, let n vary over $(0 \leq n \leq N/2 - 1)$. Then

$$A(n) = B(n) + C(n)w^n \quad (\text{valid for } 0 \leq n \leq N/2 - 1), \tag{B.65}$$

where

$$B(n) = \sum_{k=0}^{N/2-1} y(k)w^{2nk} \quad \text{and} \quad C(n) = \sum_{k=0}^{N/2-1} z(k)w^{2nk}$$

$$\text{for } n = 0, \dots, \frac{N}{2} - 1. \tag{B.66}$$

But since $B(n)$ and $C(n)$ are periodic in the half-interval, generating $A(n)$ for the second half may be done without further computing using the same $B(n)$ and $C(n)$:

$$A\left(n + \frac{N}{2}\right) = B(n) + C(n)w^n w^{N/2}$$

$$= B(n) - C(n)w^n \quad (0 \leq n \leq N/2 - 1), \tag{B.67}$$

where

$$w^{N/2} = e^{i\pi} = \cos \pi = -1. \tag{B.68}$$

The work of computing an N-point transform $A(n)$ has been reduced to computing two $N/2$ point transforms $B(n)$ and $C(n)$ and appropriate multiplicative phase factors w^n. Each of these sub-sequences $y(k)$ and $z(k)$ can be further subdivided with each step involving a further reduction in operations. These reductions can be carried out as long as the original sequence is a power of 2.

Consider $n = 8 = 2^3$. In 3 divisions we go from 1×8, to 2×4, to 4×2, to 8×1. Note: the DFT of one term is simply the term itself,

$$\text{i.e. } A(0) = \sum_{k=0} x(k)e^{-i2\pi kn/N} = x(0), \quad \text{for} \quad n = 0. \tag{B.69}$$

In the above we have assumed $N =$ power of 2. (2 is called the *radix* $= r$). One can use other values for the radix, e.g., $N = 4^2$

$$\text{Speed enhancement: } \simeq \frac{N \log_r N}{N^2}.$$

In addition to the splitting into sub-sequences, the FFT also makes use of period-icities in the exponential term w^{nk} to eliminate redundant operations.[1]

$$w^{nk} = w^{nk \bmod N} \tag{B.70}$$

e.g., if $N = 4, n = 2$ and $k = 3$, then

$$w^{nk} = w^6 = \exp\left[\left(\frac{i2\pi}{4}\right)(6)\right] = \exp(-i3\pi)$$

$$= \exp(-i\pi) = \exp\left[\left(\frac{i2\pi}{4}\right)(2)\right] = w^2 \tag{B.71}$$

$$w^2 = -w^0.$$

Note: the FFT is not an approximation but a method of computing which reduces the work by recognizing symmetries and by not repeating redundant operations.

B.10 Discrete convolution and correlation

One of the most common uses of the FFT is for computing convolutions and correlations of two time functions. Discrete convolution can be written as

$$s(kT) = \sum_{i=0}^{N-1} h(iT)r[(k-i)T] = h(kT) * r(kT). \tag{B.72}$$

According to the *Discrete Convolution Theorem*,

$$\sum_{i=0}^{N-1} h(iT)r[(k-i)T] \Longleftrightarrow H\left(\frac{n}{NT}\right)R\left(\frac{n}{NT}\right). \tag{B.73}$$

Note: the discrete convolution theorem assumes that $h(kT)$ and $r(kT)$ are periodic since the DFT is only defined for periodic functions of time. Usually, one is interested in convolving non-periodic functions. This can be accomplished with the DFT by the use of zero padding which is discussed in the next section. Reiterating, discrete convolution is only a special case of continuous convolution; discrete convolution assumes both functions repeat outside the sampling window.

---

[1] For example,

$$I \bmod m = I - \text{Int}\left(\frac{I}{m}\right) \times m$$

$$6 \bmod 4 = 6 - \text{Int}\left(\frac{6}{4}\right) \times 4 = 2$$

To efficiently compute the discrete convolution:

1. Use an **FFT** algorithm to compute $R(n)$ and $H(n)$.
2. Multiply the two transforms together, remembering that the transforms consist of complex numbers.
3. Then use the **FFT** to inverse transform the product.
4. The answer is the desired convolution $h(k) * r(k)$.

If both time functions are real (generally so) both of their transforms can be taken simultaneously. For details see Press (1992).

What about *deconvolution*? One is usually more interested in the signal $h(kT)$ before it is smeared by the instrumental response. Deconvolution is the process of undoing the smearing of the data, due to the effect of a known response function.

Deconvolution in the frequency domain consists of dividing the transform of the convolution by $R(n)$, e.g.,

$$H(n) = \frac{S(n)}{R(n)}, \tag{B.74}$$

and then transforming back to obtain $h(k)$.

This procedure can go wrong mathematically if $R(n)$ is zero for some value of n, so that we can't divide by it. This indicates that the original convolution has truly lost all information at that one frequency so that reconstruction of that component is not possible. Apart from this mathematical problem, the process is generally very sensitive to noise in the input data and to the accuracy to which $r(k)$ is known. This is the subject of the next section.

B.10.1 Deconvolving a noisy signal

We already know how to deconvolve the effects of the response function, $r(k)$ (short for $r(kT)$), of the measurement device, in the absence of noise. We transform the measured output, $s(k)$, and the response, $r(k)$, to the frequency domain yielding $S(n)$ (short for $S(n\Delta f)$) and $R(n)$. The transform, $H(n)$, of the desired signal, $h(k)$, is given by Equation (B.74).

Even without additive noise, this can fail because for some n, $R(n)$ may equal 0. The solution in this case is

$$H(n) = \frac{S(n)}{(R(n) + \epsilon)}, \tag{B.75}$$

where ϵ is very small compared to the maximum value of $R(n)$ and $\epsilon/R(n) >$ the machine precision.

Panel (a) in Figure B.10 shows our earlier result (see Figure B.4) of convolving the image of a galaxy with the response (beam pattern) of a radio telescope. Panel (b) shows the deconvolved version assuming perfect knowledge of $R(n)$ and a value of $\epsilon = 10^{-8}$. In practice, we will only be able to determine $R(n)$ to a certain accuracy which will limit the accuracy of the deconvolution. In panel (c), we have added independent Gaussian noise to (a), and panel (d) shows the best reconstruction obtained by varying the size of ϵ, which occurs for an $\epsilon \approx 0.15$. We now investigate

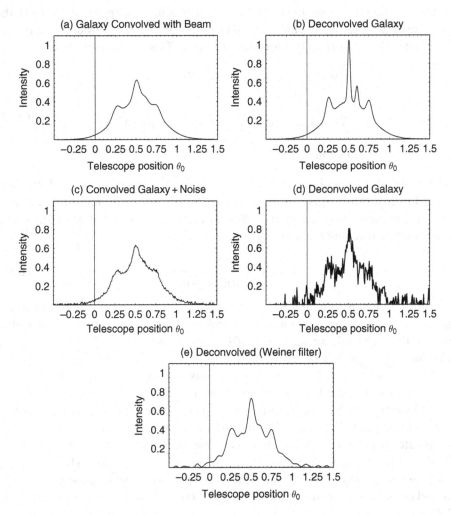

Figure B.10 Panel (a) shows the earlier result (see Figure (B.4)) of convolving a model galaxy with the point spread function of a radio telescope. Panel (b) shows the deconvolved galaxy image. Panel (c) is the same as (a) but with added Gaussian noise. Panel (d) is the best deconvolved image without any filtering of noise. Panel (e) shows the result of deconvolution using an optimal Weiner filter.

the use of an optimal Weiner filter to improve upon the reconstruction when noise is present.

B.10.2 Deconvolution with an optimal Weiner filter

If additive noise is present, the output from the measurement system is now $c(k)$, where

$$c(k) = s(k) + n(k). \tag{B.76}$$

The task is to find the optimum filter $\phi(k)$ or $\Phi(n)$, which, when applied to the $C(n)$, the transform of the measured signal $c(k)$, and then divided by $R(n)$, produces an output $\tilde{H}(n)$, that is closest to $H(n)$ in a least-squares sense. This translates to the equation

$$\sum_n \left| \frac{[S(n) + N(n)]\, \Phi(n)}{R(n)} - \frac{S(n)}{R(n)} \right|^2$$

$$= \sum_n |R(n)|^{-2} \left[|S(n)|^2 |1 - \Phi(n)|^2 + |N(n)|^2 |\Phi(n)|^2 \right] \tag{B.77}$$

$$= \text{minimum}.$$

If the signal $S(n)$ and noise $N(n)$ are uncorrelated, their cross product when summed over n can be ignored. Equation (B.77) will be a minimum if and only if the sum is minimized with respect to $\Phi(n)$ at every value of n. Differentiating with respect to Φ, and setting the result equal to zero gives

$$\Phi(n) = \frac{|S(n)|^2}{|S(n)|^2 + |N(n)|^2}. \tag{B.78}$$

The solution contains $S(n)$ and $N(n)$ but not the $C(n)$, the transform of the measured quantity $c(k)$. We happen to know $S(n)$ and $N(n)$ because we are working with a simulation. In general, we only know $C(n)$, so we estimate $S(n)$ and $N(n)$ in the following way: Figure B.11 shows the log of $|C(n)|^2, |S(n)|^2$ and $|N(n)|^2$ in panels (a), (b), (c), respectively. For small n, $|C(n)|^2$ has the same shape as $|S(n)|^2$, while at large n, it looks like the noise spectrum. If we only had $|C(n)|^2$, we could estimate $|N(n)|^2$ by extrapolating back into the signal region from high values of n. $|S(n)|^2$ is what stands above the estimate of $|N(n)|^2$ at low values of n. Finally, it is necessary to extrapolate the portion of $|S(n)|^2$ that sits above the estimated $|N(n)|^2$ to zero. Panel (d) shows the resulting optimal filter Φ given by Equation (B.78). Where $|S(n)|^2 \gg |N(n)|^2$, $\Phi(n) = 1$ and when the noise spectrum dominates, $\Phi(n) \approx 0$.

Panel (e) of Figure (B.10) shows the reconstruction $\tilde{H}(n)$ obtained using Equation (B.79):

$$\tilde{H}(n) = \frac{C(n)\Phi(n)}{R(n)}. \tag{B.79}$$

We investigate other approaches to image reconstruction in Chapter 8.

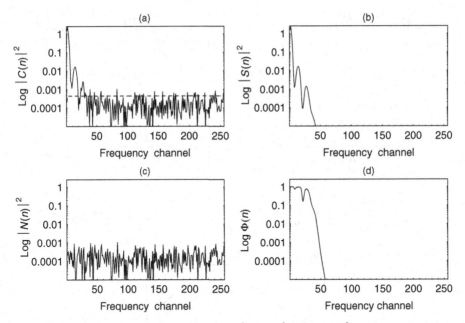

Figure B.11 The figure shows the log of $|C(n)|^2$, $|S(n)|^2$ and $|N(n)|^2$ in panels (a), (b), (c), respectively. Panel (d) shows the optimal Weiner filter.

B.10.3 Treatment of end effects by zero padding

Since the discrete convolution theorem assumes that the response function is periodic, it falsely pollutes some of the initial channels with data from the far end because of the wrapped-around response arising from the assumed periodic nature of the response function, $h(\tau)$. Although the convolution is carried out by multiplying the Fast Fourier Transforms of $h(\tau)$ and $x(\tau)$ and then inverse transforming back to the time domain, the polluting effect of the wrap-around is best illustrated by analyzing the situation completely in the time domain. Figure B.12 illustrates the convolution of $x(\tau)$, shown in panel (b), by an exponential decaying response function shown in panel (a). First, the response function is folded about $\tau = 0$, causing it to disappear from the left of panel (c). Since the DFT assumes that $h(\tau)$ is periodic, a wrap-around copy of $h(-\tau)$ appears at the right of the panel. To compute the convolution at t, we shift the folded response to the right by t, multiply $h(t - \tau)$ and $x(\tau)$ and integrate. Panel (e) shows the resulting polluted convolution.

To avoid polluting the initial samples, we add a buffer zone of zeros at the far end of the data stream. The width of this *zero padding* is equal to the maximum wrap-around of the response function (see Figure B.13). Note: if we increase N for the data stream, we must also add zeros to the response to make up the same number of samples. The wrap-around response shown in panels (c) and (d) is multiplied by zeros and so does not pollute the convolution as shown in panel (e).

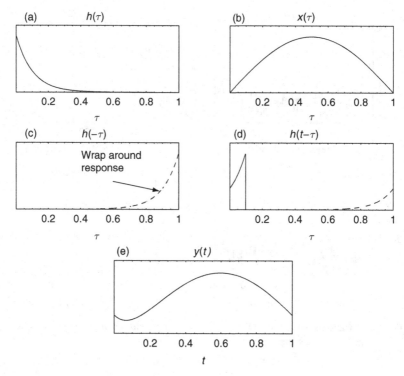

(a) $h(\tau)$

(b) $x(\tau)$

(c) $h(-\tau)$

Wrap around response

(d) $h(t-\tau)$

(e) $y(t)$

Figure B.12 Wrap-around effects in FFT convolution.

A *Mathematica* example of zero padding in the convolution of a two-dimensional image is given in the accompanying *Mathematica* tutorial.

B.11 Accurate amplitudes by zero padding

The FFT of $h(k)$ produces a spectrum $H(n)$ in which any intrinsically narrow spectral feature is broadened by convolution with the Fourier transform of the window function. For a rectangular window, this usually results in only two samples to define a spectral peak in $H(n)$ and deductions about the true amplitude of the peak are usually underestimated unless by chance one of these samples lies at the center of the peak. This situation is illustrated in Figure B.14. Panel (a) shows 12 samples of a sine wave taken within a rectangular window function indicated by the dashed lines. The actual samples are represented by the vertical solid lines. In panel (b), the DFT (vertical lines) is compared to the analytic Fourier transform of the windowed continuous sine wave. In this particular

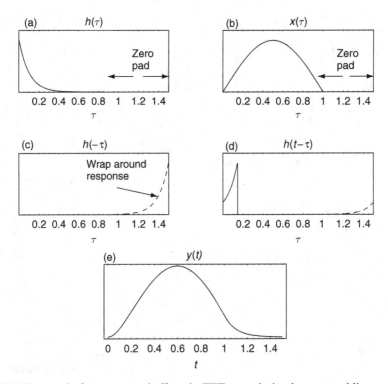

Figure B.13 Removal of wrap-around effects in FFT convolution by zero padding.

example, only two DFT components are visible; the others fall on the zeros of the analytic transform.

Doubling the length of the window function to $2 \times T_0$ causes the samples to be more closely spaced in the frequency domain $\Delta f = 1/(2T_0)$, but at the same time, the transform of the window function becomes narrower by the same factor. The net effect is to not increase the number of samples within the spectral peak. Demonstrate this for yourself by executing the section entitled "Exercise on the DFT, Zero Padding and Nyquist Sampling," in the accompanying *Mathematica* tutorial.

Consider what happens when we append $3N$ zeros to $h(k)$ which has been windowed by a rectangular function N samples long. This situation is illustrated in panels (c) and (d) of Figure B.14. There are now four times as many frequency components to define the spectrum. Even in this situation, noticeable differences between the DFT and analytic transform start to appear at larger values of f.

In the top panel, we see 12 samples of the data and to the right, the magnitude of the discrete transform. In the bottom panel, we have four times the number of points to be transformed by adding 36 zeros to the 12 original data points. Now the spectral peak remains the same in size but we have four times as many points defining the peak.

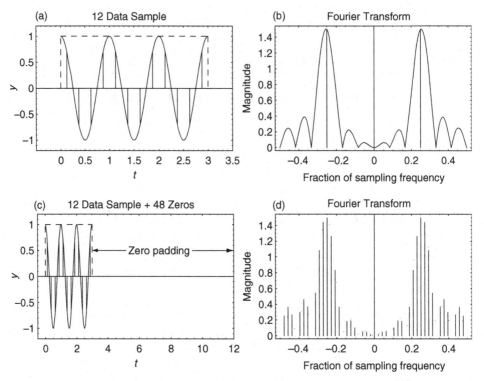

Figure B.14 How to obtain more accurate DFT amplitudes by zero padding. The frequency axis in the two right hand panels is in units of $1/T$, the sampling frequency. On this scale the Nyquist frequency $= 0.5$.

B.12 Power-spectrum estimation

The measurement of power spectra is a difficult and often misunderstood topic. Because the FFT yields frequency and amplitude information, many investigators proceed to estimate the power spectrum from the magnitude of the FFT. If the waveform is periodic or deterministic, then the correct interpretation of the FFT result is likely. However, when the waveforms are random processes, it is necessary to develop a statistical approach to amplitude estimation. We will consider two approaches to the subject in this section, and Chapter 13 provides a powerful Bayesian viewpoint of spectral analysis. We start by introducing Parseval's theorem.

B.12.1 Parseval's theorem and power spectral density

Parseval's theorem states that the energy in a waveform $h(t)$ computed in the time domain must equal the energy as computed in the frequency domain.

$$\text{Energy} = \int_{-\infty}^{\infty} h^2(t)dt = \int_{-\infty}^{\infty} |H(f)|^2 df. \tag{B.80}$$

From this equation, it is clear that $|H(f)|^2$ is an energy spectral density. Frequently, one wants to know "how much energy" is contained in the frequency interval between f and $f + df$. In such circumstances, one does not usually distinguish between $+f$ and $-f$, but rather regards f as varying from 0 to $+\infty$. In such cases, we define the one-sided energy spectral density (ESD) of the function $h(t)$ as

$$E_h(f) \equiv |H(f)|^2 + |H(-f)|^2, \quad 0 \leq f < \infty. \tag{B.81}$$

When $h(t)$ is real, then the two terms are equal, so

$$E_h(f) = 2|H(f)|^2. \tag{B.82}$$

If $h(t)$ goes endlessly from $-\infty < t < \infty$, then its ESD will, in general, be infinite. Of interest then is the one-sided ESD per unit time or power spectral density (PSD). This is computed from a long but finite stretch of $h(t)$. The PSD is computed for a function $= h(t)$ in the finite stretch which is zero elsewhere, divided by the length of the stretch used. Parseval's theorem in this case states that the integral of the one-sided PSD over positive frequency is equal to the mean-square amplitude of the signal $h(t)$.

Proof of Parseval's theorem using the convolution theorem:
FT of $h(t) \cdot h(t) = H(f) * H(f)$.
That is, $\int_{-\infty}^{\infty} h^2(t)e^{i2\pi\sigma t}dt = \int_{-\infty}^{\infty} H(f)H(\sigma - f)df$.
Setting $\sigma = 0$ yields

$$\int_{-\infty}^{\infty} h^2(t)dt = \int_{-\infty}^{\infty} H(f)H(-f)df = \int_{-\infty}^{\infty} |H(f)|^2 df. \quad \text{QED} \tag{B.83}$$

The last equality follows since

$$H(f) = R(f) + iI(f) \text{ and thus } H(-f) = R(-f) + iI(-f). \tag{B.84}$$

For $h(t)$ real, $R(f)$ is even and $I(f)$ is odd, and
$$H(-f) = R(f) - iI(f) = H^*(f). \tag{B.85}$$

B.12.2 Periodogram power-spectrum estimation

A common approach used to estimate the spectrum of $h(t)$ is by means of the periodogram also referred to as the *Schuster periodogram* after Schuster who first introduced the method in 1898.

$$\text{Let } \hat{P}_p(f) = \left(\frac{1}{L}\right) \left| \int_{-L/2}^{L/2} h(t)e^{i2\pi f t}dt \right|^2, \tag{B.86}$$

where the subscript p denotes periodogram estimate and L is the length of the data set. An FFT is normally used to compute this.

We define the power spectral density, $P(f)$, as follows:

$$P(f) = \lim_{L \to \infty} \left(\frac{1}{L}\right) \left|\int_{-L/2}^{L/2} h(t)e^{i2\pi ft}dt\right|^2. \tag{B.87}$$

We now develop another power-spectrum estimator which is in common use.

B.12.3 Correlation spectrum estimation

Let $h(t)$ be a random function of time (could be the sum of a deterministic function and noise). In contrast to a pure deterministic function, future values of a random function cannot be predicted exactly. However, it is possible that the value of the random function at time t influences the value at a later time $t + \tau$. One way to express this statistical characteristic is by means of the autocorrelation function, which for this purpose is given by

$$\phi(\tau) = \lim_{L \to \infty} 1/L \int_{-L/2}^{+L/2} h(t)[h(t+\tau)]dt, \tag{B.88}$$

where $h(t)$ extends from $-L/2$ to $+L/2$.

In the limit of $L = \infty$, the power spectral density function $P(f)$ and the autocorrelation function $\phi(\tau)$ are a Fourier transform pair:

$$\phi(\tau) = \int_{-\infty}^{+\infty} P(f)e^{-i2\pi f\tau}df \Longleftrightarrow P(f) = \int_{-\infty}^{+\infty} \phi(\tau)e^{i2\pi f\tau}d\tau. \tag{B.89}$$

Proof:
From Equation (B.87) and the correlation theorem, we have

$$P(f) \propto \left|\int_{-\infty}^{\infty} h(t)e^{i2\pi ft}dt\right|^2$$

$$= \mathrm{FT}\,[h(t)] \times \mathrm{FT}^*[h(t)]$$

$$= \mathrm{FT}\left[\int_{-\infty}^{\infty} h(t)h(t+\tau)dt\right] \tag{B.90}$$

$$= \mathrm{FT}\,[\phi(\tau)].$$

In the literature, $P(f)$ is called by many terms including: *power spectrum, spectral density function,* and *power spectral density function* (PSD). If the autocorrelation function is known, then the calculation of the power spectrum is determined directly from the Fourier transform.

Since $h(t)$ is known only over a finite interval, we estimate $\phi(\tau)$ based on this finite duration of data. The estimator generally used is

$$\hat{\phi}(\tau) = \frac{1}{L - |\tau|} \int_0^{L-|\tau|} h(t)h[t + |\tau|]dt, \quad |\tau| < L, \tag{B.91}$$

where $h(t)$ is known only over length L.

Notice that $P(f)$ cannot be calculated since $\hat{\phi}(\tau)$ is undefined for $|\tau| > L$. However, consider the quantity $w(\tau)\hat{\phi}(\tau)$, where $w(\tau)$ is a window function which is non-zero for $|\tau| \leq L$ and zero elsewhere. The modified autocorrelation function $w(\tau)\,\hat{\phi}(\tau)$ exists for all τ and hence its FT exists.

$$\hat{P}_c(f) = \int_{-\infty}^{\infty} w(\tau)\,\hat{\phi}(\tau)e^{-2\pi jf\tau}d\tau, \tag{B.92}$$

where $w(\tau) = 1$ for $|\tau| < L$ and is zero elsewhere. $\hat{P}_c(f)$ is called the *correlation* or *lagged-product estimator* of the PSD. This approach to spectral analysis is commonly referred to as the *Blackman–Tukey procedure* (Blackman and Tukey, 1958). An instrument for estimating the PSD in this way is called an *autocorrelation spectrometer*. They are very common in the field of radio astronomy and especially in aperture synthesis telescopes where it is already necessary to cross-correlate the signals from different pairs of telescopes. It is quite convenient to add additional multipliers to calculate the correlation as a function of delay to obtain the spectral information as well.

Although the periodogram and correlation spectrum estimation procedures appear quite different, they are equivalent under certain conditions. It can be shown (Jenkins and Watts, 1968) that

$$\hat{P}_p(f) = \int_{-L/2}^{+L/2} \left(1 - \frac{|\tau|}{L}\right) \hat{\phi}(\tau)e^{i2\pi f\tau}d\tau. \tag{B.93}$$

The inverse FT yields

$$\hat{\phi}_p(\tau) = \left(1 - \frac{|\tau|}{L}\right)\hat{\phi}(\tau), \quad |\tau| < L. \tag{B.94}$$

Hence, if we modify the lagged-product spectrum estimation technique by simply using a triangular (Bartlett) window function in Equation (B.92), then the two procedures are equivalent.

In spectrum estimation problems, one strives to achieve an estimator whose mean value (the average of multiple estimates) is the parameter being estimated. It can be shown (Jenkins and Watts, 1968) that the mean value of both the correlation and periodogram estimation procedures is the true spectrum $P(f)$ convolved with the frequency-domain window function:

$$E[\hat{P}_c(f)] = E[\hat{P}_p(f)] = W(f) * P(f). \tag{B.95}$$

Hence, the mean (expectation) value equals the true spectrum only if the frequency-domain window function is an impulse function (i.e., the data record length is infinite in duration). If the mean of the estimate is not equal to the true value, then we say that the estimate is biased.

B.13 Discrete power spectral density estimation

We will develop the discrete form of the PSD from the discrete form of Parseval's theorem. We start with a continuous waveform $h(t)$ and its transform $H(f)$, which are related by

$$h(t) = \int_{-\infty}^{+\infty} H(f)e^{-i2\pi ft}df \quad \text{where} \quad H(f) = \int_{-\infty}^{+\infty} h(t)e^{i2\pi ft}dt. \tag{B.96}$$

We refer to $H(f)$ as two-sided because from the mathematics, it has non-zero values at both $(+)$ and $(-)$ frequencies.

According to Parseval's theorem,

$$\text{Energy} = \int_{-\infty}^{\infty} h^2(t)dt = \int_{-\infty}^{\infty} |H(f)|^2 df. \tag{B.97}$$

Thus, $|H(f)|^2 =$ two-sided energy spectral density.

B.13.1 Discrete form of Parseval's theorem

Suppose our function $h(t)$ is sampled at N uniformly spaced points to produce the values h_k for $k = 0$ to $N - 1$ spanning a length of time $L = NT$ with $T =$ the sample interval.

$$\text{Energy} = \sum_{k=0}^{N-1} h_k^2 T = \sum_{n=0}^{N-1} |H(f_n)|^2 \Delta f$$

$$= \sum_{n=0}^{N-1} |TH_n|^2 \Delta f. \tag{B.98}$$

Note: the substitution $TH_n = H(f_n)$ comes from Equation (B.50). Thus, $|TH_n|^2 =$ two-sided discrete energy spectral density.

We note in passing that the usual discrete form of Parseval's theorem is obtained from Equation (B.98) by rewriting $\Delta f = 1/(NT)$ and then simplifying to give

$$\sum_{k=0}^{N-1} h_k^2 = \frac{1}{N} \sum_{n=0}^{N-1} |H_n|^2. \tag{B.99}$$

We will find Equation (B.98) a more useful version of the discrete form of Parseval's theorem because it makes clear that $|TH_n|^2$ is a discrete energy spectral density.

$$\text{Average waveform power} = \frac{\text{waveform energy}}{\text{waveform duration}}$$

$$= \frac{1}{NT} \sum_{k=0}^{N-1} h_k^2 T \tag{B.100}$$

$$= \sum_{n=0}^{N-1} \frac{|H_n|^2 T}{N} \Delta f$$

$$= \frac{1}{N} \sum_{k=0}^{N-1} h_k^2$$

$$= \text{mean squared amplitude.} \tag{B.101}$$

We can identify the two-sided discrete PSD with $|H_n|^2 T/N$ from the RHS of the above equation, which has units of power per cycle.

B.13.2 One-sided discrete power spectral density

Let $P(f_n) = $ the one-sided power spectral density.

$$P(f_0) = \frac{T}{N}|H_0|^2$$

$$P(f_n) = \frac{T}{N}\left[|H_n|^2 + |H_{N-n}|^2\right], \quad n = 1, 2, \ldots, (N/2 - 1) \tag{B.102}$$

$$P(f_{N/2}) = \frac{T}{N}|H_{N/2}|^2,$$

where $f_{N/2}$ corresponds to the Nyquist frequency and $P(f_n)$ is only defined for zero and positive frequencies. From Equation (B.102), it is clear that $P(f_n)$ is normalized so that $\sum_{n=0}^{N-1} P(f_n)\Delta f = $ the mean squared amplitude. Note: our expression for the one-sided discrete PSD which has units of power per unit of bandwidth differs from the one given in Press *et al.*, (1992). In particular, the definition used there, $P_{NR}(f_n)$, is related to our $P(f_n)$ by $P_{NR}(f_n) = P(f_n)\Delta f = P(f_n)/(NT)$.

B.13.3 Variance of periodogram estimate

What is the variance of $P(f_n)$ as $N \rightarrow \infty$? In other words, as we take more sampled points from the original function (either sampling a longer stretch of data, or else by resampling the same stretch of data with a faster sampling rate), how much more accurate do the estimates $P(f_n)$ become?

The unpleasant answer is that periodogram estimates do not become more accurate at all! It can be shown that in the case of white Gaussian noise[2] the standard deviation at frequency f_n is equal to the expectation value of the spectrum of f_n (Marple, 1987).

How can this be? Where did this information go as we added more points? It all went into producing estimates at a greater number of discrete frequencies f_n. If we sample a longer run of data using the same sampling rate, then the Nyquist critical frequency f_c is unchanged, but we now have finer frequency resolution (more f_n's). If we sample the same length with a finer sampling interval, then our frequency resolution is unchanged, but the Nyquist range extends to higher frequencies. In *neither* case do the additional samples reduce the variance of any one particular frequency's estimated PSD. Figure B.15 shows examples for increasing N.

As you will see below, there are ways to reduce the variance of the estimate. However, this behavior caused many researchers to consider periodograms of noisy

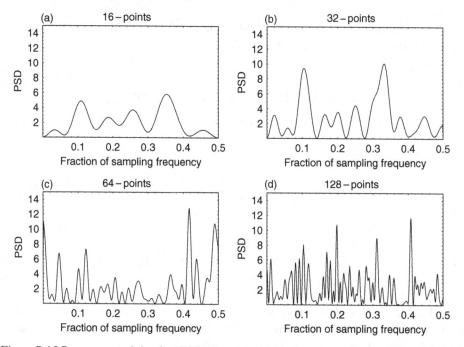

Figure B.15 Power spectral density (PSD) for white (IID) Gaussian noise for different record lengths. The frequency axis is in units of $1/T$, the sampling frequency. On this scale the Nyquist frequency $= 0.5$.

---

[2] The term *white noise* means that the spectral density of the noise is constant from zero frequency through the frequencies of interest, i.e., up to the Nyquist frequency. It is really another way of saying the noise is independent. An independent ensemble of noise values has an autocorrelation function (Equation (B.88)) which is a delta function. According to equation (B.90), the power spectral density is just the FT of the autocorrelation function, which in this case would be a constant.

data to be erratic and this resulted in a certain amount of disenchantment with periodograms for several decades. However, even Schuster was aware of the solution. This disenchantment led G. Yule to introduce a notable alternative analysis method in 1927. Yule's idea was to model a time series with linear regression analysis data. This led to the parametric methods which assume a time-series model and solve for the parameters of the random process. These include *autoregressive* (AR), *moving average* (MA) and *autoregressive-moving average* (ARMA) process models (see Priestley, 1981; Marple, 1987 for more details). In contrast, the correlation and periodogram spectral estimations are referred to as non-parametric statistics of a random process.

B.13.4 Yule's stochastic spectrum estimation model

The Schuster periodogram is appropriate to a model of a sinusoid with additive noise. Suppose the situation were more akin to a pendulum which was being hit by boys throwing peas randomly from both sides.

The result is simple harmonic motion powered by a random driving force. The motion is now affected, not by superposed noise, but by a random driving force. As a result, the graph will be of an entirely different kind to a graph in the case of a sinusoid with superposed errors. The pendulum graph will remain surprisingly smooth, but the amplitude and phase will vary continuously as governed by the inhomogeneous difference equation:

$$x(n) + a_1 x(n-1) + a_2 x(n-2) = \epsilon(n), \tag{B.103}$$

where $\epsilon(n)$ is the white noise input.

Given an empirical time series, $x(n)$, Yule used the method of regression analysis to find these coefficients. Because he regressed $x(n)$ on its own past instead of some other variable, he called it *autoregression*. The least-squares normal equations involve the empirical autocorrelation coefficients of the time series, and today these equations are called the *Yule–Walker equations*. A good example of such a time series is electronic shot noise passing through some band pass filter which rings every shot. A detailed discussion of these methods goes beyond the scope of this book and the interested reader is referred to the works of Priestley (1981) and Marple (1987).

B.13.5 Reduction of periodogram variance

There are two simple techniques for reducing the variance of a periodogram that are very nearly identical mathematically, though different in implementation. The first is to compute a periodogram estimate with finer discrete frequency spacing than you really need, and then to sum the periodogram estimates at K consecutive discrete frequencies to get one "smoother" estimate at the mid frequency of those K.[3] The variance of that summed estimate will be smaller than the estimate itself by a factor of exactly $1/K$, i.e., the standard deviation will be smaller than 100 percent by a factor $1/\sqrt{K}$. Thus, to estimate the power spectrum at $M+1$ discrete frequencies between 0 and f_c inclusive, you begin by taking the FFT of $2MK$ points (which number had better be an integer power of two!). You then take the modulus squared of the resulting coefficients, add positive and negative frequency pairs and divide by $(2MK)^2$. Finally, you "bin" the results into summed (not averaged) groups of K. The reason that you sum rather than average K consecutive points is so that your final PSD estimate will preserve the normalization property that the sum of its $M+1$ values equals the mean square value of the function.

A second technique for estimating the PSD at $M+1$ discrete frequencies in the range of 0 to f_c is to partition the original sampled data into K segments each of $2M$ consecutive sampled points. Each segment is separately FFT'd to produce a period-ogram estimate. Finally, the K periodogram estimates are averaged at each frequency. It is this final averaging that reduces the variance of the estimate by a factor of K (standard deviation by $\sqrt{K}$). The principal advantage of the second technique, how-ever, is that only $2M$ data points are manipulated at a single time, not $2KM$ as in the first technique. This means that the second technique is the natural choice for pro-cessing long runs of data, as from a magnetic tape or other data record.

B.14 Problems

1. **Exercise on the DFT, zero padding and Nyquist sampling**

   a) In the accompanying *Mathematica* tutorial, you will find a section entitled, "Exercise on the DFT, Zero Padding and Nyquist Sampling." Execute the notebook and make sure you understand each step. Do not include a copy of this part in your submission.

   b) Repeat the exercise items, but this time with

   $$\mathbf{fn} = \mathbf{Cos}[2\pi f_1 t] + \mathbf{Sin}[2\pi f_2 t].$$

   Let $f_1 = 1\,\text{Hz}, f_2 = 0.7\,\text{Hz}, T = 0.25\,\text{s}$ and the data window length $L = 3\,\text{s}$.

   c) Explain why one of the two frequencies only appeared in the real part of the analytic FT and the other only appeared in the imaginary part.

   d) What was the effect of zero padding on the DFT?

   e) Comment on the degree of agreement between the FT and the DFT in the *Mathematica* tutorial.

   f) Repeat item (b), only this time increase the window size $L = 8\,\text{s}$. What effect did this have on the spectrum? Explain why this occurred (see Figure B.7).

---

[3] Of course, if your goal is to detect a very narrow band signal, then smoothing may actually reduce the signal-to-noise ratio for detecting such a signal.

g) What is the value of the Nyquist frequency for the sampling interval used?

h) Do the two signals appear at their correct frequencies in the FT and DFT? Explain why there are low level bumps in the spectrum at other frequencies.

i) Recompute the DFT with a sample interval, $T = 0.65$ s, and a data window length $L = 65$ s. Do the two signals appear at their correct frequencies? If not, explain why.

2. **Exercise on Fourier image convolution and deconvolution**

a) In the accompanying *Mathematica* tutorial, you will find a section entitled "Exercise on Fourier Image Convolution and Deconvolution." Execute the notebook and make sure you understand each step.

b) Repeat (a) using a point spread function which is the sum of the following two multinormal distributions:

Multinormal$[\{0,0\}, \{\{1,0\}, \{0,1\}\}]$
Multinormal$[\{4,0\}, \{\{2,0.8\}, \{0.8,2\}\}]$

Appendix C
Difference in two samples

C.1 Outline

In Section 9.4, we explored a Bayesian treatment of the analysis of two independent measurements of the same physical quantity, the *control* and the *trial*, taken under slightly different experimental conditions. In the next four subsections, we give the details behind the calculations of the probabilities of the four fundamental hypotheses (C, S), $(C, \overline{S})$, $(\overline{C}, S)$ and $(\overline{C}, \overline{S})$ which arose in Section 9.4.

After determining what is different, the next problem is to estimate the magnitude of the changes. Section 9.4.4 introduced the calculation for the probability of the difference in the two means $p(\delta | D_1, D_2, I)$. The details of this calculation are given in Section C.3.

Finally, Section 9.4.5 introduced the calculation for the probability for the ratio of the standard deviations, $p(r | D_1, D_2, I)$. The details of this calculation are given in Section C.4.

C.2 Probabilities of the four hypotheses

C.2.1 Evaluation of $p(C, S | D_1, D_2, I)$

The only quantities that remain to be assigned are the two likelihood functions. The prior probability for the noise will be taken to be Gaussian.

$$p(D_1 | C, S, c_1, \sigma_1, I) \text{ and } p(D_2 | C, S, c_1, \sigma_1, I)$$
$$D_1 \equiv \{d_{11}, d_{12}, d_{13}, \ldots, d_{1N_1}\}$$

where $D_{1i} = c_1 + e_{1i}$; therefore,

$$p(D_1 | C, S, c_1, \sigma_1, I) = p(e_{11}, e_{12}, \ldots, e_{1N_1} | c_1 \sigma_1 I)$$

$$= \prod_{i=1}^{N_1} \left[\frac{1}{\sqrt{2\pi\sigma_1^2}} \exp\left(-\frac{e_i^2}{2\sigma_1^2} \right) \right] \tag{C.1}$$

$$p(D_1 | C, S, c_1, \sigma_1, I) = (2\pi\sigma_1^2)^{-\frac{N_1}{2}} \exp\left\{ -\sum_{i=1}^{N_1} \frac{(d_{1i} - c_1)^2}{2\sigma_1^2} \right\}$$

434

$$p(D_2|C, S, c_1, \sigma_1, I) = (2\pi\sigma_1^2)^{-\frac{N_2}{2}} \exp\left\{ -\sum_{i=1}^{N_2} \frac{(d_{2i} - c_1)^2}{2\sigma_1^2} \right\}. \qquad \text{(C.2)}$$

Let

$$\psi_1 = \sum_{i=1}^{N_1} (d_{1i} - c_1)^2$$
$$= \sum d_{1i}^2 + N_1 c_1^2 - 2c_1 \sum d_{1i}$$

and

$$\psi_2 = \sum_{i=1}^{N_2} d_{2i}^2 + N_2 c_1^2 - 2c_1 \sum d_{2i}$$

and

$$\psi_1 + \psi_2 = \sum_{i=1}^{N=N_1+N_2} d_i^2 + N c_1^2 - 2c_1 \sum_{i=1}^{N=N_1+N_2} d_i$$

$$= N\overline{d^2} + N c_1^2 - 2c_1 N\overline{d} \qquad \text{(C.3)}$$

$$= N c_1^2 - 2c_1 N\overline{d} + N(\overline{d})^2 - N(\overline{d})^2 + N\overline{d^2}$$

$$= N(c_1 - \overline{d})^2 + N(\overline{d^2} - (\overline{d})^2),$$

where $\overline{d}$ and $\overline{d^2}$ are the mean and mean square of the pooled data, $N = N_1 + N_2$. Therefore,

$$p(D_1|C, S, c_1, \sigma_1, I)\, p(D_2|C, S, c_1, \sigma_1, I) = (2\pi\sigma_1^2)^{-\frac{N}{2}} \exp\left\{ \frac{-N(\overline{d^2} - (\overline{d})^2)}{2\sigma_1^2} \right\}$$

$$\times \exp\left\{ \frac{-N(c_1 - \overline{d})^2}{2\sigma_1^2} \right\}. \qquad \text{(C.4)}$$

Now combine the likelihoods with the priors to obtain the posterior probability $p(C, S|D_1, D_2, I)$:

$$p(C, S|D_1, D_2, I) = K \int_{\sigma_L}^{\sigma_H} d\sigma_1 \frac{(2\pi\sigma_1^2)^{-\frac{N}{2}}}{4 R_c \sigma_1 \ln(R_\sigma)} \exp\left\{ \frac{-N(\overline{d^2} - (\overline{d})^2)}{2\sigma_1^2} \right\}$$

$$\times \int_{c_L}^{c_H} dc_1 \exp\left\{ -\frac{N(c_1 - \overline{d})^2}{2\sigma_1^2} \right\}. \qquad \text{(C.5)}$$

If the limits on the c_1 integral extend from minus infinity to plus infinity, and the limits on the σ_1 integral extend from zero to infinity, then both integrals can be evaluated in closed form. However, with finite limits, either of the two indicated integrals may be evaluated, but the other must be evaluated numerically. The integral over amplitude will be evaluated in terms of erf(x), the error function

$$\text{erf}(x) = \frac{2}{\sqrt{\pi}} \int_0^x e^{-u^2}\, du \tag{C.6}$$

by setting

$$u^2 = \frac{N(c_1 - \overline{d})^2}{2\sigma_1^2}. \tag{C.7}$$

$$\int_{c_L}^{c_H} dc_1 \exp\left\{ -\frac{N(c_1 - \overline{d})^2}{2\sigma_1^2} \right\} = \sqrt{\frac{2}{N}}\sigma_1 \int_{X_L}^{X_H} e^{-u^2}\, du$$

$$= \sqrt{\frac{2}{N}}\,\sigma_1\, \frac{\sqrt{\pi}}{2} \left[\frac{2}{\sqrt{\pi}} \int_{X_L}^{X_H} e^{-u^2}\, du \right] \tag{C.8}$$

$$= \sqrt{\frac{\pi}{2N}}\,\sigma_1\, [\text{erf}(X_H) - \text{erf}(X_L)].$$

Evaluating the integral over the amplitude, we obtain:

$$p(C, S|D_1, D_2, I) = \frac{K(2\pi)^{-N/2}\sqrt{\pi/2N}}{4R_c \log(R_\sigma)} \int_{\sigma_L}^{\sigma_H} d\sigma_1\, \sigma_1^{-N}$$

$$\times \exp\left\{ -\frac{z}{2\sigma_1^2} \right\} [\text{erf}(X_H) - \text{erf}(X_L)], \tag{C.9}$$

where

$$X_H = \sqrt{\frac{N}{2\sigma^2}}(c_{1H} - \overline{d}), \quad X_L = \sqrt{\frac{N}{2\sigma^2}}(c_{1L} - \overline{d}), \quad z = N[\overline{d^2} - (\overline{d})^2]. \tag{C.10}$$

C.2.2 Evaluation of $p(C, \overline{S}|D_1, D_2, I)$

Notice that $p(C, \overline{S}|D_1, D_2, I)$ assumes the constants are the same in both data sets, but the standard deviations are different. Thus, $p(C, \overline{S}|D_1, D_2, I)$ is a marginal probability density, where the constant and the two standard deviations were removed as nuisance parameters.

$$p(C, \overline{S}|D_1, D_2, I) = \int dc_1 d\sigma_1 d\sigma_2 \; p(C, \overline{S}, c_1, \sigma_1, \sigma_2|D_1, D_2, I)$$

$$= K \int dc_1 d\sigma_1 d\sigma_2 \; p(C, \overline{S}, c_1, \sigma_1, \sigma_2|I)$$

$$\times p(D_1, D_2|C, \overline{S}, c_1, \sigma_1, \sigma_2, I) \tag{C.11}$$

$$= K \int dc_1 d\sigma_1 d\sigma_2 \; p(C, \overline{S}|I) \; p(c_1|I) \; p(\sigma_1|I) \; p(\sigma_2|I)$$

$$\times p(D_1|C, \overline{S}, c_1, \sigma_1, I) \; p(D_2|C, \overline{S}, c_1, \sigma_2, I).$$

By analogy with Equations (C.1) to (C.3), we can evaluate

$$p(D_1|C, \overline{S}, c_1, \sigma_1, \sigma_2, I) \; p(D_2|C, \overline{S}, c_1, \sigma_1, \sigma_2, I)$$

$$= (2\pi)^{-\frac{N}{2}} \sigma_1^{-N_1} \exp\left\{-\frac{U_1}{\sigma_1^2}\right\} \sigma_2^{-N_2} \exp\left\{-\frac{U_2}{\sigma_2^2}\right\}, \tag{C.12}$$

where

$$U_1 = \frac{N_1}{2}(\overline{d_1^2} - 2c_1\overline{d_1} + c_1^2), \qquad U_2 = \frac{N_2}{2}(\overline{d_2^2} - 2c_1\overline{d_2} + c_1^2) \tag{C.13}$$

and $\overline{d_1}, \overline{d_1^2}, \overline{d_2}, \overline{d_2^2}$ are the means and mean squares of D_1 and D_2 respectively.
Substituting Equation (C.12) into (C.11) and adding the priors, we have

$$p(C, \overline{S}|D_1, D_2, I) = \frac{K(2\pi)^{-N/2}}{16R_c[\log(R_\sigma)]^2} \int_L^H dc_1 \int_{\sigma_L}^{\sigma_H} d\sigma_1 \sigma_1^{-(N_1+1)} \exp\left\{-\frac{U_1}{\sigma_1^2}\right\}$$

$$\times \int_{\sigma_L}^{\sigma_H} d\sigma_2 \sigma_2^{-(N_2+1)} \exp\left\{-\frac{U_2}{\sigma_2^2}\right\}. \tag{C.14}$$

The integrals over σ_1 and σ_2 will be evaluated in terms of $Q(r, x)$, one form of the incomplete gamma function of index r and argument x:

$$Q(r, x) = \frac{1}{\Gamma(r)} \int_x^\infty e^{-t} t^{r-1} \, dt. \tag{C.15}$$

If we let

$$t = \frac{U_1}{\sigma_1^2}, \qquad r = \frac{N_1}{2}, \tag{C.16}$$

then we can show that

$$\int_{\sigma_L}^{\sigma_H} d\sigma_1 \sigma_1^{-(N_1+1)} \exp\left\{-\frac{U_1}{\sigma_1^2}\right\} = \frac{1}{2} U_1^{-\frac{N_1}{2}} \Gamma(N_1/2)$$

$$\times \left[\left\{\frac{1}{\Gamma(N_1/2)} \int_{X_L}^{\infty} e^{-t} t^{\frac{N_1}{2}-1} dt\right\} - \left\{\frac{1}{\Gamma(N_1/2)} \int_{X_H}^{\infty} e^{-t} t^{\frac{N_1}{2}-1} dt\right\}\right] \quad \text{(C.17)}$$

$$= \frac{1}{2} U_1^{-\frac{N_1}{2}} \Gamma(N_1/2) \left[Q\left(\frac{N_1}{2}, \frac{U_1}{\sigma_H^2}\right) - Q\left(\frac{N_1}{2}, \frac{U_1}{\sigma_L^2}\right)\right].$$

Evaluating the integral over σ_1 and σ_2, one obtains

$$p(C, \overline{S}|D_1, D_2, I) = \frac{K(2\pi)^{-N/2} \Gamma(N_1/2)\Gamma(N_2/2)}{16 R_c[\log(R_\sigma)]^2} \int_L^H dc_1 U_1^{-\frac{N_1}{2}} U_2^{-\frac{N_2}{2}}$$

$$\times \left[Q\left(\frac{N_1}{2}, \frac{U_1}{\sigma_H^2}\right) - Q\left(\frac{N_1}{2}, \frac{U_1}{\sigma_L^2}\right)\right] \quad \text{(C.18)}$$

$$\times \left[Q\left(\frac{N_2}{2}, \frac{U_2}{\sigma_H^2}\right) - Q\left(\frac{N_2}{2}, \frac{U_2}{\sigma_L^2}\right)\right].$$

C.2.3 Evaluation of $p(\overline{C}, S|D_1, D_2, I)$

$$p(\overline{C}, S|D_1, D_2, I) = \int dc_1 \, dc_2 \, d\sigma_1 \, p(\overline{C}, S, c_1, c_2, \sigma_1|D_1, D_2, I)$$

$$= K \int dc_1 \, dc_2 \, d\sigma_1 \, p(\overline{C}, S, c_1, c_2, \sigma_1|I)$$

$$\times p(D_1, D_2|\overline{C}, S, c_1, c_2, \sigma_1, I) \quad \text{(C.19)}$$

$$= K \int dc_1 \, dc_2 \, d\sigma_1 \, p(\overline{C}, S, |I) \, p(c_1|I) \, p(c_2|I) \, p(\sigma_1|I)$$

$$\times p(D_1|\overline{C}, S, c_1, \sigma_1, I) \, p(D_2|\overline{C}, S, c_2, \sigma_1, I).$$

Evaluating the integrals over c_1 and c_2, one obtains

$$p(\overline{C}, S|D_1, D_2, I) = \frac{K(2\pi)^{-N/2}\pi}{8 R_c^2 \log(R_\sigma)\sqrt{N_1 N_2}} \int_{\sigma_L}^{\sigma_H} d\sigma_1 \sigma_1^{-N+1} \exp\left\{-\frac{z_1 + z_2}{2\sigma_1^2}\right\} \quad \text{(C.20)}$$

$$\times [\text{erf}(X_{1H}) - \text{erf}(X_{1L})][\text{erf}(X_{2H}) - \text{erf}(X_{2L})],$$

where

$$z_1 = N_1[\overline{d_1^2} - (\overline{d_1})^2], \qquad z_2 = N_2[\overline{d_2^2} - (\overline{d_2})^2], \quad \text{(C.21)}$$

$$X_{1H} = \sqrt{\frac{N_1}{2\sigma_1^2}}(H - \overline{d_1}), \qquad X_{1L} = \sqrt{\frac{N_1}{2\sigma_1^2}}(L - \overline{d_1}), \quad \text{(C.22)}$$

$$X_{2H} = \sqrt{\frac{N_2}{2\sigma_1^2}}(H - \overline{d_2}), \qquad X_{2L} = \sqrt{\frac{N_2}{2\sigma_1^2}}\,(L - \overline{d_2}). \qquad (C.23)$$

C.2.4 Evaluation of $p(\overline{C}, \overline{S}|D_1, D_2, I)$

$$p(\overline{C}, \overline{S}|D_1, D_2, I) = \int dc_1\, dc_2\, d\sigma_1\, d\sigma_2\; p(\overline{C}, \overline{S}, c_1, c_2, \sigma_1, \sigma_2|D_1, D_2, I)$$

$$= K \int dc_1\, dc_2\, d\sigma_1\, d\sigma_2\; p(\overline{C}, \overline{S}, c_1, c_2, \sigma_1, \sigma_2|I)$$

$$\times p(D_1, D_2|\overline{C}, \overline{S}, c_1, c_2, \sigma_1, \sigma_2, I)$$

$$= K \int dc_1\, dc_2\, d\sigma_1\, d\sigma_2\; p(\overline{C}, \overline{S}|I)\; p(c_1|I)p(c_2|I)\; p(\sigma_1|I)\; p(\sigma_2|I)$$

$$\times p(D_1|\overline{C}, \overline{S}, c_1, \sigma_1, I)\; p(D_2|\overline{C}, \overline{S}, c_2, \sigma_2, I).$$

$$(C.24)$$

Evaluating the integrals over c_1 and c_2, one obtains

$$p(\overline{C}, \overline{S}|D_1, D_2, I) = \frac{K(2\pi)^{-N/2}\pi}{8R_c^2[\log(R_\sigma)]^2\sqrt{N_1 N_2}}$$

$$\times \int_{\sigma_L}^{\sigma_H} d\sigma_1 \sigma_1^{-N_1} \exp\left\{-\frac{z_1}{2\sigma_1^2}\right\}[\mathrm{erf}(X_{1H}) - \mathrm{erf}(X_{1L})] \qquad (C.25)$$

$$\times \int_{\sigma_L}^{\sigma_H} d\sigma_2 \sigma_2^{-N_2} \exp\left\{-\frac{z_2}{2\sigma_2^2}\right\}[\mathrm{erf}(X_{2H}) - \mathrm{erf}(X_{2L})],$$

where

$$X_{1H} = \sqrt{\frac{N_1}{2\sigma_1^2}}(c_{1H} - \overline{d_1}), \quad X_{1L} = \sqrt{\frac{N_1}{2\sigma_1^2}}(c_{1L} - \overline{d_1}), \qquad (C.26)$$

$$X_{2H} = \sqrt{\frac{N_2}{2\sigma_2^2}}(c_{1H} - \overline{d_2}), \quad X_{2L} = \sqrt{\frac{N_2}{2\sigma_2^2}}(c_{1L} - \overline{d_2}), \qquad (C.27)$$

$$z_1 = N_1[\overline{d_1^2} - (\overline{d_1})^2], \qquad z_2 = N_2[\overline{d_2^2} - (\overline{d_2})^2]. \qquad (C.28)$$

C.3 The difference in the means

Section 9.4.4 introduced the calculation for the probability of the difference in the two means $p(\delta|D_1, D_2, I)$, which was expressed in Equation (9.68) as a weighted sum of $p(\delta|S, D_1, D_2, I)$ and $p(\delta|\overline{S}, D_1, D_2, I)$, the probability for the difference in means given that the standard deviations are the same (the two-sample problem) and the probability for the difference in means given that the standard deviations are different (the Behrens–Fisher problem). The details of the calculation of these two probabilities are given below.

C.3.1 The two-sample problem

$p(\delta|S, D_1, D_2, I)$ is essentially the two-sample problem. This probability is a marginal probability where the standard deviation and γ have been removed as nuisance parameters:

$$p(\delta|S, D_1, D_2, I) = \int d\gamma \, d\sigma_1 \, p(\delta, \gamma, \sigma_1|S, D_1, D_2, I)$$

$$\propto \int d\gamma \, d\sigma_1 \, p(\delta, \gamma, \sigma_1|S, I) \, p(D_1, D_2|S, \delta, \gamma, \sigma_1, I)$$

$$= \int d\gamma \, d\sigma_1 \, p(\delta|I) \, p(\gamma|I) \, p(\sigma_1|I)$$

$$\times p(D_1|S, \delta, \gamma, \sigma_1, I) \, p(D_2|S, \delta, \gamma, \sigma_1, I),$$

(C.29)

where $p(\delta|I)$ and $p(\gamma|I)$ are assigned bounded uniform priors:

$$p(\delta|I) = \begin{cases} \frac{1}{2R_c}, & \text{if } L - H \le \delta \le H - L \\ 0, & \text{otherwise} \end{cases}$$

(C.30)

and

$$p(\gamma|I) = \begin{cases} \frac{1}{2R_c}, & \text{if } 2L \le \gamma \le 2H \\ 0, & \text{otherwise.} \end{cases}$$

(C.31)

We can evaluate $p(D_1|S, \delta, \gamma, \sigma_1, I)$ by comparison with Equation (C.1), after substituting for c_1 according to Equation (9.67).

$$p(D_1|S, \delta, \gamma, \sigma_1, I) = (2\pi\sigma_1^2)^{-\frac{N_1}{2}} \exp\left\{-\frac{Q_1}{\sigma_1^2}\right\},$$

(C.32)

where Q_1 is given by:

$$Q_1 = \sum_{i=1}^{N_1} \left[d_{1i} - \frac{(\gamma + \delta)}{2}\right]^2$$

$$= \frac{N_1}{2}\left[\overline{d_1^2} - (\gamma + \delta)\overline{d_1} + \frac{\gamma^2}{4} + \frac{\gamma\delta}{2} + \frac{\delta^2}{4}\right].$$

(C.33)

Similarly, $p(D_2|S, \delta, \gamma, \sigma_1, I)$ is given by

$$p(D_2|S, \delta, \gamma, \sigma_1, I) = (2\pi\sigma_1^2)^{-\frac{N_2}{2}} \exp\left\{-\frac{Q_2}{\sigma_1^2}\right\},$$

(C.34)

where

$$Q_2 = \sum_{i=1}^{N_2} \left[d_{2i} - \frac{(\gamma - \delta)}{2} \right]^2$$

$$= \frac{N_2}{2} \left[\overline{d_2^2} - (\gamma - \delta)\overline{d_2} + \frac{\gamma^2}{4} - \frac{\gamma\delta}{2} + \frac{\delta^2}{4} \right].$$

(C.35)

The product of Equations (C.32) and (C.34) can be simplified to

$$p(D_1|S, \delta, \gamma, \sigma_1, I) \, p(D_2|S, \delta, \gamma, \sigma_1, I) = (2\pi)^{-\frac{N}{2}} \sigma_1^{-N} \exp\left\{ -\frac{V}{\sigma_1^2} \right\},$$

(C.36)

where

$$V = \frac{N}{2} \left[\overline{d^2} - 2\delta b - \gamma\overline{d} + \frac{\gamma^2}{4} + \frac{\delta^2}{4} + \frac{\delta\gamma\Delta}{2} \right],$$

(C.37)

$$\Delta = \frac{N_1 - N_2}{N}, \quad \text{and} \quad b = \frac{N_1\overline{d_1} - N_2\overline{d_2}}{2N}.$$

(C.38)

After substituting Equations (C.30) and (C.31) and (C.36) into Equation (C.29), the integral over σ_1 is evaluated in terms of incomplete gamma functions.

$$p(\delta|S, D_1, D_2, I) \propto \frac{\Gamma(N/2)}{8R_c^2 \log(R_\sigma)} \int_{2L}^{2H} d\gamma V^{-\frac{N}{2}}$$

$$\times \left[Q\left(\frac{N}{2}, \frac{V}{\sigma_H^2}\right) - Q\left(\frac{N}{2}, \frac{V}{\sigma_L^2}\right) \right].$$

(C.39)

The final integral over γ is computed numerically.

C.3.2 The Behrens–Fisher problem

The Behrens–Fisher problem is essentially given by $p(\delta|\overline{S}, D_1, D_2, I)$, the probability for the difference in means given that the standard deviations are not the same. This probability is a marginal probability where both the standard deviations and the sum of the means, γ, have been removed as nuisance parameters:

$$p(\delta|\overline{S}, D_1, D_2, I) = \int d\gamma \, d\sigma_1 \, d\sigma_2 \, p(\delta, \gamma, \sigma_1, \sigma_2|\overline{S}, D_1, D_2, I)$$

$$\propto \int d\gamma \, d\sigma_1 \, d\sigma_2 \, p(\delta, \gamma, \sigma_1, \sigma_2|\overline{S}, I) \, p(D_1, D_2|\overline{S}, \delta, \gamma, \sigma_1, \sigma_2, I)$$

$$= \int d\gamma \, d\sigma_1 \, d\sigma_2 \, p(\delta|I) \, p(\gamma|I) \, p(\sigma_1|I) \, p(\sigma_2|I)$$

$$\times p(D_1|\overline{S}, \delta, \gamma, \sigma_1, I) \, p(D_2|\overline{S}, \delta, \gamma, \sigma_2, I),$$

(C.40)

where all of the terms appearing in this probability density function have been previously assigned.

To evaluate the integrals over σ_1 and σ_2, one substitutes Equations (9.63), (C.30) and (C.31), and a Gaussian noise prior is used in the two likelihoods. Evaluating the integrals, one obtains

$$
p(\delta|\overline{S}, D_1, D_2, I) \propto \frac{\Gamma(N_1/2)\Gamma(N_2/2)}{16 R_c^2 [\log(R_\sigma)]^2} \int_{2L}^{2H} d\gamma \; W_1^{-\frac{N_1}{2}} W_2^{-\frac{N_2}{2}}
$$

$$
\times \left[Q\left(\frac{N_1}{2}, \frac{W_1}{\sigma_H^2}\right) - Q\left(\frac{N_1}{2}, \frac{W_1}{\sigma_L^2}\right) \right] \qquad (C.41)
$$

$$
\times \left[Q\left(\frac{N_2}{2}, \frac{W_2}{\sigma_H^2}\right) - Q\left(\frac{N_2}{2}, \frac{W_2}{\sigma_L^2}\right) \right],
$$

where

$$
W_1 = \frac{N_1}{2}\left[\overline{d_1^2} - \overline{d_1}(\gamma + \delta) + \frac{(\gamma + \delta)^2}{4} \right], \qquad (C.42)
$$

and

$$
W_2 = \frac{N_2}{2}\left[\overline{d_2^2} - \overline{d_2}(\gamma - \delta) + \frac{(\gamma - \delta)^2}{4} \right]. \qquad (C.43)
$$

With the completion of this calculation, the probability for the difference in means, Equation (9.68), is now complete. We now turn our attention to calculation of the probability for the ratio of the standard deviations.

C.4 The ratio of the standard deviations

Section 9.4.5 introduced the calculation for the probability for the ratio of the standard deviations, $p(r|D_1, D_2, I)$, independent of whether the means are the same or different. This is a weighted average of the probability for the ratio of the standard deviations given the means are the same, $p(r|C, D_1, D_2, I)$, and the probability for the ratio of the standard deviations given that the means are different, $p(r|\overline{C}, D_1, D_2, I)$. These two probabilities are given below.

C.4.1 Estimating the ratio, given the means are the same

The first term to be addressed is $p(r|C, D_1, D_2, I)$. This probability is a marginal probability where both σ and c_1 have been removed as nuisance parameters:

$$p(r|C, D_1, D_2, I) = \int dc_1 d\sigma \, p(r, c_1, \sigma | C, D_1, D_2, I)$$

$$\propto \int dc_1 d\sigma \, p(r, c_1, \sigma | C, I) \, p(D_1, D_2 | C, r, c_1, \sigma, I)$$

$$= \int dc_1 d\sigma \, p(r|I) p(c_1|I) \, p(\sigma|I)$$

$$\times p(D_1 | C, r, c_1, \sigma, I) \, p(D_2 | C, r, c_1, \sigma, I), \tag{C.44}$$

where the prior probability for the ratio of the standard deviations is taken to be a bounded Jeffreys prior:

$$p(r|I) = \begin{cases} 1/[2r\log(R_\sigma)], & \text{if } \sigma_L/\sigma_H \leq r \leq \sigma_H/\sigma_L \\ 0, & \text{otherwise.} \end{cases} \tag{C.45}$$

To evaluate the integral over c_1, one substitutes Equations (9.63) and (C.45), and a Gaussian noise prior probability is used to assign the two likelihoods. Evaluating the integral, one obtains

$$p(r|C, D_1, D_2, I) = \frac{(2\pi)^{-N/2} \sqrt{\pi/8w} r^{-N_1-1}}{R_c [\log(R_\sigma)]^2}$$

$$\times \int_{\sigma_L}^{\sigma_H} d\sigma \, \sigma^{-N} \exp\left\{-\frac{z}{2\sigma^2}\right\} [\text{erf}(X_H) - \text{erf}(X_H)], \tag{C.46}$$

where

$$X_H = \sqrt{\frac{w}{2\sigma^2}} [c_{1H} - v/w], \quad X_L = \sqrt{\frac{w}{2\sigma^2}} [c_{1L} - v/w], \tag{C.47}$$

$$u = \frac{N_1 \overline{d_1^2}}{r^2} + N_2 \overline{d_2^2}, \qquad v = \frac{N_1 \overline{d_1}}{r^2} + N_2 \overline{d_2}, \tag{C.48}$$

$$w = \frac{N_1}{r^2} + N_2, \qquad z = u - \frac{v^2}{w}. \tag{C.49}$$

C.4.2 Estimating the ratio, given the means are different

The second term that must be computed is $p(r|\overline{C}, D_1, D_2, I)$, the probability for the ratio of standard deviations given that the means are not the same. This is a marginal probability where σ, c_1, and c_2 have been removed as nuisance parameters:

$$p(r|\overline{C}, D_1, D_2, I) = \int dc_1 dc_2 d\sigma\, p(r, c_1, c_2, \sigma|\overline{C}, D_1, D_2, I)$$

$$\propto \int dc_1 dc_2 d\sigma\, p(r, c_1, c_2, \sigma|\overline{C}, I)\, p(D_1, D_2|\overline{C}, r, c_1, c_2, \sigma, I) \tag{C.50}$$

$$= \int dc_1 dc_2 d\sigma\, p(r|I)\, p(c_1|I)\, p(c_2|I)\, p(\sigma|I)$$

$$\times p(D_1|r, \overline{C}, c_1, \sigma, I)\, p(D_2|r, \overline{C}, c_2, \sigma, I),$$

where all of the terms appearing in this probability density function have been previously assigned.

To evaluate the integral over c_1 and c_2, one substitutes Equations (9.62), (9.63) and (C.45) and a Gaussian noise prior is used in assigning the two likelihoods. Evaluating the indicated integrals, one obtains

$$p(r|\overline{C}, D_1, D_2, I) \propto \frac{(2\pi)^{-N/2}\pi}{4R_c^2[\log(R_\sigma)]^2\sqrt{N_1 N_2}} \int_{\sigma_L}^{\sigma_H} d\sigma\, r^{-N_1}\sigma^{-N+1}$$

$$\times \exp\left\{-\frac{z_1}{2r^2\sigma^2} - \frac{z_2}{2\sigma^2}\right\}[\mathrm{erf}(X_{1H}) - \mathrm{erf}(X_{1L})] \tag{C.51}$$

$$\times [\mathrm{erf}(X_{2H}) - \mathrm{erf}(X_{2L})],$$

where

$$X_{1H} = \sqrt{\frac{N_1}{2r^2\sigma^2}}[c_{1H} - \overline{d_1}], \quad X_{1L} = \sqrt{\frac{N_1}{2r^2\sigma^2}}[c_{1L} - \overline{d_1}], \tag{C.52}$$

$$X_{2H} = \sqrt{\frac{N_2}{2\sigma^2}}[c_{2H} - \overline{d_2}], \quad X_{2L} = \sqrt{\frac{N_2}{2\sigma^2}}[c_{2L} - \overline{d_2}], \tag{C.53}$$

$$z_1 = N_1[\overline{d_1^2} - (\overline{d_1})^2], \qquad z_2 = N_2[\overline{d_2^2} - (\overline{d_2})^2]. \tag{C.54}$$

Appendix D
Poisson ON/OFF details

D.1 Derivation of $p(s|N_{on}, I)$

In Section 14.4, we explored a Bayesian analysis of ON/OFF measurements, where ON is signal + background and OFF is a just the background. The background is only known imprecisely from OFF measurement. In this appendix, we derive Equation (14.17) for $p(s|N_{on}, I)$, the posterior probability of the signal event rate.

Our starting point is Equation (14.16), which we repeat here together with some of the other relevant equations:

$$p(s|N_{on}, I) = \int_0^{b_{max}} db\ p(s, b|N_{on}, I) \tag{D.1}$$

$$p(s, b|N_{on}, I) = \frac{p(s, b|I)p(N_{on}|s, b, I)}{p(N_{on}|I)}$$
$$= \frac{p(s|b, I)p(b|I)p(N_{on}|s, b, I)}{p(N_{on}|I)} \tag{D.2}$$

$$p(b|N_{off}, I_b) = p(b|I) = \frac{T_{off}(bT_{off})^{N_{off}} e^{-bT_{off}}}{N_{off}!} \tag{D.3}$$

$$p(s|b, I) = 1/s_{max}. \tag{D.4}$$

The denominator of $p(s, b|N_{on}, I)$ in Equation (D.2) is given by

$$p(N_{on}|I) = \int_{s=0}^{s_{max}} ds \int_{b=0}^{b_{max}} db\ p(N_{on}, s, b|I)$$
$$= \int_{s=0}^{s_{max}} ds \int_{b=0}^{b_{max}} db\ p(s|b, I)p(b|I)p(N_{on}|s, b, I). \tag{D.5}$$

445

Substituting Equations (D.5), (D.4), (D.3), (D.2) and (14.15) into Equation (D.1), we obtain

$$p(s|N_{\text{on}}, I) = \frac{\int_{b=0}^{b_{\max}} db\, \frac{1}{s_{\max}} \frac{T_{\text{off}}^{(1+N_{\text{off}})} b^{N_{\text{off}}} e^{-bT_{\text{off}}}}{N_{\text{off}}!} \frac{(s+b)^{N_{\text{on}}} T_{\text{on}}^{N_{\text{on}}} e^{-(s+b)T_{\text{on}}}}{N_{\text{on}}!}}{\int_{s=0}^{s_{\max}} ds \int_{b=0}^{b_{\max}} db\, \frac{1}{s_{\max}} \frac{T_{\text{off}}^{(1+N_{\text{off}})} b^{N_{\text{off}}} e^{-bT_{\text{off}}}}{N_{\text{off}}!} \frac{(s+b)^{N_{\text{on}}} T_{\text{on}}^{N_{\text{on}}} e^{-(s+b)T_{\text{on}}}}{N_{\text{on}}!}}$$

$$= \frac{\int_{b=0}^{b_{\max}} db\, b^{N_{\text{off}}} e^{-bT_{\text{off}}} (s+b)^{N_{\text{on}}} e^{-(s+b)T_{\text{on}}}}{\int_{s=0}^{s_{\max}} ds \int_{b=0}^{b_{\max}} db\, b^{N_{\text{off}}} e^{-bT_{\text{off}}} (s+b)^{N_{\text{on}}} e^{-(s+b)T_{\text{on}}}} \tag{D.6}$$

$$= \frac{\text{Num}}{\text{Den}}.$$

D.1.1 Evaluation of Num

We start with a binomial expansion of $(s+b)^{N_{\text{on}}}$.

$$(s+b)^{N_{\text{on}}} = \sum_{i=0}^{N_{\text{on}}} \frac{N_{\text{on}}!}{i!(N_{\text{on}} - i)!} s^i b^{(N_{\text{on}} - i)}. \tag{D.7}$$

The numerator of Equation (D.6) becomes

$$\text{Num} = \int_{b=0}^{b_{\max}} db\, b^{N_{\text{off}}} e^{-b(T_{\text{on}}+T_{\text{off}})} \sum_{i=0}^{N_{\text{on}}} \frac{N_{\text{on}}!}{i!(N_{\text{on}} - i)!} s^i b^{(N_{\text{on}} - i)} e^{-sT_{\text{on}}}$$

$$= \sum_{i=0}^{N_{\text{on}}} \frac{N_{\text{on}}!}{i!(N_{\text{on}} - i)!} s^i e^{-sT_{\text{on}}} \int_{b=0}^{b_{\max}} db\, b^{(N_{\text{on}}+N_{\text{off}} - i)} e^{-b[T_{\text{on}}+T_{\text{off}}]} \tag{D.8}$$

$$= \sum_{i=0}^{N_{\text{on}}} \frac{N_{\text{on}}!}{i!(N_{\text{on}} - i)!} s^i e^{-sT_{\text{on}}} \times \text{integral}.$$

We now want to evaluate the integral in Equation (D.8), which we first rewrite in the form of an incomplete gamma function:

$$\text{Integral} = (T_{\text{on}} + T_{\text{off}})^{-(N_{\text{on}}+N_{\text{off}} - i + 1)}$$
$$\times \int_{b=0}^{b_{\max}[T_{\text{on}}+T_{\text{off}}]} d(b[T_{\text{on}} + T_{\text{off}}]) \, (b[T_{\text{on}} + T_{\text{off}}])^{(N_{\text{on}}+N_{\text{off}} - i)} \tag{D.9}$$
$$\times e^{-b[T_{\text{on}}+T_{\text{off}}]}.$$

Compare this to one form of the incomplete gamma function:

$$\gamma(n + 1, x) = \int_0^x dy\, y^n e^{-y}. \tag{D.10}$$

Thus, Equation (D.9) can be rewritten as

$$\text{Integral} = (T_{\text{on}} + T_{\text{off}})^{-(N_{\text{on}}+N_{\text{off}}-i+1)}$$
$$\times \gamma([N_{\text{on}} + N_{\text{off}} - i + 1], b_{\max}[T_{\text{on}} + T_{\text{off}}]). \tag{D.11}$$

Provided $b_{\max}[T_{\text{on}} + T_{\text{off}}] \gg [N_{\text{on}} + N_{\text{off}} - i]$, we have that

$$\gamma([N_{\text{on}} + N_{\text{off}} - i + 1], b_{\max}[T_{\text{on}} + T_{\text{off}}]) \approx \Gamma([N_{\text{on}} + N_{\text{off}} - i + 1])$$
$$= (N_{\text{on}} + N_{\text{off}} - i)! \tag{D.12}$$

Substituting Equation (D.12) into Equation (D.11), we obtain

$$\text{Integral} \approx (T_{\text{on}} + T_{\text{off}})^{-(N_{\text{on}}+N_{\text{off}}-i+1)}(N_{\text{on}} + N_{\text{off}} - i)! \tag{D.13}$$

Now substitute Equation (D.13) into Equation (D.8) to obtain

$$\text{Num} \approx \frac{N_{\text{on}}!}{(T_{\text{on}} + T_{\text{off}})^{(N_{\text{on}}+N_{\text{off}}+1)}} \sum_{i=0}^{N_{\text{on}}} \frac{(N_{\text{on}} + N_{\text{off}} - i)!}{i!(N_{\text{on}} - i)!} \frac{s^i e^{-sT_{\text{on}}}}{(T_{\text{on}} + T_{\text{off}})^{-i}}$$
$$= \frac{N_{\text{on}}!}{T_{\text{on}}(T_{\text{on}} + T_{\text{off}})^{(N_{\text{on}}+N_{\text{off}}+1)}}$$
$$\times \sum_{i=0}^{N_{\text{on}}} \frac{(N_{\text{on}} + N_{\text{off}} - i)!}{i!(N_{\text{on}} - i)!} \frac{T_{\text{on}}(sT_{\text{on}})^i e^{-sT_{\text{on}}}}{\left(1 + \frac{T_{\text{off}}}{T_{\text{on}}}\right)^{-i}}. \tag{D.14}$$

D.1.2 Evaluation of Den

The equation for denominator (Den) in Equation (D.6) is the same as Equation (D.14) for the numerator (Num) except for the additional integral over s.

$$\text{Den} = \int_{s=0}^{s_{\max}} ds \frac{N_{\text{on}}!}{T_{\text{on}}(T_{\text{on}} + T_{\text{off}})^{(N_{\text{on}}+N_{\text{off}}+1)}}$$
$$\times \sum_{i=0}^{N_{\text{on}}} \frac{(N_{\text{on}} + N_{\text{off}} - i)!}{i!(N_{\text{on}} - i)!} \frac{T_{\text{on}}(sT_{\text{on}})^i e^{-sT_{\text{on}}}}{\left(1 + \frac{T_{\text{off}}}{T_{\text{on}}}\right)^{-i}}$$
$$= \frac{N_{\text{on}}!}{T_{\text{on}}(T_{\text{on}} + T_{\text{off}})^{(N_{\text{on}}+N_{\text{off}}+1)}} \sum_{i=0}^{N_{\text{on}}} \frac{(N_{\text{on}} + N_{\text{off}} - i)!}{i!(N_{\text{on}} - i)!} \left(1 + \frac{T_{\text{off}}}{T_{\text{on}}}\right)^i$$
$$\times \int_{s=0}^{s_{\max}} d(sT_{\text{on}})(sT_{\text{on}})^i e^{-sT_{\text{on}}}. \tag{D.15}$$

The integral can be recognized as the incomplete gamma function $\gamma(i+1, s_{max} T_{on})$. Provided $s_{max} T_{on} \gg i+1$, we can write $\gamma(i+1, s_{max} T_{on}) \approx i!$, and Equation (D.15) simplifies to

$$\text{Den} \approx \frac{N_{on}!}{T_{on}(T_{on}+T_{off})^{(N_{on}+N_{off}+1)}} \sum_{i=0}^{N_{on}} \frac{(N_{on}+N_{off}-i)!}{(N_{on}-i)!} \left(1+\frac{T_{off}}{T_{on}}\right)^i. \tag{D.16}$$

Substitution of Equations (D.14) and (D.16) into Equation (D.6) yields

$$p(s|N_{on}, I) = \sum_{i=0}^{N_{on}} C_i \frac{T_{on}(sT_{on})^i e^{-sT_{on}}}{i!}, \tag{D.17}$$

where

$$C_i \approx \frac{\left(1+\frac{T_{off}}{T_{on}}\right)^i \frac{(N_{on}+N_{off}-i)!}{(N_{on}-i)!}}{\sum_{j=0}^{N_{on}} \left(1+\frac{T_{off}}{T_{on}}\right)^j \frac{(N_{on}+N_{off}-j)!}{(N_{on}-j)!}}. \tag{D.18}$$

D.2 Derivation of the Bayes factor $B_{\{s+b,b\}}$

Here, we will derive the Bayes factor, $B_{\{s+b,b\}}$, given in Equation (14.23), for the two models M_{s+b} and M_b, which have the following meaning:

$M_b \equiv$ "the ON measurement is solely due to the Poisson background rate, b, where the prior probability for b is derived from the OFF measurement."

$M_{s+b} \equiv$ "the ON measurement is due to a source with unknown Poisson rate, s, plus a Poisson background rate b. Again, the prior probability for b is derived from the OFF measurement."

$$
\begin{aligned}
B_{\{s+b,b\}} &= \frac{p(N_{on}|M_{s+b}, I_{off})}{p(N_{on}|M_b, I_{off})} \\[6pt]
&= \frac{\int_0^{s_{max}} ds \int_0^{b_{max}} db\, p(N_{on}, s, b|M_{s+b}, I_{off})}{\int_0^{b_{max}} db\, p(N_{on}, b|M_b, I_{off})} \\[6pt]
&= \frac{\int_0^{s_{max}} ds \int_0^{b_{max}} db\, p(s|b, M_{s+b}, I_{off}) p(b|M_{s+b}, I_{off}) p(N_{on}|s, b, M_{s+b}, I_{off})}{\int_0^{b_{max}} db\, p(b|M_b, I_{off}) p(N_{on}|b, M_b, I_{off})} \\[6pt]
&= \frac{\int_{s=0}^{s_{max}} ds \int_{b=0}^{b_{max}} db\, \frac{1}{s_{max}} \frac{T_{off}^{(1+N_{off})} b^{N_{off}} e^{-bT_{off}}}{N_{off}!} \frac{(s+b)^{N_{on}} T_{on}^{N_{on}} e^{-(s+b)T_{on}}}{N_{on}!}}{\int_{b=0}^{b_{max}} db\, \frac{T_{off}^{(1+N_{off})} b^{N_{off}} e^{-bT_{off}}}{N_{off}!} \frac{b^{N_{on}} T_{on}^{N_{on}} e^{-bT_{on}}}{N_{on}!}} \\[6pt]
&= \frac{\int_{s=0}^{s_{max}} ds \frac{1}{s_{max}} \int_{b=0}^{b_{max}} db\, b^{N_{off}} e^{-bT_{off}} (s+b)^{N_{on}} e^{-(s+b)T_{on}}}{\int_{b=0}^{b_{max}} db\, b^{(N_{on}+N_{off})} e^{-b(T_{on}+T_{off})}} \\[6pt]
&= \frac{\text{Num1}}{\text{Den1}},
\end{aligned}
\tag{D.19}
$$

where $I_{off} = N_{off}, I_b$, as defined in Equation (14.21). Comparing Equation (D.19) to Equation (D.6), we see that Num1 $= 1/s_{max}$ Den, which we have already evaluated in Equation (D.16). All that remains is to evaluate Den1, which we do here:

$$\text{Den1} = \int_{b=0}^{b_{max}} db \, b^{(N_{on}+N_{off})} e^{-b(T_{on}+T_{off})}$$

$$= (T_{on} + T_{off})^{-(N_{on}+N_{off}+1)} \int_{b=0}^{b_{max}[T_{on}+T_{off}]} d(b[T_{on} + T_{off}]) \qquad \text{(D.20)}$$

$$\times (b[T_{on} + T_{off}])^{(N_{on}+N_{off})} e^{-b(T_{on}+T_{off})}.$$

The integral in the above equation is the incomplete gamma function

$$\gamma([N_{on} + N_{off} + 1], b_{max}[T_{on} + T_{off}]),$$

which can be approximated as

$$\gamma([N_{on} + N_{off} + 1], b_{max}[T_{on} + T_{off}]) \approx [N_{on} + N_{off}]!, \qquad \text{(D.21)}$$

provided $b_{max}[T_{on} + T_{off}] \gg [N_{on} + N_{off}]$.

Equation (D.20) can be rewritten as

$$\text{Den1} \approx (T_{on} + T_{off})^{-(N_{on}+N_{off}+1)} [N_{on} + N_{off}]! \qquad \text{(D.22)}$$

Substituting Num1 and Den1 into Equation (D.19), and canceling quantities in common, yields

$$B_{\{s+b,b\}} \approx \frac{N_{on}!}{s_{max} T_{on} (N_{on} + N_{off})!} \sum_{i=0}^{N_{on}} \frac{(N_{on} + N_{off} - i)!}{(N_{on} - i)!} \left(1 + \frac{T_{off}}{T_{on}}\right)^i. \qquad \text{(D.23)}$$

Appendix E
Multivariate Gaussian from maximum entropy

In this appendix, we will derive the multivariate Gaussian distribution of Equation (8.59) from the MaxEnt principle, given constraint information on the variances and covariances of the multiple variables. We will start with the simpler case of only two variables, y_1 and y_2, and then generalize the result to an arbitrary number of variables. We assume that the priors for y_1 and y_2 have the following form:

$$m(y_i) = \begin{cases} \frac{1}{y_{iH}-y_{iL}}, & \text{if } y_{iL} \leq y_i \leq y_{iH} \\ 0, & \text{if } y_{iL} > y_i \text{ or } y_i > y_{iH}. \end{cases} \tag{E.1}$$

The constraints in this case are:

1. $\int_{y_{1L}}^{y_{1H}} \int_{y_{2L}}^{y_{2H}} p(y_1, y_2) dy_1 dy_2 = 1$
2. $\int_{y_{1L}}^{y_{1H}} \int_{y_{2L}}^{y_{2H}} (y_1 - \mu_1)^2 p(y_1, y_2) \, dy_1 dy_2 = \sigma_{11} = \sigma_1^2$
3. $\int_{y_{1L}}^{y_{1H}} \int_{y_{2L}}^{y_{2H}} (y_2 - \mu_2)^2 p(y_1, y_2) \, dy_1 dy_2 = \sigma_{22} = \sigma_2^2$
4. $\int_{y_{1L}}^{y_{1H}} \int_{y_{2L}}^{y_{2H}} (y_1 - \mu_1)(y_2 - \mu_2) p(y_1, y_2) \, dy_1 dy_2 = \sigma_{12} = \sigma_{21}$

Because $m(y_i)$ is a constant, we solve for $p(y_1, y_2)$ which maximizes

$$S = - \int p(y_1, y_2) \ln \left[p(y_1, y_2) \right] d^N y, \tag{E.2}$$

where $N = 2$ in this case. The problem then is to maximize $p(\{y_i\})$ subject to the constraints 1 to 4. This optimization is best done as the limiting case of a discrete problem. Let y_i and y_j (Roman typeface) represent the discrete versions of y_1 and y_2, respectively. Explicitly, we need to find the solution to

$$d \left[- \sum_{ij} p_{ij} \ln p_i - \lambda \left\{ \sum_{ij} p_{ij} - 1 \right\} - \frac{\lambda_1}{2} A_1 - \frac{\lambda_2}{2} A_2 - \frac{\lambda_3}{2} A_3 \right] = 0, \tag{E.3}$$

450

where

$$A_1 = \left\{ \sum_{ij} (y_i - \mu_i)^2 p_{ij} - \sigma_{ii} \right\}$$

$$A_2 = \left\{ \sum_{ij} (y_j - \mu_j)^2 p_{ij} - \sigma_{jj} \right\}$$

$$A_3 = \left\{ \sum_{ij} (y_i - \mu_i)(y_j - \mu_j) p_{ij} - \sigma_{ij} \right\},$$

and

$$\sum_{ij} = \sum_{i=1}^{N} \sum_{j=1}^{N}.$$

This leads to

$$\sum_{ij} \left[-\ln p_{ij} - 1 - \lambda - \frac{1}{2} \left\{ \lambda_1 (y_i - \mu_i)^2 \right\} \right.$$
$$\left. + \left[\frac{1}{2} \left\{ \lambda_2 (y_j - \mu_j)^2 + \lambda_3 (y_i - \mu_i)(y_j - \mu_j) \right\} \right] dp_{ij} = 0. \right. \tag{E.4}$$

For each *ij*, we require

$$-\ln p_{ij} - 1 - \lambda - \frac{1}{2} \left\{ \lambda_1 (y_i - \mu_i)^2 + \lambda_2 (y_j - \mu_j)^2 + \lambda_3 (y_i - \mu_i)(y_j - \mu_j) \right\} = 0, \tag{E.5}$$

or,

$$p_{ij} = e^{-\lambda_0} \times \exp \left[-\frac{1}{2} \left\{ \lambda_1 (y_i - \mu_i)^2 + \lambda_2 (y_j - \mu_j)^2 + \lambda_3 (y_i - \mu_i)(y_j - \mu_j) \right\} \right], \tag{E.6}$$

where $\lambda_0 = 1 + \lambda$.

This generalizes to the continuum assignment

$$p(y_1, y_2) = \exp\{-\lambda_0\}$$
$$\times \exp \left[-\frac{1}{2} \left\{ \lambda_1 (y_1 - \mu_1)^2 + \lambda_2 (y_2 - \mu_2)^2 + \lambda_3 (y_1 - \mu_1)(y_2 - \mu_2) \right\} \right]. \tag{E.7}$$

To simplify the notation, we will use the abbreviation $\delta y_1 = (y_1 - \mu_1)$ and $\delta y_2 = (y_2 - \mu_2)$. Then Equation (E.7) becomes

$$p(y_1, y_2) = \exp\{-\lambda_0\} \exp \left[-\frac{1}{2} \{ \lambda_1 \delta y_1^2 + \lambda_2 \delta y_2^2 + \lambda_3 \delta y_1 \delta y_2 \} \right]$$
$$= \exp\{-\lambda_0\} \exp \left\{ \left[-\frac{Q}{2} \right] \right\}, \tag{E.8}$$

where

$$Q = \lambda_1 \delta y_1^2 + \lambda_2 \delta y_2^2 + \lambda_3 \delta y_1 \delta y_2$$

$$= \lambda_1 \left(\delta y_1^2 + 2 \left(\frac{\lambda_3}{2\lambda_1} \right) \delta y_1 \delta y_2 + \frac{\lambda_3^2}{4\lambda_1^2} \delta y_2^2 \right) + \lambda_2 \delta y_2^2 - \frac{\lambda_3^2}{4\lambda_1} \delta y_2^2 \qquad \text{(E.9)}$$

$$= \lambda_1 \left(\delta y_1 + \frac{\lambda_3}{2\lambda_1} \delta y_2 \right)^2 + \left(\lambda_2 - \frac{\lambda_3^2}{4\lambda_1} \right) \delta y_2^2.$$

In Equation (E.9), we have carried out an operation called completing the squares, which will help us in our next step, evaluating λ_0 from constraint number 1.

$$\int_{y_{1L}}^{y_{1H}} \int_{y_{2L}}^{y_{2H}} p(y_1, y_2) dy_1 dy_2 = \int_{y_{1L}}^{y_{1H}} \int_{y_{2L}}^{y_{2H}} e^{-\lambda_0} \exp\left[-\frac{Q}{2} \right]$$

$$= e^{-\lambda_0} \int_{y_{2L}}^{y_{2H}} dy_2 \exp\left[-\frac{1}{2} \left(\lambda_2 - \frac{\lambda_3^2}{4\lambda_1} \right) \delta y_2^2 \right] \qquad \text{(E.10)}$$

$$\times \int_{y_{1L}}^{y_{1H}} dy_1 \exp\left[-\frac{\lambda_1}{2} \left(\delta y_1 + \frac{\lambda_3}{2\lambda_1} \delta y_2 \right)^2 \right] = 1.$$

The second integrand in Equation (E.10) is a Gaussian in dy_1, with variance $1/\lambda_1$. If the range of integration were infinite, the integral would merely be a constant (the normalization constant for the Gaussian, $\sqrt{2\pi/\lambda_1}$). With a finite range, it can be written in terms of error functions with arguments that depend on δy_2. But, as we showed in Section 8.7.4, as long as the limits y_{1H} and y_{1L} lie well outside the region where there is a significant contribution to the integral, then the limits can effectively be replaced by $+\infty$ and $-\infty$, which is what we assume here.

The first integrand in Equation (E.10) is another Gaussian in dy_2. We will also assume that range of integration is effectively infinite, so the integrand evaluates to the normalization constant, $\sqrt{2\pi/(\lambda_2 - \lambda_3^2/4\lambda_1)}$. Equation (E.10) thus simplifies to

$$e^{-\lambda_0} \frac{2\pi}{\sqrt{\lambda_1 \lambda_2 - \frac{\lambda_3^2}{4}}} = 1. \qquad \text{(E.11)}$$

The solution is

$$e^{-\lambda_0} = \frac{1}{2\pi} \sqrt{\lambda_1 \lambda_2 - \frac{\lambda_3^2}{4}}. \qquad \text{(E.12)}$$

We now make use of Equation (8.22) to evaluate the remaining Lagrange multipliers, λ_1, λ_2 and λ_3.

$$-\frac{\partial \lambda_0}{\partial \lambda_1} = \frac{1}{2} \langle (y_1 - \mu_1)^2 \rangle = \frac{\sigma_{11}}{2}, \qquad \text{(E.13)}$$

$$-\frac{\partial \lambda_0}{\partial \lambda_2} = \frac{1}{2} \langle (y_2 - \mu_2)^2 \rangle = \frac{\sigma_{22}}{2},$$ (E.14)

$$-\frac{\partial \lambda_0}{\partial \lambda_3} = \frac{1}{2} \langle (y_1 - \mu_1)(y_2 - \mu_2) \rangle = \frac{\sigma_{12}}{2}.$$ (E.15)

Note: the extra factor of 2 appearing in the denominator on the right hand side of Equations (E.13), (E.14), (E.15), when compared to Equation (8.22), arises from the factor of $1/2$ introduced in front of λ_1, λ_2 and λ_3 in Equation (E.4), which defines the meaning of these Lagrange multipliers.

The solutions to Equations (E.13), (E.14), and (E.15) are as follows:

$$\lambda_1 = \frac{\sigma_{22}}{\sigma_{11}\sigma_{22} - \sigma_{12}^2},$$ (E.16)

$$\lambda_2 = \frac{\sigma_{11}}{\sigma_{11}\sigma_{22} - \sigma_{12}^2},$$ (E.17)

$$\lambda_3 = \frac{-2\sigma_{12}}{\sigma_{11}\sigma_{22} - \sigma_{12}^2}.$$ (E.18)

Equation (E.12) for the term $e^{-\lambda_0}$ can now be expressed in terms of σ_{11}, σ_{22} and σ_{12} as follows:

$$e^{-\lambda_0} = \frac{1}{2\pi\sqrt{\sigma_{11}\sigma_{22} - \sigma_{12}^2}}.$$ (E.19)

At this point, it is convenient to rewrite Q, which first appeared in Equation (E.8), in the following matrix form:

$$Q = (\delta y_1 \delta y_2) \begin{pmatrix} \lambda_1 & \lambda_3/2 \\ \lambda_3/2 & \lambda_2 \end{pmatrix} \begin{pmatrix} \delta y_1 \\ \delta y_2 \end{pmatrix}$$ (E.20)
$$= \delta \mathbf{Y}^T \mathbf{E}^{-1} \delta \mathbf{Y}.$$

The $\mathbf{E}^{-1}$ matrix, which stands for the inverse of the $\mathbf{E}$ matrix, can be expressed in terms of σ_{11}, σ_{22} and σ_{12} as follows:

$$\mathbf{E}^{-1} = \frac{1}{\sigma_{11}\sigma_{22} - \sigma_{12}^2} \begin{pmatrix} \sigma_{22} & -\sigma_{12} \\ -\sigma_{12} & \sigma_{11} \end{pmatrix}.$$ (E.21)

Although $\mathbf{E}^{-1}$ is rather messy, the $\mathbf{E}$ matrix itself is a very simple and useful matrix.

$$\mathbf{E} = \begin{pmatrix} \sigma_{11} & \sigma_{12} \\ \sigma_{12} & \sigma_{22} \end{pmatrix}.$$ (E.22)

Now substitute Equations (E.20) and (E.19) into Equation (E.8) to obtain a final equation for $p(y_1, y_2)$.

$$p(y_1, y_2) = \frac{1}{2\pi\sqrt{\sigma_{11}\sigma_{22} - \sigma_{12}^2}} \exp\left[-\frac{1}{2}\left(\delta\mathbf{Y}^T\mathbf{E}^{-1}\delta\mathbf{Y}\right)\right]$$

$$= \frac{1}{(2\pi)^{N/2}\sqrt{\det\mathbf{E}}}\exp\left[-\frac{1}{2}\left(\delta\mathbf{Y}^T\mathbf{E}^{-1}\delta\mathbf{Y}\right)\right], \tag{E.23}$$

where $N = 2$ for two variables. Equation (E.23) is also valid for an arbitrary number of variables,[1] which we write as

$$p(\{y_i\}|\{\mu_i, \sigma_{ij}\}) = \frac{1}{(2\pi)^{N/2}\sqrt{\det\mathbf{E}}}\exp\left[-\frac{1}{2}\left(\delta\mathbf{Y}^T\mathbf{E}^{-1}\delta\mathbf{Y}\right)\right]$$

$$= \frac{1}{(2\pi)^{N/2}\sqrt{\det\mathbf{E}}}\exp\left[-\frac{1}{2}\sum_{ij}(y_i - \mu_i)[\mathbf{E}^{-1}]_{ij}(y_j - \mu_j)\right], \tag{E.24}$$

where

$$\mathbf{E} = \begin{pmatrix} \sigma_{11} & \sigma_{12} & \sigma_{13} & \cdots & \sigma_{1N} \\ \sigma_{21} & \sigma_{22} & \sigma_{23} & \cdots & \sigma_{2N} \\ \cdot & \cdot & \cdot & \cdot & \cdot \\ \sigma_{N1} & \sigma_{N2} & \sigma_{N3} & \cdots & \sigma_{NN} \end{pmatrix}. \tag{E.25}$$

The $\mathbf{E}$ matrix is called the *data covariance matrix* when each y variable describes possible values of a datum, d_i.

---

[1] ?In Equation (E.24), $\{y_i\}$ refers to a set of continuous variables.

References

Aczél, J. (1966). *Lectures on Functional Equations and their Applications.* New York: Academic Press. See also Aczel, J. (1987), *A Short Course on Functional Equations*, Dordrecht–Holland: D. Reidel.

Barber, M. N., Pearson, R. B., Toussaint, D., and Richardson, J. L. (1985). Finite-size scaling in the three-dimensional Ising model. *Physics Review B*, **32**, 1720–1730.

Bayes, T. (1763). An essay toward solving a problem in the doctrine of chances. *Philosophical Transactions of the Royal Society*, pp. 370–418.

Berger, J. O. and Berry, D. A. (1988). Statistical analysis and the illusion of objectivity. *American Scientist*, **76**, 159–165.

Berger, J. O. and Sellke, T. (1987). Testing a point null hypothesis: The irreconcilability of p-values and evidence. *Journal of the American Statistical Association*, **82**, 112–122.

Bernoulli, J. (1713). *Ars conjectandi*, Basel: Thurnisiorum. Reprinted in *Die Werke von Jakob Bernoulli*, Vol. 3, Basel: Birkhaeuser, (1975), pp. 107–286.

Blackman, R. B. and Tukey, J. W. (1958). *The Measurement of Power Spectra.* New York: Dover Publications, Inc.

Boole, G. (1854). *An Investigation of the laws of Thought.* London: Macmillan; reprinted by Dover Publications, New York (1958).

Bretthorst, G. L. (1988). *Bayesian Spectrum Analysis and Parameter Estimation.* New York: Springer-Verlag.

Bretthorst, G. L. (1990a). Bayesian analysis. I. Parameter estimation using quadrature NMR models. *Journal of Magnetic Resonance*, **88**, 533–551.

Bretthorst, G. L. (1990b). Bayesian analysis. II. Signal detection and model selection. *Journal of Magnetic Resonance*, **88**, 552–570.

Bretthorst, G. L. (1990c). Bayesian analysis. III. Applications to NMR signal detection, model selection, and parameter estimation. *Journal of Magnetic Resonance*, **88**, 571–595.

Bretthorst, G. L. (1991). Bayesian analysis. IV. Noise and computing time considerations. *Journal of Magnetic Resonance*, **93**, 369–394.

Bretthorst, G. L. (1993). On the difference in means. In *Physics & Probability Essays in honor of Edwin T. Jaynes*, W. T. Grandy and P. W. Milonni (eds.). England: Cambridge University Press.

Bretthorst, G. L. (2000a). Nonuniform sampling: Bandwidth and aliasing. In *Maximum Entropy and Bayesian Methods in Science and Engineering*, J. Rychert, G. Erickson, and C. R. Smith (eds.). USA: American Institute of Physics, pp. 1–28.

Bretthorst, G. L. (2001). Generalizing the Lomb–Scargle periodogram. In *Bayesian Inference and Maximum Entropy Methods in Science and Engineering, Paris* Ali Mohammad-Djafari (ed.). New York: American Institute of Physics Proceedings, **568**, 241–245.

Brigham, E. O. (1988). *The Fast Fourier Transform and Its Applications*, New Jersey: Prentice Hall.

Buck, B. and MaCaulay V. A. (eds.) (1991). *Maximum Entropy in Action*. Oxford Science Publication, Oxford: Clarendon Press.

Charbonneau, P. (1995). Genetic algorithms in astronomy and astrophysics. *Astrophysical Journal (Supplements)*, **101**, 309–334.

Charbonneau, P. and Knapp, B. (1995). *A User's guide to PIKAIA 1.0*, NCAR Technical Note 418 + IA. Boulder: National Center for Atmospheric Research.

Chib, S. and Greenberg, E. (1995). Understanding the Metropolis algorithm. *American Statistician*, **49**, 327–335.

Chu, S. (2003). Using soccer goals to motivate the Poisson process. *INFORMS ransactions on Education*, 3 (2) http://ite.pubs.informs.org/Vol3No2/Chu/index.php.

Cooley, J. W. and Tukey, J. W. (1965). An algorithm for the machine calculation of complex fourier series. *Mathematics of Computing*, **19**, 297–301.

Cox, R. T. (1946). Probability, frequency, and reasonable expectation. *American Journal of Physics*, **17**, 1–13.

Cox, R. T. (1961). *The Algebra of Probable Inference*, Baltimore, MD: Johns Hopkins University Press.

D'Agostini, G. (1999). *Bayesian Reasoning in High-Energy Physics: Principles and Applications*. CERN Yellow Reports.

Dayal, Hari H. (1972). *Bayesian statistical inference in Behrens–Fisher Problems*, Ph.D. dissertation, State University of New York at Buffalo, September 1972.

Dayal., Hari H. and James M. Dickey, (1976), Bayes factors for Behrens–Fisher problems. *The Indian Journal of Statistics*, **38**, 315–328.

Delampady, M. and Berger, J. O. (1990). Lower bounds on Bayes factors for multinomial distributions, with applications to chi-squared tests of fit. *Annals of Statistics*, **18**, 1295–1316.

Feigelson E. D. and Babu, G. J. (eds.) (2002). *Statistical Challenges in Modern Astronomy III*. New York: Springer-Verlag.

Fernandez, J. F. and Rivero, J. (1996). Fast algorithms for random numbers with exponential and normal distributions. *Computers in Physics*, **10**, 83–88.

Fox, C. and Nicholls, G. K. (2001). Exact MAP states and expectations from perfect sampling: Greig, Porteous and Seheult revisited. In *Bayesian Inference and Maximum Entropy Methods in Science and Engineeering*, Paris. Ali Mohammad-Djafari (ed). New York: American Institute of Physics Proceedings, **568**, 252–263.

Geyer, C. and Thompson, E. (1995). Annealing Markov chain Monte Carlo with applications to ancestral inference. *Journal of the American Statistical Association*, **90**, 909–920.

Gilks, W. R., Richardson, S. and Spiegelhalter, D. J. (1996). *Markov Chain Monte Carlo in Practice*. London: Chapman and Hall.

Goggans, P. M. and Chi, Y. (2004). Using thermodynamic integration to calculate the posterior probability in Bayesian model selection problems. In *Bayesian Inference and Maximum Entropy Methods in Science and Engineering*, Proceedings, 23rd International Workshop, G. Erickson and Y. X. Zhai (eds.). USA: American Institute of Physics, pp. 59–66.

Gregory, P. C. (1999). Bayesian periodic signal detection: Analysis of 20 years of radio flux measurements of LS I + 61° 303. *Astrophysical Journal*, **520**, 361–375.

Gregory, P. C. (2001). A Bayesian revolution in spectral analysis. In *Bayesian Inference and Maximum Entropy Methods in Science and Engineeering, Paris*. Ali Mohammad-Djafari, (ed.) New York: American Institute of Physics Proceedings, **568**, 557–568.

Gregory, P. C.(2002). Bayesian analysis of radio observations of the Be X-ray binary LS I + 61° 303. *Astrophysical Journal*, **575**, 427–434.

Gregory, P. C. and Loredo, T. J. (1992). A new method for the detection of a periodic signal of unknown shape and period. *Astrophysical Journal*, **398**, 146–168.

Gregory, P. C. and Loredo, T. J. (1993). A Bayesian method for the detection of unknown periodic and non-periodic signals in binned time series. In *Maximum Entropy and Bayesian Methods, Paris*. Ali Mohammad-Djafari and G. Demoment, (eds.). Dordrecht: Kluwer Academic Press, pp. 225–232.

Gregory, P. C. and Loredo, T. J. (1996). Bayesian periodic signal detection: Analysis of ROSAT observations of PSR 0540-693. *Astrophysical Journal*, **473**, 1059–1066.

Gregory, P. C. and Neish, C. (2002). Density and velocity structure of the Be star equatorial disk in the binary, LS I + 61° 303, a probable microquasar. *Astrophysical Journal*, **580**, 1133–1148.

Gregory, P. C. and Taylor, A. R. (1978). New highly variable radio source, possible counterpart of gamma-ray source CG 135 + 1. *Nature*, **272**, 704–706.

Gregory, P. C., Xu, H. J., Backhouse, C. J. and Reid, A. (1989). Four-year modulation of periodic radio outbursts from LS I + 61° 303. *Astrophysical Journal*, **339**, 1054–1058.

Gregory, P. C., Peracaula, M. and Taylor, A. R. (1999). Bayesian periodic signal detection: Discovery of periodic phase modulation in LS I + 61° 303 radio outbursts. *Astrophysical Journal*, **520**, 376–390.

Gull, S. F. (1988). Bayesian inductive inference and maximum entropy. In *Maximum Entropy & Bayesian Methods in Science and Engineering*, G. J. Erickson and C. R. Smith (eds.). Dordrecht: Kluwer Academic Press, pp. 53–74.

Gull, S. F. (1989a). Developments in maximum entropy data analysis. in *Maximum Entropy & Bayesian Methods*, J. Skilling (ed.), Dordrecht: Kluwer Academic Press. pp. 53–71.

Gull, S. F. (1989b). Bayesian data analysis – straight line fitting. In *Maximum Entropy & Bayesian Methods*, J. Skilling (ed.). Dordrecht: Kluwer Academic Press, pp. 511–518.

Gull, S. F., and Skilling, J. (1984). Maximum entropy method in image processing. *IEEE Proceedings*, **131**, Part F, (6) 646–659.

Hastings, W. K. (1970). Monte Carlo Sampling methods using Markov chains and their applications. *Biometrika*, **57**, 97–109.

Helene, O. (1983). Upper limit of peak area. *Nuclear Instruments and Methods*, **212**, 319–322.

Helene, O. (1984). Errors in experiments with small numbers of events. *Nuclear Instruments and Methods*, **228**, 120–128.

Holland, J. (1992). Genetic algorithms. *Scientific American*, July, 66–72.

Hutchings, J. B. and Crampton, D. (1981). Spectroscopy of the unique degenerate binary star LS I + 61° 303. *PASP*, **93**, 486–489.

James, F. (1998). *MINUIT, Function Minimization and Error Analysis Reference Manual*, Version 94.1. Computing and Network Division, CERN Geneva, Switzerland.

Jaynes, E. T. (1957). How does the brain do plausible reasoning? Stanford University Microwave Laboratory Report 421. Reprinted in *Maximum Entropy and Bayesian Methods in Science and Engineeering*, G. J. Erickson and C. R. Smith (eds.) (1988). Dordrecht: Kluwer Academic Press.

Jaynes, E. T. (1968). Prior probabilities. *IEEE Transactions on System Science & Cybernetics*, **4**(3), 227–241.

Jaynes, E. T. (1976). Confidence intervals vs Bayesian intervals. In *Foundations of Probability Theory, Statistical Inference, and Statistical Theories of Science*, 2, pp. 175–257, W. L. Harper and C. A. Hooker (eds.). Dordrecht: D. Reidel.

Jaynes, E. T. (1982). On the Rationale of Maximum Entropy Methods. *Proceedings of the IEEE*, **70**(9), 939–952.

Jaynes, E. T. (1983). *Papers on Probability, Statistics and Statistical Physics*, a reprint collection. Dordrecht: D. Reidel. Second edition, Dordrecht: Kluwer Academic Press, (1989).

Jaynes, E. T. (1987). Bayesian spectrum and chirp analysis. In *Maximum Entropy and Bayesian Spectral Analysis and Estimation Problems*, C. R. Smith and G. L. Erickson (eds.). Dordrecht: D. Reidel, pp. 1–37.

Jaynes, E. T. (1990). Probability theory as logic. In *Maximum-Entropy and Bayesian Methods*, P. F. Fougre (ed.). Dordrecht: Kluwer, pp. 1–16.

Jaynes, E. T. (2003). *Probability Theory – The Logic of Science*, G. L. Bretthorst (ed.). Cambridge: Cambridge University Press.

Jeffreys, H. (1931). *Scientific Inference*. Cambridge: Cambridge University Press. Later editions, 1937, 1957, 1973.

Jeffreys, H. (1932). On the theory of errors and least squares. *Proceedings of the Royal Society*, **138**, 48–55.

Jeffreys, H. (1939). *Theory of Probability*. Oxford: Clarendon Press. Later editions, 1948, 1961, 1967, 1988.

Jeffreys, W. H. and Berger, J. O. (1992). Ockham's razor and Bayesian analysis. *American Scientist*, **80**, 64–72.

Jenkins, G. M. and Watts, D. G. (1968). *Spectral Analysis and its Applications*, San Francisco: Holden Day.

Kirkpatrick, S., Gelatt, C. D., and Vecchi, M. P. (1983). Optimisation by simulated annealing. *Science*, **220**, 671–680.

Knuth, D. (1981). *Seminumerical algorithms*, 2nd edn, vol. 2 of *The Art of Computer Programming*. Reading, MA: Addison-Wesley.

Laplace, P. S. (1774). *Mémoire sur la probabilité des causes par les événements.* Mémoires de l'Académie royale des sciences, **6**, 621–656. Reprinted in *Laplace (1878–1912)*, vol. 8, pp. 27–65, Paris: Gauthier–Villars, English translation by S. M. Stigler (1986).

Lindley, D. V. (1965). *Introduction to Probability and Statistics*, (Part 1 – Probability and Part 2 – Inference). Cambridge: Cambridge University Press.

Liu, J. S. (2001). *Monte Carlo Strategies in Scientific Computing*. Springer Series in Statistics. New York: Springer-Verlag.

Lomb, N. R. (1976). Least squares frequency analysis of unevenly spaced data. *Astrophysical and Space Sciences*, **39**, 447–462.

Loredo, T. J. (1990). From Laplace to Supernova SN 1987A: Bayesian inference in astrophysics. *Maximum Entropy and Bayesian Methods, Dartsmouth*. P. Fougère (ed.). Dordrecht: Kluwer Academic Press, pp. 81–142.

Loredo, T. J. (1992). The promise of Bayesian inference for astrophysics. In *Statistical Challenges in Modern Astronomy*, E. D. Feigelson and G. J. Babu (eds.). New York: Springer-Verlag, pp. 275–297.

Loredo, T. J. (1999). Computational technology for Bayesian inference. In ASP Conference Series, Vol. 172, *Astronomical Data Analysis Software and Systems VIII*, D. M. Mehringer, R. L. Plante, and D. A. Roberts (eds.). San Fransisco: Astronomical Society of the Pacific, pp. 297–306.

Maddox, J. (1994). The poor quality of random numbers. *Nature*, **372**, 403.

Marple, S. L. (1987). *Digital Spectral Analysis* (Appendix 4a). Englewood Cliffs, NJ: Prentice Hall.

Metropolis, N., Rosenbluth, A., Rosenbluth, M., Teller, A., and Teller, E. (1953). Equation of state calculation by fast computing machines. *Journal of Chemical Physics*, **21**, 1087–1092.

Nedler, J. A. and Mead, R. (1965). A simple method for function minimizations. *Computing Journal*, **7**, 308–313.

Paredes, J. M., Estelle, R. and Ruis, A. (1990). Observation at 3.6 cm wavelength of the radio light curve of LS I + 61° 303. *Astronomy and Astrophysics*, **232**, 377–380.

Park, S. K. and Miller, K. W. (1988). Random number generators: good ones are hard to find. *Communications of the Association for Computing Machinery*, **31** (10), 1192–1201.

Piña, R. K. and Puetter, R. C. (1993). Bayesian image reconstruction: the Pixon and optimal image modeling. *Proceedings of the Astronomical Society of the Pacific*. **105**, 630–637.

Press, W. H., Teukolsky, S. A., Vetterling, W. T., and Flannery, B. P. (1992). *Numerical Recipes* (second edition). Cambridge: Cambridge University Press.

Priestley, M. B. (1981). *Spectral Analysis and Time Series*. London: Academic Press.

Puetter, R. C. (1995). Pixon-based multiresolution image reconstruction and the quantification of picture information content. *International Journal of Image Systems & Technology*, **6**, 314–331.

Ray, P. S., Foster, R. S., Waltman, E. B. *et al.* (1997). Long term monitoring of LS I + 61° 303 at 2.25 and 8.3 GHz. *Astrophysical Journal*, **491**, 381–387.

Roberts, G. O. (1996). Markov chain concepts related to sampling algorithms. In *Markov Chain Monte Carlo in Practice*, W. R. Gilks, S. Richardson, and D. J. Spiegelhalter (eds.). London: Chapman and Hall. pp. 45–57.

Roberts, G. O., Gelman, A. and Gilks, W. R. (1997). Weak convergence and optimal scaling of random walk Metropolis algorithms. *Annals of Applied Probability*, **7**, 110–120.

Scargle, J. D. (1982). Studies in astronomical time series analysis II. Statistical aspects of spectral analysis of unevenly sampled data. *Astrophysical Journal*, **263**, 835–853.

Scargle, J. D. (1989). Studies in astronomical time series analysis III. Autocorrelation and cross-correlation functions of unevenly sampled data. *Astrophysical Journal*, **343**, 874–887.

Schuster, A. (1905). The periodogram and its optical analogy. *Proceedings of the Royal Society of London*, **77**, 136–140.

Sellke, T., Bayarri, M. J. and Berger, J. O. (2001). Calibration of P-values for testing precise null hypotheses. *The American Statistician*, **55**, 62–71.

Seward, F. D., Harnden, F. R., and Helfand, D. J. (1984). Discovery of a 50 millisecond pulsar in the Large Magellanic Cloud. *Astrophysical Journal Letters*, **287**, L19–22.

Shannon, C. E. (1948). *Bell Systems Tech. J.*, **27**, 379, 623; these papers were reprinted in C. E. Shannon and W. Weaver, *The Mathematical Theory of Communication*, Urbana: University of Illinois Press, (1949).

Shore, J. and Johnson, R. (1980). Axiomatic derivation of the principle of maximum entropy and the principle of minimum cross-entropy. *IEEE Transactions on Information Theory*, **26**, 26–37.

Sivia, D. S. (1996). *Data Analysis: A Bayesian Tutorial*. Oxford: Clarendon Press.

Skilling, J. (1988). The axioms of maximum entropy. In *Maximum Entropy & Bayesian Methods on Science and Engineering*, Vol. 1, G. J. Erickson and C. R. Smith (eds.). Dordrecht: Kluwer Academic Press, p. 173.

Skilling, J. (1989). Classical maximum entropy. In *Maximum Entropy & Bayesian Methods*, J. Skilling (ed.). Dordrecht: Kluwer Academic Press, pp. 45–52.

Skilling, J. (1998). Probabilistic data analysis: an introductory guide. *Journal of Microscopy*, **190**, 297–302.

Skilling, J. and Gull, S. F. (1985). Algorithms and applications. In *Maximum Entropy & Bayesian Methods in Inverse Problems*, C. R. Smith and W. T. Grandy, Jr. (eds.), pp. 83–132.

Stigler, S. M. (1986). Laplace's 1774 memoir on inverse probability. Translation of Laplace's 1774 Memoir on "Probability of Causes." *Statistical Science*, **1**, 359–378.

Taylor, A. R. and Gregory, P. C. (1982). Periodic radio emission from LS I + 61° 303. *Astrophysical Journal*, **255**, 210–216.

Tierney, L. and Kadane, J. B. (1986), Accurate approximations for posterior moments and densities. *J. American Statistical Association*, **81**, 82–86.

Tinney, C. G., Butler, R. P., Marcy, G. W., Jones, H. R. A., Penny, A. J., McCarthy, C., Carter, B. D., and Bond, J. (2003). Four new planets orbiting metal-enriched stars. *Astrophysical Journal*, **587**, 423–428.

Toussaint, D. (1989). Introduction to algorithms for Monte Carlo simulations and their application to QCD. *Computational Physics Communications*, **56**, 69–92.

Tribus, M. (1969). *Rational Descriptions, Decisions and Designs*. Oxford: Pergamon Press.

Vattulainen, I., Ala-Nissila, T., and Kankaala, K. (1994). Physical tests for random numbers in simulations. *Physical Review Letters*, **73**, 2513–2516.

Wolfram, S. (1999). *The Mathematica Book* (fourth edition). Cambridge: Cambridge University Press.

Wolfram, S. (2002). *A New Kind of Science*. Champaign, IL: Wolfram Media, Inc.

Index

Terms followed by [] are *Mathematica* commands.

Index